Animal Feed Additives

NIPA® GENX ELECTRONIC RESOURCES & SOLUTIONS P. LTD.
New Delhi-110 034

Animal Feed Additives

Editors

Pankaj Kumar Singh
Assistant Professor
Department of Animal Nutrition
Bihar Veterinary College, Patna, Bihar, India

Chandramoni
Professor (Animal Nutrition)
Bihar Veterinary College, Patna, Bihar, India

Kaushalendra Kumar
Assistant Professor (Animal Nutrition)
Bihar Veterinary College, Patna, Bihar, India

Sanjay Kumar
Assistant Professor (Animal Nutrition)
Bihar Veterinary College, Patna, Bihar, India

NIPA® GENX ELECTRONIC RESOURCES & SOLUTIONS P. LTD.
New Delhi-110 034

NIPA® GENX ELECTRONIC RESOURCES & SOLUTIONS P. LTD.

101,103, Vikas Surya Plaza, CU Block
L.S.C. Market, Pitam Pura, New Delhi-110 034
Ph : +91 11 27341616, 27341717, 27341718
E-mail: newindiapublishingagency@gmail.com
www: www.nipabooks.com

For customer assistance, please contact
Phone: + 91-11-27 34 17 17
Fax: + 91-11- 27 34 16 16
E-Mail: feedbacks@nipabooks.com

ISBN: 978-81-19002-14-6

Composed and Designed by NIPA®.

भारतीय पशु–चिकित्सा अनुसंधान संस्थान

इज्जतनगर–243122, बरेली (उ.प्र.)

INDIAN VETERINARY RESEARCH INSTITUTE

Izatnagar-243122, Bareily (U.P.) INDIA

भाकृअनुप ICAR

Telefax: +91-581-2302536 (O) +91-581-2585313 (R) 91-9411008904 (M)

E.mail:dnkamra@rediffmail.com

D.N. Kamra

ICAR National Professor

Fellow NAAS

Animal Nutrition Division

FOREWORD

The livestock sector is one of the fastest growing segments of Indian agriculture, which plays an important role in socio-economic development and food security in the country. In view of increased demand for food of animal origin, driven largely by burgeoning human population, increased income and urbanization, livestock productivity in developing countries has to be increased phenomenally. Globally increase in livestock productivity in the recent past has been driven mostly by scientific and technological developments in breeding, nutrition and animal health. Maximum and economic livestock production with minimum stress to livestock and environment is the major concern of the livestock farmer. Extensive research work has been conducted for reducing the food cost encompassing judicious use of available feed resource, precise nutrient supply and augmenting nutrients utilization through feed processing, supplementation of deficient nutrients and suitable feed additives.

The feed additives are non-nutritive chemicals and microbial feed additives, also known as probiotics, are used to improve productive, reproductive and health performance of animals. The concept of using feed additives has been accepted as a potential approach to augment the health and production of livestock in economic and eco-resilient manner. In this book on '*Animal Feed Additives*' the editors have included almost all feed additives that are either being used or appear to have a potential to use them as a feed additive in livestock and poultry feed formulations. The chapters are designed in such a way that they detail the most recent scientific advances along with their valuable practical reflections and considerations. I find that the book covers theoretical and practical aspects of feed additives in a systematic and comprehensive manner to enlighten the significance of feed additives in sustainable livestock and poultry production.

I congratulate the editors and authors for their sincere effort in bringing out this publication in the present format for the benefit of science and more particularly animal nutritionists. I am sure this book will serve as a reference book for the students, teachers and researchers.

(D.N. Kamra)

Preface

Feed additives are non-nutritive substances, preparations and micro-organisms that are added to animal feed which may favourably influence characteristics of the animal feed, feed intake, gastro-intestinal flora, digestibility of the animal feeds, animal production, health, fertility, and characteristics of animal products. Modern intensive livestock and poultry production has achieved phenomenal gains in the efficient and economical production of high quality and safe animal products and by-products. Due to increasing awareness regarding environmental and food safety issues, animal nutritionists have been assigned with the responsibility to produce clean, green and wholesome animal products in a sustainable and eco-resilient manner. The use of feed additives can be an important tool of achieving the livestock production.

Despite mounting and convincing evidence that feed additives are very useful for economic and eco-friendly livestock and poultry production, most of the additives are not very popular to use. A part of the reason for this relative paucity may be the lack of complete and easily available informations on feed additives. We experienced that informations about the recent concept on different feed additives have not been presented in a single book so that readers can understand their intricacy and practical applications. However, reviews on different feed additives are distributed in diverse journals that are sometimes physically and financially out of reach of most of the students. Therefore, the need for the hour is to compile all the necessary informations pertaining to animal feed additives in a single publication and update the students, researchers and feed manufacturing agencies with current knowledge on this topic.

This book on *Animal Feed Additives* discusses the impacts of feed additives on animal metabolism, health and production in a systematic and comprehensive manner with all updated informations. In preparing this reference book, authors from major teaching and research establishments, who have an enviable tract record in their respective specialism, have been selected. The book contains chapters on antioxidants, enzymes, probiotics, prebiotics, synbiotics, antimicrobials, organic acids, coccidiostats, mycotoxin binders, immunomodulator, hen egg antibody, hormones, beta agonist, methane inhibitors, defaunating agents, essential oil and herbal feed additives. The informations presented in this book are based on the latest data available from research being conducted in the field of feed additives. The book contains good amount of the experimental evidence with references which will enable the students and research workers to obtain information quickly when necessary.

The book is dedicated to all the contributors, who made this book possible and who found their valuable time to write their comprehensive and authoritative chapters. We acknowledge and express our thanks to our family members, friends and well wishers, who have been incredibly supportive to bring this book.

We hope the book will provide a comprehensive and valuable guide on animal feed additives to the students, teachers and researchers of animal science discipline.

Pankaj Kumar Singh
Chandramoni
Kaushalendra Kumar
Sanjay Kumar

Contents

List of Contributors

A. K. Panda
Senior Scientist, Project Directorate on Poultry, Rajendranagar, Hyderabad, India

A. M. Ganai
Division of Animal Nutrition, Faculty of Veterinary Sciences and Animal Husbandry, SKUAST (K), Shuhama, Srinagar, Jammu & Kashmir, India

A.K. Srivastava
Assistant Professor, Division of LPM, College of Veterinary Science & Animal Husbandry, SDAU, Gujarat, India

Amlan K. Patra
Department of Animal Nutrition, Faculty of Veterinary and Animal Sciences, West Bengal University of Animal and Fishery Sciences, Belgachia, Kolkata, West Bengal, India

Ashutosh Mishra
Assistant Professor, College of Fisheries, G.B. Pant University of Agriculture and Technology, Pantnagar, U.S. Nagar, Uttarakhand, India

Avijit Dey
Senior Scientist, Division of Animal Nutrition & Feed Technology, Central Institute for Research on Buffaloes, Hisar- 125 001, Haryana, India

Avinash Kumar
Research Scholar, Animal Nutrition Division, Indian Veterinary Research Institute, Izatnagar, UP, India

B. K. Swain
Principal Scientist (Poultry Science), ICAR Research Complex for Goa, Ela, Goa, India

Chandramoni
Professor, Department of Animal Nutrition, Bihar Veterinary College, Patna, Bihar, India

Goutam Mondal
Senior Scientist, Dairy Cattle Nutrition Division, National Dairy Research Institute, Karnal, Haryana, India

Haidar Ali Ahmed
Assistant Professor, Division of Animal Nutrition, Faculty of Veterinary Sciences and Animal Husbandry, SKUAST (K), Shuhama, Srinagar, Jammu & Kashmir, India

Indranil Samanta
Assistant Professor, Department of Veterinary Microbiology, Faculty of Veterinary and Animal Sciences, West Bengal University of Animal and Fishery Sciences, Belgachia, Kolkata, West Bengal, India

Kamdev Sethy
Assistant Professor, Department of Animal Nutrition, College of Veterinary Science and Animal Husbandry, Orissa University of Agriculture and Technology, Bhubaneswar, Odisha, India.

Kaushal Kumar
Assistant Professor, Department of Veterinary Pathology, Bihar Veterinary College, Patna, Bihar, India

Kaushalendra Kumar
Assistant Professor, Department of Animal Nutrition, Bihar Veterinary College, Patna, Bihar, India

M.T. Banday
Professor, Division of Livestock Production and Management, Faculty of Veterinary Sciences and Animal Husbandry, SKUAST (K), Shuhama, Srinagar, Jammu & Kashmir, India

Nirbhay Kumar
Assistant Professor, Department of Pharmacology & Toxicology, Bihar Veterinary College, Patna, Bihar, India

Pankaj Kumar
Assistant Professor, Department of Veterinary and Animal Husbandry Extension, Bihar Veterinary College, Patna, Bihar, India

Pankaj Kumar Singh
Assistant Professor, Department of Animal Nutrition, Bihar Veterinary College, Patna, Bihar, India

Papori Talukdar
Ph.D Scholar, Dairy Cattle Nutrition Division, National Dairy Research Institute, Karnal, Haryana, India

Rajni Kumari
Scientist, Division of Livestock and Fishery Management, ICAR Research Complex for Eastern Region, Patna, Bihar, India

Rashmi Ranjan
Research Scholar (Microbiology), Mody Institute of Technology & Sciences, Lakshmangarh, Rajasthan, India

Ravindra Kumar
Senior Scientist, Nutrition, Feed Resources and Product Technology Division, Central Institute for Research on Goats, Makhdoom, Mathura, Uttar Pradesh, India

Rinesh Kumar
Department of Parasitology, College of Veterinary Sciences and Animal Husbandry, Rewa, Madhya Pradesh, India

S. K. Sirohi
Principal Scientist, Division of Dairy Cattle Nutrition, National Dairy Research Institute, Karnal, Haryana, India

S. Shekhar
Research Scholar, Department of Veterinary Medicine, G.B. Pant University of Agriculture and Technology, Pantnagar, Uttarakhand, India

S.B. Shudhakar
Scientist, Indian Veterinary Research Institute Regional Centre, Mukteshwar, Uttarakhand, India

S.K. Shukla
Professor, Department of Veterinary Medicine, G.B. Pant University of Agriculture and Technology, Pantnagar, Uttarakhand, India

S.V. Lal
Ph.D Scholar, Animal Biotechnology Centre, National Dairy Research Institute, Karnal, Haryana India

Sanjay Kumar
Assistant Professor, Department of Animal Nutrition, Bihar Veterinary College, Patna, Bihar, India

Shalini Vaswani
Assistant Professor, Department of Animal Nutrition, College of Veterinary Science & Animal Husbandry, DUVASU, Mathura, Uttar Pradesh, India

Shanker Dayal
Senior Scientist, Division of Livestock and Fishery Management, ICAR Research Complex for Eastern Region, Patna, Bihar, India

Suman Kumar
Department of Parasitology, College of Veterinary Sciences and Animal Husbandry, Rewa, Madhya Pradesh, India

1

An Overview of Feed Additives

Pankaj Kumar Singh

Introduction

Modern intensive livestock and poultry production has achieved phenomenal gains in the efficient and economical production of high quality and safe animal products and by-products. Furthermore, the overall performance of livestock can be increased by improving nutrients utilization, health status, fertility and efficiency of production. The use of feed additives has been an important part of achieving efficient livestock production. Feed additives are non-nutritive substances, preparations and micro-organisms that are added to animal feed to improve productive, reproductive and health performances. Any substance is considered as a feed additive when, not having a direct utilization as nutrient, is included at an optimum concentration in diet to exert a positive action over the animal health status or the dietary nutrient utilization. Hutjens (1991) defined feed additives as a group of substances that can cause a desired animal response in a non-nutrient role, such as pH shift, growth, or metabolic modifier. Feed additives are added deliberately to animal feed which may favourably influences characteristics of the animal feed, feed intake, gastro-intestinal flora, digestibility of the animal feeds, animal production, health, fertility and characteristics of animal products. Because of their chemical nature as active principles, additives are generally included in very small proportions in diet.

Global economic pressures have driven a tendency of the livestock production to produce more products per unit cost. This has led to changes to animal's environment like feeding, housing and disease control. Quality of feed nutrition is influenced not only by the content but also by some other aspects such as, feed presentation, hygiene, anti-nutritional factors, digestibility and palatability. Feed additives provide a mechanism by which such dietary deficiencies can be addressed and also benefits not only associated with the nutrition and thus the growth rate of the animal concerned, but also its health and welfare. Antimicrobial growth promoters are commonly fed to animals to prevent disease and metabolic disorders, as well as improve feed efficiency. However, in recent years, public concern about the potential for antibiotic resistant strains of bacteria, the search for alternatives to replace antibiotics growth promoters has gained increasing interest in animal nutrition. The use of phytogenic feed additives has gained momentum for their potential role as natural alternatives to antibiotic growth promoters in animal nutrition. Compared with synthetic antibiotics or inorganic chemicals, these plant-derived products have proven to be natural, less toxic, residue free, and are thought to be ideal feed additives in food animal production. There are a large number of phytogenic feed additives which have antioxidant, antibacterial, anticoccidial, antiparasitic and anti-inflammatory effect.

Additives like microbial and plant secondary metabolites offer a unique opportunity manipulating ruminal fermentation. Recent research has been greatly focused to exploit bioactive plant secondary compounds like saponins, tannins, flavonoids, essential oils to improve rumen fermentation such as enhancing protein metabolism, decreasing methane production, reducing nutritional stress like bloat, and improving animal health and productivity. Thus, feed additives can be used to improve feed intake, metabolism and effieicncy of feed utilization for economic and eco-friendly livestock production.

Types of feed additive

Common feed additives used in livestock and poultry diets include antimicrobials, antioxidants, enzymes, probiotics, prebiotics, organic acids, coccidiostats, mycotoxin binders, immunomodulators, hormones, emulsifiers, pellet binders, phytogenic herbs, essential oils and metabolic modifiers. Sometimes diets of livestock also contain additives used in diets for humans and pets such as flavour enhancers, artificial and nutritive sweeteners, colours, lubricants etc. Enzymes, antimicrobials, probiotics, prebiotics can have a very positive effect on feed intake and nutrient utilization in livestock and poultry. Enzymes increase the digestion capacity in young animal and to help to decrease the risk of digestive problems and therefore increase the health status, nutrient availability and growth performances. In older animals enzymes can be used successfully to increase the use of unconventional feedstuffs rich in dietary fibres or other nutrients with a low digestibility and to reduce environmental load with nutrients. Antibiotic growth promoters at sub-therapeutic concentrations in the diets of poultry and swine inhibit some microflora of intestinal pathogens, increasing useful intestinal microflora. Metabolic modifiers influence intermediate metabolism (Wenk, 2000). However, mechanism of action varies but positive effect of feed additives are usually expressed through better feed intake, improved feed conversion, stimulation of the immune system and increased vitality, regulation of the intestinal micro-flora, etc. Because of the fact that feed additives have different mechanisms of action, it is necessary to present every group of additive individually as follows.

Feed enzymes

Enzymes are naturally occurring proteins that function as catalysts for the vast number of chemical reactions taking place in the living being. Although animals and their associated gut microflora produce numerous enzymes, they are not necessarily able to produce sufficient quantities of specific enzymes or produce them at the right locations to facilitate absorption of all components in normal feedstuffs or to reduce anti-nutritional factors in feed that limit digestion. In today's ever increasing economic climate, commercial poultry companies are trying to help alleviate adsorbent cost even more so then ever before. The main focus on addressing the feed cost issues has been examining the benefits of adding exogenous enzymes to poultry rations. Some cereal grains (rye, barley, wheat, sorghum) have soluble non-starch polysaccharides (NSP) that can entrap large amounts of water during digestion and form very viscous gut contents. Enzymes that are harvested from microbial fermentation and added to feeds can break these bonds between sugar units of NSP and significantly reduce the gut content viscosity. Lower viscosity results in improved digestion, absorption and health (reducing moisture in manure and nutrients available for harmful gut microflora to proliferate and challenge the birds. NSP degrading enzymes have been used successfully to circumvent problems with high intestinal viscosity in broiler chicken (Bedford *et al.*, 2000). Phytates are plant storage sources of phosphorus that also bind other minerals, amino acids (proteins) and energy and reduce their availability to the bird (Singh *et al.*, 2003). Birds do not produce enzymes like cellulase, xylanase, phytase etc., required for the digestion of NSPs and phytates. Supplementation of NSPs degrading enzymes and phytase not only reduce the anti nutritive effects of NSPs and phytates, but also releases some nutrients from these, which could be utilized by the birds. Addition of enzymes such as xylanase are useful in the utilization of the non-starch polysaccharide (NSP) component of the ingredients, while others such as proteases may enhance the utilization of protein (Zakaria *et al.*, 2010). Phytase enzymes are also available commercially that significantly reduce the negative effects of phytates. An increased use of feed enzymes is expected not only from the aspect of economic gain but also from the environmental point of view as enzymes enhances nutrient utilization, thereby reducing the manure output and reducing nutrient excretion particularly excess phosphorus, nitrogen, copper and zinc.

In ruminant production system, forages are the main feed component that serve as the major source of energy available to the animal. However, due to slow or incomplete digestion of fibrous substrates only 10 to 35% of energy intake is available as net energy that significantly limits livestock performance and profits in production systems. To improve the ruminal fibre degradability many strategies have been developed to stimulate the digestion of the fibrous components in ruminant feeds. The use of supplemental exogenous enzymes to improve fibre digestibility and feed utilization was first examined for ruminants in 1960s (Beauchemin, 2004). Feeding of enzymes in ruminants was a questionable practice in past decades due to poor characterization of enzyme products, variable animal response and high cost. But now a day, with recent advancement in fermentation technology and biotechnology, economic production of large quantities of biologically active enzymes preparations as animal feed additives has become possible. Supplementation of exogenous fibrolytic enzymes (cellulose and xylanase) to ruminants has the potential to increase the digestibility of fibre and efficiency of feed utilization (Thakur *et al.*, 2010).

Antioxidants

The animal body is under constant attack from free radicals. There are a variety of sources of reactive free radicals in normal metabolism as well as those coming directly from feed ingredients. There are many lipid or fat components of feeds, which spontaneously react with atmospheric oxygen and suffer deterioration in the process of auto-oxidation. The oxidation in feed ingredients such as fats and oils decreases their nutritional value, reduces shelf life of feed ingredients and results in generation of undesirable flavours. The widespread use of fats and oils in formulation of high density diets for broiler chicken have increased incidence of auto-oxidation in poultry which ultimately reduces performances and feed utilization efficiency in poultry. Oxidative stress can disrupt normal cellular function, damage tissues and reduce health status. Antioxidants are the chemical compounds, which have the capacity of preventing the oxidation by preferentially taking up oxygen and thus prolongining the lag period of oxidation (Adams, 1997). Ethoxyquin, BHA (butylated hydroxyanisole) and BHT (butylated hydroxytoluene) are commonly used antioxidants in poultry ration. It has been reported that feeding oxidized fat and oils resulted in reduced performances and these harmful effects can be counteracted by supplementation of antioxidants (Surai, 2002).

Antibiotic feed additive

To achieve high level of economic efficiency poultry and pigs are raised under intensive production system in densely populated colonies or flocks during which they are succumbed to various kinds of stresses and diseases which adversely affect their productive performances. To prevent disease outbreaks and promote growth, low, sub-therapeutic concentrations of antibiotics are often added to the diets of poultry and swine. Antimicrobials have been used extensively in intensive poultry operations to minimise disease and improve growth and feed utilisation. Antibiotic inhibit some microflora of intestinal pathogens, increasing useful intestinal microflora, wide activity against positive gram bacteria, decreasing harmful effect of metabolites of intestinal microflora through removing them, decreasing thickness of intestinal mucosa layer to increase food absorption (Niewold, 2007).

However, human health can be affected directly through residues of an antibiotic in meat, or indirectly, through the selection of antibiotic resistance determinants that may spread to a human pathogen. In view of antibiotic resistance the European Union has moved towards a complete ban of in-feed antimicrobials. The ban on the use of antibiotics as a feed additive in animal nutrition has led to a worldwide search and implementation of alternative strategies for preventing the growth of pathogenic bacteria in farm animals, to maintain the health and performance of poultry. The industry is currently evaluating alternatives to chemical therapeutics. However, this goal can be achieved by adopting biosecurity measures as well as by the use of feed additives like probiotics, prebiotics, enzymes, antioxidants, acidifiers and phytogenic additives. These products show promise as alternatives for antibiotics as pressure to eliminate growth promotant antibiotic use increases (Yang *et al.*, 2009). They all in majority of cases demonstrated positive effect on health and performance of poultry.

Probiotics

One of the most well characterized means to reduce enteric disease is through the administration of live bacteria, or probiotics. Probiotics are individual microorganisms or groups of microorganisms which have favourable effect on host by improving the characteristics of intestinal micro-flora (Fuller, 1989). The term probiotic originates from Greek words meaning for life and contrasted with the term antibiotic, which means against life. Probiotics display several ways of action: antagonistic action towards pathogen bacteria by secretion of products which inhibit their development, such as bacteriocins, organic acids and hydrogen peroxide; the other way is competitive exclusion which represents competition for locations to adhere to the intestinal mucous membranes and in this way pathogen micro-organisms are prevented from inhabiting the digestive tract, and the third way is competition for nutritious substances (Patterson and Brukholder, 2003). In this way, they create conditions in intestines which favour useful and inhibit the development of pathogen bacteria. Probiotics not only inhibit or block the ability of pathogens to attach to epithelial surfaces, but also limit pathogen colonization through competition for nutrients and production of antibacterial substances such as bacteriocins (Edens *et al.*, 1997). Probiotic supplementation of the intestinal micro flora enhances defense, primarily by preventing colonization by pathogens and offering a greater stability of the intestinal ecosystem, and by an indirect, adjuvant-like stimulation of innate and acquired immune functions of intestine (Fuller 1989). Besides the health benefits of probiotics, it improves growth rate and feed conversion efficiency in calves (Ramaswami *et al.*, 2005), microbial protein flow and DM intake (Putnam *et al.,* 1997) particularly in poor managemental conditions.The dietary use of probiotic is gaining momentum to counteract the stresses because of their beneficial effects on live weight gain, feed conversion efficiency and reduced mortality in broilers and egg production and feed conversion in layers.

Prebiotics

Prebiotics are non-digestible feed components/ingredients which have positive effect on host in their selective growth and/or activation of certain number of bacterial strains present in intestines (Gibson and Roberfroid, 1995). The most significant compounds which belong to group of prebiotics are oligosaccharides: fructo-oligosaccharides (FOS), gluco-oligosaccharides and mannan-oligosaccharides (MOS). Their advantage compared to probiotics is that they promote growth of useful bacteria which are already present in the host organism and are adapted to all conditions of the environment *(*Yang *et al.,* 2008*).* Favourable effects of addition of probiotics reflect in presence of antagonism towards pathogens, competition with pathogens, promotion of enzyme reaction, reduction of ammonia and phenol products and increase of resistance to colonization. Prebiotics have shown promise in the prevention and control of exogenous and endogenous intestinal infections and good health of the animals (Grizard and Barthomeuf, 1999).

Synbiotics

Synbiotics are combination primarily of probiotics and prebiotics, as well as other promoting substances which together exhibit joint effect in regard to health of digestive tract, digestibility and performances of broilers. Synbiotic may improve the survival rate of probiotics during their passage through the digestive tract, thus contributing to the stabilization and/or potentiation of the probiotic effects. Marioka *et al.* (2000) found significant improvement in weight gain due to dietary supplementation of synbiotic (MOS and *Bacillus subtilis*) compared to either probiotic (*Bacillus subtilis*) or prebiotic (MOS) alone. Investigations showed that combinations used in synbiotics are often more efficient in relation to individual additives (Li *et al.*, 2008).

Organic acids

Organic acids have a long history of being utilized as food additives and preservatives for preventing food deterioration and extending the shelf life of perishable food ingredients. In recent past organic acids have been used to control microbial contamination and replace antibiotic growth promoters in poultry (Hassan *et al.,* 2010).Organic acids that have shown positive effects on growth performance in poultry include citric, formic, fumaric, and propionic acids. Acidifiers have been used in poultry nutrition for long time, in different forms and combinations. Organic acids reduce pH value of food and in this way act as conserving agents and prevent microbial contamination of feed, and this effect is exhibited also in digestive tract of poultry (Luckstadt, 2005). Organic acids are widely used to inhibit pathogens like salmonellae and in their undissociated forms are able to pass through their cell membrane. Inside the bacterial cell, the acid dissociates to produce H^+ ions, which lower the pH causing the organism to use its energy in trying to restore the normal balance. It also disrupts DNA and protein synthesis and thus the bacteria are unable to replicate or its replication slows down. Lower pH conditions thus protect the bird from infection especially at young ages. An experiment with broiler chickens also indicated that benzoic acid at concentrations of 0.2% may have a positive influence on growth (Engberg, 2001). Organic acids (mixture of fumaric acid and salt of butyric, propionic and lactic acids) supplementation @ 520ppm improved egg production, egg size, shell thickness and feed conversion efficiency (Rahman *et al*., 2008).

Hormones and beta-agonists

In growing farm animals metabolic modifiers are used to have a better partition of the energy deposition for growth in the form of protein and fat. Hormones and beta-agonists have high potential for increased protein deposition and simultaneously a reduced fat deposition (Buttary and Dawson, 1987). Hormones as additives are being added in poultry ration to help the animals gain more weight faster and fulfill consumer needs and desires. They help to reduce the waiting time and the amount of feed eaten by an animal before slaughter in meat industries. In dairy cows, hormones can be used to improve milk let down and production. The hormones approved for use in beef production are estradiol (estrogen), progesterone, testosterone, and their synthetic alternatives zeranol, melengestrol acetate, and trenbolone acetate. Melengestrol acetate, a synthetic progestin, is used as feed additive that suppresses estrus (heat or cyclic sexual activity) and improves gain and feed efficiency in beef females. While hormonal implants have been utilized for decades in cattle production, recent advancements in dietary beta-agonist supplements have offered swine and cattle producers the opportunity to further improve their production efficiency. Beta-agonist can be exploited as nutrient repartitioning agent as it diverts the nutrients from fat deposition to the muscle tissues deposition (Sillence, 2004). Ractopamine and Zilpaterol are approved â-agonists to be incorporated in cattle and hog rations to improve body weight gain and reduce the body fat content (Carr *et al.*, 2009). Recently, some developed countries banned the use of hormone implants due to their residual effect on humans.

Phytogenic feed additives

Since ancient times aromatic plants or phytobiotics have been used because of their medicinal properties. Current studies show promising results regarding the use of phytochemicals as growth and production promoters (Toghyani *et al.*, 2010). Garlic (*Allium sativum*), Turmeric (*Curcuma longa*), Thyme (*Thymus vulgaris* L.) Aloe vera (*Aloe barbadensis*), onion (*Allium sepa*), Ginger (*Zingiber officinale*, Rosc.), *Astragalus membranaceus*, neem etc., are some of the major plant additives which have been extensively reported in poultry feed for enhanced growth effect in broiler and better egg production in laying hens (Guo *et al.*, 2004; Sunder *et al.*, 2014). The positive effects of these herbs are due to the presence of essential oils, fatty acids, alkaloids, flavanoids, fats, minerals fibers, vitamins, protein and carbohydrates. Apart from the digestive and antioxidant prosperities, the herbs and plant additives may exert the beneficial influence through antioxidant, antimicrobial, immunomodulating, antiparasitic and anti-inflammatory effect (Gowda *et al.*, 2009; Khan *et al.*, 2012). Plants like Ashwagandha (*Withania somnifera*), Neem (*Azadirachta indica*), Guduchi (*Tinospora cordofolia*) and others are widely used these days due to their potent immunomodulatory and health beneficial properties (Barnes *et al.*, 2007; Tiwari *et al.*, 2014).

Metabolic modifiers in ruminants

Ruminants are provided with a microbial ecosystem (ciliate protozoa, anaerobic bacteria and anaerobic fungi) in the gastro-intestinal tract which helps in the bioconversion of lignocellulosic feeds into volatile fatty acids which are utilized by the animal as a source of energy. This mixture of organisms combines to digest the food, and it is the products of microbial digestion which form the nutrients that become available to the ruminant animal itself. This microbial biomass is also helpful in synthesis of microbial protein and B-complex vitamins for the host animals. Complex polysaccharides are converted to volatile fatty acids (acetic, butyric, and propionic acid) which provides 70% of the energy available to the rumen. Besides beneficial microbial fermentation process, there is few fermentation process which decreases efficiency of production in ruminant. Proteolysis destroys high quality protein in the feed and therefore proteolysis should be slowed. Protozoal activity is detrimental to the efficiency of microbial protein synthesis; therefore, this activity should be suppressed. Bloat and acidosis are distressing disorders which result from malfunction of microbial digestion in the rumen. Methane, a potent greenhouse gas, and ammonia, which form urinary urea, arise from normal rumen fermentation. They cause atmospheric and groundwater pollution, respectively. Since methane contains energy, its emission during rumen fermentation is considered to be a loss of feed energy that is equivalent to 2-12% of the gross energy of animal feed (Johnson and Johnson, 1995). Therefore, scientists are trying to manipulate ruminal fermentation to enhance ruminal fermentation and animal productivity. The rumen is such a delicately balanced eco-system that rumen modifiers rarely have one single effect on rumen fermentation. A number of strategies have been used to enhance ruminal fermentation.

Feed additives are used for manipulating ruminal fermentation to enhance ruminal fermentation and animal productivity. Rumen modifiers which manipulate the digestion and increase the retention of protein and energy without depressing feed intake should enhance productivity of ruminants. Direct-fed microbials offer arguably the greatest potential for manipulation of ruminal fermentation. They offer a huge spectrum of metabolic activities and enzymes as well as metabolites. Its cell envelope is quite different to that of ruminal microorganisms, which may be important in the delivery of nutrients or the absorption of toxic or undesirable compounds. It is metabolically active in the rumen but does not grow, which means that its concentration and activity can be readily controlled by its dietary inclusion level, ensuring maximum efficacy. Any compounds which can shift the fermentation from acetic acid production to propionic acid will trap more useful energy for the animal. Monensin, lasalocid salinomycin are polyether ionospheres which reduce methane production, alter volatile acid production and improved nitrogen metabolism patterns. Ionophores enhance the glucose status of dairy cows through increased production of propionate and thus improves milk production, and control ketosis and of bloat (McGufey, 2001). The process of making the rumen of animals free of rumen protozoa is called defaunation and the animal is called defaunated animal. Feed additives can be used to defaunate ruminants. Defaunation can be an important tool to improve the productivity of animals in tropical countries, where majority of livestock are maintained on sole diet on low grade roughage. Plants are part of herbivore diets, therefore, recent research has been greatly focused to exploit bioactive plant secondary compounds saponins, tannins, flavonoids, essential oils to improve rumen fermentation such as enhancing protein metabolism, decreasing methane production, reducing nutritional stress like bloat, and improving animal health and productivity (Wallace *et al.,* 2002; Patra and Saxena, 2010).

Evaluating feed additives

New feed additives are rapidly adopted by the livestock and poultry industry. There are some factors which should be used to evaluate a feed additive. These factors are 7 R's (anticipated response, economic return, available research, field responses, reliability, repeatability, and relativity) (Hutjens, 1991). Response refers to expected performance changes. The performances changes may be in terms of improved feed intake, growth rate, milk yield, milk quality, health and fertility and ultimately better economic return. Research is essential to determine if experimentally measured responses can be expected in the field. Studies should be conducted under controlled and unbiased conditions, have statistically analyzed results (to determine whether the differences are repeatable), and have been conducted under experimental designs that would be similar to field situations. Reliability is based on the research database that has been published on a feed additive. Relativity refers to other products, management changes, or on-farm practices that could replace the feed additive being used. The bottom line of evaluating criteria of a feed additive is the probability of a profitable response. Repeatability represents the statistical data results (mean and standard deviation). Each feed consultant must determine the level of risk when selecting a feed additive (Hutjens, 1991).

The future of feed additive

Although additives have been developed over the years for most species of livestock, the major uses today are with growing and finishing beef cattle, swine and poultry. Feed additives are gaining importance due to various functions such as growth promotion, controlling infectious diseases and enhancement of feed digestibility in animals. The animal feed additives market is growing at a steady pace and the market is to grow in the future due to the increasing demand for meat and meat products around the globe. Many additives have been a normal part of diets for animals. It is only recently that we have come to recognize and understand their importance in achieving high production and efficiency, maintaining health and wellbeing, improving product quality and safety and reducing the industry's impact on the environment. More work is required to further identify the positive effects of additives and minimise the negative effects they may have if not used correctly or if they interact with other additives or feed ingredients.

References

Adams, C. A. 1997. Nutriticine approach to oxidation. *Feed Mix*. 5:17-19.

Beauchemin, K. A., Colombatto, D., Morgavi, D. P., Yang, W. Z. and Rode, L. M. 2004. Mode of action of exogenous cell wall degrading enzymes for ruminants. *Can. J. Anim. Sci*. 84: 13–22.

Bedford, M. R. 2000. Exogenous enzymes in monogastric nutrition - their current value and future benefits. *Animal Feed Sci. Technol.* 86: 1-13.

Barnes, J., Anderson, L.A. and Phillipson, J.D. 2007. Herbal Medicine. 3rd Edn., Pharmaceutical Press, London. pp: 1-23.

Botsoglou, N. A., Christaki, E., Florou-Paneri, P., Giannenas, I., Papageorgiou, G., and Spais, A.B. 2004. The effect of a mixture of herbal essential oils or á-tocopheryl acetate on performance parameters and oxidation of body. *South Afr. J. Anim. Sci.,* 34: 52-61.

Buttary, P.J and Dawson, J.M. 1987. The mode of action of β-agonist as manipulators of carcass compsotion. In: β-agonist and their effects on Animal Growth and Carcass Quality (Rd. Harahan). *Elsevier Science*. pp. 29-43.

Carr, S.N., Hamilton, D.N., Miller, K.D., Schroeder, A.L. and Fernandez-Duenas, D. 2009. The effect of ractopamine hydrochloride (Paylean®) on lean carcass yields and pork quality characteristics of heavy pigs fed normal and amino acid fortified diets. *Meat Sci*. 81: 533-539.

Edens, F. W., Parkhurst, C. R., Casas, I. A. and Dobrogosz, W. J. 1997. Principles of ex ovo competitive exclusion and in ovo administration of Lactobacillus reuteri. *Poult. Science*.6:179-196.

Engberg, R.M., Hedemann, M.S. and Jensen, B.B. 2002. The influence of grinding and pelleting of feed on the microbial composition and activity in the digestive tract of broiler chickens. *British Poult Sci*. 43: 569-579.

Fuller, R. 1989. Probiotics in man and animals: a review. *Applied Bacteriology*. 66: 365- 378

Gibson, G.R. and Roberfroid, M.B. 1995. Dietary modulation of the human colonic microbiota: Introducing the concept of prebiotics. *J. Nutr*., 125: 1401-1412.

Gowda, N.K.S., Ledouxa, D.R., Rottinghausa, G.E., Bermudeza, A.J. and Chena, Y.C. 2009. Antioxidant efficacy of curcuminoids from turmeric (*Curcuma longa* L.) powder in broiler chickens fed diets containing aflatoxin B_1. *Br. J. Nutr*., 102: 1629-1634.

Grizard, D and Barthomeuf, C. 1999. Non-digestible oligosaccharides used as prebiotic agents: mode of production and beneficial effects on animal and human health. *Reproduction Nutrition Development*. 39: 563-588.

Guo, F.C., Kwakkel, R.P., Williams, B.A. Li, W.K. and Li, H.S. 2004. Effects of mushroom and herb polysaccharides, as alternatives for an antibiotic, on growth performance of broilers. *Br. Poult. Sci*., 45: 684-694.

Hassan, H. M.A., Mohamed, M.A., Youssef, A.W. and Hassan, E.R. 2010. Effect of using organic acids to substitute antibiotic growth promoters on performance and intestinal microflora of broilers. *Asian-Aust. J. Anim. Sci*., 23: 1348-1353.

Hutjens, M. F. 1991. Feed additives. *Vet. Clinics North Am. Food Animal Practice*. 7(2):525.

Johnson, K. A. and Johson, D. E. 1995. Methane emissions from cattle. *J. Anim. Sci*. 73:2483-2492.

Khan, R.U., Naz, S., Nikousefat, Z., Tufarelli, V and Laudadio, V. 2012. *Thymus vlugaris*: Alternative to antibiotics in poultry feed. *World's Poult. Sci. J.* 68: 401-408.

Khan, R.U., Nikousefat, Z., Tufarelli, V., Naz, S. Javdani, M and Laudadio, V. 2012. Garlic (*Allium sativum*) supplementation in poultry diets: Effect on production and physiology. *World's Poult. Sci. J.*, 68: 417-424.

Luckstadt, C. 2005. Synergistic acidifiers to fight *Salmonella*. *Feed Mix.* 13: (1): 28-30.

Li, X., Qiang, L., Liu and Xu, C.H. 2008. Effects of supplementation of fructooligosaccharide and/or Bacillus subtilis to diets on performance and on intestinal microflora in broilers. *Archiv fur Tierzucht.* 51(1): 64-70.

Marioka, A. Santin, E., Sugeta, S.M., Almeida, J.G. and Makari, M. 2000. Utilization of prebiotics, probiotics or synbiotics in broiler chicken diets. *Revista-Bracileria-de-Cinecia-Avicola.* 3: 75-82.

McGuffey, R. K. Richardson, L. F. and Wilkinson, J. I. D. 2001. Ionophores for Dairy Cattle: Current Status and Future Outlook . *J. Dairy Sci.* 84(E. Suppl.):E194-E203.

Niewold, T.A. 2007. The nonantibiotic anti-inflammatory effect of antimicrobial growth promoters, the real mode of action? A hypothesis. *Poult. Sci.*, 86: 605-609.

Patra, A. K and Saxena, J. 2010. A new perspective on the use of plant secondary metabolites to inhibit methanogenesis in the rumen. *Phytochemistry.* 71:1198-1222.

Patterson, J. A and Burkholder, K.M. 2003. Application of prebiotics and probiotics in poultry production. *Poult. Sci.* 82: 627-631.

Putnam, D. E., Schwab, C.G., Socha, M.T, Whitehouse, N. L., Kierstead, N.A and Garthwaite, B. D. 1997. Effect of yeast culture in the diet of early lactation dairy cows on ruminal fermentation and passage of nitrogen fractions and amino acids to the small intestine. *J. Dairy Sci.*80:374-384.

Rahman, M. S., Howlider, M. A. R., Mahiuddin, M and Rahman, M. M. 2008. Effect of supplementation of organic acids on laying performance, body fatness and egg quality of hens. *Bang. J. Anim. Sci.* 37(2): 74 - 81

Ramaswami, N., Chaudhary, L. C, Agarwal, N and Kamra, D.N. 2005. Effect of lactic acid producing bacteria on the performance of male crossbred calves fed roughage based diet. *Asian-Aust. J. Anim. Sci.*18: 1110-1115.

Singh, P. K., Khatta, V. K., Thakur, R. S., Dey, S and Sangwan, M. L. 2003. Effects of phytase supplementation on the performance of broiler chickens fed maize and wheat based diets with different levels of non-phytate phosphorous. *Asian-Aust. J. Anim. Sci.*11:1642-1649.

Sillence, M.N. 2004. Technologies for the control of fat and lean deposition in livestock. *Vet. J.*, 167: 242-257.

Sunder, J., Sujatha, T., Pazhanivel, N., Kundu, A and Kundu, M.S. 2014. Effect of *Morinda citrifolia* fruit juice and lactobacillus acidophilus on broiler duodenal morphology. *Adv. Anim. Vet. Sci.*, 2: 28-30.

Surai, P.F. 2002. Selenium in Poultry nutrition. 1. Antioxidant properties, deficiency and toxicity. *World's Poult. Sci. J.* 58:333-347.

Thakur, S.S., Verma, M.P., Ali, B., Shelke, S.K. and Tomar, S.K. 2010. Effect of exogenous fibrolytic enzyme supplementation on growth and nutrient utilization in murrah buffalo calves. *Ind. J. Anim. Sci.* 80 (12). 1217-19.

Tiwari, R., S. Chakraborty, M. Saminathan, K. Dhama and Singh, S.V. 2014. Ashwagandha (*Withania somnifera*): Role in safeguarding health, immunomodulatory effects, combating infections and therapeutic applications: A review. *J. Biol. Sci.* 14: 77-94.

Toghyani, M., Toghyani, M., Gheisari, A., Ghalamkari, G. and Mohammadrezaei, M. 2010. Growth performance, serum biochemistry and blood hematology of broiler chicks fed different levels of black seed (*Nigella sativa*) and peppermint (*Mentha piperita*). *Livestock Sci.* 129: 173-178.

Wallace, R. J., McEwan, N. R., McIntosh, F. M., Teferedegne, B and Newbold, C. J. 2002. Natural products as manipulators of rumen fermentation. *Asian-Austr. J. Anim. Sci.* 15:1458-1468.

Wenk, C. 2000. Recent advances in animal feed additives such as metabolic modifiers, antimicrobial agents, probiotics, enzymes and highly available minerals- review. *Asian-Austr. J. Anim. Sci.* 13 (1): 86-95.

Yang, Y., Iji, P. A. and Choct, M. 2009. Dietary modulation of gut microflora in broiler chickens: a review of the role of six kinds of alternatives to in-feed antibiotics. *World 's Poult. Sci. J.* 65: 97-114.

Yang, Y., Iji, P. A., Kocher, A., Mikkelsen, L.L and Choct, M. 2008. Effects of mannanoligosaccharide and fructooligosaccharide on the response of broilers to pathogenic *Escherichia coli* challenge. *Br. Poult. Sci.* 49: 550-559.

Zakaria, H. A. H; Mohammad, A. and Ishmais, M. A. A. 2010. The influence of supplemental multi-enzyme feed additive on the performance, carcass characteristics and meat quality traits of broiler chickens. *Int. J. Poult. Sci.* 9:126-133.

2

Role of Enzymes as Feed Additives in Poultry Production

B. K. Swain

Introduction

The competition between human and livestock population for the existing animal and plant resources has been of great concern to the poultry nutritionists. Today the poultry industry has already challenged by the high price of the feed ingredients. Moreover in the next century there will be a shortage of conventional feed stuffs for use in poultry production. Efforts are on to look forward for the new feed resources and to evaluate them for their inclusion in the poultry ration. However many of the feed stuffs are characterized by the presence of certain incriminating factors. The incorporation of feed stuffs containing incriminating factors may adversely affect the performance of poultry. The nutritional strategy involving the use of feed enzymes offer immense potential to overcome the problems. The use of enzymes in feed mixtures of growing poultry particularly for young chicks has been the subject of much research activity in the recent decades. The interest in the feed enzymes is a reflection in changing the attitude of the society and the economic climate of the feed industry. It has been seen that approximately 187 million metric tons of cereal grains (FAO, 2011) and 26.02 million tons oil seeds as per the trade estimate for 2011-12 are produced in India which yield non starch polysaccharides (NSP) as a part of variety of products.

Non starch (fibre) polysaccharides are found in large quantities in cell walls of plant origin feedstuffs and therefore are available for enzymatic break down. A typical cell wall consists of cellulose, pectins, hemicellulose, small amount of glycoproteins and lignin. The cellulose is nutritionally rather inert, relatively insoluble and resistant to digestion in monogastric animals. It has been suggested that these polysaccharides can reduce nutrient utilization, growth rate and efficiency of feed utilization. There may be problem of wet litter due to increased viscosity of gut lumen contents and by acting as physical barrier to endogenous enzymes. In the context of present feed situation, it is important that animals utilise the feed optimally. One possibility is the addition of substances stimulating the digestive mechanism or inhibiting unwanted microbial growth in the digestive system. There are reports in the literature with reference to enzymes, biomass, antibiotics, probiotics and other preparation such as monensin. Other possibility is the use of enzyme producing microorganisms (bio-mass) for production of feeds from unconventional substances or high fibre feedstuffs. Feed ingredients such as maize, wheat, barley, soybean meal, sunflower cake, rape seed meal, canola meal, wheat bran and deoiled rice bran etc. are used for the preparation of diet for broilers and layers. The efficiency of utilization of these feed ingredients is not optimal due to reduction in utilization of non-starch polysaccharides (NSP) and sometimes, presence of antinutritive factors. Means to utilize NSP components leads to better feed utilization, optimum performance and reduced cost of production. Enzymes can improve nutrient availability from feedstuffs lower feed costs and reduce output waste to the environment. The result of previous studies indicate that dietary addition of enzymes may be of practical importance in improving the feed value of low energy and high fibre diets for Poultry (Nagalakshmi and Devegowda, 1991). Sklan (2001) suggested that the young birds have limited types and amounts of enzymes necessary to utilize a high carbohydrate and vegetable protein diet at an early age, thus affecting nutrient digestibility. Improvements in the efficiency of poultry diets and nutrients to about 10 % have been reported due to supplementation with enzymes (Cowieson *et al.*, 2000). Use of combination of various exogenous enzymes in the form of multienzyme complex in broiler and layer diets have shown positive effects on their performance, carcass yield, meat quality, egg production, egg weight, egg quality and the economics of production in meat and egg type birds. Importance of enzyme supplementation in general and their benefit on growth performance, nutrient utilization, carcass quality, egg production and egg quality and economics of production for broilers and layers have been presented here for the use by students, researchers and feed manufacturers .

Definition of Enzymes

Enzymes are naturally occurring proteins that function as catalysts for the vast number of chemical reactions taking place in the living being. Enzymes enable metabolic and physiological processes to take place by reducing the amount of energy needed for a chemical reaction to occur, thereby speeding up the rate of the reaction (Mathews, 1990). Enzymatic activity makes life possible because vital reactions would occur millions, in some instances billions, of times more slowly if not for the catalytic activity of enzymes (Tortora and Grabowski, 1993). Enzymes catalyze reactions by binding substrates, or chemical reactants, into a region of the enzyme called the active site. Active sites may be thought of as pockets or clefts in the enzyme into which chemical substrates fit. Since every enzyme and its active site have a unique physical conformation, enzymes have a high degree of specificity for the substrates upon which they act and the type of chemical reaction they catalyze. Enzymes are not consumed by the reactions they mediate. Once an enzyme has catalyzed a chemical reaction, it releases the altered substrate from its active site and becomes available to catalyze another reaction (Mathews, 1990).

Mechanism of action

The major nutritive components of feedstuffs such as carbohydrate, proteins and fats need to be digested and broken down to sugars, amino acids and fatty acids by digestive secretions and enzymes before they can be finally assimilated and utilized by the bird. The digestive enzymes are produced in different sites of the gastrointestinal tract and act from there in response to the feed taken by the bird and these are stable and functional only at specific P^H. Enzyme application helps in improvement of feed utilization for better output in two ways;

1. The young bird in early life lacks enzymes needed for maximum utilization of feed nutrients.
2. In other occasions the endogenous enzyme is adequate but due to the nature of the diet itself further enzyme supplementation has been shown to be of significant benefit.

In general, enzymes are used in poultry feeding aiming two well defined purposes: to complement the enzymes insufficiently produced by the bird (amylases and proteases) and to provide those enzymes which are not synthesized by them (cellulases) (Fischer *et al.*, 2002). Besides, Choct (2004) and Ferket (2004) demonstrated beneficial changes on microbial intestinal population by supplementing exogenous enzymes in the diets. Such benefits occur due to higher starch, protein and fat digestion rate in the small intestine, therefore limiting substrate for pathogenic flora that eventually exists.

Effect of enzymes on gastrointestinal environment

Microorganisms in the gastrointestinal tract utilize the digesta for energy in a similar manner to the host animal. Changes in rate of passage and the type of nutrients available to the microbes influence the different microbial populations in the digestive tract. The end products of metabolism of many of the anaerobic bacteria found in the gut are volatile fatty acids which have been shown to be altered with enzyme supplementation (Choct, 1995). However, studies examining differences in specific microbial populations such as starch or xylan-degrading bacteria have yielded no significant effects (Persia *et al.*, 1999). This may be due to lack of technology to adequately examine these populations since it stands to reason that as the substrate changes so should the microorganisms that can use them. Gastrointestinal histology has also been shown to be affected by barley and wheat based diets with reduction in villi height., increased diameter and damaged villi associated with wheat and barley diets (Viveros *et al.*, 1994; Jaroni *et al.*, 1999). Enzyme supplementation of these diets counteracted some of these effects with supplemented birds having a gut morphology more similar to birds receiving enzyme supplementation. Damage to the GI tract may make the organ more susceptible to pathogenic bacterial invasion. In addition, enzyme supplemented birds had lower gut and pancreas weight. The strain of bird used also had a bearing on these results.

Utilisation of Non-starch Polysaccharides by enzyme supplementation in poultry

Most of the feed ingredients used in the poultry diet contain non-starch polysaccharides (NSPs) and phytates at different concentrations. The anti nutritional effects of NSPs have been attributed to reduce the digestibility of nutrients (Choct & Annison, 1990). Birds do not produce enzymes like cellulase, xylanase, phytase etc., required for the digestion of NSPs and phytates. Supplementation of NSPs degrading enzymes and phytase may not only reduce the anti nutritive effects of NSPs and phytates, but also releases some nutrients from these, which could be utilized by the birds. Pourreza *et al* (2007) reported that the improvement on nutrient availability attained by exogenous enzymes supplementation occurs due to NSPs hydrolysis and consequently, the negative effects reduction of those carbohydrates, resulting in a reduction of digesta viscosity. It is well known that supplemental enzymes such as β-glucanase, xylanase, protease and amylase break the non-polysaccharide complexes in to smaller compounds thereby improving their nutritional value (Cowieson and Ravindran, 2008). Addition of enzymes such as xylanase are useful in the utilization of the non-starch polysaccharide (NSP) component of the ingredients, while others such as proteases may enhance the utilization of protein (Zamora *et al.*, 2011).

Performance

Cellulase from *Trichoderma viride* supplementation @ 0.008 % to diet containing 10-20 % wheat bran significantly reduced the feed consumption with apparent effect on improving the feed to gain ratio in broilers (Nahm and Carlson, 1985). These workers also reported significant improvement in the digestibility of cell wall components (P<0.01) and calcium, phosphorous, iron, zinc and copper associated with cell walls were solubilized by cellulose. NSP-hydrolysing enzyme preparations (xylanase) when added to wheat-based diets can improve performance in broiler chickens mediated through enzyme's influence on digestion, absorption reduction in digesta viscosity and metabolism of broiler chickens (Gao *et al.*, 2008). According to results of the study conducted by Figueiredo *et al.* (2012) xylanase supplementation (a minimum of 2500 U/g of endo-1, 4-β-xylanase) with a inclusion rate of 0.5g/kg (w/w) to a wheat based diet for 36 days duration is particularly effective during the later stage of broiler growth. Xylanase supplementation during the entire production cycle may not be required and may be replaced by a restricted supplementation in the last half of the production cycle of 36 days. These observations may be of significant importance to commercial feed manufacturers and poultry producers. Xylam® (endo-1, 4-β-xylanase, 1260 U/g and amylase 800U/g) supplementation to a corn-soybean meal based diet in broilers improved the efficiency of feed utilization and reduced the abdominal fat content (Youssef *et al.*, 2011). The result of this study also indicated that addition of enzyme to corn soybean meal diets can allow the reduction in energy level of broiler diets by about 150 kcal. with increased economic efficiency. Additionally, Lazaro *et al.* (2004) suggested that supplementation of feed with β-glucanase and xylanase improved the average daily gain and feed conversion with reduction in digesta viscosity leading to improved contact between endogenous enzymes and nutrients, therby increasing digestibility.

NSP degrading enzyme (from non-genetically modified fungus *penicillium funiculosum*) containing β-glucanase, xylanase, cellulose, pectinase and protease supplemented @ 0.5g/kg to broiler diets based on corn, wheat,rapeseed meal, canola meal and soybean meal improved the feed utlisation in starter phase with increased liver weight (Nadeem *et al.*, 2005). Supplementing multi enzyme in corn-soybean meal based diet to broilers increased body weight comparatively (Kavitha Rani *et al.*, 2003). Multi enzyme supplementation to corn soy based diets though do not significantly increase the weight gain, but significantly reduces the feed intake, improve the feed efficiency and increase the blood glucose release (Balamurugan and Chandrasekaran, 2010).

Nutrient availability

Metabolizable energy (ME) of soybean is notably low for poultry due to presence of NSPs in the cell wall matrix (Pierson *et al.* 1980). Approximately 10 % of crude protein in soybean meal is located in the dry matter with the cell wall matrix (Chesson, 2001). This source of protein and energy can be made available to broiler chickens by degradation of NSPs by carbohydrates (Mandels, 1985). Digestibility of crude protein and energy increased when carbohydrases were added to the corn soybean meal diet (Mathlouthi *et al.*, 2003; Saki *et al.*, 2005). Xylanase (4 U/g) supplementation @ 0.5 % numerically increased the digestibility of protein and energy by 3 and 6 % units, respectively in broilers fed wheat-soybean meal based diet (Nian *et al.*, 2011). These workers reported significant increase in hemicellulose digestibility by xylanase addition. Reduction of crude protein content of pullet diet decreases digestibility of protein and energy and protease supplementation of diets with reduced protein enhances the digestibility of protein and energy (Yadav and Sah, 2006). Enhanced digestibility of crude protein has been reported in chickens fed raw, autoclaved and dehulled lupins due to enzyme supplementation (Brenes *et al.*, 1993). Supplementation of different combinations of enzymes like cellulose, pectinase, xylanase, glucanase, galactanase and mannanase improved the ileal digestibilities of starch and protein in broilers fed wheat, soybean meal, canola meal and peas based diet (Meng *et al.*, 2005). This beneficial effect s of combination of enzymes observed by these workers may have resulted from elimination of the nutrient encapsulating effect of the cell wall polysaccharides and, to some extent, from the reduction of intestinal viscosity.

Effect of commercial multienzyme preparations on Broiler chicken

Performance

In order to derive maximum benefit from enzyme inclusion in poultry feeds, it is necessary to ensure that the enzymes are chosen on the basis of the feed composition. The addition of a particular enzyme must match to the substrate present in the feed. Enzyme cocktail/multienzyme containing more than one enzyme will often improve the response in bird's performance compared to the use of a pure, single enzyme assuming that the cost considerations are not ignored. This is due to the fact that feedstuffs of poultry are complex compounds containing protein, fat, fibre and other complex carbohydrates. Merely targeting a specific substrate such as β-glucan may not provide maximal benefits since layers of other substrates may inherently protect some of the ß-glucan. For example, ß-glucans and arabinoxylans may be bound to peptide and protein moieties in the cell wall of the feedstuff. Therefore, enzymes capable of hydrolyzing protein may enhance the activity of pentosanases and ß-glucanases. Methods commonly used to determine the effects of enzymes on feedstuffs include determination of the NSP content of the ingredients or by measuring the viscosity of the feed digesta. According to Schang and Azcona (2003) in presence of pentosans in wheat, oligosaccharides in soybean and phytates in every vegetable ingredient limit energy, protein and phosphorous digestibility of diets. Use of specific enzymes allows the improvement in the utilization and digestibility of nutrients in these compounds which contribute to the improvement in the performance (Dale, 2000; Veira, 2003; Fernandes and Malaguido, 2004). Multienzyme (cellulase- 146 IU/g, xylanase-241 IU/g, pectinase-98 IU/g, protease-74 IU/g, amylase-778 IU/g, phytase-33 IU/g) supplementation @ 1g/kg significantly improved tibia bone ash and tibial bone phosphorous content in broilers (Bharathidharsan *et al.*, 2009). The performance of broilers as affected by multienzyme supplementation has been presented in Table 1. Broilers fed high fibre diet supplemented with multienzyme selfeed @ 0.1-0.15% had significantly higher body weight gain, better feed conversion ratio (FCR) with improved energy utilization and better nutrient retentions (Swain *et al.*, 1996). Growth performance (body weight gain and FCR) improved (Table 1) in broilers fed diet supplemented with multienzyme (Amecozyme containing amylase, protease, β-glucanase, lipase, xylanase and cellulose) and with low density rearing (7-8 birds per 0.485m^2 area) (Hassanein, 2011). Supplementation of multienzyme Avizyme @ 0.1 % improved (Table 1) the growth (1.9 % over the negative control) and reduced the FCR (2.2 % over the NC) with increase in digestibility of crude protein and amino acids (Zanella *et al.*, 1999). Broilers fed corn soybean meal diet supplemented with REAP® (Table 1) @ 0.1 % had higher body weight gain and less abdominal fat compared to those fed control diet (Shirmohammad and Mehri, 2011). Endofeed-W supplemented @ 0.05 % to broiler diets based on corn wheat and soybean meal improved body weight, FCR and energy and protein efficiencies (Hajati *et al.*, 2009). Addition of multienzyme Pliva @ 0.05 % to broiler diets based on maize, barley, soybean meal and fish meal improved body weight gain by 6.42 % and FCR by 5.52 % (Kralik *et al.*, 2002). Kemzyme-HF supplementation to diet containing maize, ground nut cake (GNC) and brewers'dried grain (BDG) improved body weight gain , FCR and net profit (Swain *et al.*, 2005). Supplementation of Endopower @0.1 % to broiler diet with low energy and 3 % tallow improved growth and relative weight of organs (Cho *et al.*, 2012). Addition of multienzyme Tomoko @ 0.05 % to corn-soybean-wheat bran (both normal density and low density) based diets in broilers improved body weight gain and feed conversion efficiency ,however, the magnitude of improvement was better in low density diets (Abudabos, 2012). These workers also reported higher eviscerated and breast yields in broilers fed enzyme supplemented diet.

Broilers fed corn-soybean meal based diet supplemented with Bergazyme P (0.025-0.005%) had higher body weight gain and better FCR with higher breast meat yield and better protein bio-availability (Abudabos, 2010). However, these workers did not find any significant differences for all the parameters studied between the two levels of Bergazyme P supplementation. Supplementation of Rovabio ® Max to broiler diet based on corn-soybean meal improved growth and FCR (Lee *et al.*, 2010). Besides, this these workers also observed that enzyme supplementation improved the tibia mineralization with reduction in the intestinal viscosity leading to better nutrient utilization. Supplementation of problend (mixture of probiotics, yeast and multienzyme-amylase, phytase, protease, cellulose, β-glucanase, lipase) @ 400mg/kg in broiler diet improved the performance in terms of better body weight gain, FCR and higher net profit (Swain and Chakurkar, 2009).

Meat quality, Carcass yield, Organ weights and Serum mineral level

Multi-enzyme containing amylase, xylanase, pectinase, glucanase, lipase, cellulase, protease and phytase @ 50g/kg feed did not improve carcass yield and serum Ca level in broilers at 6 weeks of age but the serum inorganic P level improved significantly (Singh *et al.*, 2010). Higher liver weight was recorded in broilers fed multienzyme supplemented diet. There was no difference in abdominal fat due to multienzyme supplementation (Zakaria *et al.*, 2010). Higher carcass weight, breast weight, thigh and abdominal fat weight were recorded in broilers fed multienzyme feed supplement i.e. Amecozyme (Hassanein, 2011). Broilers fed corn soybean meal diet supplemented with REAP® (Table 1) @ 0.1 % had less abdominal fat compared to those fed control diet (Shirmohammad and Mehri, 2011). Swain *et al.* (1996) observed that broilers fed low fibre diet with supplementation of multienzyme selfeed reduced the serum cholesterol content. Supplementation of Endopower @0.1 % to broiler diet with low energy and 3 % tallow improved relative weight of organs (Ho cho *et al.*, 2012). Hajati *et al.*(2009) reported higher weights of carcass and thigh in broilers fed corn soybean meal and wheat based diet supplemented with multienzyme Endofeed-W (Table 1). Significant increase in dressing per cent flavor, juiciness and tenderness of meat was observed in broilers fed diet supplemented with Roxazyme G® @ 0.3 % (Omojola and Adesehinwa, 2007).

Layers

Effect of multienzyme supplementation on the performance, egg quality and biochemical parameters has been presented in Table 2. Multienzyme supplement i.e. farmazyme (xylanase, β-glucanase, amylase, pectinase, pentosanase and hemicellulase) @ 1 or 2 g/kg diet improved the feed conversion ratio and yolk

Table 1: Effect of multienzyme supplementation on the performance of meat type of chickens

Name of commercial multienzyme	Dose	Enzyme components	Source organisms	Diets	Effect on performance	Effect on other parameters	References
Avizyme® 1500	0.1 %	Xylanase 800 µg, Protease 6000 µg, amylase 2000 µg per gram	Trichoderma longibrachiatum,	Low energy diets, 3 % tallow	Improved growth	Increased relative weight of organs	Ho cho *et al.*, 2012
REAP®	0.1 %	β-glucanase, Protease, Cellulase	-	Cornsoybean meal and fish meal	Improved body weight gain	Reduced relative weight of abdominal fat	Shirmohammad and Mehri, 2011
Endofeed-W	0.05 %	Arabinoxylanase-2250 units, β-glucanase-700 units per gram and other enzymes cellulose, hemicellulase, α-amylase and α-galactosidase	Apergillus niger	Corn wheat and soybean meal	Improved body weight gain and FCR	Energy and protein efficiency improved, carcass weight and thigh % improved	Hajati *et al.*, 2009
Pliva	0.05 %	β-glucanase, xylanase, α-amylase, protease, β-glucosidase, cellulose, chemicellulase	-	Maize, Barley, soybean meal, Fish meal	Improved body weight gain by 6.42 % and FCR by 5.52 %	-	Kralik *et al*, 2002
Endopower	0.1 %	α-galactosidase (7 U/g), galactomannase (22 U/g), xylanase (300 U/g), β-glucanase-220 U/g		Low energy diets, 3 % tallow	Improved growth	Increased relative weight of organs	Ho cho et al., 2012

Contd.

Name of commercial multienzyme	Dose	Enzyme componens	Source organisms	Diets	Effect on performance	Effect on other parameters	References
Amecozyme	0.5 %	amylase 5500 U/g, Protease2000 U/g, β-glucanase-30 U/g, lipase 50 U/g, xylanase 500 U/g and cellulose 15 U/g		Corn soybean meal	Improved body weight gain and FCR	Increased carcass yield	Hassanein, 2011
Selfeed	0.1-0.15 %	Amylase-7500U/g, cellulose-400U/g, pectinase-200 U/g, protease-1000 U/g and lipase-300 U/g		Corn soybean meal fish meal,sunflower cake, rice bran, wheat bran	Improved body weight gain, FCR, increased feed intake	Improved energy utilization and reduced serum cholesterol	Swain *et al.*,1996
Problend	400 mg/kg	Probiotics, yeast amylase-241 IU, phytase-2200 IU, protease-40000 IU, cellulose-0.15-0.25 IU, β-galactosidase-8-10 IU, lipase-0.05-1 IU per gram		Maize –soybean based diet	Improved body weight gain, FCR	Improved net profit	Swain and Chakurkar, 2009
Kemzyme-HF	0.075 %	Cellulase complex with β-glucanase		Maize-GNC and BDG based diet	Improved body weight gain and FCR	Increased net profit	Swain *et al.*,2005
Roxazyme G®	0.3 %			Maize, GNC, fish meal, wheat offals, palm kernel cake	No effect on performance	Dressing yield, flavour, juiciness and tenderness of meat improved	Omojola and Adesehniwa, 2007

Contd.

Name of commercial multienzyme	Dose	Enzyme components	Source organisms	Diets	Effect on performance	Effect on other parameters	References
Tomoko	0.05 %	Acid protease -10000 U/g, α-amylase -40 U/g, pectinase-30 U/g, phytase-10 U/g, glucoamylase-5 U/g,	Koji-feed (Aspergillus awamori)	Corn-soybean meal-wheat bran (normal density and low density diets)	Body weight gain and FCR improved but magnitude of improvement was better for low density diet	Eviscerated weight and breast yield improved	Abudabos, 2012
Bergazyme P	0.025-0.05 %	β-pentosanase, α-amylase, β-glucanase , galactomannanase		Corn-soybean meal	Body weight gain and FCR improved	Protein bioavailability improved, breast yield increased.	Abudabos, 2010
Rovabio® Max		Xylanase-1100 viscous unit/kg, β-glucanase-100 AGL units/kg,phytase-500 FTU units/kg	Penicillium funiculosum	Maize –soybean based diet	Improved body weight gain, FCR	Improved net profit	Swain and Chakurkar, 2009

Table. 2: Effect of multienzyme supplementation on the performance of laying hens

Name	Dose	Component enzymes	Source organisms	Diets	Effect on Performance	Effect on other parameters	References
Multienzyme	0.05 %	Acid protease -10000 U/g, α-amylase -40 U/g, pectinase-30 U/g, phytase-10 U/g, glucoamylase-5 U/g,	Koji-feed (Aspergillus awamori)	Corn-soybean meal-wheat bran (normal density and low density diets)	Body weight gain and FCR improved but magnitude of improvement was better for low density diet	Eviscerated weight and breast yield improved	Abudabos, 2012
Multienzyme	0.35 %	Xylanase150 U/kg, amylase-200 U/kg, protease-2000 U/kg, phytase-200 U/kg, β-glucanase-300 U/kg, hemicellulase-1000 U/kg, pectinase-1500 U/kg		Corn-soybean meal	Numerical increase in egg production and egg mass	Improvement in egg shell quality and reduction in no. of broken eggs	Malekian *et al.*, 2013
Superzyme-OM Canadian Bio-systems Inc., Calgary, Alberta, Canada	Units/ Kg	Pectinase-1100 U, cellulose-50 U, xylanase-1000 U, glucanase-600 U, mannanase-400 U, galactanase-50 U, amylase-2500 U	Saccharomyces cerevisiae	Corn, broken rice, wheat,soybean meal, sunflower meal, rapeseed meal, fish meal	Improved egg production, egg weight and egg mass	-	Khan *et al.*, 2011
Kemzyme ®	0.5 %	Protease, α-amylase, lipase, cellulose and β-glucanase			Improved egg production		Shehata *et al.*, 2000
Farmazyme2010TM	0.1-0.2 %	Xylanase, β-glucanase, α-amylase, pectinase, pentosanase and hemicellulase		Corn soybean meal	Decreased feed intake, improved FCR	Increased serum albumin and yolk index	Yoruk *et al.*, 2006

index in laying hens fed corn-soybean meal based diets which improved the net profit also (Yoruk *et al.*, 2006). Multi-enzyme supplementation @250-350g/ton feed had a positive effect on egg shell thickness but no effect on egg production, egg weight or egg quality during 26 to 34 weeks age (Daramani *et al.*, 2010). Multi-enzyme complex (xylanase, amylase, protease, phytase, β-glucanase, hemicellulase and pectinase) supplementation numerically improved the egg weight and egg production but the improvement in egg shell quality was significant which led to decrease in the proportion of cracked and broken eggs (Malekin *et al.*, 2013). Multienzyme (Phytase, xylanase,cellulose, acid protease and amylase) supplementation @ 0.1 % to the diet of laying hens fed maize-soybean meal based diet (Table 2) increased the activities of digestive enzymes and nutrient retentions without any effect on performance (Wen *et al.*, 2012). Layers fed diet supplemented with Kemzyme @ 0.5 % improved egg production (Shehata *et al.*, 2000). Supplementation of superzyme - OM to layer diets containing mixture of corn, broken rice , wheat soybean meal, sunflowere cake, rape seed meal and fish meal improved egg production, egg weight and egg mass (Khan *et al.*, 2011).

Prospects of microbial phytase in utilization of phytate-P by poultry

Up to about 85% of the total phosphorus found in feedstuffs of vegetable origin, particularly the cereals, cereal by-products and oil cakes is present in the form of phytic acid (inositol hexa-hosphate). Under most dietary conditions the phytate phosphorus is poorly utilized by monogastrics including poultry and, consequently, excreted via the feces. The bird's need for phosphorus, therefore, has to be met through supplementing diet with inorganic phosphorus sources such as defluorinated phosphate or the dicalcium phosphate. The traditional practice of adding inorganic phosphorus in poultry ration adds to the cost of feeding besides constraints in finding the right kind of stuff for regular use. Considerable research emphasis has been placed on reducing the phosphorus (P) content of poultry manure (Paik, 2003; Maguire *et al.*, 2004; Smith *et al.*, 2004). Rezaei *et al.* (2007) reported the use of phytase as a way to reduce phosphorous emission to the environment. The reason for this is that long term application of poultry manure to agricultural land often leads to soil P accumulation which has the potential to accelerate P transfer and runoff into water bodies (Maguire *et al.*, 2004). Presently in India, about 12 million tones of compounded poultry feed are being manufactured annually. Considering the minimum level of 1% addition in diet, about 0.12 tones of dicalcium phosphate of worth USD400, 000 is required. Besides being an indigestible constituent the phytate phosphorus tends to act as an anti-nutritional factor in diet in that it possesses strong chelating properties thereby markedly reducing the bioavailability of several multivalent cations mainly the Ca^{2+}, Mn^{2+}, Zn^{2+},

Fe^{2+} and Fe^{3+} by forming insoluble, phytate-metal, complexes. Moreover, the phytate also decreased protein digestibility through formation of indigestible protein-phytate complex and strongly inhibit the activity of amylases. Most feed ingredients of plant origin contain from 55-85% of the total phosphorus in phytate form (Johri, 2006) (Table 3). The cereal by- products like wheat or rice brans and oilseed meals usually possess higher levels of total phosphorus than that in whole grains and with a larger proportion bound as phytate. Hence, supplementation of phytase enzyme to poultry feed not only improves the utilization of phosphorous but also increases the availability of minerals like Ca, Mn, Zn and Fe. In addition to this phytase enzyme improves the digestibility of protein, fat, amino acids and energy in poultry.

Factors affecting the efficacy of phytase enzyme and phosphorus bioavailability

The practical poultry rations are largely composed of vegetable oil cakes, meals, grains, and their byproducts. Phosphorus bioavailability from such feedstuffs is only about 30%. The hydrolysis and absorption of phytate phosphorus by monogastric animals are complex processes that are affected by certain factors like Ca bioavailability and inorganic phosphorus supplements, age of the bird (Kornegay, 1996), available phosphorus levels, amount of vitamin D3 present (Mohammad *et al.*, 1991; Edwards, 1993), Ca to total P ratio (Liu *et al.*, 1998), weight of the bird (Punna and Roland, 1999) and the level of phytase in the diet (Shirley and Edwards, 2003). The values reported in the literature on the availability in most of the single and many feed mixtures are variable. In feedstuffs of plant origin, the P occurs in association with phytates and the variation in its availability may party be due to the presence of the endogenous phytase which hydrolyses phytate and affects P bioavailability. High dietary levels of calcium have been reported to have a slightly negative effect on phytate P utilization in pullets and laying hens. It has been reported that a chick's ability to retain phytate P increases when the diet contained inadequate levels of non-phytin phosphorus (NPP) compared to normal phosphorus levels. Phytate P is more efficiently retained when chickens are fed suboptimal levels of NPP. It is generally conceded that older birds hydrolyze phytate phosphorus to a greater extent than chicks, the basis of this being that there is more of the phytase activity present in the gastrointestinal tract of older birds. Limited reports indicate that there may be breed and strain differences in the utilization of phytate phosphorus. Study on the relative ability of three genotype (White Leghorn, Kadaknath and Aseel) on phytate phosphorus utilization from diets varying in calcium and phosphorus has indicated that the chicks of different genotypes differed in their response to utilize phytate phosphorus (Johri, 2006).

In view of such undesirable effects of phytic acid, it is preferred to either remove it altogether or reduce its amount in poultry feed or ingredients. Efforts have been made in different laboratories to either eliminate or reduce phytic acid content in plant feedstuffs through chemical methods, solid state fermentation technology, autolysis or by the use of phytase enzyme in diet. Of these, supplementation of diet with microbial phytase appears to be more promising.

Effect of microbial phytase on the performance of meat type birds

Predictably, the addition of phytase to P-inadequate diets has been consistently shown to enhance growth performance. Phytase supplementation @ 250 or 500 U/kg in corn-soybean meal diets yielded significant effects on growth in broilers fed corn-soybean meal diet (Libert *et al.*, 2008). Supplementation of phytase @ 500U/Kg to low NPP (0.30 %) diets improved the growth performance of broilers and allowed complete, safe and economical replacement of dietary P (di-calcium phosphate) to reduce feed cost per kg live weight gain of broilers (Singh *et al.*, 2003). Considerable work has been done in India in recent years on the effectiveness of supplemental microbial phytase in improving the P bioavailibity in poultry. Microbial phytase supplementation @ 250-500 units/kg in diet increased body weight gain and feed intake in broiler chicken. However, as a result of simultaneous increase in body weight gain and feed intake, several reports have indicated that microbial supplementation had no significant effect on feed to gain ratio in broiler chicken. Improvement in chick performance, however, has been related to an improved utilization of dietary phosphorus. The phytase supplementation of diet was found associated with a significant increase in plasma inorganic phosphorus concentrations, tibia ash percentages, as well as improved absorption and reduced excretion of dietary phosphorus. However, contrary to these reports, the addition of dietary phytase to corn-soya diet containing less phosphorus than the NRC (1994) recommendation did not improve either body weight gain or feed intake, but it did increase toe and tibia ash and plasma inorganic phosphorus in broiler chickens, growing quails and guinea fowl (Johri, 2006). The apparent cause of dissimilar results may partly be attributed to the complex nature of factors influencing the bio-efficacy of phytase and utilization of phytate phosphorus.

Nonetheless, the phytase supplementation to a standard broiler diet can allow the reduction of the usual addition of inorganic phosphate in the diet and also reduce the amount of P excretion into the environment. Enhanced level of vitamin D_3 can also be used as an alternative to phytase to improve phytate P availability in low Ca and NPP diets. Phytase addition to broiler diet with low phytate barley increased the availability of P due to which it may be possible to reduce the amount of inorganic P used when formulating diets with low phytate barley compared with the levels needed when formulating diets with normal phytate barley (Thacker *et al.*, 2009).

Effect of microbial phytase on the performance of laying birds

As layer feed differs from broiler feed in respect of among other things, very high calcium content, the supplemental phytase may just not be as effective in presence of high dietary calcium (3.5%) as in case of broilers (1.0%) (Johri, 2006). A number of studies have been conducted on phytase supplementation in layers with more conventional inorganic P than tricalcium phosphate (TCP) (Boling *et al.*, 2000a; Keshavarz, 2003; Sohali and Roland, 2000; Ceylan *et al.*, 2003; Francesh *et al.*, 2005 and Panda *et al.*, 2005). The thrust of these studies in which diets were based mainly on maize and soybean meal is that phytase supplementation of layer diets permits reduction in non-phytate P (NPP) for layers consequently P excretion without compromising performance and egg quality. It is worthy to mention that phytase inclusion levels (250-350 FTU/kg) in layer diets are generally lower than that for broilers which may be related to larger retentions of feed in the crop, which facilitates phytate degradation. It is possible that phytase enhances Ca availability (Sohail and Roland, 2000) and Ca influences phytase efficacy (Scott *et al.*, 1999). Hence, the possibility to reduce additional nutrient specifications is associated with phytase supplementation can be exploited in formulation of layer feed (Scott *et al.*, 2001) which could be economically advantageous. Phytase supplementation improved egg production and reduced percentage of broken and soft eggs and P excretion (Lim *et al.*, 2003). However, it was concluded that dietary levels of Ca and NPP could significantly influence the effects of phytase supplementation. Phytase supplementation to diet with low NPP increased hen housed egg production, hen day egg production(%), fertility (%) and FCR of broiler breeders (Plumstead, 2007). Contrary to these findings, some workers did not find any improvement in the production performance of laying birds. Panda *et al.* (2005) found that addition of phytase @ 500U/Kg to a diet containing 1.8g/Kg NPP had no additional advantage on shell weight, shell thickness, shell strength and tibia strength. However, it has been demonstrated that phytase supplementation @ 250 units/kg in layer eliminated inorganic P supplementation without affecting laying performance. Microbial phytase supplementation (300U/kg diet) was not effective in improving egg production, egg weight in addition to utilization of P, N and ME in laying hens fed corn-soybean meal based diets with low NPP (Liebert *et al.*, 2005). No significant difference between the enzyme treated group and control was found for egg shell quality in laying hens fed diet supplemented with phytase (Cabuk *et al.* 2004; Casartelli *et al.*, 2005).

Effects of microbial phytase on mineral utilization

Phytate is capable of forming complexes with proteins and inorganic cations such as calcium, magnesium, iron and zinc. The use of phytase not only releases the bound phosphorous but also these essential minerals and nutrients which led to higher nutritional value for the feed ingredients (Ceylan *et al.*, 2003; Panda *et al.*, 2005). Rezaei *et al.* (2007) reported the use of phytase as a way to reduce phosphorous emission to the environment. Phytic acid (PA) has strong chelating potential and forms a variety of complexes with cations and proteins, rendering, these nutrient biologically unavailable. Theoretically, when PA is hydrolysed by microbial phytase, all minerals bound to it should be released. Phytase supplementation to low NPP (0.30 %) improved the relative retention of nutrients (N, Ca, P) and minerals (Ca, P) status of blood and bone in broiler chickens fed maize based diets (Singh *et al.* 2003). Relatively, very little work has been done in India on the effect of phytase supplementation on the availability of minerals. Reports on this aspect suggested that contents of ash and minerals in tibia such as Ca, P, Mg, Zn and Fe were significantly higher in phytase supplemented layer diets (Johri, 2006). Woyengo *et al.* (2010) found that phytase supplementation improved ileal phytate and total P digestibilities and retention of total P due to liberation of P from the PA molecule , which has also been reported by several other studies (Dilger *et al.,* 2004; Olukosi *et al.*, 2007; woyengo *et al.*, 2008).

The efficacy of phytase with regard to hydrolysis of PA is, however, limited, in part, by presence of divalent cations like Ca in practical poultry diets, which form insoluble complexes with PA at P^H found in small intestine (Selle and Ravindran, 2007). Addition of organic acids to phytase-supplemented diets can improve the efficacy of phytase because the organic acids chelate multivalent cations like Ca, thereby reducing the amount of cations available for binding PA (Boling *et al.*, 2000). Furthermore, the organic acids reduce the P^H of the digesta (Radcliffe *et al.*, 1998) which can result in increased dissociation between PA and minerals (Maenz *et al.*, 1999) and thus increased phytase efficacy due to more acidic P^H (Simon and Igbasan, 2002).

Table 3: Phytate phosphorus contents and phytase activities of plant feed ingredients

Ingredient	Phytate P (%)	PhytateP (as % of total P)	Phytase activity (units/kg)
Cereals			
Maize	0.24 (0.17-0.29)	72 (66-85)	15(0-46)
Barley	0.27 (0.19-0.33)	64(56-70)	582(408-882)
Wheat	0.27 (0.17-0.38)	69(60-80)	1193(915-1581)
Oats	0.29 (0.22-0.35)	67 (59-78)	42(0-108)
Rye	0.22 (0.20-0.23)	61(56-66)	5140 (4132-6127)
Sorghum	0.24 (0.21-0.28)	66 (64-69)	24(0-76)
Foxtail millet	0.19	70	-
Finger millet	0.14	58	-
Rice	0.27 (0.25-0.28)	77 (74-81)	-
Rice polished	0.09 (0.04-0.17)	51 (49-55)	-
Cereal By-Products			
Rice bran	10.31(10.02-10.09)	80 (72-86)	122(108-135)
Wheat bran	0.92 (0.88-0.96)	71 (70-72)	2957(1180-5208)
Rice polishing	20.42	89	
Oilseed Meal			
Soybean meal	0.39 (0.37-0.42)	60 (57-61)	8(0-20)
Cottonseed meal	0.84 (0.75-0.90)	70 (70-71)	-
Peanut meal	0.48	80	3(0-8)
Rapeseed meal	0.70 (0.54-0.78)	59 (43-70)	16(0-36)
Sunflower meal	0.89	77	62(0-185)
Coconut meal	0.29 (0.26-0.33)	49 (43-56)	24(0-80)
Sesame meal	10.18 (10.03-10.46)	81 (77-84)	-

Effects of phytase on nutrient digestibility

It has been suggested that the de novo formation of binary protein-phytase complexes in the gastrointestinal tract, which are refractory to pepsin activity may be key mechanism whereby phytate depresses the digestibility of dietary amino acids (Selle *et al.*, 2000). The other possible mode of action is that phytate may induce increases in endogenous amino acid flows (Ravindran *et al.*, 1999a; Cowieson *et al.*, 2004). Both mechanisms would depress apparent ileal digestibility of amino acids in poultry diets which should be countered, at least in part, by phytase supplementation. Ravindran *et al.* (1999a) and Rutherfurd *et al.* (2002) found that phytase increased the ileal digestibility of essential amino acids in wheat (9.2 and 13.4 %) to a greater extent than in maize (3.3 and 3.9 %). In another study, phytase increased the average digestibility of 14 amino acids by 5.1 % (0.839 % vs 0.800) in diets based on wheat (8 % inclusion) and casein (Ravindran *et al.*, 1999b). It is possible that phytate is more likely to complex wheat protein than maize protein as discussed by Selle *et al.* (2000). Significant effect of phytase addition@500U/kg on N utilization was established in broilers fed corn-soybean meal diet from 7-42days (Liebert *et al.*, 2008). Relative retention of N was improved in broilers fed maize based diets supplemented with phytase @ 500U/kg (Singh *et al*, 2003). Phytase has been shown to increase the co-efficient of apparent ileal digestibility (CAID) (Ravindran *et al.*, 1999a) and co-efficient of true ileal digestibility (CTID) (Rutherfurd *et al.*, 2002) of amino acids of individual feed ingredients in broilers to a marked extent. Phytate-Ca forms insoluble complex with starch and fatty acids in the gastrointestinal tract of broilers which depress the digestibility of carbohydrates and fats (Rama and Reddy, 2001). Therefore, supplemental phytase had positive effects on dry matter digestibility by releasing bound organic nutrients like starch (Knuckles and Betschart, 1987) and protein and amino acids (Ravindran *et al.*, 2000).

Summary

Numerous studies over the past ten years have demonstrated improvements in feed utilization with enzyme supplementation. Use of β-glucanases in barley diets and pentosanases in feed ingredients high in NSP is now common practice. Reduction in viscosity of the digesta may enhance nutrient utilization by the bird by "normalizing" the histology of the gut when NSP diets are fed. Evidence also exist that enzyme cocktails containing proteases may enhance the beneficial effects associated with these enzymes. As enzyme technology improves we have also seen benefits in areas not traditionally associated with digestive inefficiencies such as energy and protein utilization from soy and other feed ingredients. Tradition tells us that the use of enzymes in corn/soy diets is not efficacious. However, recent advances indicate that this is not the case. Future developments in enzyme technology will likely focus on more thermo-tolerant enzyme preparations, greater enzyme activity and enzymes which function optimally at low gastric pH values. Additionally, as more is known of the chemical nature of our feed ingredients, better methods of degrading these compounds may be found. In reporting studies on enzyme efficacy it is becoming increasingly important to ensure that the units of activity of the enzyme, as well as the use rate, be reported to improve our understanding of efficacy. Commercial enzymes can come from a variety of source microorganisms. The organisms produce enzymes that may have different pH and temperature optima. Because of this, different enzyme manufacturers may use other units of activity to describe an enzyme. Hence, the focus of enzyme supplementation should be on source, units of expression, heat stability for better storage, facilities to determine the activities of different enzymes for targeting the specific substrate/ component of the feed to derive maximum benefit.

References

Abudabos, A. 2010. Enzyme supplementation of corn-soybean meal diets improves performance in broiler chickens. *Int. J. Poult. Sci.*9:292-297.

Abudabos, A. M.2012.Effect of enzyme supplementation to normal and low density broiler diets based on corn-soybean meal. *Asian J. Anim Vet. Adv.*7:139-148.

Balamurugan, R. and D. Chanrasekaran. 2010. Effect of multienzyme supplementation on weight gain, feed intake, feed efficiency and blood glucose in broiler chickens. *Indian J. Sci. Technol.* 3:193-195.

Bharathidharsan, A., D. Chandrasekaran, A. Natarajan, R. Ravi and S. Ezhilvalavan. 2009. Effect of enzyme supplementation on carcass quality, intestinal viscosity and ileal digestibilities of broilers to nutrient reduced diet. *TamilNadu J. Vet. Anim. Sci.* 5: 239-245.

Boling, S. D., M. W. Douglas, M. L. Johnson, X. Wang, C. M. Parsons, K. W. Koelkebeck, R. Zimmerman. 2000a. The effects of dietary available phosphorous levels and phytase on performance of young and older laying hens. *Poult. Sci.* 79:224-230.

Brenes, A., M. Smith, W. Guenter and R. R. Marquardt. 1993. Effect of enzyme supplementation on the performance and digestive tract size of broiler chicken fed wheat and barley based diets. *Poult. Sci.* 72:1731-1739.

Cabuk,, M., M. Bozukurt, F. Kirpinarand and O. Ozku. 2004. Effect of phytase supplementation of diets with different levels of phosphorous on performance and egg quality of laying hens in hot climatic condition. *South African J. Anim. Sci.* 34: 13-17.

Casartelli, E. M., O. M. Junqueira, A. C. Laurentiz, J. Lucas junior and L. F. Araujo. 2005. Effect of phytase in laying hens diet with different phosphorous sources. *Brazilian J. Poult. Sci.* 7:93-98.

Ceylan, N., S. E. Scheideler and H. L. Stilborn. 2003. High available phosphorous corn and phytase in layer diets. *Poult. Sci.* 82:789-795.

Chesson, A. 2001. Non starch polysaccharide degrading enzymes in poultry diets: influence of ingredients on the selection of activities. *World Poult, Sci. J.* 57:251-263.

Cho, J.H., P.Y.Zhao and I. H. Kim. 2012. Effect of emulsifier and multi-enzyme in different energy density on growth performance, blood profile and relative organ weight in broiler chickens. *J.Agric. Sci.* 4:161-168.

Choct, M. 1995. Chicken Meat Research and Development Council-Final Report on Project CSN 2 CM, CSIRO Division of Human Nutrition, Adlaide, South Australia.

Choct, M. and G. Annison.1990. Anti-nutritive activity of wheat pentosans in broiler diets. *Br. Poult. Sci.* 31:811-821.

Choct, M. 2004. Effect of organic acids, prebiotics and enzymes on control of necrotic enteritis and performance of broiler chickens. School of Rural Science and Agriculture, University of New England.

Cowieson, A. J., T. Acamoic and M. R. Ford. 2000. Enzyme supplementation of diets containing *Camelina sativa* meal for poultry. *Br. Poult. Sci.* 41:689-690.

Cowieson, A. J., T. Acamovic and M. R. Bedford. 2004. The effects of phytase and phytic acid on the loss of endogenous amino acids and minerals for broiler chickens. *Br. Poult. Sci.* 45: 101-108.

Cowieson, A. J. and V. Ravindran. 2008. Effect of exogenous enzymes in maize-based diets varying in nutrient density for young broilers growth performance and digestibility of energy, minerals and amino acids. *Br. Poult. Sci.* 49: 37-44.

Dale, N. 2000. Soy products as energy sources for poultry. In: Drackley, J.K. Soy in Animal Nutrition, Illinois, Savoy, IL, pp. 283-288.

Dilger, R. N., E. M. Onyango, J. S. Sands and O. Adeola.2004. Evaluation of microbial phytase in broiler diets. *Poult. Sci.* 83:962-970.

Edwards, H. M. Jr.1993. Dietary 1, 25-dihydroxycholecalciferol supplementation increases natural phytate phosphorous utilization in chickens. *J. Nutr.* 123:567-577.

FAO, FAOSTAT.2011. In http://faostat.fao.org/site/291/default.aspx, Food and Agriculture Organization, Rome.

Ferket, P. R. 2004. Alternatives to antibiotics in poultry production: Responses, practical experience and recommendations. In: Alltech's Annual Symposium, 20th, 2004, Lexington: Alltech, pp.54-67.

Figueiredo, A. A., B. A. Correia, T. Ribeiro, P. I. P. Ponte, L. Falcao, J. P. Freire, J. A. M. Prates, L. M. A. Ferreira, C. M. G. A. Fontes and M. M. Lordelo. 2012. The effects of restricting enzyme supplementation in wheat-based diets of broilers. *Anim. Feed Sci. Technol.* 172: 194-200.

Fischer, G., J. C. Maier and F. Rutz.2002. Performance of broilers fed diets based on corn and soybean meal , with or without addition of enzymes. *Brazilial J. Anim. Sci.* 31:402-410.

Francesch, M. J. Broz and J. Brufau. 2005. Effects of an experimental phytase on performance, egg quality, tibia ash content and phosphorous bioavailability in laying hens fed on maize or barley based diets. *Br. Poult. Sci.* 46:340-348.

Gao, F., Y. Jiang, G.H., Zhou and Z.K. Han. 2008. The effects of xylanase supplementation on performance, characteristics of gastrointestinal tract, blood parameters and gut microflora in broilers fed on wheat-based diets. *Anim. Feed Sci. Technol.* 142:173-184.

Hassanein, H. H. M. 2011. Growth performance and carcass yield of broilers as affected by stocking density and enzymatic growth promoters. *Asian J. Poult. Sci.* 5:94-101.

Hajati, H., M. Rezaei and H. Sayyahzadeh. 2009. The effects of enzyme supplementation on performance, carcass characteristics and some blood parameters of broilers fed on corn-soybean meal-wheat diets. *Int. J. Poult. Sci.* 8:1199-1205.

Jaroni, D., S. E. Scheideeler, M. M. Beck, C. Wyatt. 1999. The effect of dietary wheat middlings and enzyme supplementation II. Apparent digestibility, digestive tract size, gut viscosity, and gut morphology in two strain of Leghorn hens. *Poult. Sci.* 78:1664-1674.

Johri, T. S. 2006. Poultry Nutrition Research in India and its perspective. 3. Prospects of microbial phytase in phosphorous utilization by poultry. www.fao.org/doc.rep/ ARTICLE/ AGRIPPA/ 659_en-10.htm visited on March 16, 2013.

Karamani, K., M. S. Ranjbari, F. Fatahnia and Y. Mohamadi. 2010. Effects of multi-enzyme supplementation on performance and egg quality of laying hens fed diets with high levels of wheat bran. *Anim. Sci. J.* 23:84-91.

Kavitha Rani, B., J. Desai, D. Reddy and P. M. Radhakrishna. 2003. Effect of supplementation of enzymes for non starch polysaccharides in corn-soya diet in broilers. *Indian J. Anim. Nutr.* 20:63-69.

Keshavarz, K. 2003. Effects of continuous feeding of low-phosphorous diets with and without phytase during the growing and laying periods of two strains of leghorns. *Poult. Sci.* 82: 1444-1456.

Khan, S. H., A. Muhammad, N. Mukhtar, A. Rahman and G. Fareed. 2011. Effects of supplementation of multi-enzyme and multi-species probiotic on production performance, egg quality, cholesterol level and immune system in laying hens. *J. Appl. Anim. Res.* 39: 386-398.

Knuckles, B. E. and A. A. Betschart.1987. Effect of phytate and other myo-inositol phosphate esters on alpha-amylase digestion of starch. *J. Food. Sci.* 52:719-721.

Kornegay, E. T. 1996. Nutritional, environmental and economic considerations for using phytase in pig and poultry diets. Pages 277-299 in Nutrient Management of Food Animals to Enhance and Protect the Environment. E. T. Kornegay, ed. Lewis Publishers, New York.

Kralik, G, I. Bogut, Z. Skrtic and G Kusec. 2002. The effect of multienzyme preparation on the growth performance of broilers. *Acta Agraria Kaposvariensis*. 6:245-252.

Lazaro, R., M. A. Latorre, P. Medel, M. Garcia and G. G. Mateos. 2004. Feeding regimen and enzyme supplementation to rye based diets for broilers. *Poult. Sci.* 83:152-160.

Lee, S.Y., J. S. Kim, J. M. Kim, B. Ki An and C.W. Kang. 2010. Effects of multiple enzyme (Rovabio ® Max) containing carbohydralases and phytase on growth performance and intestinal viscosity in broiler chicks fed corn-wheat-soybean meal based diets. *Asian-Aust. J. Anim. Sci.* 23:1198-1204.

Liebert, F., J. K. Htoo and A. Sunder. 2005. Performance and nutrient utilization of laying hens fed low phosphorous corn-soybean and wheat-soybean diets supplemented with microbial phytase. *Poult. Sci.* 84:1576-1583.

Liebert, F., J. K. Htoo and A. Sunder. 2008. Microbial phytase and nutrient utilization in low phosphorous chicken diets. J. *Poult. Sci.* 45:255-264.

Lim, H. S., H. Namkung and I. Paik. 2003. Effects of phytase supplementation on the performance, egg quality, and phosphorous excretion of laying hens fed different levels of dietary calcium and non-phytate phosphorous. *Poult. Sci.* 82: 92-99.

Liu, J., D. W. Bollinger, D. R. Ledoux and T. I. Veum. 1998. Lowering the dietary calcium to phosphorous ratio increases phosphorous utilization in low phosphorous corn-soybean meal diets supplemented with microbial phytase for growing finishing pigs. *J. Anim. Sci.* 76:808-813.

Maenz, D. D., C.M. Engele-Schaan, R.W. Newkirk and H. I. Classen. 1999. The effect of minerals and mineral chelators on the formation of phytase- resistant and phytase-susceptible forms of phytic acid in solutionand in slurry of canola meal. *Anim. Feed Sci. Technol.* 81:177-192.

Maguire, R.O., J.T.Sims, W.W.Saylor, B.L.Turner, R. Angel and T.J. Applegate. 2004. Influence of phytase addition to poultry diets on phosphorous forms and solubility in litters and amended soils. *J. Environ. Qual.* 33: 2306-2316.

Malekin, G, A.K.Zamani Moghaddam and F. Khajali. 2013. Effect of using enzyme complex on productivity and hatchability of broiler breeders fed a corn-soybean meal diet. *Poult. Sci. J.* 1:36-45.

Mandels, M. 1985. Application of cellulases. *Biochem Soc. Transactions.* 13:414-415.

Mathews, C. K. and K. E. van Holde. 1990. Biochemistry. Redwood City, CA: The Benjamin/ Cummings Publishing Company.

Mathlouthi, N., M. A. Mohamoed and M. Larbier. 2003. Effect of enzyme preparation containing xylanase and β-glucanase on performance of laying hens fed wheat/barley or maize/soybean meal-based diets. *Br. Poult. Sci.* 44:60-66.

Meng, X., B. A. Slominski, C. M. Nyachoti, L. D. Campbell and W. Guenter. 2005. Degradation of cell wall polysaccharides by combinations of carbohydrase enzymes and their effect on nutrient utilization and broiler chicken performance. *Poult. Sci.* 84:37-47.

Mohammad, A. M., M. J. Gibney and T. G. Taylor.1991. The effects of dietary levels of inorganic phosphorous, calcium and cholecalciferol on the digestion of phytate phosphorous by the chick. *Br. J. Nutr.* 66: 251-259.

Nadeem, M. A., M. I. Anjum, A. G. Khan and A. Azim. 2005. Effect of dietary supplementation of non-starch polysaccharide degrading enzymes on growth performance of broiler chicks. *Pakistan Vet. J.* 25: 183-188.

Nagalakshmi, R. and G. Devegowda. 1991. Effect of supplementation of enzymes in different diets on broiler performance. Department of Poultry Science, M. V. Sc. Thesis submitted to Univ. Agric. Sci. Bangalore, India.

Nahm, K. H. and C. W. Carlson.1985. Effects of cellulose from *Trichoderma viride* on nutrient utilization by broilers. *Poult. Sci*.64:1536-1540.

Nian, F., Y. M. Guo, Y. J. Ru., F. D. Ali and A. Peron.2011. Effect of exogenous xylanase supplementation on the performance, net energy and gut microflora of broiler chickens fed wheat based diets. *Asian-Aust. J. Anim. Sci*. 24:400-406.

National Research Council. 1994. Nutritional Requirements of Poultry, nineth revised ed. National Academic Press, Washington, DC.

Olukosi, O. A., A.J. Cowieson and O. Adeola. 2007. Age related influence of a cocktail of xylanase, amylase and protease or phytase individually or in combination in broilers. *Poult. Sci*. 86:77-86.

Omojola, A. B. and A. O. K. Adesehinwa. 2007. Performance and carcass characteristics of broiler chickens fed diets supplemented with graded levels of Roxazyme G®. *Int. J. Poult. Sci*. 6:335-339.

Paik, I.2003. Application of phytase, microbial or plant origin, to reduce phosphorous excretion in poultry production. *Asian Aust. J. Anim. Sci*. 16:124-135.

Panda, A. K., S. V. Rama Rao, M. V. L. N. Raju and S. K. Bhanja. 2005. Effect of microbial phytase on production performance of white Leghorn layers fed on a diet low in non-phytate phosphorous *Br. Poult. Sci*. 46:464-469.

Persia, M. E., B. A. Dehority and M. S. Liburn. 1999. Cereal bacterial concentrations in turkey hens fed corn or wheat based diets with and without supplemental enzymes. *Poult. Sci*. 78 (Suppl. 1):16.

Pierson, E. E. M., L. M. Potter and R. D. Brown. 1980. Amino acid digestibility of dehulled soybean meal by adult turkeys. *Poult. Sci*. 59:845-848.

Plumstead, P.W.2007. Strategies to reduce fecal phosphorous excretion in the broiler industry without affecting performance, P.D. faculty of North Carolina State University.

Poureza, J., A. H. Samie and E. Rowghani. 2007. Effect of supplemental enzyme on nutrient digestibility and performance of broiler chicks fed diet containing triticale. *Int. J. Poult. Sci*. 6:115-117.

Punna, S. and D. A. Ronald, Sr. 1999. Variation in phytate phosphorous utilization within the same broiler strain. *J. Appl. Poult. Res*. 8:10-15.

Rama Rao, S. V. and V. Ramasubba Reddy.2001. Phytic phosphorous for eco-friendly products. *Poult. International*. 40: 46-52.

Radcliffe, J. S., Z. Zhang and E. T. Kornegay. 1998. The effects of microbial phytase, citric acid and their interaction in acorn-soybean meal-based diet for weanling pigs. *J. Anim. Sci*. 76:1880-1886.

Ravindran, V., S. Cabahug, G. Ravindran and W. L. Bryden. 1999a. Influence of microbial phytase on apparent ileal amino acid digestibility in feedstuffs for broilers. *Poult. Sci*. 78:699-706.

Ravindran, V., P. H. Selle and W. L. Bryden. 1999b. Effects of phytase supplementation, individually and in combination, on nutritive value of wheat and barley. *Poult. Sci*. 78:1588-1595.

Ravindran, V., S. Cabahug, G. Ravindran, P. H. Selle and W. L. Bryden. 2000. Response of broiler chickens to microbial phytase supplementation as influenced by dietary phytic acid and non-phytate phosphorous levels. II. Effects on apparent metabolizable energy, nutrient digestibility and nutrient retention. *Br. Poult. Sci*. 41: 193-200.

Rezai, M., S. Borbor and M. Zaghari. 2007. Effect of phytase supplementation on nutrients availability and performance of broiler chicks. *Int. J. Poult. Sci*. 6:55-58.

Rutherfurd, S. M., T. K. Chung and P. J. Moughan.2002. The effect of microbial phytase on ileal phosphorous and amino acid digestibility in broiler chickens. *Br. Poult. Sci*. 44:598-606.

Saki, A. A., M. T. Mazugi and A. Kamyab. 2005. Effect of mannanase on broiler performance,ileal and in vitro protein digestibility, uric acid and litter moisture in broiler feeding. *Int. J. Poult. Sci.* 4:21-26.

Schang, M. J. and J. O. Azcona. 2003. Natural enzyme applications to optimize animal performance. In: Alltech's Annual Symposium, 19, Lexington. Proceedings. Lexington Alltech. pp. 54-57.

Scott, T. A., R. Kampen and F. G. Silversides. 1999. The effects of phosphorous, phytase enzyme, and calcium on the performance of layers fed corn-based diets. *Poult. Sci.* 78: 1742-1749.

Scott, T. A., R. Kampen and F. G. Silversides. 2001. The effect of adding exogenous phytase to nutrient-reduced corn and wheat based diets on performance and egg quality of two strains of laying hens. *Can. J. Anim. Sci.* 81: 393-401.

Selle, P. H., V. Ravindran, R. A. Caldwell and W. L. Bryden. 2000. Phytate and phytase: consequences for protein utilization. *Nutr. Res. Rev.* 13: 255-278.

Selle, P. H. and V. Ravindran. 2007. Microbial phytase in poultry nutrition. A review. *Anim. Feed Sci. Technol.* 135:1-41.

Shehata, A. A. M.2000. Using some aquatic plants in feeding chicks.Ph.D. Thesis, Faculty of Agriculture, Zagazig University, Egypt.

Shirley, R. B. and H. M. Edwards, Jr. 2003. Graded levels of phytase past industry standards improves broiler performance. *Poult. Sci.* 82:671-680.

Shirmohammad, F. and M. Mehri.2011. Effect of dietary supplementation of multi-enzyme complex on the energy utilization in rooster and performance of broiler chicks. *African J. Biotechnol.* 45: 9200-9206.

Simon, O. and F. Igbasan. 2002. In vitro properties of phytase from various microbial origins. *Int. Food Sci. Technol.* 37:813-822.

Singh, P. K., V. K. Khatta, R. S. Thakur, S. Dey and M. L. Sangwan.2003. Effects of phytase supplementation on the performance of broiler chickens fed maize and wheat based diets with different levels of non-phytate phosphorous. *Asian-Aust. J. Anim. Sci.*11:1642-1649.

Singh, J., C. P. Verma and R. K. Sharma. 2010. Effect of period of supplementation of multienzyme on carcass yield serum minerals in broilers during humid weather. *Haryana Vet.* 49:59-61.

Sklan, D. 2001. Development of the digestive tract of poultry. *World. Poult. Sci.* 57:415-428.

Smith, D.R., P. A. Moore, D. M. Miles, B. E. Haggard and T.C. Daniel.2004.Decreasing phosphorous runoff loses from land applied poultry litter with dietary modifications and alum addition. *J. Environ. Qual.* 33: 2210-2216.

Sohali, S. S. and D. A. Roland. 2000. Influence of phytase on calcium utilization in commercial layers. *J. Appl. Poult. Res.* 9: 81-87.

Swain, B. K., T. S. Johri, A. K. Shrivastav and S. Majumdar. 1996. Performance of broilers fed on high or low fibre diets supplemented with digestive enzymes. *J. Appl. Anim. Res.*10:95-102.

Swain, B. K., R. N. S. Sundaram and S. B. Barbuddhe. 2005. Effect of feeding brewery dried grain (BDG) with enzyme supplement on the performance of broilers. *Indian J. Poult. Sci.,* 40: 26-31.

Swain, B. K. and E. B. Chakurkar. 2009. Assessment of problend supplementation on the performance and economics of production of Vanaraja chickens. *J. Applied Anim. Res.,* 36:203-206.

Thacker, P. A., I. Haq, B. P. Willing and A. B. Laytem. 2009. The effects of phytase supplementation on performance and phosphorous excretion from broiler chickens fed low phosphorous containing diets based on normal or low phytic acid barley. *Asian-Aust. J. Anim. Sci.* 22:404-409.

Tortora, G. J. and S. R. Grabowski. 1993. Principles of Anatomy and Physiology. 7th ed. Harper Collins College Publishers, New York.

Viveros, A., A. Brenes, M. Pizaro and M. Castano. 1994. Effect of enzyme supplementation of a diet based on barley, and autoclave treatment on apparent digestibility, growth performance and gut morphology of broilers. *Anim. Feed Sci. Technol.* 48: 237-251.

Wen, C., L. C. Wang, Y. M. Zhou, Z. Y. Jiang and T. Wang. 2012. Effect of enzyme preparation on egg production, nutrient retention, digestive enzyme activities and pancreatic enzyme messenger RNA expression of late phase laying hens. *Anim. Feed Sci. Technol.* 172:180-186.

Woyengo, T. A., W. Guenter, J. S. Sands, C. M. Nyachoti and M. A. Mirza. 2008. Nutrient utilization and performance responses of broilers fed a wheat-based diet supplemented with phytase and xylanase alone or in combination. *Anim. Feed Sci. Technol.* 146:113-123.

Woyengo, T. A., B. A. Slominski and R. O. Jones. 2010. Growth performance and nutrient utilization of broiler chickens fed diets supplemented with phytase alone or in combination with citric acid and multicarbohydrase. *Poult. Sci.* 89:2221-2229.

Yadav, J. L. and R. A. Sah. 2006. Supplementation of corn –soybean based layers diets with different levels of acid protease. *J. Int. Agric. Anim. Sci.* 27:93-102.

Yoruk, M. A., Gul, M., Hayirli, A. and Karaoglu, M. 2006. Multi-enzyme supplementation to peak producing hens fed corn-soybean meal based diets. *International* J. *Poult. Sci.* 5: 374-380.

Youssef, A. W., H. M. A. Hassan, H. M. Ali and M. A. Mohamad. 2011. Performance, abdominal fat and economic efficiency of broilers fed different energy levels supplemented with xylanase and amylase from 14 to 40 days of age. *World J. Agric. Sci.* 7: 291-297.

Zakaria, H. A. H., A. R. Mohammad and M. A. A. Ishmais. 2010. The influence of supplemental multi-enzyme feed additive on the performance, carcass characteristics and meat quality traits of broiler chickens. *Int. J. Poult. Sci.* 9:126-133.

Zamora, V., J. L. Figueroa, L. Reyna, J. L. Cordero, M. T. Sanchez-Torres and M. Martinez. 2011. Growth performance, carcass characteristics and plasma urea nitrogen concentration of nursery pigs fed low-protein diets supplemented with glucomannans or protease. *J. Appl. Anim. Res.* 39: 53-56.

Zanella, I., N. K. Sakomura, F. G. Silversides, A. Fiqueirdo and M. Pack. 1999. Effect of enzyme supplementation of broiler diets based on corn and soybeans. *Poult. Sci.*78:561-568.

3

Scope of the use of Antioxidants in Poultry Production

A. K. Panda

Introduction

During the intensive system of poultry production birds are exposed to several types of stresses. The birds can tolerate mild stress but when the stress is severe, there is imbalance between oxidants and antioxidants which lead to oxidative stress and lipid peroxidative damage of tissues, ultimately results in poor performance. Nutrition has a strong impact on oxidative stress. Insufficient intake of antioxidants, a high intake of prooxidants, or both may lead to oxidative stress. The detrimental effects of oxidation process can be prevented by addition of antioxidant compounds in the feed. Vitamin E, carotenoids and selenium (Se) are among important antioxidant components of poultry diets. It is important to understand the role of antioxidants in poultry nutrition in general and in production and reproduction in particular. The chapter, here in will discuss the stressors in poultry production, oxidative stress and possible alleviation through dietary antioxidant supplementation.

Stressors in Poultry Production

In poultry, the embryo develops outside the body in a close system (inside the egg). Thus, the proper development of embryo depends on the nutritional status of the egg as well as the breeder hen. Once the egg is collected, it has to go the process of incubation and hatching (21 days) to get a chick. After the chick is hatched, it is placed in the intensive system of production where several challenges like vaccination, overheating/chilling, transportation to growing site are happened. All these conditions inevitably lead to stress. Some of the stressors are defined below.

- **Incubation Temperature and humidity** – If proper incubation temperature and humidity is not maintained it may affect embryonic development, oxidation and phosphorylation in embryonic tissues leading to free radical production, and embryonic death.
- **Stages of embryonic development** – The later stage of embryonic development (Day 19) is important as the risk of lipid peroxidation is very high. The chick tissues are characterized by comparatively high levels of PUFAs during this stage. Low antioxidant status in combination with high temperature, humidity, and PUFAs could increase susceptibility to lipid peroxidation.
- **Delay in collecting chicks from hatche** – As not all chicks are hatched at the same time, some would be in the hatcher for 2-12 hrs longer than others. This puts pressure on antioxidant defense capacity.
- **Delay in access to feed and water** – It has been reported that feed and water deprivation beyond 24 hours of hatch lead to stress and has a long term negative effect on performance of the bird.
- **Improper Brooding and growing management** – If proper management conditions are not provided (like high density rearing, sub-optimal temperatures in the poultry house, high levels of ammonia and CO_2 in the poultry house, improper vaccination, imbalanced feed) will lead to stress and this could substantially decrease antioxidant system efficiency.
- **High oil content of poultry feed** – Poultry feeds are usually supplemented with oil to increase the energy content of diet. These oils are rich in PUFAs making them more susceptible for lipid peroxidation.
- **Mycotoxin contamination** – Contamination of mycotoxin by poultry feed can substantially decrease antioxidant assimilation from the feed and increase their requirement to prevent damaging effects of free radicals.
- **Excess concentration of vitamins** – It has been found that Vitamin A excess in the diet is shown to cause an oxidative stress decreasing vitamin E and carotenoid concentrations in tissues and increasing tissue susceptibility to lipid peroxidation.
- **Induced molting with feed withdrawal** – When birds are subjected to induced molting with feed withdrawal practice, inevitably lead to stress.

- **Excessive preventive medication** – Excessive administration of coccidiostats have been reported to cause hepatic lipid peroxidation with decreasing concentrations of vitamin E.

Oxygen is the primary oxidant in metabolic reactions designed to obtain energy from the oxidation of a variety of organic molecules. Oxidative stress results from the metabolic reactions that use oxygen, and it has been defined as a disturbance in the equilibrium status of pro-oxidant/anti-oxidant systems in intact cells. When additional oxidative events occur, the pro-oxidant systems outbalance the anti-oxidant, potentially producing oxidative damage to lipids, proteins, carbohydrates, and nucleic acids, ultimately leading to cell death in severe oxidative stress. Mild, chronic oxidative stress may alter the anti-oxidant systems by inducing or repressing proteins that participate in these systems, and by depleting cellular stores of anti-oxidant materials such as glutathione and vitamin E. Let us try to understand the phenomena of oxidation and why oxidative stress occurs.

Oxidation

Nutrients obtained from food are oxidized to generate energy for metabolic processes and to transform dietary nutrients into body tissue, where the body tissues consume oxygen. Therefore, oxidation is an important tissue process in the normal metabolism. However, many different uncontrolled oxidation reactions, which usually termed, as auto-oxidation is a destructive process, which result in destruction of dietary nutrients, damage of cellular tissue and cause metabolic diseases (Surai, 2000). Thus controlling auto-oxidation is very important in feed and food to preserve the quality of feed and poultry products and prevent disease in chicken.

Phases of Oxidation

The oxidation processes are fairly complex which consists of three separate phases as illustrated in figure 1.

The first phase is the **Initiation** – The production of free radicals that can be generated in many different ways through catalytic effects of metals such as iron and copper, by sunlight, by enzymes and by radiation. During this phase there are no obvious gross changes in the composition of fat.

The second phase is **Propagation**, which occurs when these free radicals combine with oxygen to produce peroxides.

The third phase is **Termination**, which comprises of various species of free radical to produce stable end products.

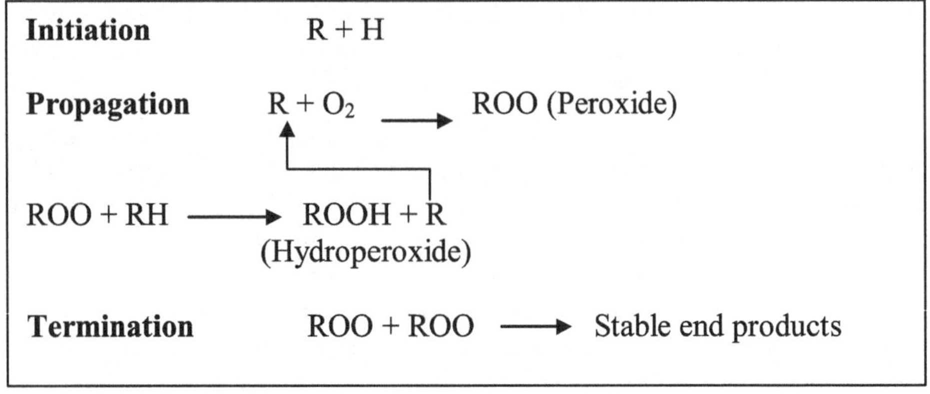

Panda *et al.* (2008)

Fig. 1: Phases of oxidation process

Generation of ROS and free radical formation

Oxygen is the primary oxidant in metabolic reactions designed to obtain energy from the oxidation of a variety of organic molecules. It is required for normal respiration in animals, however, at the same time oxygen is potentially a toxic substance and this has frequently been described as the oxygen paradox (Miller and Brzezinska, 1993).

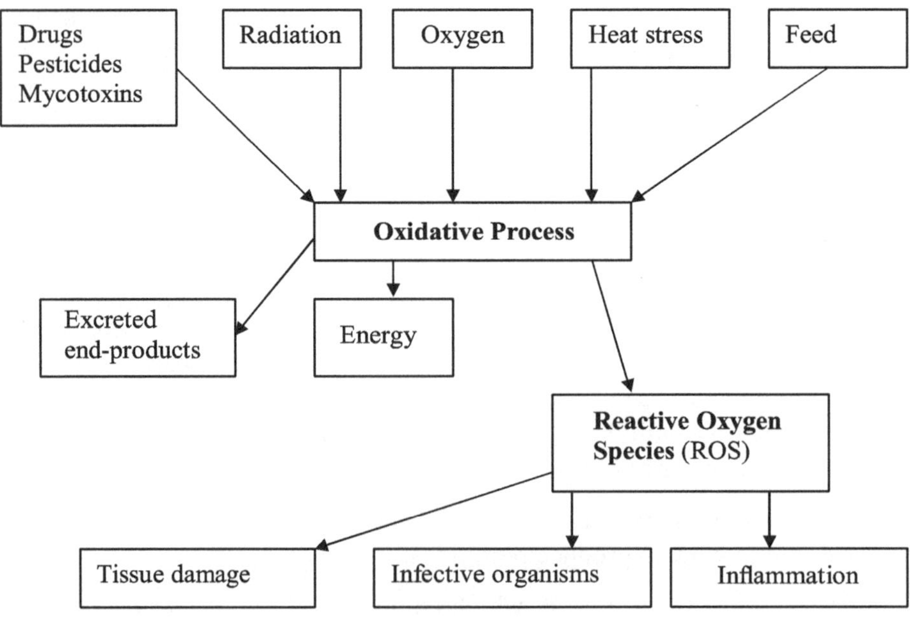

Fig. 2: Production of reactive oxygen species in living tissue

During the normal respiration processes of animals oxygen is progressively reduced to yield water. However, the incomplete reduction of oxygen during the process leads to the formation of chemical entities that have powerful oxidizing properties. These are known as reactive oxygen species (ROS) and basal cellular metabolism in the body of an animal continuously produces ROS. Oxidative stress results from the metabolic reactions that use oxygen, and it has been defined as a disturbance in the equilibrium status of pro-oxidant and anti-oxidant systems in intact cells (Voljc *et al.*, 2011). When the pro-oxidant systems outbalance the anti-oxidant, oxidative damage to lipids, proteins, carbohydrates, and nucleic acids occurs, ultimately leading to cell death in severe oxidative stress (Tavarez *et al.*, 2011). Several conditions lead to ROS formation in living tissue (Fig. 2).

Free radicals (FR) are atoms, molecules or any compounds containing one or more unpaired electrons. They are highly unstable and reactive, and are capable of damaging biologically relevant molecules such as DNA, proteins, lipids and carbohydrates. The FR is often known as Reactive oxygen species (ROS) and is defined as oxygen-containing, reactive chemical species. There are two types of ROS, those of free radicals, which contain one or more unpaired electron(s) in their outer molecular orbital such as superoxide, nitric oxide and hydroxyl radicals, and non-radical ROS, which do not have unpaired electron(s) but are chemically reactive and can be converted to radical ROS such as hydrogen peroxide, ozone, peroxynitrate and hydroxide (Trachootham *et al.* 2009). The rate of ROS generation depends on the concentration of oxygen in the tissues, though other factors, such as metabolic rate, which may also play an important role (Turrens, 2003). The most important effect of free radicals on the cellular metabolism is due to their participation in lipid peroxidation reactions.

Table 1: Reactive oxygen species

Radicals	Non-radicals
Alkoxyl, RO*	Hydrogen peroxide H_2O_2
Hydroperoxyl, HOO*	Hypochlorous acid, HOCl
Hydroxyl, *OH	Ozone, O_2
Peroxyl, ROO*	Singlet $^1O_2$
Superperoxide, ROO*	Peroxinitrite, $ONOO^-$
Nitrioxide, NO*	Nitroxyl anion, NO^-
Nitrogen dioxide, NO_2*	Nitrous acid, HNO_2

(Halliwell and Gutteridge, 1999)

Three levels of antioxidant defense

All the living organisms have specific antioxidant defense mechanism to deal with ROS (Halliwell and Gutteridge, 1999). They are nothing but the natural antioxidants in living organisms which make them possible to survive in oxygen rich environment (Halliwell, 1994). This mechanism is generally known as antioxidant system and includes natural antioxidant vitamins (vitamin E, A, carotenoids, C etc.), antioxidant enzymes (glutathione peroxidase, catalase and superoxide dismutase) and thiol redox system containing glutathione system (glutathione reductase and peroxidase) and thioredoxin system (thioredoxin peroxidase or reductase). The antioxidant system of living cell consists of three major levels of defense (Surai, 1999) as described below.

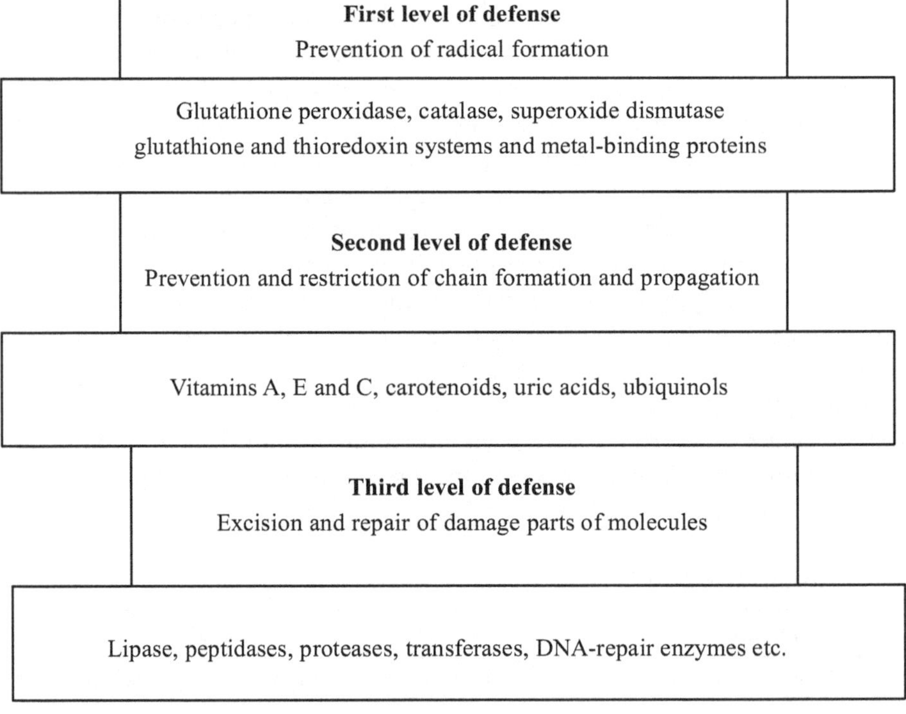

Fig. 3: Three line of antioxidant defense in animal system (Surai, 1999)

Stress and prooxidant-antioxidant balance

The economic and nutritional demands of food from poultry necessitate the raising of large number of birds in relatively small areas with high rates of productivity. During this intensive system of production, birds are exposed to considerable stress. These conditions stimulate free radical generation by a decrease of coupling of oxidation and phosphorylation in the mitochondria that result in an increased electron leakage and overproduction of superoxide radical (Voljc *et al.*, 2011). Cells can tolerate mild oxidative stress by additional synthesis of various antioxidants to restore antioxidant-prooxidant balance. Once the free radical production exceeds the ability of the antioxidant system to neutralize them, lipid peroxidation develops and causes damage to unsaturated lipids in cell membranes, amino acids in proteins, and nucleopeptides in DNA resulting in disruption of membrane and cell integrity (Surai, 2000). These results in poor efficiency of absorption of nutrients leading to imbalance of vitamins, amino acids, inorganic elements and nutrients in the organism (Panda *et al.*, 2010). All these result in adverse affect on the productive and reproductive performance of the animal. Thus, it is important to understand the limitations to poultry growth and productivity from nutritional point of view (Fig. 4). Poultry production must be on a greater understanding of the relationship between diet and health. The purpose is to reduce stress on the bird by improving feed utilization, increasing resistance against enteric and infectious diseases, avoiding oxidative stress and non infectious diseases.

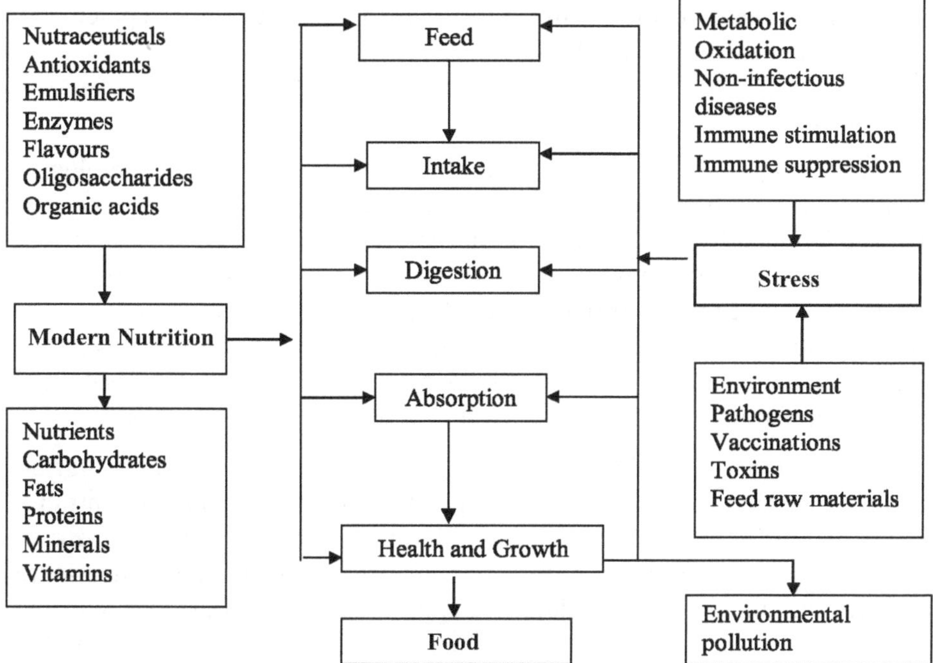

Fig. 4: Relationship between nutrition, stress and food production

Antioxidants

Several compounds in nature possess antioxidant properties. Some of them are synthesized in the body (ascorbic acid, glutathione) while others are supplemented with feed (vitamin E, carotenoids, Se etc.). Antioxidant enzymes those are synthesized in the body require metal co-factors (Se for glutathione peroxidases (GSH-Px) and thioredoxin reductase (TR); Zn, Cu and Mn for superoxide dismutases (SOD); Fe for catalase). The synthesis of the antioxidant enzymes required sufficient concentrations of the element in the diet. Deficiency of any of the above metals causes oxidative stress and allows damage to biological molecules and membranes. The antioxidant/prooxidant balance in the body are responsible for maintaining health, productive and reproductive performances of birds. Sub-optimal concentration of the nutrients in the diet may adversely affect the antioxidant-prooxidant balance resulting in oxidative stress. Poultry in intensive systems of farming frequently exposed to oxidative stress which may result in damage to body proteins, lipids and DNA (McCall and Frie, 1999) and can lead to poor performance and health (Lykkesfeldt and Svendsen, 2007). Therefore, optimizing the dietary intake of antioxidant nutrients is an important step in balancing oxidative damage and antioxidant defense in the animal body. Some of the antioxidants which play a potent role in antioxidant defense in poultry are discussed below.

Vitamin E

Vitamin E (α-tocopheryl acetate), a fat soluble vitamin is an excellent biological chain-breaking antioxidant that protects cells and tissues from lipoperoxidative damage induced by free radicals. Eight naturally occurring substances have been found to have vitamin E activity: four tocopherols (α-, β-, γ-, δ-tocopherols) and four tocotrienols (α-, β-, γ-, δ--tocotrienols). The β-, γ- and δ-tocopherols can act as an antioxidant but not retained well in the body tissues and biologically less active than α-tocopherol. One IU of vitamin E is defined as 1 mg of all-rac-α-tocopherol acetate (all-rac-α-tocopheryl acetate).

The avian embryo develops in a close system using the nutrients that are available within the egg before being laid. Therefore, optimum concentration of all the nutrients should be deposited by mother hen into the egg to get a healthy chick. Several factors like egg composition, incubation temperature, relative humidity, egg storage duration, egg storage temperature, antioxidants, etc. affect the chick embryo development (Surai, 2003). Thus, anything that affects the growth and development during embryonic period can have a marked effect on overall performance during the post hatch period. During chicken embryo development, considerable accumulation of highly PUFAs occurs within the embryonic tissue (Noble and Speake, 1997) and the rate of oxidative metabolism increases dramatically over the hatching period (Freeman and Vince, 1974). At hatching, the chick is suddenly exposed to atmospheric oxygen and has a dramatic increase in metabolic rate. Thus oxidative stress may be a problem during the last days of pre-hatch and first day of post-hatch life. It has been reported that vitamin E content in newly hatched chicks was dramatically depleted during the first nine days to about 5% of the day one (Surai and Ionov, 1994). This indicates that antioxidant may play a vital role during embryonic stage to initial few days of chick's life. Vitamin E is a major lipid soluble antioxidant which breaks the chain reaction of LPO and is transported from the yolk to the embryonic tissues during embryonic development (Gaal *et al.*,1995; Surai *et al.*, 1996). It prevents lipid peroxidation of polyunsaturated fatty acids within the cell, thus protecting the cell against the toxicity of free radicals (Khan, 2011). Surai *et al.* (1999) reported that supplementation of higher vitamin E in the maternal diet (365 ìg vitamin E/g feed) can substantially increase the concentration of vitamin E in the embryonic tissue of the chick and significantly decrease their susceptibility to lipid peroxidation. The same worker (Surai, 2000) studied the effect of various concentrations of Se (0.2, 0.4 mg/kg) and vitamin E (40,100, 200 mg/kg) in maternal laying hen diet and concluded that higher concentration of Se and vitamin E increased the concentration of antioxidant enzymes and decreased susceptibility of lipid peroxidation in day old chick. Thus, nutritional status of laying hen (maternal diet composition) is a major determinant of antioxidant development in chick during embryogenesis and in early post natal development (Cherian and Sim, 2003).

Table 2: Effect of dietary supplementation of the hen with vitamin E on the concentration of vitamin E in the initial yolk and of tissues of the developing chick

	μg TBARS/g tissue/h		P value
	Control diet	Supplemented diet	
Initial yolk	46.6	20.2	0.01
Yolksac membrane	30.1	18.3	0.05
Liver	21.1	16.6	0.05
Brain	110.2	19.7	0.001
Lungs	24.6	18.4	0.05

(Surai *et al.*, 1999)

The time from hatching to the onset of receiving nutrition is also a critical period in the development of hatchling poultry. However, in commercial poultry operations, the poultry hatch over a 2-day period and get transferred from the incubator only when majority of them clear the shell. Following removal from the incubator, other practices such as sexing, vaccination and packaging are carried out before they are boxed and transported to farms. Thus, under practical condition the time between placements of chick right from hatch to growing site is a stressful condition (Panda *et al.*, 2009). To achieve high level of economic efficiency poultry are again raised under intensive production system in densely populated colonies or flocks which causes stress. The birds are additionally stressed by several other factors like transportation, chilling/over heating, vaccination, etc.. With so many challenges, the bird has to grow and produce optimum performance within the shorter span of 35 days which is the marketable age of broiler now. Nutrition has been identified as one source of early exposure that might affect early development and later phenotype performance (Cherian, 2011).

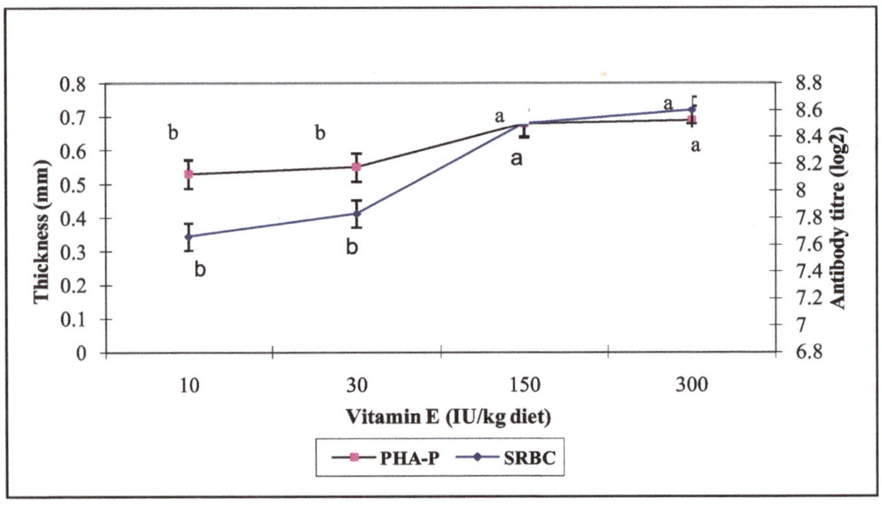

Panda *et al.* (2009)

Fig. 5: Vitamin E supplementation on immune response of broilers

Vitamin E is added in the poultry diet to improve performance, increase immunity and alleviating stress (Sunder *et al.,* 1997, Flachowsky, 2000; Paneri *et al.,* 2006; Panda *et al.,* 2009). Vitamin E affects the development and maintenance of immune competence through multiple functions, by acting directly on the immune cells or by indirectly altering metabolic and endocrine parameters, which in turn influence immune function (Greshwin *et al.*, 1985). The antioxidant property of vitamin E has been considered to have role in the development of immune response in chickens. Vitamin E has been reported to protect cells involved in immune response, such as lymphocytes, macrophages and plasma cells against oxidative damage and to enhance the function and proliferation of these cells (Franchini *et al.*, 1991).

Supplementation of vitamin E is beneficial in lowering the oxidative stress in broilers (Iqbal *et al.,* 2001; Sahin *et al.,* 2002). Panda *et al.* (2009) studied the effect of higher supplementation of vitamin E (10, 30, 150 and 300 IU/kg diet) in broiler chicks and observed a linear a decrease in the lipid peroxidase activity and linear increase in glutathione peroxidase (GSH-Px) activity by enhancing the levels of vitamin E from 10 to 300 IU per kg diet. Thus it is suggested that higher level of vitamin E in the diet of young chicks could be beneficial in alleviating oxidative stress and thereby improving performance and immunity in broilers. He further concluded that vitamin E requirement for better immunity and antioxidant status is higher than growth performance.

Table 3: Effect of vitamin E supplementation on antioxidant status of broiler chickens

Vitamin E (IU/kg diet)	Lipid peroxidase (nmol/MDA/mg protein)	Glutathione peroxidase (U/ml)	RBC catalase (U/g/Hb)
10	1.81[a]	217[d]	284
30	1.71[b]	235[c]	279
150	1.59[c]	269[b]	290
300	1.31[d]	289[a]	288
SEM	0.02	3.20	6.49

[a, b, c, d] Means with different superscripts in a column differ significantly ($P<0.05$)

(Panda *et al*., 2009)

Selenium

Selenium is a trace element that plays a key role in the antioxidant defense as one selenium atom is absolutely required at the active site of all seleno-proteins in the form of selenocysteine. These selenoproteins include glutathione peroxidase (GSH-Px) and thioredoxin reductase, which have important antioxidant and detoxification functions. Selenium as a component GSH-Px is an integral part of antioxidant system present in most mammalian cells. Its main role is to prevent the formation of peroxides from unsaturated fatty acids and removal of hydrogen peroxides (H_2O_2) from the bio-system.

Vitamin E and Se interrelationship

Selenium as a component of GSH-Px in concert with vitamin E undoubtedly plays an important role in protecting biological membrane from peroxidative damage. Both vitamin E and GSH-Px guard against the excessive accumulation of organic hydroperoxidase but by different mechanisms. Vitamin E prevents the formation of these toxic products and GSH-Px convents them to less harmful alcohol. Se and vitamin E has gat a sparing effect on each other.

Vitamin E is known to reduce the Se requirement by two ways:

- By maintaining the body Se in active form.
- By preventing destruction of membrane lipids.

Similarly Se can spare vitamin E in atleast three ways:

- By maintaining the integrity of pancreas.
- Reducing vitamin E requirement for maintaining integrity of lipid membranes.
- In some unknown ways-by retention of vitamin E in plasma.

Selenium supplementation in poultry has long been associated with energy metabolism, increased feed efficiency, improved reproduction, and improved immune responses. A concentration of 0.15 mg of Se/kg of diet is recommended for broiler chickens throughout the growth period (NRC, 1994). Although this requirement often can be met by natural feedstuffs in the diets, it is common practice to supplement broiler diets with Se. Traditionally, the primary form of supplemental Se has been inorganic sodium selenite (NaSe). Recently, organic sources of Se, such as Se yeast, have been gaining popularity as an alternative to inorganic supplementation (Payne and Southern, 2005). Organic selenium supplementation in chicken diet increases GSH-Px activity in liver and blood plasma as well as erythrocyte (Arai *et al.*, 1994). In the investigation of Yang *et al.* (2012), Se supplementation caused an increase of GSH-Px activity in blood by about 60% in two and four week old chickens and about 20% at six weeks of age. Smaller elevation of GSH-Px activity in six week old chickens is possibly related to intensive increase of body mass and muscular tissue which is the main site where selenium is stored in the organism (Schrauzer, 2000).

Table 4: Effect of selenium sources on antioxidant status of broiler chickens

Selenium (U/ml)	Superoxide dismutase (U/L)	Glutathione peroxidase (U/ml)	Alkaline (IU/kg diet) phosphatase
0.3 ppm (inorganic)	124.9[b]	2075[b]	1068
0.3 ppm (organic)	131.6[a]	5310[a]	1103

[a, b] Means with different superscripts in a column differ significantly ($P<0.05$)

Source: Yang *et al.*, (2012)

A Se-deficiency affects the total antioxidant system because the mineral in the inorganic forms and be a pro-oxidant (Edens, 2002) and in the organic form (SeMet) it can be an anti-oxidant as well (Sies and Arteel, 2000). This lack of antioxidant protection can lead to an increase in oxidative stress and alterations in redox signaling (Gladyshev *et al.*, 1999). Selenium supplementation is also known to affect the antioxidant defense of chicken spermatozoa (Edens, 2002). When cockerels were fed on a basal diet containing 0.28 ppm Se without additional dietary supplementation on this trace element, the percentage of normal spermatozoa was only 57.9% and two major abnormalities were seen (bent mid piece and corkscrew head). When the diet was supplemented with 0.2 ppm Se in the form of sodium selenite, the percentage of normal spermatozoa increased to 89.4 % and abnormalities were significantly reduced.

Selenium enriched products are believed to be most valuable as they determine human health (Rayman, 2000) and adequate supplementation of Se in layer and broiler chicken diets can enhance Se concentration in eggs and meat (Leeson *et al.*, 2008). The organic form of Se from yeast (Seleno-methionine) was more efficient than selenite Se for its deposition in eggs (Paton *et al.*, 2002; Utterback *et al.* 2005)). Within the egg, Se along with other minerals is preferentially deposited in yolk than in albumen, primarily due to mineral binding activity of lipoproteins during the formation of egg yolk (Pappas *et al.*, 2005). The maternal effect of Se supplementation persists in progeny for several weeks after hatching (Pappas *et al.*, 2005), indicating the importance of Se level in parental diet. The inclusion of Se-methionine in the layer diet (0.3 ppm) enhanced GPx activity in the egg yolk and albumen, and maintained higher egg quality (Surai, 2002). In addition, organic Se increases the concentration of tissue Se and maintains high GPx activity compared to inorganic source (Payne *et al.,* 2005), which has relevance on the egg and meat quality during storage and transportation. Similarly, higher supplementation of vitamin E in diet produced linear increase in alpha-tocopherol retention in egg yolk based on its level of consumption (Barreto *et al.*, 1999). The requirement of vitamin E could be less for productive parameters, but for oxidative stability of meat higher levels would be useful. Optimal combination of Se and vitamin E can be even more beneficial in preventing lipid peroxidation, membrane deterioration, peroxide accumulation and maintaining meat freshness and quality (Surai, 2002).

The primary function of selenium in animal systems is as a component of glutathione peroxidase (GSH-Px), a plasma and cytosolic enzyme responsible for destroying potentially damaging hydrogen peroxide and organic hydroperoxides in animal tissues (Burk, 1989; Omaye and Tappel, 1974; Rotruck *et al.*, 1973). Beyond its role as a component of GSH-Px, selenium is also involved in several enzyme systems regulating energy metabolism, spermatozoa function, prostaglandin synthesis and essential fatty acid metabolism, purine and pyrimidine base synthesis, and animal immunity (McDowell, 1992). A discussion of selenium would be incomplete without mentioning the interrelationship between selenium and vitamin E. Vitamin E, is the first line of defense against cellular damage resulting from membrane phospholipid peroxidation while Se, as a component of GSH-Px, acts as a second line of defense against cellular peroxide damage due to inability of vitamin E to destroy all metabolic peroxides. For the most part, vitamin E and selenium are mutually replaceable (except at low levels) and each acts as a sparing mechanism for the other (McDowell, 1992; Combs, 1981).

Highest selenium concentrations are found in kidney, liver, spleen, and pancreatic tissues. Several factors influence tissue storage and retention of selenium, including animal selenium status and supplemental selenium source. Cantor *et al.* (1982) and Osman and Latshaw (1976) showed that selenium concentrations in the gizzard, breast muscle, and pancreas were increased when turkey poults and chicks were fed selenomethionine instead of sodium selenite. Furthermore, selenium is absorbed and retained more efficiently in Se-deficient animals (McDowell, 1992). Selenium, as a component of intracellular GSHPx, acts with vitamin E for reduction of cellular oxidative stress. Selenium, then, may also contribute to maintaining meat quality. Several reports have suggested a link between antioxidant (i.e., vitamin E and Se) intake and increased or maintained meat quality parameters for pork, poultry, and fish (Nielsen and Rasmussen, 1979; Mahan *et al.*, 1999; Higgins *et al.*, 1998; Combs and Regenstein, 1980; Mahan, 1996; Bartov, 1977). Beyond the role of Se in protection from cellular damage and meat quality maintenance, dietary selenium source seems to affect meat quality, with particular reference to excessive muscle tissue water loss. Mahan *et al.* (1999) and Edens (1997) reported a reduction in pork and poultry meat quality (increased drip loss), respectively, when supplemental dietary sodium selenite was replaced with organic selenium in the form of Se-enriched yeast (Sel-PlexTM).

Carotenoids

Carotenoids are large lipophilic molecules composed of a chain of 40 carbon atoms joined with alternating single and double bonds. In most carotenoids, this linear hydrocarbon skeleton is cyclized at both ends, and these end rings are often substituted by different functional groups. Carotenoids are primarily derived from plants and are found in roots, leaves, shoots, seeds, fruit, and flowers. It is well known that carotenes are precursor of vitamin A and this function is well defined. Literature reports suggest that carotenoids such as β-carotene can have a positive effect on health and well-being in a variety of species. Of the 600 or more naturally occurring carotenoids, β-carotene is probably best known as one of 50 or so that possess provitamin A activity. In addition to their provitamin A activity, carotenoids have been considered to have many other biological functions (Mascio *et al.*, 1989). They are proposed to be efficient scavengers of free radicals (Surai, 1999), and they have also been shown to protect low density lipoproteins (LDLs) against oxidation in vitro. Interestingly, canthaxanthin, a diketo-carotenoid better known for its pigmenting properties, has also been reported to possess pro-vitamin A activity in birds (Olson, 1983). Canthaxanthin, like β-carotene, is a naturally occurring carotenoid produced by certain bacteria and algae (Goodwin, 1980). Both β-carotene and canthaxanthin have been shown to enhance resistance to pathogenic agents (Seifter *et al.*,1981; Gruner *et al.*, 1986) and to significantly enhance immune response (Mathews-Roth, 1985; Bendich and Shapiro, 1986). Bendich and Shapiro (1986) reported that β -carotene and canthaxanthin significantly enhanced splenic T and B lymphocyte proliferation in response to mitogen stimulation. It is generally regarded that carotenoids can provide protection against certain diseases through an antioxidant affect (Palozza and Krinsky, 1992).

The antioxidant properties of carotenoids include scavenging singlet oxygen and peroxyl radicals (Krinsky, 1989b; Terao *et al.*, 1992), sulphur radicals (Chopra *et al.*, 1993) as well as thiyl, sulphonyl and NO_2 radicals (Everett *et al.*, 1996) and provide protection for lipids against superoxide and hydroxyl radical attack (Krinsky and Deneke, 1982). The mechanism of protection of biological systems against damage due to $^1O_2$ by carotenoids includes both a physical component as well as a chemical reaction between the carotenoid and the reactive oxygen molecule (Krinsky, 1989a). The efficiency of the antioxidant defence provided by carotenoids depends on many factors including stress conditions, method of oxidative stress detection, concentrations of carotenoid used, oxygen tension and interaction with other antioxidants (Rock, 1997; Rice-Evans *et al.*, 1997; Edge *et al.*, 1997). It has been suggested that depending on the redox potential of the carotenoid molecules and oxygen tension, carotenoid concentration and interactions with other antioxidants these pigments could show antioxidant or pro-oxidant properties (Palozza, 1998). When a Se and vitamin E deficient chicken diet was supplemented with canthaxanthin (0.5 g CX/kg diet), liver homogenates exhibited significantly (P=0.02) decreased formation of thiobarbituric acid-reactive substances over time in ferrous ion-induced peroxidation conditions (Mayne and Parker, 1989).

Ascorbic acid (Vitamin C)

Vitamin C is chemically known as L-ascorbic acid. It is not considered as dietary essential in the diet of chicken because under normal conditions the body synthesis of this vitamin is adequate to meet the requirements. However, under certain conditions (under stress) the bird is not able to synthesize sufficient vitamin C to meet the requirements and hence needs dietary supplementation. Vitamin C is necessary for various biosyntheses (collagen, 1, 25-dihydroxyvitamin D and adrenaline) as well as for regulation of diverse reactions (secretion of corticocosterones, regulation of body temperature and activation of immune system) in the body. Vitamin C has been demonstrated to enhance the antioxidant activity of vitamin E by reducing the tocopheroxy radicals back to their active form of vitamin E (Jacob, 1995) or by sparing available vitamin E (Retsky and Frei, 1995). Adult poultry under normal conditions are able to synthesize vitamin C to meet the requirement. However, it has been reported that vitamin C requirement is higher during stress (McDowell, 1989) and several reports have documented a beneficial effect of supplementing the feed with ascorbic acid (De Faria *et al.*, 2001, Kutlu, 2001). Vitamin C synthesis is inadequate under stress conditions (McDowell, 1989) and thus birds need dietary source of vitamin C to sustain production during heat stress.

Heat stress leads to generation of free radicals (O_2-, HO.) which can damage cell membrane by inducing lipid peroxidation of polyunsaturated fatty acids in the cell membrane (Laudicina and Marnett, 1990), resulting in abnormal membrane integrity. Endogenous enzymatic antioxidants like RBCC, GSH-PX and GSHR play a vital role in scavenging oxidative radicals (Spurlock and Savage, 1993) and are considered as markers for evaluation of oxidative stress. According to the antioxidant theory, when the concentrations of antioxidant vitamins decrease, lipid peroxidation increase in plasma and tissues leading to damage of cell membranes. Lipid peroxidation products increase in plasma and tissues during heat stress as result of reduction in synthesis of vitamin C and E. We also observed similar findings in our study (Panda *et al.*, 2008). The activities of LP, an oxidative enzyme which causes lipid peroxidation increased and activity of antioxidant enzymes (RBCC, GSH-PX and GSHR) decreased in the birds fed diet with lowest concentration of vitamins E and C in the diet. However, vitamin E at 125 IU/kg and C 200 mg/kg diet had an additive effect on reducing the activity of oxidative enzyme LP and increasing the activity of antioxidant enzyme GSHR. Vitamin E at 125 IU/kg and vitamin C at 200 mg/kg diet independently enhanced the activity of antioxidant enzyme such as RBCC and GSH-PX. Our findings suggested a higher requirement of vitamins E (125 IU/kg) and C (200 mg/kg) to counteract the negative effect of heat stress on lipid peroxidation. In another experiment conducted in broilers during summer, we have also observed significant improved in performance (body weight gain and feed conversion efficiency) and reduction in lipid peroxidase activity in serum of broilers due to dietary supplementation of 200mg/kg vitamin C in diet (Panda *et al.*, 2007). On the other hand the antioxidant enzymes activity increased in serum by vitamin C supplementation.

Table 5: Effect of dietary vitamin C supplementation on performance and antioxidant enzyme profile of commercial broilers at 5 weeks of age during summer stress

Vitamin C (mg/kg)	Performance		Antioxidant enzyme activity	
	Body weight (g)	Feed conversion ratio	Lipid peroxidase (nM. MDA. ml^{-1})	Glutathione peroxidase (mM. ml^{-1})
0	1397[c]	1.84[c]	12.44[d]	1.33[c]
100	1424[b]	1.82[b]	7.86[c]	1.45[b]
200	1458[a]	1.76[a]	4.26[b]	1.79[a]
400	1462[a]	1.77[a]	4.1[a]	1.83[a]

(Panda *et al.*, 2007)

Summary

The above review explains how antioxidants are beneficial in alleviating oxidative stress in poultry besides maintaining health, productive and reproductive performance. All the antioxidants in the body work in concert as a team called the antioxidant system responsible for prevention of the damaging effects of free radicals and toxic products of metabolism. The efficiency of the antioxidant defence provided by the antioxidants depend on many factors including stress conditions, antioxidant status of the body, stages of production, age and health of the birds. With continuous increase in genetic potential of the birds for productive and reproductive parameters over the years, the author beliefs more important roles of antioxidants in poultry production systems in the year to come.

References

Arai, T., M. Sugawara, N. Sako, S. Motoyoshi, T. Shimura, N. Tsutsui, T. Konno. 1994. Glutathione peroxidase activity in tissues of chicken supplemented with dietary selenium. *Com. Biochem. Physiol.* 107A: 245-248.

Barreto, S.L.T., W.M Ferreira, and T. Moraes. 1999. Protein and vitamin E levels for broiler breeder hens. 1. Effects on broiler breed performance, egg composition and performance of their progeny . *Arquiv. Brasil. Med. Vet. Zootec.* 51: 183-192.

Bartov, I. 1977. Pro- and antioxidants in the diets of broilers and their effects on carcass quality: copper, selenium and acidulated soybean-oil soapstock. *Poult.Sci.* 56: 829-835.

Burk, R.F. 1989. Recent developments in trace element metabolism and function: newer roles of selenium in nutrition. *J. Nutr.* 119: 1051-1054.

Cantor, A.H., P.D. Moorhead and M.A. Musser. 1982. Comparative effects of sodium selenite and selenomethionine upon nutritional muscular dystrophy, selenium-dependent glutathione peroxidase and tissue selenium concentrations of turkey poults. *Poult. Sci.* 61:478-484.

Combs, Jr., G.F. 1981. Influences of dietary vitamin E and selenium on the oxidant defense system of the chick. *Poult. Sci.* 60: 2098-2105

Bendich, A., and S. S. Shapiro, 1986. Effect of b-carotene and canthaxanthin on the immune responses of the rat. *J. Nutr.* 116:2254–2262.

Cherian, G. 2011. Essential fatty acids and early life programming in meat type birds. *World's Poult. Sci.* J.90: 599-614.

Cherian, G. and J.S. Sim. 2003. Maternal and post-hatch dietary polyunsaturated fatty acids alter tissue tocopherol status of chicks. *Poult. Sci.* 82: 681-686.

Chopra, M., R. L. Willson and D. I. Thurnham, (1993) Free radical scavenging activity of lutein *in vitro*, in: Cranfield, L.M., Olson, J.A. & Krinsky, N.I. (Eds) *Annals of the New York Academy of Sciences,* 691: 246-249.

Combs, Jr. G. F. and J. M. Regenstein. 1980. Influence of selenium, vitamin E and ethoxyquin on lipid peroxidation in muscle tissues from fowl during low temperature storage. *Poult. Sci.* 59: 347-351.

Edens, F.W. 1997. Potential for organic selenium to replace selenite in poultry diets. *Zootech. Int.* 20: 28-31.

De faria, D. E., O. M. Junqueira, P. A. Souza and E. A. I Titto. 2001. Performance, body temperature and egg quality of laying hens fed vitamins D and C under three environmental temperatures. *Brasil. J. of Poul. Sci.* 3: 49-56.

Di Mascio P, S. Kaiser, H. Sies. 1989. Lycopene as the most efficient biological carotenoid singlet oxygen quencher. *Arch. Biochem. Biophys.* 274*:532–538.*

Edens, F.W. 2002. Practical applications for selenomethionine: broiler breeder reproduction. In: Biotechnology in the Feed Industry. Proceedings of the 18th Alltech's Annual Symposium, Edited by Lyons, T.P. and Jacques, K.A., Nottingham University Press, Nottingham, UK. pp. 29-42.

Edge, R., D.J. Mcgarvey and T.G. Truscott.1997. The carotenoids as anti-oxidants - a review. *J. Photochem. Photobiol.* 41B: 189-200

Everett, S. A., M. F. Dennis, K. B.Patel, S. Maddix, S.C. Kundu, and R. L. Wilson.1996. Scavenging of nitrogen dioxide; thiyl and sulfonyl free radicals by the nutritional antioxidant β-carotene. *J. Biol. Chem.* 271: 3988-3994.

Flachowsky, G. 2000. Vitamin E transfer from feed into pig tissues. *J. Appl. Anim. Res.* 17: 69-80.

Franchini, A. M., M. Canti, G. Manfreda, and G. S. Bertuzzi, 1991. Vitamin E in viral inactivated vaccines. *Poult. Sci.*, 70: 1709-1715.

Freeman, B.M. and Vince M.A. 1974. Development of the avian embryo. Chapman and Hall, London. pp 249-260.

Gaal, T., M. Mezes, R. C.Noble, J. Dixon, and B. K. Speake. 1995. Development of antioxidant capacity in tissues of the chick embryo. *Com. Biochem. Physiol.* 112B: 711–716.

Gladyshev, V. N., T. C. Stadtman, D. L. Hatfield, and K. T. Jeang. 1999. Levels of major selenoproteins in t cells decrease during hiv infection and low molecular mass selenium compounds increase. *Proc. Nat. Acad. Sci. USA.* 96: 835-839.

Goodwin, T. W. 1980. The Biochemistry of the Carotenoids. Vol. 1. Plants. Chapman and Hall, London, UK.

Greshwin, M., R. Beach, and L. Hurley. 1985. The potent impact of nutritional factors on immune response. *Nutr. Immun.* 1: 1-7.

Gruner, S., H. D. Volk, P. Falck, and R. Von Baehr, 1986. The influence of phagocytic stimuli on the expression of HLADR antigens; role of reactive oxygen intermediates. *Eur. J. Immunol.* 16:212–215.

Halliwell, B. and J. M. C. Gutteridge, 1999. Free radicals in biology and medicines. Third edition, Oxford University Press, Oxford.

Halliwell, B. 1994. Free radicals and antioxidants. A personal view. *Nutr. Rev.* 52: 253-265.

Higgins, F.M., J.P. Kerry, D.J. Buckley and P.A. Morrissey. 1998. Assessment of α-tocopheryl acetate supplementation, addition of salt and packaging on the oxidative stability of raw turkey meat. *Bri. Poult. Sci.* 39: 596-600.

Iqbal, M., D. Cawthon, JR. R. F. Wideman, and W. G. Bottje, 2001. Lung mitrochondrial dysfunction in pulmonary hypertension syndrome. II. Oxidative stress and inability to improve function with repeated additions of adenosine diphosphate. *Poult. Sci.* 76: 1505-12.

Jacob, R. A. 1995. The integrated antioxidant system. *Nutr. Res.*, 15: 755-766.

Khan, R. U. 2011. Antioxidants and poultry semen quality. *World's Poult. Sci.* J. 67: 297-308.

Krinsky, N. I. 1989a. Antioxidant functions of carotenoids. *Free Radical Biol.* and *Med.* 7: 617-635.

Krinsky, N.I. 1989b. Carotenoids in medicine, in: Krinsky, N. I., Mathews-roth, M.M. and Taylor, R.F. (Eds) *Carotenoids: Chemistry and Biology,* pp. 279-291.

Krinsky, N. I. and S. M. Deneke. 1982. Interaction of oxygen and oxygen radicals with carotenoids. *J. Nat. Can. Inst.* 69: 205-209.

Kutlu, H.R. 2001. Influences of wet feeding and supplementation with ascorbic acid on performance and carcass composition of broiler chicks exposed to a high ambient temperature. *Arch. Anim. Nutr.* 54: 127-139.

Laudicina, D.C., and , L.J. Marnett. 1990. Enhancement of hydro-peroxide dependent lipid peroxidation in rat liver microsomes by ascorbic acid. *Arch. Biochem. Biophy.* 278: 73-80.

Leeson, S., H. Namkung, L. Caston, S. Durosoy and P.Schlegel. 2008. Comparison of selenium levels and sources and dietary fat quality in diets for broiler breeders and layer hens. *Poult. Sci.* 87: 2605-2612.

Mahan, D.C. 1996. How organic selenium may help reduce drip loss. *Misset World Poultry*. 12: 19-21.

Mahan, D.C., T.R. Cline and B. Richert. 1999. Effects of dietary levels of selenium-enriched yeast and sodium selenite as selenium sources fed to growing-finishing pigs on performance, tissue selenium, serum glutathione peroxidase activity, carcass characteristics and loin quality. *J. Anim. Sci.* 72: 2172-2179.

Mathews-Roth, M.M., 1985. Carotenoids and cancer prevention-experimental and epidemiological studies. *Pure Appl. Chem.* 57:717–722.

Mayne, S. T. and R. S. Parker. 1989. Antioxidant activity of dietary canthaxanthin. *Nutr. Canc.* 12: 225-236.

McDowell, L. R. 1992. Selenium. In: Minerals in Animal and Human Nutrition. Academic Press, Inc., San Diego, CA, pp. 294-311.

Miller, J. K. and E. Brzezinska. 1993. Oxidative stress, antioxidants, and animal function. *J. Dairy Sci.* 76: 2812-2823.

National Research Council. 1994. Nutrient requirements of poultry, 9th rev. edn. National Academy Press, Washington, DC.

Nielsen, H.E. and O.K. Rasmussen. 1979. The influence of selenium on performance, meat production and the quality of some edible tissues in pigs. *Acta Agriculturae Scandinavica Supplementum*. 21: 246-257

Omaye, S.T. and A.L. Tappel. 1974. Effect of dietary selenium on glutathione peroxidase in the chick. *J. Nutr.* 104: 747-753.

Noble, R. C. and B. K. Spaeke. 1997. Observation on fatty acid uptake and utilization by the avian embryo. *Prenat. Neonat. Med.* 2: 92-100.

Olson, J. A. 1983. Formation and function of vitamin A. Pages 371–412 *in*: Biosynthesis of isoprenoid compounds. J. W. Porter and S. L. Spurgeon, ed. Wiley and Sons, New York,

Osman, M. and J.D. Latshaw. 1976. Biological potency of selenium from sodium selenite, selenomethionine and selenocystine in the chick. *Poult. Sci.* 55:987-994.

Rotruck, J. T., A. L. Pope, H. E. Ganther, A. B. Swanson, D. G. Hafeman and W. G. Hoekstra. 1973. Selenium: biochemical role as a component of glutathione peroxidase. *Sci.* 179:88-590.

Palozza, P. 1998. Prooxidant actions of carotenoids in biologic systems. *Nutr. Rev.* 56: 257-265.

Palozza, P. and N. I. Krinsky, 1992. Antioxidant effects of carotenoids in vivo and *in vitro*: an overview. *Methods Enzymol.* 213: 403–420.

Panda, A. K., M. V. L. N. Raju, and S. V. Rama Rao. 2010. Nutraceuticals in Poultry Production. Pp 45-53. Proceedings of the Poultry Expo and Conference held at Namakkal, Tamilnadu, 8-10 January.

Panda, A. K., S.V. Rama Rao and M. R. Reddy, 2008. Growth Promoters in Poultry. Published by IBDCo. Lucknow, India. pp. 64-68.

Panda, A. K., S. V. Rama Rao, M. V. L. N. Raju and R. N. Chatterjee. 2008. Effect of vitamins E and C supplementation on production performance, immune response and antioxidant status of White Leghorn layers during summer stress. *Bri. Poult. Sci.* 49: 592-599.

Panda, A.K., M. V. L. N. Raju, and S. V. Rama Rao. 2007. Effect of vitamin C supplementation on performance, immune response and antioxidant status of heat stressed White Leghorn layers. *Ind.* J. *Poult. Sci.* 42: 169-173.

Panda, A. K., S. V., Rama Rao, Raju, M.V. L. N., G. Shyam Sunder and M. R. Reddy, M. R. 2009. Effect of higher concentration of vitamin E supplementation on growth performance, immune competence and antioxidant status in broilers. *Ind. J. Poult. Sci.* 44: 187-190.

Paneri, P., I. Giannenas, E. Christaki, A. Govaris and N. Botsoglou. 2006. Performance of chickens and oxidative stability of the produced meat as affected by feed supplementation with oregano, vitamin C and their combinations. *Arch. Fur Geflugel.* 70: 232-40.

Pappas, A. C., T. Acamovic, N. H. C. Sparks, P. F. Surai and R. M. Mc Devitt. 2005. Effects of supplementing broiler breeder diets with organic selenium and polyunsaturated fatty acids on egg quality during storage. *Poult. Sci.* 84: 865-874.

Paton, N. D., A. H. Cantor, A. J. Pescatore, M. J. Ford and C. A. Smith. 2002. The effect of dietary selenium source and level on the uptake of selenium by developing chick embryos. *Poult. Sci.* 81: 1548-1554.

Payne, R. L., and L. L. Southern. 2005. Comparison of inorganic and organic selenium sources for broilers. *Poult. Sci.* 84:898–902.

Payne, R. L., T. K. Lavergne and L. L. Southern. 2005. Effect of inorganic versus organic selenium on hen production and egg selenium concentration. *Poult. Sci.* 84: 232-237.

Rayman, M. P. 2000. The importance of selenium to human health. *Lancet* 15: 233-41.

Retsky, K. L. and Frie, B. (1995). Vitamin C prevents metal ion dependent initiation and propagation of lipid peroxidation in human low density lipoprotein. *Bicochem. Biophy. Acta.* 1257: 279-287.

Rice-Evans, C.A., Samson, J., Bramley, P.M. and Holloway, D.E. 1997. Why do we expect carotenoids to be antioxidants *in vivo*? *Free Rad. Res.* 26: 381-398.

Rock, C. L. 1997. Carotenoids: Biology and treatment. *Pharmac. Therap.* 75: 185-197.

Sahin, K., N. Sahin and M. Onderci. 2002. Vitamin E supplementation can alleviate negative effects of heat stress on egg production, egg quality, digestibility of nutrients and egg yolk mineral concentrations of Japanese quails. *Res. Vet. Sci.* 73: 307-312.

Schrauzer, G. N. 2000. Selenomethionine: a review of its nutritional significance, metabolism and toxicity. *J. Nutr.* 130: 1653-1656.

Seifter, E., G. Rettura, and S. M. Levenson, 1981. Carotenoids and cell-mediated immune responses. Pages 335–347 *in*: The Quality of Foods and Beverages. Chemistry and Technology. G. Charalambous and G. E. Inglett, ed. Academic Press, New York, NY.

Sies, H., and G. E. Arteel. 2000. Interaction of peroxynitrite with selenoproteins and glutathione mimics. *Free Radic Biol. Med.* 28: 1451-1455.

Spurlock, M. E. and J. E. Savage. 1993. Effects of dietary protein and selected antioxidants on fatty haemorrhagic syndrome induced in Japanese quails. *Poult. Sci.* 72: 2095-2105.

Sunder, A., G. Richter and G. Flachowsky. 1997, Influence of different concentrations of vitamin E in the feed of laying hens on the vitamin E transfer to eggs. *Proc. Soc. Nutr. Physiol.* 6: 14-152.

Surai, P. and Ionov, I. 1994. Vitamin E in the liver of developing embryos. *J. Ornithology*. 135: 85.

Surai, P.F. 1999. Vitamin E in avian reproduction. *Poultry and Avian Biology Reviews*, 10: 1–60.

Surai, P. F. 2000. Effect of selenium and vitamin E content of the maternal diet on the antioxidant system of the yolk and the developing chick. *Bri. Poult. Sci.* 41: 235–243.

Surai, P.F. 2003. Natural Antioxidants in Avian Nutrition and Reproduction. Nottingham University Press, Nottingham.

Surai, P.F., R.C. Noble, and B.K. Speake. 1996. Tissue-specific differences in antioxidant distribution and susceptibility to lipid peroxidation during development of the chick embryo. *Biochem. Biophysic. Acta* 1304: 1–10.

Surai, P. F., R.C. Noble, and B. K. Speake. 1999. Relationship between vitamin E content and susceptibility to lipid peroxidation in tissues of the newly hatched chick. *Bri. Poult. Sci.* 40: 406-410.

Tavarez, M.A., D. D. Boler, K. N. Bess, J. Zhao, F. Yan, A. C. Dilger, F. K. McKeith and J. Killefer. 2011. Effect of antioxidant inclusion and oil quality on broiler performance, meat quality and lipid oxidation. *Poult. Sci.* 90: 922-930.

Terao, J., P. L.Boey, F. Ojima, A. Nagao, T. Suzuki and K. Takama. 1992. Astaxanthin as a chain-breaking antioxidant in phospholipid peroxidation, in: Yagi, K., Kondo, M., Niki, E. & Yoshikawa, T. (Eds) Oxygen Radicals. Elsevier Science Publishers, New York. pp. 657-660.

Trachootham, D., J. Alexandre and P. Huang. 2009. Targeting cancer cells by ROS-mediated mechanisms: a radical therapeutic approach? *Nature Rev. Drug Discov.* 8: 579-591.

Turrens. J. F. 2003. Mitochondrial formation of reactive oxygen species. *J Physiol.* 552: 335–344.

Voljc, M., T. Frankic, A. Levart, M. Nemec and J. Salobir. 2011. Evaluation of different vitamin E recommendations and bioactivity of α-tocopherol isomers in broiler nutrition by measuring oxidative stress in vivo and the oxidative stability of meat. *Poult. Sci.* 90: 1478-1488.

Yang, Y. R., F. C. Meng, P. Wang, Y. B. Jiang, Q. Q. Yi and J. X. Liu. 2012. Effect of organic and inorganic selenium supplementation on growth performance, meat quality and antioxidant property of broilers. *African J. Biotech.* 11: 3031-3036.

4

Role of Probiotics in Poultry Production

Pankaj Kumar Singh and Chandramoni

Introduction

The term "probiotic" is derived from Greek and means pro: for and bios: life (for life) in contradiction to antibiotic which means: against life. The term probiotic was first introduced by Lilly and Stillwel (1965) to describe growth-promoting factors produced by microorganisms. Parker (1974) first specified designation "probiotic". He defined probiotics as microorganisms or substances, which contribute to the balance of the intestinal micro flora. Crawford (1979) defined probiotics as a culture of specific living microorganisms, primarily *Lactobacillus* spp. that are implanted in the organism and ensure the rapid and effective establishment of a beneficial intestinal population. Fuller (1989) discussed the definition given by Parker (1974) and considered it too broad, as cultures, cells, and metabolites are also included in antibiotic preparations. He redefined "probiotic" as a live microbial feed additive, which beneficially affects the animal by improving its microbial balance. Havenaar *et al.* (1992) pointed out that the definition of "probiotic" made by Fuller (1989) was restricted to feed supplements, animals, and their intestinal tract. Therefore, they generalized Fuller's definition of "probiotic" as a mono or mixed culture of living microorganisms, which beneficially affect the host by improving the properties of the indigenous micro-flora. Through 1989, United States Department of Agriculture (USDA) advised manufactures to use the term: "direct-fed microbial" (DFM) instead of "probiotic" (Miles and Bootwalla, 1991). The USFDA defined DFM as a source of live naturally occurring microorganisms, including bacteria, fungi, and yeast. Vanbelle *et al.* (1990) pointed out that most researchers considered "probiotic" for selected and concentrated viable counts of lactic acid bacteria. Koh *et al.* (1992) pointed out that as a biological product for newly hatched chicks a bacterial culture producing acetic acid could be used. Such a culture might be supplied to the chicks either through their drinking water or by the feed. For controlling the biological balance in the chicken's intestinal tract different probiotics may be used. The US National Food Ingredient Association presented, probiotics (direct fed microbial) as a source of live naturally occurring microorganisms and this includes bacteria, fungi and yeast (Miles and Bootwalla, 1991). According to the currently adopted definition by FAO, probiotics are: "live microorganisms which when administered in adequate amounts confer a health benefit on the host" (FAO/ WHO, 2001). More precisely, probiotics are live microorganisms of nonpathogenic and nontoxic in nature, which when administered through the digestive route, are favorable to the host's health (Guillot, *et al.*, 1998; Thomke and Elwinger, 1998,).

Characteristics of an ideal probiotics

The perceived desirable traits for selection of functional probiotics are many. The probiotic bacteria must fulfill the following conditions:

- Be of host origin
- Non-pathogenic
- Withstand processing and storage
- Resist gastric acid and bile
- Adhere to epithelium or mucus
- Persist in the intestinal tract
- Produce inhibitory compounds
- Modulate immune response
- Alter microbial activities

Probiotic must be a normal inhabitant of the gut, non-pathogenic, and it must be able to adhere to the intestinal epithelium to overcome potential hurdles, such as the low pH of the stomach, gastric juice, the presence of bile acids in the intestines, and the competition against other micro-organisms in the gastro-intestinal tract (Nurmi *et al,* 1983; Chateau *et al.*, 1993). Many *in vitro* assays have been developed for the pre-selection of probiotic strains (Ehrmann *et al.*, 202; Morelli *et al.*, 2000; Koenen *et al.*, 2004). The competitiveness of the most promising strains selected by *in vitro* assays was evaluated *in vivo* for monitoring of their persistence in chickens (Garriga *et al.*, 1998). In addition, potential probiotics must exert its beneficial effects (*e.g.*, enhanced nutrition and increased immune response) in the host. Finally, the probiotic must be viable under normal storage conditions and technologically suitable for industrial processes (*e.g.*, lyophilized).

The species currently being used in probiotic preparations are varied and many. These are mostly *Lactobacillus bulgaricus, Lactobacillus acidophilus, Lactobacillus casei, Lactobacillus helveticus, Lactobacillus lactis, Lactobacillus salivarius, Lactobacillus plantarum, Streptococcus thermophilus, Enterococcus faecium, Enterococcus faecalis, Bifidobacterium spp.* and *scherichia coli.* With two exceptions, these are all intestinal strains. The two exceptions, *Lactobacillus bulgaricus* and *Streptococcus thermophilus*, are yoghurt starter organisms (Fuller, 1989). Some other probiotics are microscopic fungi such as strains of yeasts belonging to *Saccharomyces cerevisiae* species (Guillot, *et al.*, 1998).

Methods of administration of probiotics

There are different methods for administering probiotics:

Treatment of individual birds

Practically there are four different ways of treating birds individually:

a) Introducing the treatment material into the crop by tube and syringe,

b) Introducing the treatment material into the beak using a hypodermic syringe fitted with a beaded needle,

c) Allowing each chick to drink from the tip of a pipette,

d) Dipping the beak of the bird in the treatment material.

Intubation of probiotics into the crop can be used in laboratory trials especially when precise control of the treatment dose is important (Nurmi and Rantala (1973). Mead *et al.* (1989) recommended the method, which allows chicks to drink from the tip of pipette. Administration of probiotics via the beak is also commonly used in laboratory trials and beak dipping may be appropriate in some circumstances.

For mass application of probiotics in poultry flock, following methods can be used:

a) Administration via drinking water

b) Droplet and spray application

c) Administering through the feed

a) Administration via drinking water

This method of probiotics application was introduced by Rantala (1974). The first field application of this method showed 11% reduction in the incidence of *Salmonella typhimurium* var. *copenhagen* (Seuna *et al*., 1978). Better results were reported from field trials in Sweden (Wierup *et al*., 1988). Practical application of competitive exclusion preparations through the first drinking water of the hatched chicks while the feed is withheld is not always optimal. Sometimes some of the chicks refuse to drink and the probiotics preparation spreads unevenly among the flock. The viability of the anaerobic organisms shows a rapid decline especially in chlorinated water (Seuna *et al*., 1978).

b) Droplet and spray application

Pivnick and Nurmi (1982) applied first the method of administering competitive exclusion cultures by using aerosols. Goren *et al*., (1984) developed a spray application method for treating of newly hatched chicks, either in the hatchers themselves or in the delivery boxes. Newly hatched chicks were treated with a homogenate of either crop or caecal material or a mixture of both cultures of aerobically and anaerobically cultured intestinal microorganisms from adult hens. Spray application of competitive exclusion cultures in the hatchers followed by drinking water administration on the farm was described by Blankenship (1992). This method is highly effective for the control of *Salmonellae* (Schneitz *et al*., 1990). Schneitz (1992) studied the method of automated spray application of competitive exclusion (CE) preparations, an automated cabin for the vaccination against Infectious Bronchitis were used. Ghadban *et al*. (1998) and Ghadban (1999) reported that spray application of competitive exclusion preparations when 50–60% of the chicks were hatched followed by treatment of the chicks through their first drinking water on the farm was a highly effective method in controlling *Salmonellae* and *E. coli* and in improving growth performance of treated chicks.

c) Administering through the feed

Classical probiotics like *Lactobacillus* or *Streptococcus* rarely produce optimum results in the pelleted feed usually fed to broilers. This seems to be due to the fact that the lactic acid bacteria are destroyed partly or totally by the current pelleting process. The optimum viability temperature of lactic acid bacteria is around 35–38°C (Crawford, 1979), while pelletization may increase the temperature of finished feed up to 80°C. Gould and Hurst (1969) reported that spores of Bacillus are well known for being able to survive high temperatures.

Modes of action of probiotics

It is believed by most investigators that there is an unsteady balance of beneficial and non-beneficial bacteria in the tract of normal, healthy, non-stressed poultry. When a balance exists, the bird performs to its maximum efficiency, but if stress is imposed, the beneficial floras, especially lactobacilli, have a tendency to decrease in numbers and an overgrowth of the non-beneficial ones seems to occur. The protective flora which establishes itself in the gut is very stable, but it can be influenced by some dietary and environmental factors. The three most important factors are excessive hygiene, antibiotic therapy and stress. Enhancement of colonization, resistance and/or direct inhibitory effects against pathogens is important factors where probiotics have reduced the incidence and duration of diseases. Probiotic strains have been shown to inhibit pathogenic bacteria both *in vitro* and *in vivo* through several different mechanisms. Some of the proposed modes of action of probiotics in poultry include:

i) Modifying intestinal microflora by competitive exclusion and antagonism
ii) Altering metabolism by increasing digestive enzyme activity and decreasing bacterial enzyme activity
iii) Improving feed intake and digestion
iv) Decrease ammonia and urea excretion
v) Stimulating the immune system

(i) Competitive exclusion (CE)

Competitive exclusion implies the prevention of entry of one entity into a given environment by occupying the available space. As a result of competitive exclusion, native micro-flora competitively excludes undesired bacterial contamination from the intestinal tract of poultry (Voltera, 1928). Probiotic and competitive exclusion approaches have been used as one method to control endemic and zoonotic agents in poultry. Competitive exclusion in poultry has implied the use of naturally occurring intestinal microorganisms in chicks and poults that were ready to be placed in brooder house. Nurmi Metchnikoff (1907) recognized the beneficial effects of probiotics, based on his observations on the longevity of Bulgarian pheasants who consumed large amounts of milk fermented with *Lactobacillus acidophilus*. He specified that detrimental microbe in the intestinal tract produce harmful substances to the host, which could be neutralized by beneficial organisms in yoghurt. Later, it was assumed that the beneficial effects were due to the colonization of the gut by *L. acidophilus* (Rettger and Chaplin, 1921). Tortuero (1973) used for poultry preparations containing living bacteria and observed that using of lactobacilli resulted in performances similar to those obtained when using antibiotics.

Millner and Shaffer, (1952) recognized that in chickens natural resistance to *Salmonella* infections developed with the establishment of a mature intestinal flora. The significance of normal indigenous intestinal micro-flora, especially the anaerobes, in protecting the host against pathogenic transient bacteria such as *Salmonella typhimurium* was demonstrated in mice (Bonhoff *et al.*, 1954) and with Vibrio cholera in guinea pigs (Freter, 1955). Royal and Mutimer (1972) were the first to demonstrate that caecal cultures inhibited the growth of *Salmonella typhimurium* in vitro, and they forecasted the use of these cultures as a preventive in vivo treatment. Nurmi and Rantala (1973) introduced the method of "competitive exclusion" (CE) to increase the resistance of young chicks to Salmonella infection by inoculating them orally with intestinal content from adult birds. They demonstrated that orally inoculation of 1–2 day old chicks with a 1:10 dilution of normal intestinal contents from healthy adult birds one day prior to oral challenge with *S. infantis* resulted in 77% of birds free from infection. This study was the basis for further development of the competitive exclusion methods. During the past two decades, many studies on the efficiency of CE against pathogenic bacteria (i.e. Salmonella, Escherichia coli, and Campylobacter) have been carried out. Undefined preparations of cultured fecal or caecal micro-flora generally reduce the prevalence of infected chicks following challenge with a standard dose of *Salmonella* under laboratory conditions (Goren *et al.*, 1984; Blankenship *et al.*, 1993). In contrast, results under field conditions have been more variable (Stavric and D'Aoust, 1993). Defined cultures are less effective than undefined cultures under laboratory conditions. The potency of defined cultures has been found to decrease gradually during cold storage and during repeated laboratory manipulation of the bacterial isolates (Stavric *et al.*, 1991; Mead, 1989). Lactic acid bacteria have been the component of the used defined cultures. Fuller (1977) reported that host-specific *Lactobacillus* strains 59 and 74/1 were able to decrease E. coli in crop and small intestine, but not in caeca of gnotobiotic chickens. Muralidhara *et al.* (1977) found that the homogenates of washed intestinal tissues dosed with *L. lactis* had higher numbers of attached Lactobacilli and lower E. coli counts than control birds.

Jin *et al.* (1996 c) found that only 26% of the isolates of *Lactobacillus* spp. from chicken intestines were able to attach moderately or strongly to the ileal epithelial cells of chickens. The ability of *Lactobacillus* to adhere in vitro to intestinal epithelial cells varies considerably among species and among strains of the same species (Barrow *et al.*, 1980; Kleeman and Klaenhammer, 1982; Jin *et al.*, 1996c). Sissons (1989) suggested that *Lactobacilli* compete with pathogens for sites of adherence on the intestinal surface. Attachment is necessary for proliferation and for reducing the rate of removal of organisms from specific sites in the gastrointestinal tract due to the movement of digesta caused by peristalsis. Light and electron microscopic studies confirmed that invading Salmonellae penetrate through the laminar surface of the epithelial cells of the intestine (Turnbull and Richmond, 1978). Electron microscopic examination revealed that *Salmonellae* adhere firmly to the mucosa of the caeca and, in the absence of other micro-flora are able to colonize any point along the gastrointestinal tract (Soerjadi *et al.*, 1982).

The adherent bacteria of the intestinal flora are interconnected with fibers to the mucosal surface. Bacterial colonization on the epithelial wall of the caeca is more extensive in treated than in untreated chicks. Maximum colonization of chick's intestinal flora occurs at 48–72 hours after treatment (Soerjadi *et al.*, 1982). The early colonization of the intestinal wall by a dense mat of micro-flora plays an important role in the initial protection of chicks against Salmonella infection. This supports the theory that direct competition for attachment sites is probably the primary mechanism of competitive exclusion (Muralidhara *et al.*, 1977; Cownway *et al.*, 1987; Stavric, 1987). Competitive exclusion is a very effective measure to protect newly hatched chicks, turkey poults, quails and pheasants and possibly other game birds, too, against *Salmonella* and other enteropathogens (Schneitz, 2005).

(ii) Antagonistic activity

The antagonistic activity of lactic acid bacteria against different pathogenic microorganisms can be related to the production of bacterial substances as bacteriocins, organic acids, and hydrogen peroxides. Bacteriocins are defined as compounds produced by bacteria, which have a biologically active protein moiety and a bactericidal action (Tagg *et al.*, 1976). Lactobacilli have been extensively studied for the production of antagonistic substances, which are referred to as bacteriocins. These include well-characterized bacteriocins (DeKlerk and Smith, 1967; Barefoot and Klaenhammer, 1983; Joerger and Klaenhammer, 1986), bacteriocins like substances (Vincent *et al.*, 1959), and other antagonistic substances not necessarily related to bacteriocins (Shanhani *et al.*, 1976). DeKlerk and Smith (1967) found that out of the studied 121 strains of *L. fermentum*, 25 produced bacteriocin-like substances, which were not affected by pH or catalase, were non-dialyzable, and were precipitated by ammonium acetate. Bacteriocins from *L. helveticus*, known as lactocin-27, have been identified and characterized by Uperti and Hindsdill (1973, 1975) and by Joerger and Klaenhammer (1986).

(iii) Improving digestive enzyme activity

Lactobacillus spp. has been shown to produce digestive enzymes in vitro and the enzymes may enrich the concentration of intestinal digestive enzymes. Szylt *et al.* (1980) reported that two of five strains of Lactobacillus isolated from male chicks showed alpha-amylase activity. The lactobacilli colonizing the intestine may secrete the enzyme, thus increasing the intestinal amylase activity (Duke, 1977; Sissons, 1989). *Lactobacillus* spp., which are used in cheese products were found to have amylolytic, lipolytic and proteolactic activities (Moon and Kim, 1989; Lee, 1990). Jin *et al.* (1996 b) reported that all 12 *Lactobacillus spp.* isolated from chicken's intestine were found to secret amylase, protease, and lipase both extracellularly and intracellularly. They also found that amylase activity in small intestine increased when Lactobacillus cultures were fed to the broilers, but there was no effect on lypolytic and proteolytic activities (Jin *et al.*, 1996a). Collington *et al.* (1990) found that the inclusion of a probiotic (mixture of multiple strains of *L. plantarum, L. acidophilus, L. casei,* and *S. faecium*) in the diet resulted in significantly higher carbohydrase activities in the mucosal tissue of pigs.

The effect of *Aspergillus oryzae* on macronutrients metabolism in laying hens was observed (Schneitz, 2005), of which findings might be of practical relevance. They postulated that active amylolytic and proteolytic enzymes residing in *Aspergillus oryzae* may influence the digested nutrients. Similarly, it was reported that an increase in the digestibility of dry matter was closely related to the enzymes released by yeast (Han *et al.*, 1999).

(iv) Decreasing bacterial enzyme activity

Goldin and Gordbach (1977) reported that the activities of nitroreductase, azoreductase and beta-glucuronidase in the gut of rates could be reduced by feeding supplements of L. acidophilus. The same results were observed in humans (Goldin and Gorbach, 1984a). A similar reduction in beta-glucuronidase has been detected in chickens fed 40% yoghurt in the drinking water (Coloe *et al.*, 1984) and in pigs (Cole *et al.*, 1987).

(v) Suppressing ammonia production

Suppressing ammonia production and urea's activity can be beneficial for improving animal health and enhancing growth as ammonia can cause damage to the surface of cells. Probiotics may contribute to the improvement of health status of birds by reducing ammonia production in the intestines. Chiang and Hsien (1995) reported that probiotics containing L. acidophilus, S. faecium, and *B. subtilis* reduced the concentration of ammonia in the excreta and litter of broilers. Increasing feed intake and digestion. The intestinal bacterial flora of domestic animals has an important role in the digestion and absorption of feed. It participates in the metabolism of dietary nutrients such as carbohydrates, proteins, lipids, and minerals and in the synthesis of vitamins. Nahanshon *et al.* (1992, 1993, 1994, and 1996) determined that addition of *Lactobacillus* cultures in maize/soybean or maize/barely/soybean diets stimulated appetite and increased fat, nitrogen, calcium, phosphorus, copper, and manganese retention in layers.

(vi) Enterotoxin neutralization

A substance produced by a probiotic may neutralize enterotoxins produced by pathogenic bacteria. Different studies with L. bulgaricus showed that this microorganism produces a metabolite that has a neutralizing effect on enterotoxins released from coli-forms (Mitchell and Kenworthy, 1976; Stuart *et al.*, 1978; Schwab *et al.*, 1980). Stimulation of immune system Immunity resulting from gut exposure to a variety of antigens, such as pathogenic bacteria and dietary protein, is important in the defense of young animals against enteric infections (Perdigon *et al.*, 1990 and 1995). Dunham *et al.* (1993) reported that birds treated with *L. reuteri* exhibited longer ileal villi and deeper crypts, which are a response, associated with enhanced T cell function and increased production of anti-Salmonella IgM antibodies. Nahanshon *et al.* (1994) found that Lactobacillus supplementation of layers diets increased cellularity of Peyer's patches in the ileum indicating a stimulation of the mucosal immune system that responded to antigenic stimuli by secreting immunoglobulin (IgA).

(vii) Stimulating the immune system

The manipulation of gut microbiota via the administration of probiotics influences the development of the immune response (McCracken and Gaskins, 1999). The exact mechanisms that mediate the immunomodulatory activities of probiotics are not clear. However, it has been shown that probiotics stimulate different subsets of immune system cells to produce cytokines, which in turn play a role in the induction and regulation of the immune response (Massan *et al.*, 2000; Christensen *et al.*, 2002). Stimulation of human peripheral blood mononuclear cells with *Lactobacillus rhamnosus* strain GG *in vitro* resulted in the production of interleukin 4 (IL-4), IL-6, IL-10, tumor necrosis factor alpha, and gamma interferon (Schultz *et al.*, 2003). Other studies have provided confirmatory evidence that Th2 cytokines, such as IL-4 and IL-10, are induced by lactobacilli (Christensen *et al.*, 2002; Rakoff-Nahoum *et al.*, 2004). The outcome of the production of Th2 cytokines is the development of B cells and the immunoglobulin isotype switching required for the production of antibodies. The production of the mucosal IgA response is dependent on other cytokines, such as transforming growth factor β (Lebman and Edmiston, 1999). Importantly, various species and strains of lactobacilli are able to induce the production of transforming growth factor β, albeit to various degrees (Blum *et al.*, 2000). Probiotics, especially lactobacilli, could modulate the systemic antibody response to antigens in chickens (Kabir *et al.*, 2004; Haghighi *et al.*, 2005; Huang *et al.*, 2004).

Effects of probiotics on poultry production

Growth performances

Studies on the beneficial impact on poultry performance have indicated that probiotic supplementation can have positive effects. Increased live weight gain in probiotic fed birds were reported by various research workers (Jin *et al.*, 1998; Kalavathy *et al.*, 2003; Kabir *et al.*, 2004; Khaksefidi *et al.*, 2006; Mountzouris *et al.*, 2007). On the other hand, Lan *et al.* (2003) found higher (P<0.01) weight gains in broilers subjected to two probiotic species. Huang *et al* (2003) demonstrated that inactivated probiotics, disrupted by a high-pressure homogenizer, have positive effects on the production performance of broiler chickens when used at certain concentrations. In addition, Torres-Rodriguez *et al.* (2007) reported that administration of the selected probiotic (FM-B11) to turkeys increased the average daily gain and market BW, representing an economic alternative to improve turkey production. However, Karaoglu and Durdag (2005) used *Saccharomyces cerevisiae* as a dietary probiotic to assess performance and found no overall weight gain difference. Kabir *et al.*(2004) reported the occurrence of a significantly (P<0.01) higher carcass yield in broiler chicks fed with the probiotics on the 2nd, 4th and 6th week of age both in vaccinated and non vaccinated birds. Although Mahajan *et al.* (1999) recorded in their study that mean values of giblets, hot dress weight, cold dress weight and dressing percentage were significantly (P<0.05) higher for probiotic (Lacto-Sacc) fed broilers. On the other hand, Mutus *et al.* (2006) investigated the effects of a dietary supplemental probiotic on morphometric parameters yield stress of the tibia and they found that tibiotarsi weight, length, and weight/length index, robusticity index, diaphysis diameter, modulus of elasticity, yield stress parameters, and percentage Ca content were not affected by the dietary supplementation of probiotic, whereas thickness of the medial and lateral wall of the tibia, tibiotarsal index, percentage ash, and P content were significantly improved by the probiotic.

The supplementation of either mixture of Lactobacilli cultures or preparations of Lactobacilli and other bacteria in chickens' feed has given variable results. Kim *et al.* (1988) reported that addition of commercial probiotic (*L. sporogenes*) increased weight gain of chicks fed a diet containing 10% mouldy maize at 2 or 6 weeks of age. Similar improvement in body weight gain of chickens fed a culture of *L. sporogenes* have also been reported by Kalbande *et al.* (1992) and Mohan Kumar and Christopher (1988). Han *et al.* (1984) supplemented chicks' diets with an aerobic spore-former (*L. sporogenes*) and Clostridium butyricum and found significant improvements in weight gains and feed conversion rates. Similar results were obtained by Meluzzi *et al.* (1986) with male chicks fed diets containing a mixture of *L. lactis* and *Streptococcus thermophilus* from day 1 to day 60 of age. Tortuero *et al.* (1989) demonstrated that weight gain and feed efficiency increased significantly ($P < 0.05$) when chickens were fed with a diet containing 30% beans supplemented with a mixture of *L. acidophilus* and *S. faecium* ($2x10^9$ CFU/kg) for 5–8 weeks. Mohan *et al.* (1995) reported that body weight gain could vary by 5% to 9% when chickens were supplemented with probiotic containing a mixture of *L acidophilus*, *L. casei, Bifidobacterium bifidum, Aspergillus oryzae, and Torulopsis*. Owings *et al.* (1990) observed that feed efficiency and body weight were significantly ($P < 0.05$) improved in broilers treated with S. faecium in the feed and in the water compared to birds supplemented antibacterial products. Jin *et al.* (1996a) treating two hundred 10 days–old broiler chicks with commercial Lactobacilli added to their diets under a hot and humid environment found that weight gain was significantly higher and feed gain ratio was significantly lower than that of control birds ($P < 0.05$). Jin *et al.* (1997) reported that adding to the feed from 0 to 6 weeks of age of either a single strain of *L. acidophilus* I 26 or a mixture of *Lactobacillus* (adherent Lactobacillus cultures isolated from the intestine of the chickens) significantly improved body weight and feed gain ratios of broilers ($P < 0.05$). Yeo and Kim (1997) reported that feeding a diet containing probiotic (L. casei) significantly increased average daily gain during the first 3 weeks ($P < 0.05$). Addition of probiotic Streptococcus faecium M-74 to broiler diet (0.5 millon CFU/g and 1.0 million CFU/g) from 14th to 21st day of age increased body weight, improved feed conversion ratio, and decreased mortality of the treated chickens (Gerendai and Gippert, 1988). Including the probiotic Lactosacc and S. faecium JMB 52 cultures (400 million CFU/g) to the feed of broilers leads to the improvement of their productivity, consistent improvement has also been reported by Koudela *et al.* (1992). Nguyen *et al.* (1988a) fed broiler chicks with five diets supplemented with probiotics presented in crumbles and pellets, and reported that both probiotics acted as potential growth promoters and did not affect microbial quality of carcasses. Van Wambeke and Peeters (1995) studied the effect of probiotic added to broiler's diet and determined that it had no significant effect on the chickens' body weight and did not improve feed conversion ratio. Grashorn (1998) studied the possibilities of using probiotics and/or digestion enhancers instead of antibiotics (Virginamycin) in the feed of heavy turkeys. He observed no negative effects on turkey performance. Ghadban (1999) observed the insufficiency of probiotic Toyocerin to overcome high temperatures during pelletization and reported that spray application and drinking water administration were more effective methods for probiotic treatment to broiler-chicks resulting in a significantly better growth performance.

Mohan-Kumar and Christopher (1988) observed that lactic acid bacteria exert an influence on the synthesis of vitamins of complexes B_{12} and K. Those bacteria also were effective against different strains of pathogenic microorganisms such as E. coli, Salmonella, *Shigella, Pasteurella, and Staphylococcus.* Basing on the in vitro activity against some pathogenic strains several authors considered the possibilities of lactic acid bacteria to be used for controlling and inhibiting the growth of non-useful microflora at the gastrointestinal tract of animals and humans (Vicent *et al.*, 1959; Gilliland *et al.*, 1977; McCormick *et al.*, 1983). Application of probiotic to poultry resulted in 5–6% less mortality through the first week, fully suppressing the growth of E. coli, improving daily gain and feed conversion ratio (Kabakchiev *et al.*, 1994). Bilgili and Moran (1990) used a probiotic bacterial culture from *Bifidobacterium pseudulongum, Bifidobacterium thermophilium,* and *L. acidophilus* in dose of $6.8x10^6$ for obtaining safe and healthy poultry products. Vladimirova and Sourdjiyska (1996) reported that a combined supplementation of probiotics, Yeasacc, and Lactosacc to diets in doses of 0.1% from 1st to 28th day of live and 0.05% from 29th to 42nd day of live increased live weight significantly by 11% in the groups and by 12% when probiotics were used combined. Ghadban (1998) studied the effect of spray application of probiotics when chicks were 50–70% hatched followed by drinking water administration and reported that treated chicks had significantly higher weight gain ratio at the end of the 6th week of age, less mortality by 55–70 %, and better feed conversion ratio compared to chickens from the control group.

There are many other investigations lacking positive effects of probiotics. Watkinz and Kratzer (1984) using a strain from the specific host (KTM, 74/1 and 59), *Lactobacilli*, and commercial products containing *Lactobacilli* did not find any improvement in growth performance of treated chicks. Maiolino *et al.* (1992) did not find any significant differences in live weights of chickens fed diets containing L. acidophilus and S. faecium from 8 to 60 days of age. The effect of probiotic against Salmonella contamination Spray application of probiotic to newly hatched chicks followed by administration via the chicks' first drinking water has been a very efficient method for controlling of the intestinal Salmonella colonization in poultry. Goren *et al.* (1988) treated 284 flocks with resuspended freeze-dried intestinal homogenates from specific pathogen free (SPF) birds. The incidence of Salmonella infected flocks was reduced from 38.6% to 7.6% and the incidence of infected broilers within the positive flocks was reduced from 38.6% to 5.2 %. Blankenship *et al.* (1993) reported that Salmonella prevalence in caeca and in processed carcasses was significantly reduced from 41% in control flocks to 10% in treated flocks. This shows that treating chickens with probiotics can serve as useful means to reduce Salmonella contaminations. The same authors, using mucosal competitive exclusion (MCE) for the treatment of newly hatched chicks through spraying followed by water administration reported that initial feed, water, and litter contamination was at a low frequency (<10 %). Eggshell fragment and chicks paper pads were frequently contaminated (>50 %). After 3 weeks growth, contamination of litter, skin with feathers, and caeca were significantly ($P < 0.05$) reduced in treated flocks as compared to control ones. Ghadban *et al.* (1998) reported that spray application of probiotic followed by water administration showed a clear tendency for elimination of Salmonella (38.8% in control group to 9.72% in treated groups) and E. coli (51.4% in control group to 22.2% in treated group) in the organism of treated chickens and it was concluded that the probiotic was highly effective in preventing intestinal colonization by pathogenic bacteria. The same authors found that the main challenged organism in birds of the control group grew heavily at the pure culture, while the same organism grew only in a single colony in the experimental groups. Bolder *et al.* (1995) studied the effect of spray application of Broilact1 followed by its administration by drinking water in broilers in a field trial with 2.4×10^6 day old chicks from 40 farms during six consecutive rearing periods. They determined that application of this probiotic led to a significantly lower Salmonella incidence in cloacal swabs at 4 weeks of age and to a significantly lower Campylobacter incidence in the caecal content at the moment of slaughtering. The Salmonella incidence of both swabs and caeca contents were significantly correlated with the Salmonella status of the one-day-old chicks. The same authors reported that the Broilact1 treatment significantly reduced the incidence at 4 weeks of age of initially Salmonella positive one-day-old chickens from 58.3% to 38.1 %. Stern *et al.* (1998) tested a Mucosal Starter Culture TM (MSC) with over 100.000 broiler chicks.

Salmonella incidence in treated chicks was significantly reduced both on the farm and after processing with 9% and 31 %, respectively. Untreated control birds were 2% and 7.5% Salmonella positive, respectively, on the farm and after processing compared to 0% and 3.5% after the probiotic treatment (Bailey *et al.*, 1994). Stern *et al.* (1998) have tested the MSC in a field trial involving over 160.000 birds. Salmonella on final processed carcasses was reduced from 5% positive in untreated control birds to 0% in carcasses of treated birds. Administration of probiotic to chickens protected them from *S. enteritides* colonization and reduced the incidence of Salmonella in litter from 52% to 13% (Hinton *et al*, 1991).

Probiotic is a generic term, and products can contain yeast cells, bacterial cultures, or both that stimulate microorganisms capable of modifying the gastrointestinal environment to favor health status and improve feed efficiency. Mechanisms by which probiotics improve feed conversion efficiency include alteration in intestinal flora, enhancement of growth of nonpathogenic facultative anaerobic and gram positive bacteria forming lactic acid and hydrogen peroxide, suppression of growth of intestinal pathogens, and enhancement of digestion and utilization of nutrients (Yeo and Kim, 1997). Therefore, the major outcomes from using probiotics include improvement in growth (Yeo and Kim, 1997) reduction in mortality (Kumprecht and Zobac, 1998) and improvement in feed conversion efficiency (Yeo and Kim, 1997). These results are consistent with previous experiment of Tortuero and Fernandez (1995), who observed improved feed conversion efficiency with the supplementation of probiotic to the diet.

Probiotic and Disease control

Despite the introduction of the most stringent quality control and hygiene measures, absolute biosecurity is impossible especially under commercial conditions. Competitive exclusion (CE) has proved very effective in controlling both vertical and horizontal infection of Salmonella. Administration of probiotic is one of the most effective means of controlling Salmonella infection in poultry under correct hygienic conditions and requirements.

It was shown that treating chickens with probiotics can serve as useful means to reduce Salmonella contaminations. Goren *et al.* (1988) treated 284 flocks with resuspended freeze dried intestinal homogenates from specific pathogen free (SPF) birds. The incidence of Salmonella infected flocks was reduced from 38.6% to 7.6% and the incidence of infected broilers within the positive flocks was reduced from 38.6% to 5.2 %. Blankenship *et al.* (1993) reported that Salmonella prevalence in caeca and in processed carcasses was significantly reduced from 41% in control flocks to 10% in treated flocks. The same authors, using mucosal competitive exclusion (MCE) for the treatment of newly hatched chicks through spraying followed by water administration reported that initial feed, water, and litter contamination was at a low frequency (<10 %). Eggshell fragment and chicks paper pads were frequently contaminated (>50 %). After 3 weeks growth, contamination of litter, skin with feathers, and caeca were significantly ($P < 0.05$) reduced in treated flocks as compared to control ones. Ghadban *et al.* (1998) reported that spray application of probiotic followed by water administration showed a clear tendency for elimination of Salmonella (38.8% in control group to 9.72% in treated groups) and E. coli (51.4% in control group to 22.2% in treated group) in the organism of treated chickens and it was concluded that the probiotic was highly effective in preventing intestinal colonization by pathogenic bacteria. The same authors found that the main challenged organism in birds of the control group grew heavily at the pure culture, while the same organism grew only in a single colony in the experimental groups. Bolder *et al.* (1995) studied the effect of spray application of Broilact followed by its administration by drinking water in broilers in a field trial. They determined that application of this probiotic led to a significantly lower Salmonella incidence in cloacal swabs at 4 weeks of age and to a significantly lower Campylobacter incidence in the caecal content at the moment of slaughtering. The Salmonella incidence of both swabs and caeca contents were significantly correlated with the Salmonella status of the one-day-old chicks. The same authors reported that the Broilact1 treatment significantly reduced the incidence at 4 weeks of age of initially Salmonella positive one-day-old chickens from 58.3% to 38.1 %. Stern *et al.* (1998) tested a Mucosal Starter Culture (MSC) with over 100.000 broiler chicks. Salmonella incidence in treated chicks was significantly reduced both on the farm and after processing with 9% and 31 %, respectively. Untreated control birds were 2% and 7.5% Salmonella positive, respectively, on the farm and after processing compared to 0% and 3.5% after the probiotic treatment (Bailey *et al.*, 1994). Stern *et al.* (1998) have tested the MSC in a field trial. Salmonella on final processed carcasses was reduced from 5% positive in untreated control birds to 0% in carcasses of treated birds. Administration of probiotic to chickens protected them from *S. enteritides* colonization and reduced the incidence of Salmonella in litter from 52% to 13% (Hinton *et al*, 1991).

Probiotic and Immune response

Kabir *et al.* (2004) evaluated the dynamics of probiotics on immune response of broilers and they reported significantly higher antibody production ($P<0.01$) in experimental birds as compared to control ones. They also demonstrated that the differences in the weight of spleen and bursa of probiotics and conventional fed broilers could be attributed to different level of antibody production in response to SRBC. Similarly, Khaksefidi and Ghoorchi (2006) reported that the antibody titer in the 50 mg/kg probiotic supplemented group was significantly higher at 5 and 10 days of post immunization (PI) compared to control, when SRBC was injected at 7 and 14 days of age. Haghighi *et al* (2005) demonstrated that administration of probiotics enhances serum and intestinal natural antibodies to several foreign antigens in chickens. On the other hand, Dalloul *et al.* (2005) examined the effects of feeding a *Lactobacillus*-based probiotic on the intestinal immune responses of broiler chickens over the course of an *E. acervulina* infection and they demonstrated that the probiotic continued to afford some measure of protection through immune modulation despite a fairly overwhelming dose of *E. acervulina*. They also suggested a positive impact of the probiotic in stimulating some of the early immune responses against *E. acervulina*, as characterized by early IFN-γ and IL-2 secretions, resulting in improved local immune defenses against coccidiosis. Brisbin *et al.* (2008) investigated spatial and temporal expression of immune system genes in chicken cecal tonsil and spleen mononuclear cells in response to structural constituents of *L. acidophilus* and they found that cecal tonsil cells responded more rapidly than spleen cells to the bacterial stimuli, with the most potent stimulus for cecal tonsil cells being DNA and for splenocytes being the bacterial cell wall components. They also discovered that in both splenocytes and cecal tonsil cells, STAT2 and STAT4 genes were highly induced and the expression of STAT2, STAT4, IL-18, MyD88, IFN-alpha, and IFN-gamma genes were up-regulated in cecal tonsil cells after treatment with *L. acidophilus* DNA. Simultaneously, several investigators demonstrated the potential effect of probiotic on immunomodulation (Matsuzaki and Chin, 2000; Dalloul *et al.*, 2003; Koenen *et al.*, 2004; Haghighi *et al.*, 2005). On the other hand, Midilli *et al.* (2008) showed the ineffectiveness of additive supplementation of probiotics on systemic IgG.

Probiotic and intestinal microbiota and intestinal morphology

Kabir *et al* (2005) attempted to evaluate the effect of probiotics with regard to clearing bacterial infections and regulating intestinal flora by determining the total viable count (TVC) and total lactobacillus count (TLC) of the crop and cecum samples of probiotics and conventional fed groups at the 2nd, 4th and 6th week of age. Their result revealed competitive antagonism. The result of their study also evidenced that probiotic organisms inhibited some non-beneficial pathogens by occupying intestinal wall space. They also demonstrated that broilers fed with probiotics had a tendency to display pronounced intestinal histological changes such as active impetus in cell mitosis and increased nuclear size of cells, than the controls. This results of histological changes support the findings of Samanya and Yamauchi (2002) and they indicated that birds who were fed dietary *B. subtilis* var. *natto* for 28 days had a tendency to display greater growth performance and pronounced intestinal histologies, such as prominent villus height, extended cell area and consistent cell mitosis, than the controls. On the other hand, Chichlowski *et al.* (2007) compared the effects of providing a direct-fed microbials (DFM) with the feeding of salinomycin on intestinal histomorphometrics, and microarchitecture and they found less mucous thickness in DFM-treated chickens and the density of bacteria embedded in the mucous blanket appeared to be lower in DFM-treated chickens than in the control in all intestinal segments. Watkins and Kratzer (1983) reported that chicks dosed with *Lactobacillus* strains had lower numbers of coliforms in cecal macerates than the control. Francis *et al.* (1978) also reported that the addition of *Lactobacillus* product at 75 mg/kg of feed significantly decreased the coliform counts in the ceca and small intestine of turkeys. Using gnotobiotic chicks, Fuller (1977) found that host-specific *Lactobacillus* strains were able to decrease *Escherichia coli* in the crop and small intestine. Kizerwetter-Swida and Binek (2009) demonstrated that *L. salivarius* strain reduced the number of *Salmonella enteritidis* and *Clostridium perfringens* in the group of chickens treated with Lactobacillus. Watkins *et al.* (1982) similarly observed that competitive exclusion of pathogenic *E. coli* occurred in the gastrointestinal tract of gnotobiotic chicks dosed with *L. acidophilus*. It was demonstrated that probiotic species belonging to *Lactobacillus*, *Streptococcus*, *Bacillus*, *Bifidobacterium*, *Enterococcus*, *Aspergillus*, *Candida*, and *Saccharomyces* have a potential effect on modulation of intestinal microflora and pathogen inhibition (Yaman *et al.*, 2006; Mountzouris *et al.*, 2007; Higgins *et al.*,2007).

Evaluating probiotic effects on meat quality

Kabir *et al.* (2005) evaluated the effects of probiotics on the sensory characteristics and microbiological quality of dressed broiler meat and reported that supplementation of probiotics in broiler ration improved the meat quality both at pre freezing and post freezing storage. Mahajan *et al.* (2000) stated that the scores for the sensory attributes of the meat balls appearance, texture, juiciness and overall acceptability were significantly higher and those for flavour were lower in the probiotic (Lacto-Sacc) fed group. Simultaneously, Mahajan *et al.* (2000) reported that meat from probiotic (Lacto-Sacc) fed birds showed lower total viable count as compared to the meat obtained from control birds. On the other hand, Loddi *et al.* (2000) reported that neither probiotic nor antibiotic affected sensory characteristics (intensity of aroma, strange aroma, flavour, strange flavour, tenderness, juiciness, acceptability, characteristic colour and overall aspects) of breast and leg meats. On the other hand, Zhang *et al.* (2005) conducted an experiment with 240, day-old, male broilers to investigate the effects of *Saccharomyces cerevisiae* (SC) cell components on the meat quality and they reported that meat tenderness could be improved by the whole yeast (WY) or *Saccharomyces cerevisiae* extract (YE).

Influence of antibiotics on the effectiveness of probiotics

Antibiotic treatment of chicks at sub-therapeutic levels is effective in different situations, but it has negative effect on the lactic acid bacteria, which are the main bacteria of the anaerobic micro-flora of the gastrointestinal tract of the birds. Lactic acid bacteria are very sensitive to different antibiotics (Tortuero, 1973; Dilworth and Day, 1978; Szylit and Charlet, 1981; Watkins and Kratzer, 1984; Roth and Kirchgessner, 1986; Bougon *et al.*, 1987; Ghadban, 1999). The influence of a therapeutically antibiotic treatment of broilers with Furazolidone and Trimetoprim/Sulfamethoxasol on the stability and the efficacy of probiotic were tested (Bolder and Palmu, 1995). Probiotic treatment of broilers resulted in an increased competitive potency against colonization with Salmonella. The IF (Infection Factor) in the probiotic treated groups or groups treated with a combination of probiotics and antibiotics was lower than in the other groups. Treatment of chicks with Furazolidone after probiotic treatment initially damaged the intestinal micro-flora, but after two weeks, the broilers appeared to be able to "self repair" and re-establish the exclusion of Salmonella from the gut. When Trimethoprim/ Sulfamethoxasol were given after probiotic, Infection factor of broilers increased although the differences were not significant.

The probiotic treatment showed a protection against Salmonella colonization of broilers and led to lower Salmonella CFU counts in the Salmonella positive broilers (Bolder *et al.*, 1992 and Goren *et al.*, 1984). Mohan *et al.*, (1995) studied the effect of probiotic and combined probiotic and antibiotic supplementation on growth, nitrogen utilization, and serum cholesterol content of broiler chickens. In the first experiment chicks treated with probiotic had a better weight gain, retained significantly more nitrogen than the control birds, and serum cholesterol content was lower in the probiotic-supplemented birds. In the second experiment, the probiotic plus antibiotic supplemented group had the highest weight gain. Nitrogen utilization was greatest in the antibiotic treated group.

The widespread use of antibiotics as therapeutic agents and growth promoters resulted in the development of resistant populations of bacteria, which made subsequent use of antibiotics for therapy difficult. Over the past twenty years, there has been increasing concern on the possible contribution of antibiotics used routinely at sub-therapeutic levels to the reservoir of resistant Salmonellae and to the occurrence of resistant Coliforms to which susceptible human beings may be exposed and to the occurrence of antibiotics residues in the poultry products (DuPont and Steele, 1987). The possibility of antibiotics ceasing to be used as growth stimulants for farm animals and the concern about the side effect of their use as therapeutic agents has produced a climate in which both consumer and manufacturer are looking for alternatives. Probiotics are being considered to fill this role and farmers are using them in preference to antibiotics (Fuller, 1989).

Future of Probiotics

Probiotics appear to be the feed additives of the future, especially under the politics of banning of growth promoters (feed antibiotics). In view of specificity to meet the challenge from different strains of pathogen, it will be necessary to develop specific probiotics with high inhibitory activity. The probiotic preparations have to be of adequate concentration and sufficiently stable both in storage and during administration to the birds, especially for those administered through the feed to be able to support high temperature of feed pelletization. The complex micro-flora present in the gastrointestinal tract of the birds is effective in providing resistance to disease, but this protective flora can be altered by dietary and environmental influences. The establishment of desirable microorganisms may depend on environmental factors such as the composition of the diet, the existing gut micro-flora at the time of administration, the general health status of the birds, and the different hygienic conditions. Therefore, in most types of probiotics, the extent to which adequate levels of colonization in the digestive tract of the whole birds in the flock can be achieved has to be determined.

Summary

Probiotics is a live microbial feed additive, which beneficially affects the host animal by improving its intestinal microbial balance. The probiotics could be successfully used as nutritional tools in poultry feeds for promotion of growth, feed efficiency, modulation of intestinal micro flora and pathogen inhibition, immunomodulation and promoting meat quality of poultry. Probiotics offer immense potential to become an alternative to feed antibiotics as they do not result in the development and spread of microbial resistance.

References

Barefoot S. F. and Klaenhammer, T. R. 1983. Detection and activity of Lactacin B, a bactariocin produced by *L. acidophilus*. *Applied and Environmental Microbiology*. 45: 1808–1815.

Barrow P. A, Brooker, B. E., Fuller, R. and Newport, M. J. 1980. The attachment of bacteria to the gastric epithelium of the pig and its importance in the micro-ecology of the intestine. *J. Appl. Bactriol.* 48:147–154.

Blankenship L. C. 1992. Report at International Poultry Exposition in Atlanta, USA. pp. 22–24.

Blankenship L. C., Bailey, J. S., Cox, N. A., Stern, N. J., Brewer, R. and Williams, O. 1993. Two-step mucosal competitive exclusion flora treatment to diminish Salmonella in commercial broiler chickens. *Poult. Sci.* 72: 1667–1762.

Blum, S., Haller, D., Pfeifer, A., Schiffrin, E.J. 2002. Probiotics and immune response. *Clin. Rev. Allergy Immunol.* 22: 287-309.

Bolder, N. M. and Palmu, L. 1995. The effect of antibiotic treatment of broilers on the efficacy of Broilact. In: Proceedings of XII European Symposium on the Quality of Poultry Meat and VI European Symposium on the Quality of Eggs and Egg Products, 25–29 September, Zaragoza, Spain, pp. 199–204.

Bolder, N. M., van Lith, L. A. , Putirulan, F. F., Jacobs-Reitsma, W. F. and Mulder, R. W.1992. Prevention of colonization by Salmonella enteritidis PT4 in broiler chickens. *Int. J. Food Microbiol.* 15: 313–317

Bonhoff , M., Drake, B. L. and Miller, C. P. 1954. Effect of streptomycin on susceptibility of intestinal tract to experimental Salmonella infection. *Proc. Soc. Exp. Biol. Med.* 86: 132.

Bougon M., Menec, M. Le and Launay, M. 1987. Influence des probiotiques sur les performances des poulets. *Bulletin d'information, Station Experimentale d'Aviculture de Ploufragan.* 27 (4): 120–125.

Brisbin, J.T., Zhou, H., Gong, J., Sabour, P., Akbari, M.R., Haghighi, H.R., Yu, H., Clarke, A.,Sarson, A.J., Sharif, S. 2008. Gene expression profiling of chicken lymphoid cells after treatment with *Lactobacillus acidophilus* cellular components. *Dev. Comp. Immunol.*32: 563-574.

Chichlowski, M., Croom, W.J., Edens, F.W., McBride, B.W., Qiu, R., Chiang, C.C., Daniel, L.R., Havenstein, GB., Koci, M.D. 2007. Microarchitecture and spatial relationship between bacteria and ileal, cecal, and colonic epithelium in chicks fed a direct-fed microbial, primalac, and salinomycin. *Poult. Sci.* 86: 1121-1132.

Christensen, H.R., Frokiaer, H., Pestka, J.J. 2002. Lactobacilli differentially modulate expression of cytokines and maturation surface markers in murine dendritic cells. *J. Immunol.* 168:171-178.

Cole, C. B., Anderson., Philips, P. H. S. M., Fuller , R. and Hewitt, D. 1984. The effect of yogurt on the growth, lactose-utilizing gut organisms and b-glucuronidase activity of caecal contents of a lactose-fed, lactose-deficient animal. *Food Microbiol.*1: 217–222.

Cole C. B., Fuller, R. and Newport, M. J. 1987. The effect of diluted yogurt on the gut microbiology and growth of piglets. *Food Microbiol.* 4: 83–85.

Collington, G. K., Parker, D. S. and Armstrong, D. G. 1990. The influence of inclusion of either an antibiotic or a probiotic in the diet on the development of digestive enzyme activity in the pig. *British J. of Nutrition.* 64: 59-70.

Coloe P. J., T. J. Bagust and L. Ireland, 1984. Development of the normal gastrointestinal micro-flora of specific pathogenfree chickens. *Journal of Hygiene* 92: 79–87.

Crawford J. S., 1979. Probiotics in animal nutrition. Proceedings of the Arkansas Nutrition Conference, Arkansas, USA, pp. 45–55.

Dalloul, R.A., Lillehoj, H.S., Shellem, T.A., Doerr, J.A. 2003. Enhanced mucosal immunity against *Eimeria acervulina* in broilers fed a *Lactobacillus*-based probiotic. *Poult. Sci.* 82: 62-66.

Dalloul, R.A., Lillehoj, H.S., Tamim, N.M., Shellem, T.A., Doerr, J.A. 2005. Induction of local protective immunity to *Eimeria acervulina* by a *Lactobacillus*-based probiotic. *Comp. Immun.Microbiol. Infect. Dis.* 28: 351-361.

Deklerk H. C. and J. A. Smith, 1967. Properties of Lactobacillus ferment bacteriocin. *J. General Microbiol.* 48: 309–316.

Dibner, J. J., and J. D. Richards. 2005. Antibiotic growth promoters in agriculture: history and mode of action. *Poult. Sci.* 84: 634-643.

Dilworth B. C. and E. J. Day, 1978. Lactobacillus cultures in broiler diets. *Poult. Sci.* 57: 1101.

Duke, G.E. Avian digestion. 1997. In *Physiology of Domestic Animals*, 9th ed., Duke, G.E., Ed.,Cornell University Press: Ithaca, NY, USA. pp. 313-320.

Dupont H. L. and J. H. Steele, 1987. Use of anti-microbial agents in animal feeds: Implication for human heath. *Rev. Infect. Dis.* 9: 447–460.

FAO/WHO. 2001. Health and nutritional properties of probiotics in food including powder milk with live lactic acid bacteria. *Report of a Joint FAO/WHO Expert Consultation on Evaluation of Health and Nutritional Properties of Probiotics in Food Including Powder Milk with Live Lactic Acid Bacteria*; FAO/WHO: Amerian Córdoba Park Hotel, Córdoba, Argentina. pp. 1-34.

Francis, C., Janky, D.M., Arafa, A.S., Harms, R.H. 1978. Interrelationship of *Lactobacillus* and zinc bacitracin in diets of turkey poults. *Poult. Sci.*, 57: 1687-1689.

Freter, R., 1955. Experimental enteric shigella and vibrio infection in mice and guinea pigs. *J. Experim. Med.* 104: 411–418.

Fuller, R., 1977. The importance of Lactobacilli in maintaining normal microbial balance in the crop. *British Poult. Sci.* 18: 85–94.

Fuller, R. 1989. Probiotics in man and animals: A review. *J. Applied Bacteriol.* 66: 365–378.

Fuller, R. 2001. The chicken gut microflora and probiotic supplements. *J. Poult. Sci.*, 38:189-196.

Ghadban G, 1999. Studying on productivity of chickens broilers treated by biological products, Ph.D. thesis, Thracian University, Stara Zagora, Bulgaria.

Gilliland S. E. and M. L. Speck, 1977. Antagonistic action of (A&A) L. acidophilus toward intestinal and food-borne pathogens in associative cultures. *J. Food Protection.* 40: 820–823.

Ghadban G., 1998. Investigation on the efficacy of early probiotic treatment on the performance of broiler chicks. Proceedings of 10th European Poultry Conference, June 21–26, Jerusalem, Israel, II: 305–310.

Ghadban G., M. Kabakchiev and A. Angelov, 1998. Efficacy of different methods of probiotic treatment in preventing infection of broiler chicks with Salmonella typhimurium and E. coli O7. Proceedings of 10th European Poultry Conference, June 21–26, Jerusalem, Israel, I: 305–310.

Goldin B. R. and Gorbach, S. L. 1977. Alternations in fecal micro-flora enzymes related to diet, age, Lactobacillus supplements and dimethylhydrazine. *J. National Cancer Institute.* 40: 2421–2426.

Goldin B. R. and Gorbach, S. L. 1984a. The effect of milk and Lactobacillus feeding on human intestinal bacteria enzyme activity. *American J. Clinical Nutrition.* 39: 756–761.

Goldin B. R. and Gorbach, S. L. 1984b. Alternations of the intestinal micro-flora by diet, oral antibiotic and Lactobacillus: Decreased production of free amines from aromatic nitro compounds, azo dyes and glucuronides. *J. National Cancer Institute.* 73: 689–695.

Goren E., W. A. de Jong, P. Doornenbal, J. P. Koopman and H. M. Kennis, 1984. Protection of chicks against Salmonella infection induced by spray application of intestinal micro-flora in the hatchery. *Vet. Q.* 6: 73–79.

Goren E., W. A. De Jong, P. Doornenbal, N. M. Bolder, R. W. A. W. Mulder and A. Jansen, 1988. Reduction of Salmonella infection of broilers by spray application of intestinal microflora: a longitudinal study. *Vet. Q.* 10: 249–255.

Gould G. W. and A. Hurst, 1969. The bacterial spore. Academic press – London and New York.

Guillot, J.F. 1998. Les probiotiques en alimentation animale. *Cah. Agric.* 7: 49-54.

Haghighi, H.R., Gong, J., Gyles, C.L., Hayes, M.A., Sanei, B., Parvizi, P., Gisavi, H., Chambers, J.R., Sharif, S. 2005. Modulation of antibody-mediated immune response by probiotics in chickens. *Clin. Diagn. Lab. Immunol.* 12: 1387-1392.

Han, S.W., Lee, K.W., Lee, B.D., Sung, C.G. 1999. Effect of feeding Aspergillus oryzae culture on fecal microflora, egg qualities, and nutrient metabolizabilities in laying hens. *Asian Aust. J. Anim. Sci.*, 12: 417-421.

Havenaar R., B. T. Brink, J. H. H. Huis Veld and R. Fuller, 1992. Selection of strains for probiotics use. In: Probiotics: The Scientific Basis (Ed. Fuller R.), Chapman and Hall, London, pp. 209–224

Higgins, J.P., Higgins, S.E., Vicente, J.L., Wolfenden, A.D., Tellez, G., Hargis, B.M. 2007. Temporal effects of lactic acid bacteria probiotic culture on *Salmonella* in neonatal broilers. *Poult. Sci.* 86:1662-1666.

Hinton M., G. C. Mead and Impey, C. S. 1991. Protection of chicks against environmental challenge with Salmonella enteritidis by "competitive exclusion" and acid treated feed. *Lett. Appl. Microbiol.* 12: 69–71.

Huang, M.K., Choi, Y.J., Houde, R., Lee, J.W., Lee, B., Zhao, X. 2004. Effects of lactobacilli and an acidophilic fungus on the production performance and immune responses in broiler chickens. *Poult. Sci.* 83: 788-795.

Jin L. Z., Y. W. Ho, A. Ali N. Abdullah, B. K. Ong and Jalaludin, S. 1996c. Adhesion of Lactobacillus isolates to the intestinal epithelial cells of chicken. *Letters in Applied Microbiol.* 22: 229–232.

Jin L. Z., Y. W. Ho, N. Abdullah and Jalaludin, S. 1996a. Influence of dried Bacillus subtilis and Lactobacilli culture on intestinal micro-flora and performance in broilers. *Asian-Australian J. Animal Sci.* 9: 397–404.

Jin L. Z., Y. W. Ho, N. Abdullah and Jalaludin, S. 1996b. Effect of Lactobacillus culture on the digestive enzymes in chicken intestine. Proceedings of the 8th Animal Science Congress, Tokyo, Chiba, Japan, pp. 224–225.

Jin L. Z., Y. W. Ho, N. Abdullah and Jalaludin, S. 1997. Probiotics in poultry: modes of action. *World's Poult. Sci. J.* 53: 352–368.

Jin, L.Z., Ho, Y.W., Abdullah, N., Jalaludin, S. 2000. Digestive and bacterial enzyme activities in broilers fed diets supplemented with *Lactobacillus* Cultures. *Poult. Sci.* 79: 886-891.

Jin, L.Z., Ho, Y.W., Abdullah, N., Jalaludin, S. 1998. Growth performance, intestinal microbial populations and serum cholesterol of broilers fed diets containing *Lactobacillus* cultures. *Poult. Sci.* 77: 1259-1265.

Joerger M. C. and Klaenhammer, T. R.1986. Characterization and purification of helvetin J and evidence for a chromosomally determined bacteriocin produced by Lactobacillus helveticus 481. *J. Bacteriology* 167: 439–446.

Kabir, S.M.L., Rahman, M.M., Rahman, M.B. 2005. Potentiation of probiotics in promoting microbiological meat quality of broilers. *J. Bangladesh Soc. Agric. Sci. Technol.* 2: 93-96.

Kabir, S.M.L., Rahman, M.M., Rahman, M.B., Hosain, M.Z., Akand, M.S.I., Das, S.K. 2005. Viability of probiotics in balancing intestinal flora and effecting histological changes of crop and caecal tissues of broilers. *Biotechnology.* 4: 325-330.

Kabir, S.M.L., Rahman, M.M., Rahman, M.B., Rahman, M.M., Ahmed, S.U. 2004. The dynamics of probiotics on growth performance and immune response in broilers. *Int. J. Poult. Sci.* 3: 361-364.

Kalavathy, R., Abdullah, N., Jalaludin, S., Ho, Y.W. 2003. Effects of *Lactobacillus* cultures on growth performance, abdominal fat deposition, serum lipids and weight of organs of broiler chickens. *Br. Poult. Sci.* 44:139-144.

Kalbande V. H., M. A. Gaffer and S. V. Deshmukh, 1992. Effect of probiotic and nitrofurin on performance of growing commercial pullets. *Indian J. Poultry Sci.* 27: 116–117.

Karaoglu, M., Durdag, H. 2005. The influence of dietary probiotic (*Saccharomyes cerevisiae*) supplementation and different slaughter age on the performance, slaughter and carcass roperties of broilers. *Int. J. Poult. Sci.*,4: 309-316.

Khaksefidi, A., Ghoorchi, T. 2006. Effect of probiotic on performance and immunocompetence in broiler chicks. *J. Poult. Sci.* 43: 296-300.

Kim C. J., H. Namkung, M. S. An and I. K. Paik, 1988. Supplementation of probiotics to the broiler diets containing moldy corn. *Korea J. Animal Sci.* 30: 542–548.

Kleeman E. G. and Klaenhammer, T. R. 1982. Adherence of Lactobacillus species to human fetal intestinal cells. *J. Dairy Sci.* 65: 2036–2069.

Koenen, M.E., Kramer, J., van der Hulst, R., Heres, L., Jeurissen, S.H.M., Boersma, W.J.A. 2004. Immunomodulation by probiotic lactobacilli in layer and meat-type chickens. *Br. Poult. Sci.* 45: 355-366.

Koh, K., E. Ibradolaza, and Isshuki, Y. 1992. Effect of administered lactic acid bacteria on feed utilization in chickens. *Japanese Poult. Sci.* 29: 242–246.

Kumprecht, I., Zobac, P. 1998. The effect of probiotic preparations containing *Saccharomyces cerevisiae* and *Enterococcus faecium* in diets with different levels of B-vitamins on chicken broiler performance. *Zivocisna Vyroba.* 43: 63-70.

Lan, P.T.N., Binh, L.T., Benno, Y. 2003. Impact of two probiotic *Lactobacillus* strains feeding on fecal lactobacilli and weight gains in chicken. *J. Gen. Appl. Microbiol.* 49: 29-36.

Lebman, D.A., Edmiston, J.S. 1999. The role of TGF-beta in growth, differentiation, and maturation of B lymphocytes. *Microbes Infect.* 15: 1297-1304.

Lee S. Y. and Lee, B. H. 1990. Esterolytic and lipolytic activities of Lactobacillus casei-subsp-casei LLG. *J. Food Sci.* 55: 119– 122.

Lilly D. M. and Stillwell, R. H. 1965. Probiotics: Growth promoting factors produced by microorganisms. *Science.* 147: 747–748.

Loddi, M.M., Gonzalez, E., Takita, T.S., Mendes, A.A., Roca, R.O., Roca, R. 2000. Effect of the use of probiotic and antibiotic on the performance, yield and carcass quality of broilers. *Rev. Bras.Zootec.*, 29: 1124-1131.

Maassen, C.B., van Holten-Neelen, C., Balk, F., den Bak-Glashouwer, M.J., Leer, R.J., Laman,J.D., Boersma, W.J., Claassen, E. 2000. Strain dependent induction of cytokine profiles in the gut byorally administered *Lactobacillus* strains. *Vaccine.* 18: 2613-2623.

Mahajan, P., Sahoo, J., Panda, P.C. 2003. Effect of probiotic (Lacto-Sacc) feeding, packaging methods and season on the microbial and organoleptic qualities of chicken meat balls during refrigerated storage. *J. Food Sci. Technol. Mysore.* 37: 67-71.

Mahajan, P., Sahoo, J., Panda, P.C. 1999. Effects of probiotic feeding and seasons on the growth performance and carcass quality of broilers. *Indian J. Poult. Sci.* 34: 167-176.

Matsuzaki, T and Chin, J. 2000. Modulating immune responses with probiotic bacteria. *Immunol. Cell Biol.* 78: 67-73.

McCracken, V.J., Gaskins, H.R. 1999. Probiotics and the immune system. In *Probiotics, a Critical Review*; Tannock, G.W., Ed., Horizon Scientific Press: Norfolk, UK. pp. 85-112.

Mead, G. C., P. A. Barrow, M. H. Hinton, F. Humbert, C. S. Impey, C. Lahellec, R. W. A. W. Mulder, S. Stavric and Stern, N. J. 1989. Recommended assay for treatment of chicks to prevent Salmonella colonization by "competitive exclusion". *J. Food Protect.* 52: 500–502.

Miles, R.D., Bootwalla, S.M. 1991. Direct-fed microbials in animal production. In *Direct-Fed Microbials in Animal Production. A Review*; National Food Ingredient Association: West Des Monies, Iowa, USA. pp. 117-132.

Milner K. C. and Shaffer, M. F. 1952. Bacteriologic studies of experimental Salmonella infections in chicks. *J. Infection and Diseases*. 90: 81–96.

Mitchell, I. G and Kenworthy, R. 1976. Investigation on a metabolite from Lactobacillus bulgaricus which neutralizes the effect of enterotoxin from Escherichia coli pathogenic for pigs. *J. Applied Bacteriol*. 41: 163–174.

Mohan, B., R. Kadirvel, A. Natarajan and Bhaskaran, M. 1995. Effect of probiotic supplementation on growth, nitrogen utilization and serum cholesterol in broilers. *British Poultry Sci.* 37: 395–401.

Mohan–Kumar O. R., and K. J. Christopher, 1988. The role of Lactobacillus Sporogenes (Probiotic) as Feed additive. *Poult. Guide*. 25 (10): 37–40.

Moon , Y. I. and Kim, Y. K. 1989. Study on the proteolytic action of intracelluar protease of Lactobacillus bulgaricus CH-18. *Korean J. Dairy Sci.* 11: 34–41.

Mountzouris, K.C., Tsirtsikos, P., Kalamara, E., Nitsch, S., Schatzmayr, G., Fegeros, K. 2007. Evaluation of the efficacy of probiotic containing *Lactobacillus*, *Bifidobacterium*, *Enterococcus*, and *Pediococcus* strains in promoting broiler performance and modulating cecal microflora composition and metabolic activities. *Poult. Sci.* 86: 309-317.

Muralidhara, K. S., G. G. Sheheby, P. R. Elliker, D. C. England and Sandine, W. E. 1977. Effect of feeding Lactobacilli on the Coliform and Lactobacillus flora of intestinal tissue and feces from piglets. *J. Food Prot*. 40: 288.

Mutu°, R., Kocabagli, N., Alp, M., Acar, N., Eren, M., Gezen, S.S. 2006. The effect of dietary probiotic supplementation on tibial bone characteristics and strength in broilers. *Poult. Sci.* 85: 1621-1625.

Nahanshon, S. N., H. S. Nakaue and Mirosh, L. W. 1992. Effect of direct-fed microbials on nutrient retention and parameters of laying pullets. *Poult. Sci.* 71: 1: 111.

Nahanshon , S. N., H. S. Nakaue and Mirosh, L. W. 1993. Effect of direct-fed microbials on nutrient retention and parameters of Single Comb White Leghorn pullets. *Poult. Sci.* 72:87.

Nahanshon, S. N., H. S. Nakaue and Mirosh, L. W. 1994. Performance of Single Comb White Leghorn laying pullets fed diets supplemented with direct – fed microbials. *Poult. Sci.* 73: 1699–1711.

Nahanshon, S. N., H. S. Nakaue and Mirosh, L. W. 1996. Performance of Single Comb White Leghorn layers fed with a live microbial during the growth and egg laying phases. *Animal Sci. and Technology*. 57: 25–38.

Nurmi, E. and Rantala, M. 1973. New aspects of Salmonella infection in broiler production. *Nature* (London) 244: 210-211.

Parker, R. B. 1974. Probiotics, the other half of the antibiotics story. *Anim. Nutrition Health*. 29: 4–8.

Patterson, J. A. and Burkholder, K. M. 2003. Application of Prebiotics and Probiotics in Poultry Production. *Poult. Sci.*. 82: 627–631

Perdigon, G., S. Alavares, M. E. N. De Macias, M. E. Roux and De Ruiz Holgado, A. A. P. 1990. The oral administration of lactic acid bacteria increases the mucosal intestinal immunity in response to enteropathogens. *J. Food Protection*. 53: 404–410.

Perdigon, G., S. Alavares, M. Rachid, G. Aguero and Gobbato, N. 1995. Immune system stimulation by probiotics. *J. Dairy Sci.* 78: 1597–1606.

Pivnick, H. and Nurmi, E.1982. The Nurmi Concept and its role in the control of Salmonella in poultry. In: R. Davies (Ed.): Developments in food microbiology. Applied Science Publishers London, pp. 41–70.

Rakoff-Nahoum, S., Paglino, J., Eslami-Varzaneh, F., Edberg, S and Medzhitov, R. 2004. Recognition ofcommensal microflora by toll-like receptors is required for intestinal homeostasis. *Cell*.118: 229-241.

Rantala, M. 1974. Cultivation of a bacterial flora able to prevent the colonization of Salmonella infantis in the intestines of broiler chickens, and it use. *Acta. Pathol. Microbiol, Scand.* 82: 75–80.

Rantala, M. 1976. The influence of the intestinal flora and some antimicrobial drugs on the colonization of Salmonella infantis in the intestine of the chicken. Ph.D. thesis, College of Veterinary Medicine, Helsinki, Finland.

Rettger L. R. and Chaplin, H. A. 1921. Treatise on the transformation of the intestinal flora with special reference to the implantation of Bacillus acidophilus. Yale University Press, New Haven, Connecticut.

Roth, F. and Kirchgessner, M. 1986. Zur nutritiven Wirksamkeit von Streptococcus faecium (Stamm M 74) in der Ku¨kenmast. *Arch. Geflu¨gelk.* 50: 225–228.

Royal, W. A. and Mutimer, M. D. 1972. Inhibition of Salmonella typhimurium by fowl caecal cultures. *Res. Vet. Sci.* 13: 184.

Samanya, M and Yamauchi, K.2002. Histological alterations of intestinal villi in chickens fed dried *Bacillus subtilis* var. *natto. Comp. Biochem. Physiol. Physiol.* 133: 95-104.

Schneitz C., L. Nuotio, G. Mead and E. Nurmi, 1992. Competitive exclusion in the young bird. Challenge models, administration and reciprocal protection. *Int. J. Food Microbiol.* 15: 241–255.

Schneitz, C., M. Hakkinen, L. Nuotio, E. Nurmi and Mead, G. 1990. Droplet application for protecting chicks against Salmonella colonization by competitive exclusion. *Vet. Record* 126: 510.

Schneitz, C. 2005. Competitive exclusion in poultry-30 years of research. *Food Control.* 16: 657-667.

Seuna, E., M. Raevuori and Nurmi, E. 1978. An epizootic of Salmonella typhimurium var. copenhagen in broilers and the use of cultured chicken intestinal flora for its control. *British Poultry Sci.* 19: 309–314.

Shanhani, K. M., J. R. Vakil and Kilara, A. 1976. Natural antibiotic activity of Lactobacillus acidophilus and L. bulgaricus. 1. Cultural conditions for the production of antibiotics. *Cultured Dairy Production. J.* 11: 117.

Sissons, J. W. 1989. Potential of probiotic organisms to prevent diarrhea and promote digestion in farm animals; a review. *J. Food and Agriculture Sci.* 49: 1–13.

Soerjadi, A. S., R. Rufner, G. H. Snoeyenbos and Weinack, O. M. 1982. Adherence of Salmonellae and native gut microflora to the gastrointestinal mucosa of chick. *Avian Diseases.* 26: 576–584.

Stavric S and D'Aoust, J. Y. 1993. Undefined and defined bacterial preparations for the competitive exclusion of Salmonella in poultry – a review. *J. Food Protection.* 56: 173–180.

Stavric, S. 1987. Microbial colonization control of chicken intestine using defined cultures. *Food Technology.* 43: 93–98

Stavric, S., Buchnan, B. and Gleeson, T. M. 1991. Competitive exclusion of Escherichia coli 0157 :H7 from chicks with anaerobic cultures of faecal microflora. *Lett. Appl. Microbiol.* 14: 191–193.

Stuart R. L., H. C. Surprise and Davis, L. W. 1978. Response of growing rats to diets supplemented with a liquid nonviable Lactobacillus fermentation product. *J. Animal Sci.* 47: 322.

Szylit O. and Charlet, G. 1981. Energy and protein retention in holoxenic, axenic and gnotobiotic chickens mono-associated with Lactobacillus spp. *British Poultry Sci.* 22: 305–315.

Szylit O., M. Champ, N. Ait-Abdelkader and Raibaud, P.1980. Role of five Lactobacillus strains on carbohydrate degradation in monoxenic chickens. *Reproduction Nutrition Development,* 20: 1701–1706.

Tagg J. R., A. S. Dajani and Wannamaker, L. W. 1976. Bacteriocins of gram-positive bacteria. *Bacteriological Reviews.* 40: 722–756.

Thomke, S., Elwinger, K. 1998. Growth promotants in feeding pigs and poultry. III. Alternatives to antibiotic growth promotants. *Ann. Zootech.*, 47: 245-271.

Torres-Rodriguez, A., Donoghue, A.M., Donoghue, D.J., Barton, J.T., Tellez, G., Hargis, B.M. 2007. Performance and condemnation rate analysis of commercial turkey flocks treated with a *Lactobacillus* spp.-based probiotic. *Poult. Sci.* 86: 444-446.

Tortuero, F. 1973. Influence of implantation of Lactobacillus acidophilus in chicks on the growth, feed conversion, malabsorption of fats syndrome and intestinal flora. *Poult. Sci.* 52: 197–203.

Tortuero F., Rodriguez, L. M. and Barrera, J. 1989. Lactic acid bacteria and beans in the diets for chickens. *Archivos de Zootecnia.* 141: 151–165.

Tortuero, F and Fernandez, E. 1995. Effect of inclusion of microbial culture in barley-based diets fed to laying hens. *Anim. Feed. Sci. Tec.,* 53: 255-265.

Turnbull P. C. B. and Richmond, J. E. 1978. A model of Salmonella enteritidis: The behavior of Salmonella enteritidis in chick intestine studied by light and electron microscopy. *British J. Experimental Pathology*. 59: 64–75.

Upreti G. C. and Hindsdill, R. D. 1973: Isolation and characterization of bacteriocin from a homofermentive Lactobacillus. *Antimicrobial Agents and Chemotherapy.* 4: 487–494.

Upreti G. C. and Hindsdill, R. D. 1975: Production and mode of action of Lactocin 27: Bacteriocin from a homofermentive Lactobacillus. *Antimicrobial Agents and Chemotherapy.* 7:139-145.

Vanbelle M., E. Teller and Focant, M.1990. Probiotics in animal nutrition: a review. *Arch. Animal Nutrition (Berlin).* 40: 543–556.

Voltera, V. 1928. Variation and fluctuations of the number of individuals in an animal species living together. *J. Cons. Int. Explor. Mer.* 3: 3.

Watkins B. A. and F. H. Kratzer, 1984. Drinking water treatment with commercial preparation of a concentrated Lactobacillus culture for broiler chickens. *Poult. Sci.* 63: 1671–1673.

Watkins, B.A., Kratzer, F.H. 1983. Effect of oral dosing of *Lactobacillus* strains on gut colonization and liver biotin in broiler chicks. *Poult. Sci.* 62: 2088-2094.

Watkins, B.A., Miller, B.F., Neil, D.H. 1982. *In vivo* effects of *Lactobacillus acidophilus* against pathogenic *Escherichia coli* in gnotobiotic chicks. *Poult. Sci.* 61: 1298-1308.

Wierup M., Wold-Troell, M., Nurmi, E. and Hakkinen, M. 1988. Epidemiological evaluation of the Salmonella-controlling effect of a nationwide use of a competitive exclusion culture in poultry. *Poult. Sci.* 67:1026–1033.

Yaman, H., Ulukanli, Z., Elmali, M., Unal, Y. 2006. The effect of a fermented probiotic, the kefir, on intestinal flora of poultry domesticated geese (*Anser anser*). *Revue. Méd. Vét.* 157: 379-386.

Yang, Y., Iji, P. A. and Choct, M. 2009. Dietary modulation of gut microflora in broiler chickens: a review of the role of six kinds of alternatives to in-feed antibiotics. *World's Poult. Sci.* J. 65: 97-114.

Yeo, J. and Kim, K. 1997. Effect of feeding diets containing an antibiotic, a probiotic, or yucca extract on growth and intestinal urease activity in broiler chicks. *Poult. Sci.* 76: 381-385.

Zhang, A.W., Lee, B. D., Lee, S.K., Lee, K.W., An, G.H., Song, K.B., Lee, C.H. 2005. Effects of yeast (*Saccharomyces cerevisiae*) cell components on growth performance, meat quality, and ileal mucosa development of broiler chicks. *Poult. Sci.*, 84: 1015-1021.

5

Application of Prebiotics in Poultry Production

Pankaj Kumar Singh, Kaushalendra Kumar, Chandramoni and Sanjay Kumar

Introduction

Prebiotics are non-digestible food ingredient which beneficially affects the host by selectively stimulating the growth and/or activity of one or a limited number of health-promoting bacteria in the intestinal tract, thus improving the host's microbial balance (Gibson and Roberfroid, 1995). Normally, prebiotics contain some type of a non-digestible carbohydrate fraction (Rehman *et al.*, 2009). Almost 90% of oligosaccharides escape digestion in the small intestine and reach the colon where they selectively stimulate the growth and/or activity of beneficial bacteria like *Bifidobacteria, Lactobacillus and Streptococcus*, providing prebiotic properties (Stanton *et al.*, 2003). These bacteria populate the gut filling any niches were pathogenic bacteria may attempt to take hold. This concept is better known as competitive exclusion. Prebiotics function by lowering the gut pH through lactic acid production, inhibiting/preventing colonization of pathogens, modifying metabolic activity of normal intestinal flora, and stimulation of the immune system (Doyle and Erickson, 2006). Studies have demonstrated that these bacteria reduce pathogenic bacteria like *Salmonella* spp., *Clostridium diffile* and *Campylobacter* spp. (Hopkins and Macfarlane, 2003; Stern *et al.* 2000), by diverse mechanisms including competitive exclusion of nutrients or adhesion sites to intestinal mucosa (Stern *et al.* 2000) and fatty acid of short chain production, like propionic and butyric acids, which modifies the acidic conditions and inhibits the growth of microorganisms that cause infections (Collins and Gibson, 1999).

The bacteria that feed on fermentable carbohydrates produce many beneficial substances, including short chain fatty acids (SCFAs) and certain B-vitamins. Additionally, there is some evidence that they may promote further absorption of some minerals that have escaped the small intestine, including calcium and magnesium (Patterson and Burkholder, 2003; Hajati and Rezaei, 2010). Other possible benefits include: lower cholesterol, lower triglycerides, improved insulin sensitivity and glucose metabolism and improved immune system function (Niness, 1999).

Mechanism of actions of prebiotics

Prebiotics can either directly bind the pathogens or increasing the osmotic value in the intestinal lumen. However, they have indirectly effects through metabolites that are generated by intestinal flora while utilizing prebiotics compounds for their own metabolism. Mechanism of actions of prebiotics can be listed as followed:

- Lowering the gut pH through lactic acid production (Chio *et al.*, 1994; Gibson and Wang, 1994).
- Inhibiting/preventing colonization of pathogens (Morgan *et al.*, 1992; Bengmark, 2001).
- Modifying metabolic activity of normal intestinal flora (Demigne *et al.*, 1986).
- Stimulation of immune system (Monsan and Paul, 1995).

Substances used as prebiotics

Many of the non-digestible sources that are used for prebiotics are derived from plants. Non-digestible carbohydrates (oligo and polysaccharides), some peptides, proteins and certain lipids are candidate of prebiotic. Oligosaccharides, polysaccharides, and lactose are non-digestible carbohydrate sources typically used in poultry as the foundation for prebiotics (Hajati and Rezaei, 2010); however, dietary fibers and certain fungi can also be used. The most common are carbohydrate sources that cannot be broken down by the chicken but can be utilized by the beneficial microflora of the intestinal tract. Lactic acid based prebiotics are very popular due in part to the fact that birds are unable to enzymatically digest lactose. Many native species of microflora in the gastrointestinal tract utilize the lactose as a nutrient source. Lactose is a disaccharide consists of glucose and galactose, which has prebiotic effect in chickens. Since chickens does not have lactase enzyme, lactose enters to the lower segment of the intestine and caeca, where hydrolyzed by microbial activity. The dominant prebiotics are fructo resistance oligosaccharide products (Fructo-oligosaccharides, oligufroctose, inulin); gluco-oligosaccharides, stachyose, maltobalance). Oligosaccharides and oligochitosan have also been investigated in broiler chickens (Jiang *et al.*, 2006; Huang *et al.*, 2007). Fructo-oligosaccharides (FOS) and mannan-oligosaccharides (MOS) are the two most common prebiotics available for use in poultry.

Characteristics of an ideal prebiotics (Simmering and Blaut, 2001)

- Be neither hydrolyzed or absorbed by mammalian enzymes or tissues
- Selectively enrich for one or a limited number of beneficial bacteria
- Beneficially alter the intestinal microbiota and their activities
- Beneficially alter luminal or systemic aspects of the host defense system

Table 1: Major oligosaccharide candidates for prebiotics (Hazati and Rezaei, 2010)

Oligosaccharides	Structure	Linkages	Process	Origin
Xylo-oligosaccharides	(Glu)n	β -1,4	Hydrolysis	Cereals
Lactulose	Gal-Fru	β -1, 4	Isomerisation	Lactose
Isomalto-oligosaccharides	(Glu)n	α-1, 6	Hydrolysis	Algae
Gluco-oligosaccharides	(Glu)n	α-1,2 and α-1,6	Synthesis	Sucrose
Galacto- oligosaccharides	(Gal)n-Glu	β -1,4 and β -1,6	Synthesis	Lactose
Fructo- oligosaccharides	(Fru)n Glu	(β-2,1)- α-1,2	Synthesis	Sucrose
Oligofructose	(Fru)n-(Fru)n-Glu	(β-2,1)	Hydrolysis	Inulin

Fructo-oligosaccharides (FOS)

The general term fructo-oligosaccharides (FOS) includes all non-digestible oligosaccharides composed of fructose and glucose units, but specifically FOS refers to short chains of fructose units bound by β (2-1) linkages attached to a terminal glucose unit (Swanson and Fahey, 2002). All of these non digestible carbohydrates are expressed as non digestible polysaccharides because they are not hydrolyzed by endogenous enzyme in the small intestine, but hydrolyzed by colonic bacteria in the large intestine. However, all of these could not be classified as prebiotics but rather colonic food because the process of colonic fermentation in most of these substances is nonspecific (Gibson and Roberfroid, 1995).

The predominant FOS that has been studied is inulin, oligofructose, and short-chain fructo-oligosaccharides (scFOS). Generally, FOS range in length from 2 to 60 fructose units with or without a glucose molecule residing at the end of each fructose chain (Flickinger *et al.*, 2003). Due to the β (2-1) linkages, these compounds are resistant to hydrolytic digestion in the upper part of the intestinal tract of monogastric animals (Fishbein *et al.*, 1988), making them highly digestible substrates for fermentation by cecal microbes, and any caloric contribution to the host is obtained from lactate and short-chain fatty acids (SCFA) produced by these microbes (Ellegard *et al.*, 1997). Lactate, which is produced from saccharolytic bacteria such as lactobacilli and bifidobacteria, decreases the pH in the lumen, thus creating an unfavorable environment for the growth of pathogenic organisms that are not acid tolerant such as *Salmonella* spp., *E. coli,* or *Clostridium perfringens.* The main SCFAs that are produced from bacterial fermentation are acetate, propionate, and butyrate, and they can provide several benefits to the host animal.

Fructo-oligosaccharides are found naturally in plants sources such as onions, Jerusalem artichokes, chicory, and bananas. Commercially they are made by the hydrolysis of inulin (a class of plant fiber) or by enzymatic synthesis from sucrose and lactose. FOS is a type of fiber that is indigestible by poultry but fermentable by the microflora of the gastrointestinal tract. The bacteria in the intestinal tract ferment the fiber and turn it into short chain fatty acids. These short chain fatty acids acidify in the large intestine allowing for beneficial bacteria like *Bifidobacteria* and *Lactobacillus* to grow (Hajati and Rezaei, 2010). The epithelial cells which line the mucosa of the intestine can utilize the short chain fatty acids for energy. This helps to preserve intestinal health and also contributes to enhanced nutrient absorption which can decrease feed conversion. FOS is fermentable by most strains of helpful bacteria such as *Bifidobacteria. Bifidobacteria* are gram-positive, anaerobic (not requiring oxygen) bacteria naturally found in the gastrointestinal tract of humans and many warm-blooded animals (Doyle and Erickson, 2006). These bacteria suppress the growth of pathogens such as *E. coli* by producing natural antimicrobial compounds by lowering the gut pH through volatile fatty acid production.

FOS has been postulated to enhance growth responses in livestock. Ammerman *et al.* (1988) investigated the growth response of broiler chickens to diets containing 2.5 or 5.0 g/kg supplemental oligofructose when fed for 46 days. The addition of oligofructose did not result in a significant increase in body weights, but both concentrations of oligofructose improved feed efficiency compared to the control containing no oligofructose. In a second study performed by Ammerman *et al.* (1989), higher concentrations of oligofructose

(3.75 or 7.5 g/kg) were included in the diet to see if there were similar effects on growth performance. The authors found that at 3.75 g/kg oligofructose, broilers had increased weight gain, increased hot and chilled carcass weights, and a higher percentage of breast meat. For 7.5 g/kg oligofructose, there was also increased weight gain but none of the other effects were observed. Wu *et al.* (1999) also observed beneficial effects on body weight gain and feed efficiency when 2.5 or 5.0 g/kg FOS were included in the diet. Xu *et al.* (2003) found that the addition of 4.0 g/kg FOS significantly increased average daily weight gain of broilers, while diets containing 2.0 or 8.0 g/kg FOS had no significant effect. Feed to gain ratios were also decreased in the chicks fed 2.0 and 4.0 g/kg FOS, while there was no improvement when 8.0 g/kg FOS was supplemented. Other authors, however, have reported no effects of FOS on growth in poultry (Stanczuk *et al.*, 2005; Waldroup *et al.*, 1993), which indicates that the effects on growth performance may not be consistent.

The primary benefit of FOS, however, is that they selectively stimulate indigenous populations of bifidobacteria and lactobacilli, which are considered beneficial species (Yang *et al.*, 2009), usually at the expense of E. coli and Salmonella species. Bifidobacteria are anaerobic, gram-positive bacteria and they are found in the gastro-intestinal tract of human infants and adults as well as various warm-blooded animals. Bifidobacteria are saccharolytic organisms and all strains of fermented glucose, galactose and fructose. Glucose is fermented via the fructose-6-phosphate shunt to acetic and L lactic acids. Bifidobacteria do not produce CO_2, butyric or propionic acid. The optimum growth temperature of bifidobacteria is 37-43°C and optimum pH for growth is 6.5-7.1 (Scardovi, 1986). Bifidobacteria populations in the gastrointestinal tract of piglets range from 10^4-10^6/g chyme in the stomach to 10^8/g chyme in the ileum (Stewart *et al.*, 1993) and 10^8 - 10^9 in the large intestine (Borg Jensen, 1993). Several studies showed that bifidobacteria and lactobacillus may be beneficial and the dominant bacteria in the colon. Bifidobacteria have an antibacterial effect because they can suppress potential pathogens like *E. coli.* They do this by producing antimicrobials like bacteriocin or by lowering pH through the rapid production of volatile fatty acids especially acetate and lactate. The undissociated acid which is presented in a higher proportion where pH decreases in the gut can function as antibacterial agent (Eklund, 1983). It is the use of nitrogenous compounds for the growth of bifidobacteria which will lead to less proteinous substances used for energy. Thus, when there is saccharolytic fermentation, the bacteria does not use as much protein for energy. This may result in less amines and branched chain fatty acids. Non-specific immune activity can be increased by feeding fermented milk products with *bifidobacterium bifidum* (Schiffrin *et al.*, 1995). In an in vitro study, FOS and xylo oligosaccharides are converted to acids at a high rate by most strains of bifidobacterium, at a lower rate by most lactobacilli, most bacteroides, but not used by eubacteriaceae, most clostridia (except *Clostridium butyricum*), E. coli and staphylocus (Wada *et al.*, 1987).

Bifidobacteria and lactobacilli are selectively enhanced because of their ability to ferment FOS due to the presence of â-fructosidase (Swanson and Fahey, 2002). These results can be achieved with very low dosages in several species. Niness (1999) reported that oligofructose intakes ranging from 5 to 20 g per day in humans resulted in altered colonic microbiota, especially bifidobacteria populations. In a study conducted with male beagles fed 3, 6, or 9 g/kg supplemental oligofructose for 18 days, the dogs fed the highest concentration of oligofructose tended ($P < 0.10$) to have higher numbers of bifidobacteria when compared to the negative control group (Flickinger *et al.*, 2002). This was also shown in a study conducted with rats where higher doses of 6% oligofructose or scFOS included in the diet showed an increase in bifidobacteria populations (Campbell *et al.*, 1997). Sparkes *et al.* (1998) were one of the first to examine the effects of FOS in cats. When cats were fed a diet containing 7.5 g/kg FOS, increased lactobacilli and bacteroides concentrations were seen compared to the control group, and *Clostridium perfringens* and *E. coli* concentrations tended to decrease. In broiler chickens fed diets containing 0, 2, 4, or 8 g/kg FOS up to 49 days, the cecal concentrations of bifidobacteria were increased when birds were fed 4 g/kg FOS, and concentrations of lactobacilli were increased, whereas concentrations of *E. coli* were decreased when birds were fed 2 or 4 g/kg FOS compared with the control (Xu *et al.*, 2003). There was no effect observed on cecal microbes when 8 g/kg FOS were fed.

Mannan-oligosaccharides (MOS)

Mannan-oligosaccharides (MOS) are another type of oligosaccharide that may positively influence gut microbial populations and immune function. These compounds are special when compared to other oligosaccharides because of their proposed mode of action when influencing microbial populations in the GIT. Mannose, the main component of MOS, is a unique sugar because many enteric bacteria have receptors that bind to it (Griggs and Jacob, 2005). These receptors, called Type-1 fimbriae, are involved in the attachment of bacteria to host cells, and this attachment is critical for pathogens if they want to establish colonization in the host. MOS are derived from yeast cell wall and work slightly differently than fructo-oligosaccharides. The major difference between fructo-oligosaccharides and mannan-oligosaccharides is that MOS products do not selectively enrich for beneficial bacteria. The binding and removing of pathogens while stimulating the immune system are the primary modes of action for MOS products (Patterson and Burkholder, 2003). The most common strain of yeast used on the market is *Saccharomyces cervisiae* which is obtained from fruits and grains. The yeast cell wall gets broken down into mannan, glucan, and protein (Hajati and Rezaei, 2010). While the composition remains consistent between yeast strains the interaction among the components has a tendency to vary. MOS is theorized to work against type 1 gram-negative pathogenic bacterium by binding their fimbriae. Fimbriae are used by bacteria to attach to their host; type 1 gram-negative bacteria use fimbriae to bind to mannose receptor sites found in the mucosal lining of the gastrointestinal tract. *E.coli* and *Salmonella* are examples of type 1 gram-negative bacteria. MOS essentially works by attaching to the fimbriae of the bacteria preventing them from binding to the epithelial lining so they pass through the host. The major difference between fructo-oligosaccharides and mannan-oligosaccharides is that MOS products do not selectively enrich for beneficial bacteria. Research has shown that mannan-oligosaccharide based products made from *Saccharomyces cervisiae* or *Saccharomyces boulardi* have decreased the colonization of *Salmonella* and *Campylobacter (*Hofacre *et al.*, 2003).

Mannan-oligosaccharides contain a high affinity ligand for bacteria and offer a competitive binding site, so pathogens, which possess these fimbriae, attach to the MOS instead of attaching to the intestinal wall and, therefore, move through the intestine without colonization (Newman, 1994). *Escherichia coli, Salmonella* spp. and *Clostridia* are a few types of pathogenic bacteria that exhibit this mannan-binding behavior.

In several experiments conducted with broiler chickens, it was determined that supplemental MOS increased lactobacilli and bifidobacteria populations in the ceca compared to diets containing no antibiotic growth promoters (Newman *et al.*, 1994; Fernandez *et al.*, 2002). Mannan-oligosaccharide supplementation of turkey poults increased the concentration of bifidobacteria, while *Clostridium* spp. were decreased from 4.22 log10 CFU/g in the control to 2.98 log 10 CFU/g in the MOS supplemented diet (Finucane *et al.*, 1999). Mannan-oligosaccharides may also have an effect on growth performance, along with the intestinal changes discussed above. Based on a meta-analysis from 1993-2003 of 44 broiler pen research trials where MOS was fed as a supplement, Hooge (2004a) concluded that birds fed MOS (ranged from 0.05 to 0.3% of the diet) showed improved growth performance and feed conversion ratios in 79.5% of cases compared with those fed diets containing no antibiotic growth promoters. In a similar meta-analysis conducted by the same author, turkeys fed MOS as a supplement showed increased body weight in 70.4% of cases and reduced mortality in 56.3% of cases (Hooge, 2004b). In contrast, studies by Stanczuk *et al.* (2005) and Waldroup *et al.* (2003) reported no significant changes in body weight gain or feed conversion when MOS was added to the diet of either turkeys or broilers, respectively.

Gluco-oligosaccharides and Galacto-oligosaccharides

Two other types of oligosaccharides with prebiotic activity are gluco-oligosaccharides and galacto-oligosaccharides. These compounds have been studied very little in livestock. Gluco-oligosaccharides are synthesized from the transfer of glucose molecules from sucrose to maltose by the enzymatic action of glucosyltransferase, which is derived from *Leuconostoc mesenteroides* (Vallete *et al.*, 1993). Galacto-oligosaccharides are typically generated from lactose by the enzymatic action of β-galactosidase, an enzyme synthesized by *Aspergillus oryzae* (Kan *et al.*, 1989). The concentrations of SCFAs produced by fermentation of these compounds are similar to that of FOS.

Gluco-oligosaccharides and galacto-oligosaccharides positively affect beneficial types of intestinal bacteria, including bifidobacteria and lactobacilli. *In vitro* and *in vivo* studies performed by Djouzi *et al.* (1995) with gnobiotic rats demonstrated that gluco-oligosaccharides served as a substrate for enhanced growth of bacteroides, bifidobacteia, and *C. butyricium* strains. Transgalacto-oligosaccharides (TOS) have been patented in Japan and are marketed as bifidogenic agents for humans. *In vivo* experiments utilizing rats inoculated with human microbiota and fed 5% TOS resulted in a significant increase in total cecal anaerobes including lactobacilli and bifidobacteria (Rowland and Tanaka, 1993). Smiricky-Tjardes *et al.* (2003) obtained similar results when 6% TOS was fed to pigs.

Lactose

Lactose cannot be considered a "true" prebiotic by definition; however, lactose fits within the prebiotic concept for poultry because birds cannot digest it because of their lack of lactase activity (Denbow, 2000); therefore, it becomes an available substrate for microflora in the hindgut. Initially, the interest in the prebiotic effect of lactose was focused on reducing *Salmonella* colonization in the bird (Corrier *et al.*, 1990; Hinton *et al.*, 1990); however, more recent research has evaluated its potential to improve growth rate and efficiency. Douglas *et al.* (2003) demonstrated that inclusion of 2 or 4% galactose or lactose to the diet increased weight gain from 0 to 21 days post-hatch in broiler chicks. Simoyi *et al.* (2006) reported that when turkeys were fed diets containing lactose levels from 0 to 8%, body weight gain significantly improved after a six-week feeding period, with the optimal response at the 2% inclusion level. McReynolds *et al.* (2007) evaluated the effects of dietary lactose levels ranging from 0 to 4.5% on the control of necrotic enteritis in broiler chicks. All of the control birds

(100%) challenged with *Clostridium perfringens* had intestinal lesions compared to only 30% of the birds fed 2.5% lactose.

Effects of prebiotics on poultry production

Favourable effects of addition of prebiotics may be the result presence of antagonism towards pathogens, competition with pathogens, and promotion of enzyme reaction, reduction of ammonia and phenol products and increase of resistance to colonization. Also prebiotics application in poultry result in improvement in intestinal microbial balance, improvement in performance, better nutrient utilization (eg. amino acids and proteins) and reduced environmental pollution (Peric *et al.*, 2009; Khksar *et al.*, 2008; Midilli *et al.*, 2008; Ghiyasi *et al.*, 2007).

Effects of MOS on broiler production resulted in daily body weight gain of 4-8% (Peterson, 1998; Kumprecht *et al.*, 1998). Prebiotics have also been shown to decrease incidences of necrotic enteritis. Necrotic enteritis is due to pathogenic *Clostridium perfringens*. In many European countries, after the banning of sub-therapeutic antibiotics in 2006, subclinical necrotic enteritis has been reported at percentages as high as 40%. *Hofacre et al* (2003) conducted a study comparing non-antibiotic feed additives (prebiotics, probiotics, herbal supplements, organic acids, and competitive exclusion products) and their affects on birds challenged with necrotic enteritis. The results indicated that the use of competitive exclusion products alone or in combination with MOS products can effectively reduce mortality and feed efficiency in birds suffering from subclinical necrotic enteritis. If a pathogen cannot bind to a receptor site nor grow as fast as the environmental passage rate, then the constant flow of digesta will essentially wash the pathogen out of the animal. Decreased feed conversion and increased weight gain have both been scientifically proven results from the usage of either FOS or MOS.

It has been shown in various animal species that Salmonella colonization of the gut is decreased when the bifidobacterial population is increased, either by administration of bifidobacteria as probiotic strains, or by addition of certain types of oligosaccharides that stimulation proliferation of these bacteria in the gut (Asahara *et al.*, 2001; Buddington *et al.*, 2002; Bovee-Oudenhoven *et al.*, 2003; Silva *et al.*, 2004; Thitaram *et al.*, 2005). When the caecal Bifidobacterium population in broilers was increased by isomalto-oligosaccharide addition to the feed, and the animals were infected with a high dose of *Salmonella typhimurium*, large reductions in caecal colonization were observed (Thitaram *et al.*, 2005).

Future Scope

As with any feed additive prebiotics offer both advantages and disadvantages to the producer. Prebiotics provide competition against pathogens, improve gut health, decrease production costs, reduce ammonia emissions, enhance nutrient utilization, and are effective at low inclusion rates (Hajati and Rezaei, 2010; Patterson and Burkholder, 2003). The lack of consistent research results is the largest hurdle prebiotics face in the poultry industry. While most prebiotic research has focused on humans, the major factors for variation in commercial poultry production are theorized to be differences between strains, sex, age, diet composition, diet inclusion levels, and a very limited understanding of the immune system and gastrointestinal tract of poultry.

The use of prebiotics to modulate microbial populations, as well as overall gut health, has been studied exclusively in human and companion animal nutrition. The use of these compounds in livestock has gained increased interest because of their potential prebiotic properties and their potential to maintain gut health and animal performance in the absence of antibiotics growth promoters. However, more research needs to be performed to fully understand how these additives may improve gut health and maintain performance.

Summary

It can be concluded that the use of prebiotics is a promising approach for enhancing the role of endogenous beneficial organisms in the gut of simple stomached animals. Prebiotics can be used as potential alternatives to growth promoting antibiotics.

References

Ammerman, E., C. Quarles, and Twining, P. V. 1989. Evaluation of fructoligosaccharides on performance and carcass yield of broilers. *Poult. Sci.* 68 (Suppl.):167.

Ammerman. E., C. Quarles, and Twining, P. V. 1988. Broiler response to the addition of dietary fructooligosaccharides. *Poult. Sci.* 67 (Suppl.):1.

Asahara, T., Nomoto, K., Shimizu, K., Watanuki, M. and Tanaka, R. 2001. Increased resistance of mice to Salmonella enterica serovar Typhimurium infection by symbiotic administration of Bifidobacteria and transgalactosylated oligosaccharides. *J. Appl. Microbiol.*, 91:985-996.

Bengmark, S., 2001. Pre, pro and symbiotics. *Current opinion in clinical nutrition and metabolic car.* 4:571-579.

Borg Jensen, B. 1993. The possibility of manipulation of the microbial activity in the digestive tract of monogastric animals. In: Proceedings of the 44th Annual Meeting of European Association for Animal production, Aarus, Denmark. pp. 49.

Bovee-Oudenhoven, I.M., Ten Bruggencate, S.J., Lettink-Wissink, M.L., Van Der Meer, R. 2003. Dietary fructo-oligosaccharides and lactulose inhibit intestinal colonisation but stimulate translocation of Salmonella in rats. *Gut.* 52:1572-1578.

Buddington, K.K., Donahoo, J.B. and Buddington, R.K. 2002. Dietary oligofructose and inulin protect mice from enteric and systemic pathogens ant tumor inducers. *J. Nutr.,*132: 472-477.

Campbell, J. M., G. C. Fahey Jr., and Wolf, B. W. 1997. Selected indigestible oligosaccharides affect large bowel mass, cecal and fecal short-chain fatty acids, pH, and microflora in rats. *J. Nutr.* 127: 130-136.

Chio, K.H., H. Namkung and Paik, I.K. 1994. Effects of dietary fructo-oligosaccharides on the suppression of intestinal colonization of *Salmonella typhimurium* in broiler chickens. *Korean J. Anim. Sci.,* 36: 271-284.

Collins, M. D., and G. R. Gibson. 1999. Probiotics, prebiotics, and synbiotics: approaches for modulating the microbial ecology of the gut. *Am. J. Clin. Nutr.* 69 (Suppl. 1):1042S–1057S.

Corrier, D. E., A. Hinton, Jr., R. L. Ziprin, and DeLoach, J. R. 1990. Effect of dietary lactose on *Salmonella* colonization of market-age broilers. *Avian Dis.* 34: 668-676.

Denbow, D. M. 2000. Gastrointestinal anatomy and physiology. Pages 299-625 in Sturkie's Avian Physiology. 5th ed. G. Causey Whittow, ed. Acad. Press, San Diego, CA.

Demigne, C., C. Yacoub, C. Remezy and Fafournoux, P. 1986. Effects of absorption of large amounts of volatile fatty acids on rat liver metabolism. *J. Nutr.,* 116: 77-86.

Djouzi, Z., C. Andrieux, V. Pelenc, S. Somarriba, F. Popot, F. Paul, P. Monsan, and Szylit, O. 1995. Degradation and fermentation of a gluco-oligosaccharide by bacterial strains from human colon: In vitro and in vivo studies in gnotobiotic rats. *J. Appl. Bacteriol.* 79:117-127.

Douglas, M. W, M. Persia, and Parsons, C. M. 2003. Impact of galactose, lactose, and Grobiotic-B70 on growth performance and energy utilization when fed to broiler chicks. *Poult. Sci.* 82: 1596-1601.

Doyle, M. and Erickson, M, 2006. Reducing the carriage of foodborne pathogens in livestock and poultry. *J. Poult. Sci.* 85: 960-973.

Eklund, T., 1983. The antimicrobial effect of dissociated and undissociated sorbic acid at different pH levels. *J. Appl. Bacteriol.,* 54: 383.

Ellegard, L., H. Anderson, and Bosaeus, I. 1997. Inulin and oligofructose do not influence the absorption of cholesterol and the excretion of cholesterol, Fe, Ca, Mg, and bile acids, but increases energy excretion in man. A blinded, controlled cross-over study in ileostomy subjects. *Eur. J. Clin. Nutr.* 51: 1-5. 29

Fernandez, F., M. Hinton, and Van Gils, B. 2002. Dietary mannan-oligosaccharides and their effect on chicken caecal microflora in relation to *Salmonella enteritidis* colonization. *Avian Pathol.* 31: 49-58.

Finucane, M. C., K. A. Dawson, P. Spring, and Newman, K. 1999. Effects of mannan-oligosaccharides on composition of the gut microflora of turkey poults. *Poult. Sci.* 78 (Suppl.1): 77 (abstr.).

Fishbein, L., M. Kaplan, and Gough, M. 1988. Fructooligosaccharides: A review. *Vet. Hum. Toxicol.* 30: 104-107.

Flickinger, E. A., J. Van Loo, and Fahey Jr, G. C. 2003. Nutritional responses to the presence of inulin and oligofructose in the diets of domesticated animals: A review. *Crit. Rev. Food. Sci. Nutr.* 43: 19-60.

Flickinger, E. A., T. F. Hatch, R. C. Wofford, C. M. Grieshop, S. M. Murray, and G. C. Fahey Jr. 2002. *In vitro* fermentation properties of selected fructooligosaccharide containing vegetables and *in vivo* colonic microbial populations are affected by diet in healthy human infants. *J. Nutr.* 132: 2188-2194.

Ghiyasi, M., M. Rezaei and H. Sayyahzadeh, 2007. Effect of prebiotic (Fermacto) in low protein diet on performance and carcass characteristics of broiler chicks. *Int. J. Poult. Sci.,* 6: 661-665.

Gibson, G.R. and X. Wang, 1994. Bifidogenic properties of different types of fructooligosaccharides. *Food Microbial.,* 11: 491-498.

Gibson, G.R., and M. B. Roberfroid. 1995. Dietary modulation of the human colonic microbiota: Introducing the concept of prebiotics. *J. Nutr.* 125: 1404-1412.

Griggs, J. P., and J. P. Jacob. 2005. Alternatives to antibiotics for organic poultry production. *J. Appl. Poult. Res.* 14:750-756.

Hajati, H. and M. Rezaei. 2010. The application of prebiotics in poultry production. *Int. J. Poult. Sci.* 9(3): 298-304.

Hinton, A., Jr., D. E. Corrier, G. E. Spates, J. O. Norman, R. L. Ziprin, R. C. Beier, and J. R. DeLoach. 1990. Biological control of *Salmonella typhimurium* in young chicks. *Avian Dis.* 34: 626-633.

Hofacre, C.L., T. Beacorn, S.Collett, and Mathis, G. 2003. Using competitive exclusion, mannan-oligosaccharide and other intestinal products to control necrotic enteritis. *J. Appl. Poult. Res.,* 12:60-64.

Hooge, D. M. 2004a. Meta-analysis of broiler chicken pen trials evaluating dietary mannanoligosaccharide. *Int.* J. *Poult. Sci.* 3: 163-174.

Hooge, D. M. 2004b. Turkey pen trials with dietary mannan oligosaccharide: Meta-analysis, 1993-2003. *Int.* J. *Poult. Sci.* 3: 179-188.

Hopkins, M.J. and G.T. Macfarlane. 2003. Nondigestible Oligosaccharides Enhance Bacterial Colonisation Resistance against *Clostridium difficile. App. Environ. Microbiol.* 69 (4): 1920-1927.

Huang, R.L., Y.L. Yin and M.X. Li, 2007. Dietary oligochitosan supplementation enhances immune status of broilers. *J. Sci. Food. Agric.,* 87: 153-159.

Jiang, H.Q., L.M. Gong., Y.X. Ma., Y.H. He., D.F. Li and H.X. Zhai, 2006. Effect of stachyose supplementation on growth performance, nutrient digestibility and caecal fermentation characteristics in broilers. *Br. Poult. Sci.,* 47: 516-522.

Kan, T., Y. Kobayashi, and K. Matsumoto. 1989. Characterization of galacto-oligosaccharides and application for food. *New Food Industry.* 31: 25-30.

Khksar, V., A. Golian, H. Kermanshahi, Movasseghiand and A. Jamshidi, 2008. Effect of prebiotic fermacto on gut development and performance of broiler chickens fed diet low in digestible amino acids. *J. Anim. Vet. Adv*., 7: 251-257.

Kumprecht, I., Zobac, P., Siske, V., Sefton, A.E. Spring, P. 1998. Effect of dietary mannanoligoschharude level on performance and nutrient utilization of broilers. . *Proc. Alltechs 14th Annual Symposium Enclosure code,* 016C.

Midilli, M., M. Alp, N. Kocabagli, Ö.H. Muglali, N. Turan, H. Yilmaz and S. Çakir, 2008. Effects of dietary probiotic and prebiotic supplementation on growth performance and serum IgG concentration of broilers. *South Afr. J. Anim. Sci*., 38: 12-16.

Monsan, P.F. and F. Paul, 1995. Oligosaccharide feed additives. In: Biotechnology in Animal Feeds and Feeding, Wallace, R.J. and A. Chesson (Eds.). VCH Verlagsgesellschaft, Weinheim and New York, pp: 233-245.

Morgan, A., A.J. Mul, G. Beldman and A.G.J. Voragen, 1992. Dietary oligosaccharides. New insights. *Agro. Food Ind. Hi-tech*., 11: 35-38.

McReynolds, J. L., J. A. Byrd, K. J. Genovese, T. L. Poole, S. E. Duke, M. B. Farnell, and D. J. Nisbet. 2007. Dietary lactose and its effects on the disease condition of necrotic enteritis. *Poult. Sci.* 86: 1656-1661.

Newman, K. 1994. Manna-oligosaccharides: Natural polymers with significant impact on the gastrointestinal microflora and the immune system. In Biotechnology in the Feed Industry. Proc. of Alltech's 10th Annual Symposium. T. P. Lyons and K. A. Jacques (ed). Nottingham University Press, Nottingham, UK. pp. 167-174.

Niness, K. R. 1999. Inulin and oligofructose: What are they? *J. Nutr.* 129: 1402S-1406S.

Patterson, J and Burkholder, K. 2003. Application of prebiotics and probiotics in poultry production. *J. Poult. Sci.* 82: 627-631.

Petersen, C.B.1998. Comparative effect of Zoolac, Bio-Pro on performance of broilers upto 36 days. Proc. Alltechs 14th Annual Symposium Enclosure code: 51.160.

Rehman, H., W. Vahjen, A. Kohl-Parisini, A. Ijaz, and J. Zentek. 2009. Influence of fermentable carbohydrates on the intestinal bacteria and enteropathogens in broilers. *World's Poult. Sci. J.* 65: 75-89.

Rolfe, R. D. 2000. The role of probiotic cultures in the control of gastrointestinal health. *J. Nutr.* 130: 396S-402S.

Rowland, I. R., and R. Tanaka. 1993. The effects of transgalactosylated oligosaccharides on gut flora metabolism in rats associated with a human fecal flora. *J. Appl. Bacteriol.* 74: 667-674.

Scardovi, V. 1986. Bifidobacterium. In: Bergey's Manual of Systematic Bacteriology, Sneath, P.H., N.S. Mair, M.E. Sharpe and J.G. Holt (Eds.). 9th Edn., Williams and Wilkins Publishers, Baltimore, MD.

Schiffrin, E.J., F. Rochat, H. Link-Amster, J.M. Aeschlimann and A. Donnet-Hughes. 1995. Immunomodulation of human blood cells following the ingestion of lactic acid bacteria. *J. Dairy Sci.,* 78:491.

Silva, A.M., Barbosa, F.H., Duarte, R., Vieira, L.Q., Arantes, R.M. and Nicoli, J.R. 2004. Effect of Bifidobacterium longum ingestion on experimental salmonellosis in mice. *J.Appl. Microbiol.,* 97, 29-37.

Simmering, R., and M. Blaut. 2001. Pro- and prebiotics—the tasty guardian angles? *Appl. Microbiol. Biotechnol.* 55:19–28.

Simoyi, M. F., M. Milimu, R. W. Russell. R. A. Peterson and P. B. Kenney. 2006. Effect of dietary lactose on the productive performance of young turkeys. *J. Appl. Poult. Res.* 15: 20-27.

Smiricky-Tjardes, M. R., C. M. Grieshop, E. A. Flickinger, L. L. Bauer, and G. C. Fahey, Jr. 2003. Dietary galactoligosaccharides affect ileal and total-tract nutrient digestibility, ileal and fecal bacterial concentrations, and ileal fermentative characteristics of growing pigs. *J. Anim. Sci.* 81: 2535-2545.

Sparkes, A. H., K. Papasouliotis, G. Sunvold, G. Werrett, E. A. Gruffydd-Jones, K. Egan, T. J. Gruffydd-Jones. and G. Reinhart. 1998. Effect of dietary supplementation with fructooligosaccharides on fecal flora of healthy cats. *Am. J. Vet. Res.* 59: 436-440.

Stanczuk, J., Z. Zdunczyk, J. Juskiewicz, and J. Jankowski. 2005. Indices of response of young turkeys to diets containing mannanoligosaccharide or inulin. *Veterinarija ir Zootechnika* 31: 98-101.

Stanton, C.C., Desmond, G. Fizgerald and R.P. Ross. 2003. Probiotic health benefits-reality or myth? *Aust. J. Dairy Tech.* 58: 107.

Stern, N.J., N.A. Cox, J. S. Bailey, M. E. Berrang and M.T. Musgrove. 2000. Comparison of mucosal competitive exclusion and competitive exclusion Treatment to Reduce *Salmonella* and *Campylobacter* spp. Colonisation in Broiler Chickens. *Poult. Sci.,* 80: 156-160.

Stewart, C.S., K. Hillman, F. Maxwell, D. Kelly and T.P. King, 1993. Recent Advances in Probiotics for Pigs. *In*: Recent Advances in Animal Nutrition, Garnsworthy, P.C. and D.J.A. Cole (Eds.). Nottingham University Press. pp: 197-220.

Swanson, K. S. and G. C. Fahey, Jr. 2002. Prebiotics in companion animal nutrition. In: Biotechnology in the Feed Industry. T. P. Lyons and K. A. Jacques (ed). Nottingham University Press, Nottingham, UK. pp. 461-473.

Thitaram, S.N., Chung, C.H., Day, D.F., Hinton, A.Jr., Bailey, J.S., Siragusa, G.R. 2005. Isomaltooligosaccharides increases cecal Bifidobacterium population in young broilerchickens. *Poult. Sci.,* 84: 998-1003.

Vallete, P., V. Pelenc, Z. Djouzi, C. Andrieux, F. Paul, P. Monsan, and O. Sztlit. 1993. Bioavailability of new glucooligosaccharides in the intestinal tract of gnotobiotic rats. *J. Sci. Food Agric.* 62: 121-127.

Wada, M., H. Fujita and H. Itikawa, 1987. Genetic suppression of a temperature-sensitive groES mutation by an altered subunit of RNA polymerase of Escherichia coli K-12. J. *Bacteriol.,* 169: 1102-1106.

Waldroup, A. L., J. T. Skinner, R. E. Hierholzer, and Waldroup P.W. 1993. An evaluation of fructooligosaccharides in diets for broiler chickens and effects on salmonellae contamination of carcasses. *Poult. Sci.* 72: 643-650.

Wu, T. X., X. J. Dai, and L. Y. Wu. 1999. Effects of fructoligosaccharides on the broiler production. *Acta Agric. Zhejiangensis.* 11:85-87

Xu, Z. R., C. H. Hu, M. S. Xia, X. A. Zhan, and Wang, M. Q. 2003. Effects of dietary fructooligosaccharides on digestive enzyme activities, intestinal microflora, and morphology of male broilers. *Poult. Sci.* 82: 1030-1036.

Yang, Y., P. A. Iji, and Choct, M. 2009. Dietary modulation of gut microflora in broiler chickens: a review of the role of six kinds of alternatives to in-feed antibiotics. *World's Poult. Sci. J.* 65: 97-114.

6

Significance of Synbiotic in Animal Production

Kamdev Sethy

Introduction

Synbiotics refer to nutritional supplements combining probiotics and prebiotics in form of a synergism.The synbiotic concept was first introduced as "mixtures of probiotics and prebiotics that beneficially affect the host by improving the survival and implantation of health-promoting bacteria. Probiotics are live bacteria which are intended to colonize the large intestine and confer physiological health benefits to the host. A prebiotic is a food or dietary supplement product that enhances the activity of probiotics. Synbiotics are not drugs and the effects are due to changes in bacteria population or enhancement of activity of the bacteria. A synbiotic may contain fiber, but all fiber is not necessarily a synbiotic.

Using prebiotics and probiotics in combination is often described as synbiotic, but the United Nations Food & Agriculture Organization (FAO) recommends that the term "synbiotic" be used only if the net health benefit is synergistic. A synbiotic contains fiber such as fructose oligosaccharide, galactose oligosaccharide, *etc.* and is intended to stimulate the microflora in the large intestine. The combination of probiotics and prebiotics work separately in the small and large intestine, but synergistically increase the overall gut health. Prebiotic has shown to increase the population and/or function of the probiotic it is paired with, as the probiotic is an external species, whereas prebiotics stimulate the flora which is already present.

The manipulation of composition of the gut microbiota in animals and birds through dietary supplementation is possible by probiotic/prebiotic/synbiotic therapies. Synbiotics products containing potentially beneficial bacteria and oligosaccharides make up an important part of maintaining intestinal health. The probiotics use the prebiotics as a food source, which enables them to survive for a longer period of time inside the animal digestive system. Synbiotics [probiotics + prebiotics] enable to improve the viability of probiotics and to deliver specific health benefits. Characteristics synbiotic include antimicrobial, anticarcinogenic, antidiarrheal, antiallergenic qualities, osteoporosis prevention, reduction in serum fats and blood sugars, regulation of the immune system and treatment of liver-related brain dysfunction. Clinical indications of synbiotic therapies include severe surgical cases, liver transplantation, biliary cancer, acute pancreatitis, asthma, atopic dermatitis and animals under emergency care. Synbiotics are strengthening their roles in the reduction of antibiotic associated diarrhea, management of rotavirus diarrhea, and diarrhea due to various causes; exhibiting positive effect in the management of lactose intolereance, irritable bowel syndrome, inflammatory bowel disease, and research is in progress in various nongastrointestinal diseases. The effects of synbiotic microorganisms were found strain specific. Synbiotics have potential as replacements for antibiotics in pigs and poultry. Synbiotics products offer the potential to develop prebiotics targeted at specific probiotic strains to optimize health benefits.

Pharmaceutical medicine has been unable to stop the increasing global morbidity and mortality both in acute and chronic diseases. In view of this, there is a growing awareness of the preventative and therapeutic potential of alternative agents in the pharmaceutical field. Synbiotics amongst other agents fall into this category and can have both direct and indirect effects on the pathogenesis and progress of disease. First it was probiotics and then it was prebiotics. Now combining probiotics and prebiotics into "synbiotics" have potential to further enhance the immunosupportive effects billions of bacteria inhabit the digestive system forming over a kilo of body weight and these bacteria are referred to as the gut flora. The gut flora consists of more than 400 different species, or types, of beneficial/harmful bacteria. The most numerous probiotic bacteria normally inhabiting the small intestine and colon are species of *Lactobacilli* and *Bifidobacteria* respectively. Most probiotic products consist of one or more species of bacteria from one or both of these types. The largest group of probiotic bacteria in the intestine is lactic acid bacteria. Feed supplements containing prebiotics, probiotics, or synbiotics are attracting special interest by consumers because they have a health benefit beyond basic nutrition. Examples are *DanActive*™, *Danimals® and Activia*™ are low-fat yogurt that contains *Bifidus regularis*, a probiotic culture. Example of single strain symbiotic products available in market is Align and Culturelle. Align contains the probiotic bacteria

Bifantis (*Bifidobacterium infantis* 35624) and has been found in clinical studies to help build and maintain a healthy digestive system. Align has been recommended for symptoms such as constipation, diarrhoea, abdominal discomfort, gas and bloating. Culturelle contains the probiotic bacteria *Lactobacillus casei* which is thought to be one of the best probiotic bacteria strains for surviving the harsh acidic environment of the stomach, therefore, subsequently able to colonize the lower gastrointestinal tract.

Probiotics acts as synbiotics

Probiotic means "for/in favor of life". It contrasts directly with "anti" "biotic" or "killing life". Probiotic is a functional food that is essential for good health. It is to be the "medicine" of the twenty first century. According the FAO/ WHO, probiotics are defined as mono or mixed cultures of "live microorganisms which, when administered in adequate amounts confer a health benefit on the host". They are also called "friendly bacteria" or "good bacteria" and can be used as complementary and alternative medicine, a group of diverse medical and health care systems, practices, and products that are not presently considered to be part of conventional medicine. In other words, they are defined as the microbial food supplement with nearly 20 known species, which beneficially affect the host by improving its intestinal microbial balance. Various types of synbiotic bacteria include: *Lactobacillus* species [*L acidophilus, L reuteri, L plantarum, L casei, L salivarius, L bulgaricus, L fermentum, L gasseri, L johnsonii, L lactis, L paracasei*], *Bifidobacterium* species [*B bifidum, B infantis, B lactis, B longum, B breve, B adolescentis*], *Saccharomyces* species [*S boulardii*] and *Streptococcus* species [*S thermophilus*]. These organisms are known as digestive bioregulators or direct-fed microbials (DFMs) (WHO, 1994).Thorough research into probiotics for human use began at the onset of the 20th century when a Russian biologist Elie Metchnikoff studied the mystery of the high life expectancy of Cossacks in Bulgaria. He proposed that the high life expectancy in human due to consumption of fermented milk products. He named the microorganism as *Bacillus bulgaricus*, later classified as *Lactobacillus bulgaricus*, which was used to treat scours and gastrointestinal diseases in humans as early as the 1920s.

Common organisms used as synbiotic and their possible roles in animal body

Probiotics	Possible Roles
Bacillus coagulans	Improve abdominal pain and bloat. May increase immune response to a virus.
Bifidobacterium longum	Possible relief from abdominal pain/discomfort, bloating and constipation.
Lactobacillus acidophilus	Reduce the side effects of antibiotic therapy
Lactobacillus paracasei	Reduction of diarrhea in animals
Lactobacillus johnsonii	Reduce incidence of gastritis
Lactobacillus acidophilus & Lactobacillus casei	Inhibition of *Listeria*, *Escherichia*, *Staphylococcus*, *Enterococcus*. May reduce symptoms of lactose intolerance and immune stimulation
Lactobacillus plantarum & Lactobacillus paracasei	Reduces common cold infections in human being
Saccharomyces cerevisiae	Prevents adhesion of pathogenic microbes

Role of synbiotic organisms in animals

The exact impact of synbiotic organisms on animals has not been completely elucidated and is still under investigation. However, some of the common mechanisms of synbiotic strains are as follows:

Maintenance of microbiological equilibrium in the alimentary tract

Synbiotics plays an important role to maintain microbial balance in the alimentary tract. Immediately after birth the alimentary tract of animals is sterile and can easily be colonized by the microorganisms including the pathogenic groups like *E.coli* and *Salmonella* that may cause diseases in the animals. Synbiotic strains compete with pathogenic microorganisms for adhesion and colonization of biological membranes (Nousiainen *et al.,* 2004).

Synbiotic bacteria form thin durable layers known as biofilm while adhering to the alimentary tract. The biofilm consists of bacteria and exopolymers. The exopolymers include polysaccharides, proteins, nucleic acids and phospholipids. This biofilm prevents the attachment of pathogenic microorganisms to the digestive tract (Czaczyk, 2003).

Metabolites of lactic acid bacteria play a pivotal role to control the intestinal micro flora. The organic acids, especially lactic acid, acetic acid, as well as hydrogen peroxide and bacteriocins produced by lactic acid bacteria inhibits the growth of the pathogenic microorganisms (Mercenier *et al.,* 2003).The organic acids exert their antibacterial effect by rapid reduction of the pH values for the microorganisms i.e. 6-7 that is beyond the optimum level for their growth and also inhibit the biochemical processes in the microorganisms (Messens & de Vuyst, 2002).Weak acids (lactic and acetic acid) pass into the cytoplasm in an un dissociated form and reduce the pH inside the cell and disrupt the process of proton movement through the outer membrane and increases its tension. Consequently there is increase in the permeability and denaturation of proteins of the pathogenic strains (Caplice & Fitzgerald, 1999). The dissociation constant of lactic acid is 3.08, and in case of acetic acid it equals 4.87. Acetic acid, due to higher pKa, shows stronger antimicrobial activity than lactic acid (Cherrington *et al.,* 1991). Thus, lactic acid is mainly responsible for lowering pH of the environment, while the acetic acid acts as an antimicrobial factor (Ouwehand and Vesterlund, 2004). It should also be taken into account that lactic acid not only lowers pH, but also functions as a factor causing increased permeability of the outer membrane of gram-negative bacteria, and thus it may increase the effectiveness of other antagonistic substances (Alakomi *et al.,* 2000).

Further the lactic acid producing bacteria like Lactobacillus convert the oxygen to hydrogen peroxide (H2O2). Hydrogen peroxide may inhibit the growth or kill other microbes that do not possess H_2O_2 degrading enzymes such as catalase and peroxidase. Studies also confirmed that hydrogen peroxide inhibits various bacteria like *Staphylococcus aureus, Salmonella Typhimurium, Escherichia coli, Clostridium perfringens, Clostridium butyricum* and Pseudomonas etc. (Tomas *et al.,* 2004).

Certain diversified protein like bacteriocins produced by the lactic acid bacteria cause poration of bacterial cytoplasmatic membrane, which leads to the dispersal of the transmembrane potential and induces leakage of K^+ ions, ATP and amino acids from the cytoplasm of affected cells causing cell lysis (Grajek & Sip, 2004).

Detoxification of mycotoxins synbiotic microorganisms

Mycotoxins like aflatoxins (B_1, B_2, G_1 and G_2), ochratoxins, trichothecenes, zearalenone and fumonisins is mainly influenced by substrate composition, moisture, temperature and the presence of competitive microflora (Fink-Gremmels, 1999). Mycotoxins impair the process of tissue protein reconstruction by interfering with the metabolic pathways by blocking the co-factors responsible for protein synthesis and causing abnormal multiplication of the cellular genetic code. So in order to detoxify them, many microbes like lactic acid bacteria and yeasts like *Saccharomyces cerevisiae* are used (Shetty and Jespersen, 2006). It has been seen that different strains of lactic acid bacteria can inhibit the biosynthesis of aflatoxins (Coallier-Ascah and Idziak, 1985).

Stimulation of immune system

Synbiotics micro flora plays a major role to stimulate the immune system which is highly essential for the development of the lymphoid structures of the system. Intestinal micro flora, including the probiotic bacteria exerts an immune modulating effect by the following phenomenon (Isolauri *et al.,* 2001):

i. Induction and maintenance of immune tolerance to environmental antigens.

ii. Induction and control of immune responses against pathogens of bacterial and viral origin

iii. Inhibition of autoimmune and allergic reactions.

iv. Animals receiving synbiotics have increased level of natural antibodies (IgA, IgG and IgM) in the blood serum and more reactive against toxins like tetanus and alpha toxin of *Clostridium* (Haghighi *et al.,* 2006).

Yeast as synbiotic

Selected strains of the yeast such as *Saccharomyces cerevisiae* have been tested for their efficacy in the digestive tract and propagated in pure culture. Products consisting of live yeast cells and their dried culture media were developed to be used as synbiotics. *Saccharomyces cerevisiae* contain high amount of glucan and mannan content in their cell wall and so have affinity for certain bacterial adhesions. It prevents adhesion of pathogenic bacteria, such as *Escherichia coli* or *Salmonella* spp (Gedek, 1999).

Synbiotic substrate

Synbiotic substrates are substances that can promote the growth of beneficial microorganisms, mainly in the intestinal tract, and will modify the colonic microbiota. These substrates are commonly known as prebiotics. A prebiotic is defined as 'a non-digestible food ingredient that beneficially affects the host by selectively stimulating the growth and/or activity of one or a limited number of bacteria in the colon and thus improves host health' (Gibson and Roberfroid, 1995). The stimulated bacteria should be of a beneficial nature, namely bifidobacteria and lactobacilli (Gibson *et al.,* 1999). These substrates are selectively fermented ingredient that allows specific changes, both in the composition and/or activity in the gastrointestinal microflora that confers benefits upon host well-being and health (ISAPP, 2008).

However many food components, especially many food oligosaccharides and polysaccharides (including dietary fibre), have been claimed to have synbiotic activity without due consideration to the criteria required. These criteria for a prebiotics are:

i. Resistance to gastric acidity, hydrolysis by mammalian enzymes,

ii. Resistance to gastrointestinal absorption

iii. Fermentation by intestinal microflora

iv. Selective stimulation of the growth and/or activity of those intestinal bacteria that contribute to health and well-being. (Gibson *et al.* 2004).

Resistance, in the first criterion, does not necessarily mean that the prebiotic is completely indigestible, but it should guarantee that a significant amount of the compound is available in the intestine (especially the large bowel) to serve as a fermentation substrate.

Symbiotic substrates are reported to be particularly suited to the growth and activities of the *bifidobacteria and lactobacilli* (Rowland and Tanaka, 1993). These are classed as beneficial microorganisms because species within these groups have been reported to exert therapeutic and prophylactic influences on the health of young animals (Goldin and Gorbach, 1992).

Types of Synbiotics Substrate

Prebiotic Oligosaccharides

The production of prebiotic oligosaccharides is usually achieved through the following three general processes (Grizard and Barthomeuf, 1999):

i. Direct extraction of natural oligosaccharides from plants

ii. Controlled hydrolysis of natural polysaccharides and

iii. Enzymatic synthesis using hydrolases and/or glycosyl transferases from plant or of microbial origin

Important prebiotics of oligosaccha ride groups includes fructo oligosaccharides (FOS), mannanoligosac charides (MOS), inulin, galacto-oligosaccharides (GOS), xylo oligosaccharide (XOS), isomaltooligosaccharide (IMO) and lactulose, which alter the microflora, increasing the level of bifidobacteria and/or lactobacilli.

Fructans

Fructans is used to name molecules that have a majority of fructose residues, (Roberfroid, 2005). Fructans can vary with respect to the sources, chain composition, linkage, degree of polymerisation and functions. The sources may be Plant, bacteria, and fungi. The linkage found in fructans is mostly 2, 1 and 2, 6. Plant fructans do not exceed degree of polymerisation maximum up to 200 units; where as bacterial fructans can exceed 100000 units of polymerisation (Roberfroid, 2005)

Fructo- Oligosaccharides (FOS)

These are the subclass of the group Fructans. These have low molecular weight and degree of polymerisation is less than 10 (Roberfroid, 2005). The oligofructose subgroup can be further subdivided into the group called short-chain fructo-oligosaccharides (scFOS). Commercially, scFOS consists of low-molecular-weight linear chains synthesized by enzymatic fermentation from sucrose; however, the short chains also exist in nature. Owing to differences in structure, it is important to characterize and understand the collective nutritional, chemical, and food science properties of scFOS as a separate fructan subgroup. scFOS is present in selected foods that include onion, artichoke, garlic, wheat, and banana. scFOS is manufactured by a bioenzymatic (or fermentation) process, using sucrose from sugar beet or cane sugar as the starting raw material. scFOS has been included in a wide variety of foods and supplements globally that have been marketed for animal use.

Physiological effects of sc-FOS

i. Both *in vivo* and *in vitro* models demonstrate that scFOS is not digested between the mouth and small intestine. This is because neither the pancreas nor the small intestine mucosa secretes enzymes capable of hydrolyzing the β–(1$\rightarrow$2) fructosylfructose linkages.

ii. Bacteria utilize FOS, by the production of short chain fatty acids (Hosoya *et al.,* 1988). Selective utilization of scFOS by intestinal bacteria has been demonstrated *in vitro* using pure cultures of selected bacterial species or using mixed fecal flora inoculations, and also in animal and human studies by measuring the bacterial composition of the feces.

iii. Piglets fed with sc-FOS in milk replacer develop resistance against *E. coli* & don't develop diarrhoea (Bunce *et al.,* 1995).

iv. Bouhnik *et al.* (2004) demonstrated that feeding of sc-FOS at 10 g/day not only increases counts of bifidobacteria but also the percentage of the bacteria among total anaerobes.

Studies suggest that scFOS may be a better substrate for intestinal bacteria than oligofructose or inulin due to its shorter and more specific degree of polymerisation. (McKellar *et al.,* 1993).

Inulin type Fructans

Inulin is a carbohydrate that is extremely widespread in nature. It occurs in plants mainly as an energy reserve and as a cryoprotectant. Different plant species typically contain inulin with varying chain lengths. Wheat, onions and bananas have short-chain inulins (maximal degree of polymerisation is 10); dahlia tubers, garlic and Jerusalem artichoke have medium-chain inulins (maximal degree of polymerisation is 40); and globe artichoke and chicory typically contain long-chain inulin molecules (maximal degree of polymerisation is 100) (Van Loo *et al.,* 1995). Other plants, such as certain types of lily (*Urginea maritima*) and blue agave, and certain bacteria (e.g. *Streptococcus mutans*) produce high degree of polymerisation (more than 1000). Inulin resists hydrolysis by acid and digestive enzymes due to presence of glycosidic bond (Bjorck, 1991).

Physiological effects of Inulin

i. Inulin is bifidogenic in nature when given @ 15gm/day (Bouhnik *et al.,* 2004).

ii. Inulin does not appear to be absorbed significantly in the small intestine, except possibly the very short oligosaccharides (Ma *et al.,* 1995). However, even if such small oligosaccharides are absorbed, they are not hydrolysed inside the body and are excreted as such in the urine.

iii. Inulin may influence the digestion process and thus affect metabolic responses; especially digestion of disaccharides, glucose transport, and absorption of leucine and proline *etc* (Buddington *et al.,* 1999).

iv. More slowly fermented long-chain inulin is fermented in distal parts of the intestine. Inulin modifies the composition of the intestinal flora, but, due to the slower and less intense fermentation as compared to FOS, impact on composition of the intestinal flora is less pronounced (Harmsen *et al.,* 2002)

v. Mixture of the long-chain inulin and short-chain oligofructose @ 1:1 may be more potent than oligofructose alone in promoting Ca absorption. (Coudray *et al.*, 2003)

vi. The development of chemically induced cancer in animal was suppressed by feeding diet containing inulin (Taper and Roberfroid, 1999). This indicates that systemic effects were involved and the cancer-preventing action was carried out by continuous feeding of inulin.

As antibiotics are being banned as growth promotors in animal feed, prebiotics such as inulin offers a good alternative. Young animals are typically fed relatively high doses of inulin [up to 2% (w/w) in feed] directly or indirectly to improve health of the animals.

Galacto Oligosaccharides (GOS)

The GOS are composed of oligosaccharides ranging from disaccharides to octasaccharides. GOS have attracted particular attention because they have certain similarities to oligosaccharides occurring in milk and modulate the microbial population in the gut. They affect different gastrointestinal activities and have the potential to influence inflammatory and immunological processes (Shin and Yang, 1998).

Physiological effects of GOS

i. GOS is potential to stimulate the growth of bifidobacteria and lactobacilli at 1% (w/v) in animals (Tanaka *et al.,* 1983).

ii. GOS also supports growth of several enterobacteriaceae and streptococci.

iii. GOS has been shown to be protective against the development of induced colorectal tumors in rats (Taper and Roberfroid, 1999).

Functional Disaccharides

Lactulose

Lactulose is the first true prebiotic recognized for its effects on the gut flora. The disaccharide lactulose was first manufactured from lactose. It has also been shown that small amounts (10 to 15 g twice a day) may induce tonic contractions of the colon leading to the known anti constipating effect. Lactulose has also been used to reduce the rate of *Salmonella* carriage in chronic carriers. Mineral absorption, particularly calcium and magnesium, have been shown to be enhanced by ingestion of lactulose (Schumann, 2002).

Lactitiol

Lactitol is also derived from lactose through hydrogenation of the parent compound. Feeding of Lactitiol increases both *bifidobacteria* and *Lactobacilli*, but decreases the no of *bacteroids* and *clostridia* species. Combinations of sucrose and lactitol were significantly increase bifidobacteria concentration and Ca absorption (Ballongue *et al.*, 1997).

Effect of Synbiotics Substrate

Effect on Gut:

- Synbiotics substrate helps in selective growth of some selective bacteria like Lactobacillus.
- They may also inhibit pathogens through a multiple mechanisms. Some may also compete with receptor sites on the gut wall and inhibit pathogen persistence and thus reduce the potential risk of infection. They may also compete effectively for nutrients with pathogens
- It may reduce the risk of colon cancer through stimulating apoptosis. (Scheppach,1996)
- Feeding of synbiotics substrate develops resistance to gastric acidity and enhances GI absorption. (Gibson *et al.,* 2004).

Effect on Immune System:

- Synbiotics substrate may influence the immune system directly or indirectly as a result of intestinal fermentation and promotion of growth of certain members of the gut microbiota.
- Presence of increased numbers of a particular microbial genus or species, or a related decrease of other microbes, may change the collective immuno-interactive profile of the microbiot.
- Microbial products such as SCFA may interact with immune cells and enterocytes and modify their activity.

Effect on Gastro intestinal Disorder:

- Synbiotics substrates are able to manipulate the microbial intestinal environment and subsequently prevent the occurrence of infectious disease.
- A number of substrate has been demonstrated to be effective in the manipulation of the microbiota (Guarner, 2007).
- A mixture of various synbiotics substrates decrease the incidence of bloody diarrhoea and mucosal injury (Kanauchi *et al.,* 1998)
- Synbiotics substrate may interfere with the process of carcinogenesis. Increasing the proportion of lactic acid bacteria in the gut may lead to decrease in certain bacterial enzymes involved in the synthesis or activation of carcinogens.

Effect on Mineral Absorption:

Synbiotics substrate increase Ca and Mg absorption in various species of animals. Substrates are resistant to hydrolysis by small intestinal digestive enzymes. The colonic fermentation produces SCFA and other organic acids that contribute to lower luminal pH in the large intestine which, in turn improves the passive diffusion (Lopez *et al.*, 2000). Fe and Zn balance can be improved by consumption of prebiotics.

Researchs related to synbiotics:

- Supplementation of Synbiotics reported decreased population of E.coli and increased population of Lactobacilli in piglets (Shim *et al.,* 2005).
- Supplementation of Lactobacillus and FOS causes reduction in cholesterol activity (Liong and Shah, 2005)
- Feeding of Lactobacillus and FOS improved digestibility of dry matter and crude protein in pigs (Shim *et al.,* 2005).
- Supplementation of Lactobacillus and FOS improved body weight and absorption of Ca and P in pigs (Shim *et al.,* 2005).
- Improved immunity in rats supplemented with Lactobacillus rhamnosus and innulin (Roller *et al.,* 2004).

Combination of Lactobacillus, Biofidobacterium and oligosaccharides used as therapeutic agents to prevent infection (Nomoto *et al.,* 2004)

Summary

Synbiotic act by supporting the dynamic equilibrium of the intestinal micro flora. It causes high performance of an animal by improving feed conversion and daily weight gain. It can improve digestive problems and general health of an animal. By ensuring high performances and averting problems, synbiotics contribute to an environmentally friendly animal husbandry.

References

Alakomi, H.L., Skytta, E., Saarela, M., Mattila-Sandholm, T., Latva-Kala, K. and Helander, I.M. 2000. Lactic acid permeabilizes gram-negative bacteria by disrupting the outer membrane. *Applied and Environmental Microbiology*. 66 (5): 2001- 2005.

Ballongue, J., Schumann, C. and Quignon, P. 1997. Effects of lactulose and lactitol on colonic microflora and enzyme activity. *Scandanavian Journal of Gastroenterology.* 222: 41-48

Bjorck, I. 1991. On the possibility of acid hydrolysis of inulin in the rat stomach. *Food Chemistry.* 41: 243–250.

Bouhnik, Y., Raskine, L., Simoneau, G., Vicaut, E., Neut, C., Brouns, F. and Bornet, F. 2004. The capacity of non digestible carbohydrates to stimulate fecal bifidobacteria in healthy humans: A double-blind, randomized, placebo-controlled, parallel-group, dose-response relation study. *American Journal of Clinical Nutrition*. 6:1658–1664.

Buddington, R. K. 1999. Influence of fermentable fibre on small intestinal dimensions and transport of glucose and proline in dogs. *American Journal of Veterinary Research*. 60: 354–358

Bunce, T.J., Howard, M.D., Kerley, M.S., Allee, G.L., and Pace, L.W. 1995. Protective effect of fructooligosaccharide (FOS) in prevention of mortality and morbidity from infectious E. coli K:88 challenge, Abstract for the American Society of Animal Science & American Dairy Science Association Annual Meeting.

Caplice, E. and Fitzgerald, G.F. 1999. Food fermentations: role of microorganisms in food production and preservation. *International Journal of Food Microbiology*. 50(1-2): 131-149.

Cherrington, C.A., Hinton, M., Pearson, G.R. and Chopra, I. 1991. Short-chain organic acids at pH 5.0 kill *Escherichia coli* and *Salmonella* spp. without causing membrane perturbation. *Journal of Applied Bacteriology*. 70(2):161-165.

Coallier-Ascah, J. and Idziak, E.S. 1985. Interaction between *Streptococcus lactis* and *Aspergillus flavus* on production of aflatoxin. *Applied and Environmental Microbiology*. 49(1): 163-167.

Coudray, C., Demigne, C. and Rayssiguier, Y. 2003. Effects of dietary fibers on magnesium absorption in animals and humans. *Journal of Nutrition*. 133: 1–4.

Czaczyk, K. 2003. Tworzenie biofilmów bakteryjnych-istota zjawiska i mechanizmy oddzia³ywañ. Biotechnologia, The creating biofilms of bacteria - the creature the phenomenon and mechanisms of influences. *Biotechnology*. 3:180-192.

Fink-Gremmels, J. 1999. Mycotoxins: their impact on human and animal health. *Veterinary Quarterly*. 21: 115-120.

Fuller, R. 1992. Probiotics. The scientific basis, Chapman & Hall, London.

Gedek, B.R. 1999. Adherence of *Escherichia coli* sero group O 157 and the *Salmonella typhimurium* mutant DT 104 to the surface of *Saccharomyces boulardii*. *Mycoses*. 42(4): 261-264.

Gibson, G. R. and Roberfroid, M. B. 1995. Dietary modulation of the human colonic microbiota: introducing the concept of prebiotics. *Journal of Nutrition*. 125: 1401–1412.

Gibson, G. R., Rastall, R. A. and Roberfroid, M. B. 1999. Prebiotics. In: Colonic Microbiota: Nutrition and Health. G. R. Gibson and M. B. Roberfroid (Eds). Kluwer Academic Press, Dordrecht, pp. 101–124.

Gibson, G.R., Probert, H.M., Van Loo, J.A.E., Roberfroid, M.B. 2004. Dietary modulation of the human colonic microbiota: Updating the concept of prebiotics. *Nutrition Research Review*. 17: 257–259.

Goldin, B. R. and Gorbach, S.L. 1992. Probiotics for humans. In: Probiotics. The Scientific Basis. Fuller R. (Ed.). Chapman and Hall, London, pp. 355–376.

Grajek, W. and Sip, A. 2004. Biological fixation food with utilization of lactic acid bacteria metabolites In: Lactic acid bacteria. Classification, metabolism, genetics, *Politechnikalodzka*. 8: 103–197.

Grizard, D. and Barthomeuf, C. 1999. Enzymatic synthesis and structure determination of NEO-FOS. *Food Biotechnology*. 13: 93–105.

Guarner, F. 2007. Prebiotics in inflammatory bowel diseases. *British Journal of Nutrition*. 98 (Suppl 1): 85–89.

Haghighi, H.R., Gong, J. Gyles, C.L., Hayes, M.A., Zhou, H., Sanei, B., Chambers, J.R. and Sharif, S. 2006. Probiotics stimulate production of natural antibodies in chickens. *Clinical and Vaccine Immunology*. 13(9): 975-980.

Harmsen, H. J., Raangs, G. C., Franks, A.,Wildeboer-Veloo, A. C. and Welling, G.W. 2002. The effect of the prebiotic inulin and the probiotic *Bifidobacterium longum* on the fecal microflora of healthy volunteers measured by FISH and DGGE. *Microbial Ecology Health and Diseases.* 14: 219-237.

Hosoya, N., Dhorranintra, B. and Hidaka, H. 1988. Utilization of [U-14C] fructooligosaccharides in man as energy resources. *Journal of Clinical Biochemistry and Nutrition.* 5: 67–74.

ISAPP 2008. 6th Meeting of the International Scientific Association of Probiotics and Prebiotics, London, Ontario.

Isolauri, E., Sütas, Y.; Kankaanpää, P., Arvilommi, H. and Salminen, S. 2001. Probiotics: effects on immunity. *The American Journal of Clinical Nutrition.* 73: 444S-450S.

Kanauchi, O., Nakamura, T., Agata, K., Mitsuyama, K. and Iwanaga, T. 1998. Effects of germinated barley foodstuff on dextran sulfate sodium-induced colitis in rats. *Journal of Gastroenterology.* 33(2): 179–88, 1998.

Liong, M.T. and Shah, N.P. 2005. Optimization of Cholesterol Removal by Probiotics in the Presence of Prebiotics by Using a Response Surface Method. *Applied and environmental Microbiology.* 71 : 1745-1753.

Lopez, H.W., Coudray, C. and Levrat-Verny, M.A. 2000. Fructooligosaccharides enhance mineral apparent absorption and counteract the deleterious effects of phytic acid on mineral homeostasis in rats. *Journal of Nutrition and Biochemistry.* 11: 500–508.

Ma, T. Y., Hollander, D., Erickson, R. A., Truong, H., Nguyen, H. and Krugliak, P. 1995. Mechanism of colonic permeation of inulin: is rat colon more permeable than small intestine? *Gastroenterology.* 108: 12–20.

McKellar, R.C., Modler, H.W. and Mullin, J. 1993. Characterization of growth and inulinase production by *Bifidobacterium spp.* on fructo oligosaccharides, Bifidobacteria. *Microflora.* 12(2):75–86.

Mercenier, A., Pavan, S. and Pot, B. 2003. Probiotics as biotherapeutic agents: Present knowledge and future prospects. *Current Pharmaceutical Design.* 9(2):175-191.

Messens, W. and de Vuyst, L. 2002. Inhibitory substances produced by *Lactobacilli* isolated from sourdough - a review. *International Journal of Food Microbiology.* 72(1):31-43.

Nomoto, K. 2008. Prevention of post operative microbial infection by synbiotics, Indian *Journal of Experimental Biology.* 46: 557-561

Nousiainen, J., Jaranainen, P., Setala, J. and von Wright, A. 2004. Lactic acid bacteria as animal probiotics. *Food Science and Technology.* 139: 547-580.

Ouwehand, A.C. and Vesterlund, S. 2004. Antimicrobial components from lactic acid bacteria. *Food Science and Technology.*139: 375-396.

Roberfroid, M. 2005. Inulin: A fructan, in Inulin-Type Fructans: Functional Food Ingredients, Roberfroid, CRC Press, Boca Raton, FL, USA.

Roller, M., Rechkemmer, G. and Watzi, B. 2004. Prebiotic inulin enriched with oligofructose in combination with the probiotics Lactobacillus rhamnosus and Bifidobacterium lactis modulates intestinal immune functions in rats. *Journal of Nutrition.*134: 153-156.

Rowland, I. R. and Tanaka, R. 1993. The effects of transgalactosylated oligosaccharides on gut flora metabolism in rats associated with human faecal microflora. *Journal of Applied Bacteriology.* 74: 667–674.

Scheppach, W. 1996. Treatment of distal ulcerative colitis with short-chain fatty acid enemas. A placebo-controlled trial. *Digestive Disease Science.* 41: 2254–2259.

Schumann, C.(2002). Medical, nutritional and technological properties of lactulose. An update. *European Journal of Nutrition.* 41(Suppl.): 1-17.

Shetty, P.H. and Jespersen, L. 2006. *Saccharomyces cerevisiae* and lactic acid bacteria as potential mycotoxin decontaminating agents. *Trends in Food Science and Technology.*17: 48-55.

Shim, S.B. 2005. Effects of prebiotics, probiotics and synbiotics in the diet of growing pigs. Wageninngen Institute of Animal Sciences, Wageninngen, Netherlands.

Shin, H. J. and Yang, J.W. 1998. Enzymatic production of galactooligosaccharide by *Bullera singularis* β-galactosidase. *Journal of Microbiology and Biotechnology.* 8: 484–489.

Tanaka, R., Takayama, H., Morotomi, M., Kuroshima, T., Ueyama, S., Matsumoto, K., Kuroda, A. and Mutai, M. 1983. Effects of administration of TOS and Bifidobacterium breve 4006 on the human fecal flora. *Microflora.* 2:17–24.

Taper, H. S. and Roberfroid, M. 1999. Influence of inulin and oligofructose on breast cancer and tumor growth. *Journal of Nutrition*, 129 (Suppl.): 1488S–1491S.

Van Loo, J., Coussement, P., De Leenheer, L., Hoebregs, H. and Smits, G. 1995. On the presence of inulin and oligofructose as natural ingredients in the Western diet. *Critical Review on Food Science and Nutrition*. 35: 525–552.

WHO. 1994. Environmental health criteria, No.136. *Environmental aspects.* World Helath Organisation, Geneva.

7

Use of Dietary Organic Acids in Poultry Production

Pankaj Kumar Singh, Sanjay Kumar, Chandramoni and Kaushalendra Kumar

Introduction

With the perception that antibiotics should no longer be used as animal growth promoters, there has been widespread interest in natural methods of inhibiting detrimental bacteria. Their removal from the poultry diets is therefore brings with it consequences. This has lead to somewhat of a renaissance in research activity on the discovery and application of non antibiotic chemical compounds capable of either killing microorganisms outright or at the very least retarding growth sufficiently to limit their dissemination. Organic acids have a long history of being utilized as food additives and preservatives for preventing food deterioration and extending the shelf life of perishable food ingredients. In recent past organic acids have been used to control microbial contamination and replace antibiotic growth promoters in poultry (Ricke, 2003; Hassan *et al.*, 2010; Adil *et al.*, 2011).

Organic acid as Feed Acidifiers

Organic acids are organic carboxylic acids, including fatty acids and amino acids, of the general structure R-COOH referred to as fatty acids. As a group these compounds primarily include the saturated straight-chain mono-carboxylic acids and their respective derivatives (unsaturated, hydroxylic, phenolic, and multi-carboxylic versions) and are often generically weak or carboxylic acids (Cherrington *et al.*, 1991). Several of these organic acid compounds are used as direct additives incorporated into human foods or can accumulate over time as a consequence of the fermentation activity of indigenous or starter cultures added to certain dairy, vegetable, and meat products. In addition, acid sprays have been incorporated as sanitizers during meat processing (Cherrington *et al.*, 1991; Dickson, 1992; Dorsa, 1997). Some organic acids, particularly the short-chain fatty acids (SCFA), acetate, propionate, and butyrate, are produced in millimolar quantities in the gastrointestinal tracts of food animals and humans and characteristically occur in high concentrations in regions where strictly anaerobic microflora are predominant. In the food animal industry, organic acids were originally added to animal feeds to serve as fungistats (Paster, 1979; Dixon and Hamilton, 1981), now different organic acids and their various combinations have also been examined for potential bactericidal activity in feeds and feed ingredients contaminated with food borne pathogens, particularly *Salmonella* spp. (McHan and Shotts, 1992; Berchieri and Barrow, 1996; Thompson and Hinton, 1997). The short chain acids (C_1-C_7) are associated with antimicrobial activity. They are either simple monocarboxylic acids such as formic, acetic, propionic and butyric acids or carboxylic acids with hydroxyl group such as lactic, malic, tartaric and citric acids or short chain carboxylic acids containing double bonds like fumaric and sorbic acids. Organic acids are weak acids and are only partly dissociated. Most organic acids with antimicrobial activity have a pKa (the pH at which the acid is half dissociated) between 3 and 5.

Organic acids have been used for decades in feed preservation, protecting feed from microbial and fungal destruction or to increase the preservation effect of fermented feed. Feed acidifiers are acids included in feeds in order to lower the pH of the feed, gut, and microbial cytoplasm thereby inhibiting the growth of pathogenic intestinal microflora. This inhibition reduces the microflora competing for the host nutrients and results in better growth and performance of the chicken. They also act as mold inhibitors. Most acids are efficacious and their effect remains as long as the acid is not volatilized. Organic acids have been used extensively for more than 25 years in swine production and more recently in poultry. The antimicrobial effect of organic acid ions in controlling bacterial populations in the upper intestinal tract leads to beneficial effects. Comparative studies of six organic acids showed that the inhibiting effect of the acids was more pronounced in stomach contents than in content from the small intestine, probably due to the lower pH in the stomach content. The bactericidal effect of the organic acids is:

Benzoic acid >Fumaric acid >Lactic acid > Butyric acid > Formic acid > Propionic acid.

Commercial preparations appear to enhance digestibility and diet palatability, thus improving feed conversion and growth of animals, including pigs and poultry. The antibacterial mechanism(s) for organic acids vary depending on physiological status of the organism and the physicochemical characteristics of the external environment.

Table 1: Some physicochemical properties of most common organic acids

Name	**Physical form**	**Mol. wt (g/mol)**	**GE (MJ /kg)**	**Density (g/ml)**	**Disso-ciation Con-stant (pKa)**	**Corrosive rate Odour**	**Odour**
Formic acid	Liquid	46.03	5.7	1.220	3.75	High	Pungent
Acetic acid	Liquid	60.05	14.6	1.049	4.76	Medium	Pungent
Propionic acid	Liquid	74.08	20.6	0.993	4.88	Medium	Pungent
Butyric acid	Liquid	88.12	24.8	0.958	4.82	Low	Rancid
Lactic acid	Liquid	90.08	15.1	1.0206	3.86	Low	Sour milk
Sorbic acid	Solid	112.14	27.8	1.204	4.76	Low	Mildy acrid
Fumaric acid	Solid	116.07	11.5	1.635	4.38	Negligible	Odourless
Malic acid	Liquid	134.09	10.0	1.601	5.10	Low	Apple
Citric acid	Solid	192.14	10.2	1.665	6.4	Negligible	Odourless

Mode of action of organic acids

Although the antibacterial mechanism(s) for organic acids are not fully understood, they are capable of exhibiting bacteriostatic and bactericidal properties depending on the physiological status of the organism and the physicochemical characteristics of the external environment. Several mechanisms through which dietary organic acids may produce beneficial effects on health status and growth performance have been proposed (Partanen and Mroz, 1999; Diebold and Eidelsburger, 2006; Tung and Pettigrew, 2006), however, the following appear to be the most prominent:

1. Reduction of gastric pH
2. Reduction of buffering capacity of diets
3. Increase of proteolytic enzymes activity/Improvement of pancreatic secretions

4. Stimulating the activity of digestive enzymes
5. Increase of nutrient digestibility
6. Promotion of beneficial bacterial growth
7. Reduced survival of pathogens through balancing the microbial population
8. Direct killing of bacteria
9. Alterations in the nutrient transport and synthesis within the bacterium
10. Depolarization of the bacterial membrane.

The antimicrobials activity of organic acids is related to reduction in pH and its ability to dissociate, which is determined by the pKa value of the respective acid and the pH of the surrounding environment. The organic acids are lipid soluble in undissociated form. The more the undissociated form of organic acids the better the efficacy. Hence, the selection of organic acid is very critical in determining the efficacy of the product. The undissociated form of the organic acid penetrates the cell membrane of bacteria. Inside the cell the acid dissociates according to internal pH:

Organic acid $\rightarrow$ Anion + H^+ (proton)

The H^+ (proton) decreases pH in the cell. The bacteria use its energy resources trying to remove the protons and dies. Anions of organic acids deactivate RNA transferase enzyme, which damage nucleic acid multiplication process and eventually result in death of the organism. In general, potential bacterial targets of biocidal compounds include the cell wall, cytoplasmic membrane, and specific metabolic functions in the cytoplasm associated with replication, protein synthesis, and function (Denyer and Stewart, 1998; Davidson, 2001). Given the weak acid nature of most of these compounds, pH is considered a primary determinant of effectiveness because it affects the concentration of undissociated acid formed (Davidson, 2001). It has been traditionally assumed that undissociated forms of organic acids can easily penetrate the lipid membrane of the bacterial cell and once internalized into the neutral pH of the cell cytoplasm dissociate into anions and protons (Salmond *et al.*, 1984; Cherrington *et al.*, 1991; Davidson, 2001). Generation of both of these species potentially presents problems for bacteria that must maintain a near neutral pH cytoplasm to sustain functional macromolecules. Export of excess protons requires consumption of cellular adenosine triphosphate (ATP) and may result in depletion of cellular energy (Davidson, 2001). Anions of organic acids deactivate RNA transferaze enzyme, which damage nucleic acid multiplication process and eventually result in death of the organism.

Other toxicity mechanisms have been proposed that attribute membrane uncoupling capabilities for organic acids. It has been speculated that organic acids interfere with cytoplasmic membrane structure and membrane proteins such that electron transport is uncoupled and subsequent ATP production is reduced or that organic and electrical gradients across cell membranes (Salmond *et al.*, 1984; Russell, 1992; Axe and Bailey, 1995; Davidson, 2001). Russell (1992) has hypothesized that anion accumulation is the primary toxic effect of organic acids and that some organisms are more resistant to organic acids because they are capable of allowing their internal pH to decline. Less direct antibacterial activities have also been attributed to organic acids and include interference with nutrient transport, cytoplasmic membrane damage resulting in leakage, disruption of outer membrane permeability, and influencing macromolecular synthesis (Cherrington *et al.*, 1991; Denyer and Stewart, 1998; Alakomi *et al.*, 2000; Davidson, 2001). The mechanisms associated with these activities have been difficult to establish due to the complex nature of the interaction between energy dissipation and disruption of ATP generation capabilities of the bacteria. Consequently, bactericidal concentrations of organic acids may be due to the combination of dissipation of proton-motive force and inability to maintain internal pH followed by denaturation of acid-sensitive proteins and DNA. Sub lethal concentrations may elicit their effects on overall cell physiology and lead to responses such as enlargement of bacterial cell size (Thompson and Hinton, 1996).

Antibacterial effect of organic acids

It has been shown that acid conditions favor the growth of lactobacilli in the stomach, which possibly inhibits the proliferation of E. coli and produces lactic acid and other metabolites which lower the pH and inhibit E. coli. The antibacterial effect of organic acids has been reported by many researchers. Sofos *et al.* (1985) reported that the broilers on sorbic acid-containing feed had lower coliform counts in the duodenum, lower yeast and mold counts in the caeca, and higher Bacteroides counts in the caeca. Humphrey and Lanning (1988) observed that 0.5% formic acid resulted in significant reduction in the isolation rate of Salmonella from laying hens, and also, a reduction in the incidence of infection in newly hatched chicks. Use of organic acid mixture containing formic and propionic acid decreased the Salmonella and lactic acid producing bacteria counts in hen's crop (Thompson and Hinton, 1997). Alp *et al.* (1999) reported that inclusion of antibiotic and an organic acid mixture containing lactic, fumaric, propionic, citric and formic acid separately or in combination reduced the Enterobacteriacae count in the ileum of broilers. Ramarao *et al.* (2004) studied the efficacy of gut acidifier in broiler diets at the rate of 300 g/100 kg feed vis-à-vis use of antibacterial compounds. They observed that the total bacterial, coliform, and Escherichia coli counts in crop and caecal contents were low in broilers fed gut acidifier and opined that gut acidifier can safely replace antibacterial compounds in broiler chicken diets with beneficial effects on the intestinal bacterial colonization and resistance to E. coli challenge. Moharrery and Mahzonieh (2005) described that malic acid have the potential for reduction of E. coli population in the intestines of broiler chicken. Thirumeignanam *et al.* (2006) reported a decrease in the total bacterial load with concomitant increase in lactobacilli load as a result of dietary acidification. Organic acids mixture at the level of 0.2% in the diet of broilers significantly decreased total bacterial and gram negative bacterial counts compared to the basal diet (Gunal *et al.,* 2006). Fumaric and sorbic acid lowered the numbers of lactic acid bacteria and Coliforms in the ileum and caeca (Pirgozliev *et al.,* 2008). Due to antimicrobial effect, organic acids result in inhibition of intestinal bacteria leading to the reduced bacterial competition with the host for available nutrients and diminution in the level of toxic bacterial metabolites as a result of lessened bacterial fermentation resulting in the improvement of protein and energy digestibility; thereby ameliorate the performance of bird.

The antimicrobial effects of six different organic acids (formic, propionic, butyric, lactic, benzoic, and fumaric acid) were compared in stomach content at pH 4.5 and in small intestinal content at pH 5.5. In contrast to lactic acid bacteria, coliform bacteria were unable to grow in stomach content at the pH values investigated. A pH-dependent inhibition of bacterial growth was demonstrated both in the content of the stomach and the small intestine. The comparative studies of the six organic acids showed that the inhibiting effect of the acids was more pronounced in stomach content than in content from the small intestine, properly due to the lower pH in the stomach content. The killing effect of the organic acids increased in the following way: propionic acid < formic acid < butyric acid < lactic acid < fumaric acid < benzoic acid. Benzoic acid was superior to the other acids tested in exhibiting a bactericidal effect on coliform as well as lactic acid bacteria in both stomach and small intestinal content (Maribo *et al.,* 2000).

Trophic effects of organic acids

Organic acids have direct stimulatory effect on the gastro-intestinal cell proliferation as has been reported by various workers with short chain fatty acids. The short chain fatty acids are believed to increase plasma glucagon-like peptide 2 (GLP-2) and ileal pro-glucagon mRNA, glucose transporter (GLUT2) expression and protein expression, which are potential signals mediating gut epithelial cell proliferation (Tappenden and McBurney, 1998).

LeBlay *et al* (2000) and Fukunaga *et al* (2003) also reported that short chain fatty acids can accelerate gut epithelial cell proliferation, thereby increase intestinal tissue weight and changing mucosal morphology. Improved proliferation of the epithelial cell lining is another positive effect that has been observed for a number of organic acids. This feature results in an improved villi length and villus-crypt ratio. The beneficial effect of organic acids on the gastro-intestinal tract was observed by Denli *et al* (2003) reported that organic acids resulted in remarkable increase in the intestinal weight and length of broiler chicken. Senkoylu *et al.* (2007) reported increased villus height with combination of formic and propionic acids in broilers but found no effect on the thickness of lamina muscularis mucosae. Paul *et al.* (2007) also reported that the propionate, formate and lactate supplementation improved duodenal villus height. Similar results were observed by Garcia *et al.* (2007) who found improved villus height with formic acid and also greater crypt depth but villus surface area was not influenced. Abdel-Fattah *et al.* (2008) reported that the addition of any level and source of organic acids increased feed digestion and absorption as a result of increased small intestine density which is an indication of the intestinal villi dimension. The increased villus height in the small intestines has been related to a higher absorptive intestinal surface, which facilitates the nutrient absorption and hence, has a direct impact on growth performance.

Other effects of organic acids

Dietary supplementation of organic acids has been found to reduce the pH of crop, gizzard and duodenal contents. Abdel-Fattah *et al.* (2008) reported that the pH values in different gastro-intestinal tract segments were decreased with supplementation of organic acids irrespective of type and dose used. The lowered pH is conducive for the growth of favourable bacteria while it simultaneously hampers the growth of pathogenic bacteria which grow at relatively higher pH. This reduces bacterial competition with the host for available nutrients, thereby increasing the nutrient absorption. The acid anion has been shown to complex with calcium, phosphorus, magnesium and zinc, which improves the digestibility of these minerals (Li *et al.*, 1988). Organic acids serve as substrates in the intermediary metabolism (Kirchgessner and Roth, 1998) and lower chyme pH, consequently, enhancing the protein digestion (Gauthier, 2002). Besides its antimicrobial function, pancreatic enzyme secretion can also be increased by dosing organic acids in piglet feeds (Harada *et al*, 1986). Reduction in gastric pH occurs following organic acid feeding which may increase the pepsin activity (Kirchessner and Roth, 1982) and the peptides arising from pepsin proteolysis trigger the release of hormones, including gastrin and cholecystokinin, which regulate the digestion and absorption of protein (Hersy, 1987).

Use of organic acid in poultry feed

Enhancement of growth, feed efficiency and prevention of disease transmission are critical factors in modern poultry production. Feed materials may be contaminated with microorganisms, mostly bacteria and fungi, at any time during growing, harvesting, processing and storage. Food-producing animals (e.g. chickens, pigs) are the main reservoirs for many of these microorganisms, which include *Salmonella enterica*, *Campylobacter* species, Shiga toxin producing strains of *Escherichia coli*, and *Yersenia enterocolitica* (Mead, 2000). The microflora found in feed materials comes from a variety of ecological niches e.g. soil and the animals' gastrointestinal (GI) tract. The GI tract pathogens can be introduced into food chain by animals defecating in the farm environment or by fertilization of crops with manures (Maciorowski *et al.*, 2007), consequentially making feed a carrier for animal and human pathogens. When pathogenic bacteria contaminate feed, it becomes a potential route of transmission of disease to animal and human populations, and is consequently of great concern to producers and consumers (Crump *et al.*, 2002). Heat treatment, usually during conditioning, pelleting or extrusion has been shown to be an effective way to reduce microbial loads in feed materials and compound feed. Reduction of the bacterial contamination by heat is dependent on the temperature and treatment time. However, these methods do not prevent a recontamination of feed materials and compound feed afterwards (Çelik *et al.* 2003; Maciorowski *et al.*, 2007).

Dietary acidification with organic acids has been shown to contribute to environmental hygiene preventing feed raw materials and compound feed from microbial and fungal deterioration. Moreover, constant treatment with organic acids has a residual protective effect in feed, which helps to reduce recontamination and also reduce the contamination of milling and feeding equipment as well. Supplementation of organic acids in feed tends to decrease the feed pH, buffer capacity and to prevent undesirable microbes' growth. Dietary acidification is important to create unfavorable conditions for microorganisms and for reduction of pH and stimulation of GI tract enzymes. Optimum pH is needed for enzyme activation, for example, pepsinogens are rapidly activated at pH 2, but very slow at pH 4. Pepsin has optimum pH between 2 and 3.6, and remains inactive at pH 6 (Kidder and Manners, 1978). Thus, acidifiers are powerful tools to maintain animal health and improve their performance, as well as to control feed and environmental hygiene (Reddy, 2004).

Factors affecting efficiency of organic acid

The efficacy of some individual organic acids or their salts enhance the growth and feed to gain ratio of weaned and growing-finishing pigs, although there is a relatively large variation in responses due to different factors as follows:

- Chemical properties of organic acid.
- Type and doses of supplemented organic acids;
- Type of diets and their acid-base or buffering capacity;
- Level of intraluminal production of SCFA in particular segments of the gastrointestinal tract by inhabiting microflora;
- Quantity of fermentable carbohydrate substrates in the diet for bacterial growth, colonization and activity resulting in SCFA production;
- Presence or absence of maternal vaccination against pathogens;
- Presence or absence of receptors for bacterial colonization on the epithelial villi;
- Existing level of performance of birds
- Hygiene and welfare standards (density per pen, ventilation intensity and area, cleaning frequency etc.).

Chemical forms, molecular weight, pKa value, animal species, kind of micro-organisms are major factors affecting efficiency of organic acid in feed (Reddy, 2004). Another factor affecting the response of acidifiers are differences in type of diets and their acid-base or buffering capacity (B-value) of the diets. By definition, the buffering capacity (B-value) is the change in the pH value of a defined volume or mass after the addition of a strong acid. A more practical definition in the feed industry is that the B-value is the amount of 1 M hydrochloric acid (HCl) solution which needs to be added to a 10% slurry of feed or a feed ingredient in 100 ml of water in order to obtain a pH-value of 5, in some cases a pH value of 4 or 3. This definition is the reason why we find different values for the same expression in practical applications. For example, various studies reported B-values ranging from 380 to 700 mEq per kg feed (Makkink, 2001). Acid-buffering capacity is lowest in cereals and cereal by-products, intermediate or high in protein feedstuffs and very high in minerals. High protein and mineral content of feed ensures rapid animal growth, but generates high buffering capacity, thus reducing levels of HCl in the stomach. Results of some studies demonstrated that high B-value of the diet increased gastric pH and resulted in decreased amino acids digestibility (Jung and Bolduan, 1986; Lawlor *et al.* 1994; Blank *et al.*, 1999). In young animals, capacity to secrete gastric juice is limited. High B-value may pose problems. Pathogenic bacteria multiply in the digestive tract. The recommended B-value for poultry is about1-10 for 1-10 days age and 10-20 for 10-30 days age. Lowering dietary buffering capacity, via acidification with organic acids, has been shown to inhibit luminal growth of enterotoxigenic microflora and to enhance animal performance (Gabert and Sauer, 1994).

Organic acid and the performance of poultry

Growth Performances

Organic acids are beneficial in practical studies. The efficacy of poultry digestion depends on microorganisms, which live naturally in the digestive tract. Various studies revealed that body weight gain, feed intake, feed conversion rate, carcass weight, abdominal fat weight, abdominal fat percentage, intestinal weight were affected significantly by giving organic acids. Patten and Waldroup (1998) reported that addition of 0.5 or 1.0% fumaric acid significantly improved body weights of broilers but did not influence feed utilization. Skinner *et al.* (1997) reported that addition of 0.125% fumaric acid significantly improved 49-day body weight of females and average weight gain of both sexes with no effect on feed utilization. Also, higher body weight gain, feed intake and better feed efficiency due to organic acid supplementation has been reported (Denli *et al.*, 2003; Sheikh *et al.*, 2011). Christian *et al.* (2004) observed that organic acid blend (3 kg inclusion rate per ton of feed) increased the growth of broiler chicken under controlled conditions without the use of anti-biotic growth promoters. The body weight of broilers at 6th week of age was significantly higher in the groups fed diet containing mixture of organic acid at 1 kg/ton or 1.5 kg/ton with a better feed conversion ratio noticed in the group containing organic acid at 1 kg/ton (Thirumeignanam, *et al.,* 2006). Paul *et al.* (2007) reported that ammonium formate or calcium propionate at the level of 3 g/kg feed increased the live weight and live weight gain and feed conversion ratio at day 21 in broiler chicken. Formic acid at the rate of 5,000 ppm and 10,000 ppm in the diet of chicken significantly improved feed conversion ratio and both levels were beneficial for improving the growth traits (Garcia *et al.*, 2007). Fumaric acid improved growth performances and feed efficiency in broilers (Vogt and Matthews, 1981; Patten and Waldroup, 1988; Skinner *et al.*, 1991). Vogt *et al.* (1982) studied malic, sorbic, and tartaric acids (0.5 to 2%) in broilers. They reported increases in weight gain, with optimal levels of 1.12 and 0.33% for sorbic and tartaric acids, respectively. Sorbic and malic acids also tended to improve feed efficiency. Versteegh and Jongbloed (1999) tested the effect of dietary lactic acid on performance of broilers from 0 to 6 wk of age. Body weight gains tended to be greater, whereas feed-to-gain ratios were significantly improved when birds were fed 2% lactic acid. Vieira *et al.* (2008) reported improved body weight and feed conversion ratio with diets supplemented with a blend of organic acids (40% lactic, 7% acetic, 5% phosphoric and 1% butyric). Owens *et al.* (1997) reported 12 % increase in total live weight gain and about 9 % improvement in gain feed ratio with diets supplemented with dietary organic acids. Improvement in live body weight, body weight gain and feed conversion ratio by organic acid supplementation (containing acetic acid, citric acid and lactic acid, each at 1.5 and 3.0 % in the diet) was also observed by Abdel-Fattah *et al.* (2008). Benzoic acid is extensively used as food preservative in human nutrition. An experiment with broiler chickens also indicated that benzoic acid at concentrations of 0.2% may have a positive influence on growth (Engberg, 2001). Addition of organic acids in broiler diets increased the villus height in all the segments of small intestines (Sheikh *et al.*, 2011). Organic acids (mixture of fumaric acid and salt of butyric, propionic and lactic acids) supplementation @ 520ppm improved egg production, egg size, shell thickness and feed conversion efficiency (Rahman *et al.*, 2008).

Control of Salmonella Infection

Organic acids exert their antimicrobial action both in the feed and in the GI-tract of the animal. In poultry, pathogen bacteria e. g. *Salmonella* enters the gastro-intestinal tract via the crop. The environment of the crop with respect to microbial composition and pH seems to be very important in relation to the resistance to pathogens. High amounts of lactobacilli and low pH in the crop have shown to decrease the occurrence of Salmonella in the crop (Hinton *et al.*, 2000). Also the antibacterial effect of dietary organic acids in chickens is believed to take mainly place in the upper part of the digestive tract (crop and gizzard). A study on the metabolism of dietary added propionic acid reveals that only little if any dietary propionic acid reaches the lower digestive tract and the caeca (Hume *et al.*, 1993).

Antimicrobial effect of organic acids helps in prevention of the growth of bacteria (especially Gram negative bacterial species, like *Salmonella spp.* and *E. coli*), yeasts and moulds. Inclusion of organic acids reduced pH in crop and gizzard but not in intestinal tract. Organic acids in crop reduce *salmonella* populations and toxic components by bacteria and a change in the morphology of the intestinal wall and reduce colonization of pathogens on the intestinal wall, thus preventing damage to the epithelial cells. Various organic acid alone or in combinations were tested to prevent *Salmonella* contamination through the ration in chicken. Waldroup *et al.* (1995) studied the use of citric acid at dietary concentrations up to 1% in relation to the caecal colonization with *Salmonella typhimurium* and carcass contamination following oral challenge.

The number of birds colonised with *Salmonella typhimurium* was actually increased following supplementation with citric acid as compared to control, which indicates that citric acid may not be reliable with respect to the prevention of Salmonella colonization of the caeca. In broilers the inclusion of 0.4% and 0.8% propionic acid decreased the number of *coliforms* and *E. coli* in the small intestine without any effect on intestinal pH (Izat *et al.*, 1990a). Following periodic dosage of *Salmonella typhimurium* the same authors observed a reduced number Salmonella on post chill broiler carcasses when 0.4% propionic acid was added to the diet. In an experiments with broiler chickens, increasing amounts of dietary lactic acid (0.25, 0,5, 1,0 and 2.0%) or fumaric acid (0.5, 1,0 and 2.0%) did not offer protection from caecal Salmonella colonization or carcass contamination following oral challenge with *Salmonella typhimurium* (Waldroup *et al.,* 1995).

In poultry formic acid alone or a combination of formic acid with propionic acid (68% formic acid and 20% propionic acid) at concentrations of 0.6% were effective with respect to the prevention of infection with *Salmonella kedougou* (Hinton and Linton, 1988) and *Salmonella gallinarum* (Berchieri and Barrow, 1996). The acid mixture was reported to reduce the concentrations of lactic acid in the crop of laying hens, indicating a growth inhibition of lactic acid bacteria at this location. *In vitro* experiments with hens′ crop contents, where dietary addition of acid mixture was simulated, showed a bactericidal effect for *Salmonella enteritidis* (Thompson and Hinton, 1997). In an experiment with broiler chickens infected with *Salmonella typhimurium*, the addition of 0.36% calcium formate and 0.25% formic acid significantly reduced levels of *Salmonella* on pre chill carcasses. In caecal contents the number of Salmonella was reduced following addition of either 0.36% calcium formate or 0.5% formic acid (Izat *et al.*, 1990b). However, Waldroup *et al.* (1995) found a reduction of ceacal pH in relation to an addition of a formic acid/propionic acid blend in concentrations of 1%. However, the authors found that the formic acid/propionic acid blend did not offer a reliable protection regarding the caecal colonization with *Salmonella typhimurium* following oral challenge. It was observed that a mixture of formic (70%) and propionic acid (30%), at a 0.8% dose, was efficient to eliminate *S. enteritidis* and *S. thyphimurium*, in broiler chicken (Vale *et al.*, 2004). It was concluded that dietary supplementation of organic acids may effectively replace AGPs as a growth promoter in broiler diets (Samanta *et al.*, 2010; Hassan *et al.,* 2010).

Other effects of organic acid

There are other benefits resulting from the antimicrobial activity of organic acids. Runho *et al.* (1997) observed higher metabolizable energy in broilers fed 0.5 to 1% fumaric acid. Eckel *et al.* (1992) and Eidelsburger *et al.* (1992) describe a significant reduction in ammonia in the stomach, small intestine, and cecum of weaned piglets fed 1.25% formic acid. This finding could be due to reduced microbial deamination of amino acids, which would then be available for absorption, resulting in increased nitrogen digestibility and reduced ammonia excretion as observed in swine fed organic acids. Reduction in microbial synthesis of ammonia has been proposed as part of the mechanism of the growth response associated with feeding antibiotics (Visek, 1978). Eckel *et al.* (1992) also reported reduced concentrations of biogenic amines in the small intestines of these animals fed 0.6% formic acid (Eckel *et al.*, 1992). These and other microbial metabolites may exert a growth-depressing effect (Bonem *et al.,* 1976).

Another effect associated with antibiotics that appears to be a direct result of antimicrobial activity is the often-reported thinning of the intestinal tract (Franti *et al.,* 1972). Orally ingested antibiotics have an antimicrobial effect that includes the gut microflora. This reduction in micro flora, and its consequences, may be the underlying mechanism for beneficial effects of antibiotics (Bedford, 2000).

Organic acid as an antibiotic growth replacer

There are various reports in which it was shown that organic acid can replace antibiotic growth promoters without affecting growth performance. Runho *et al.* (1997) compared 0.25 to 1.0% fumaric acid with an antibiotic growth promoter (Nitrovin) in Hubbard broilers. It was reported that growth was not affected, but feed consumption was reduced, resulting in a significant improvement in feed to gain. Feed efficiencies for 0.5, 0.75, and 1% fumaric acid were comparable to the antibiotic control. Effects of buffered propionic acid in the presence and absence of bacitracin or roxarsone were reported by Izat *et al.* (1990a), who found a significant increase in dressing percentage for female broilers and a significant reduction in abdominal fat for males at 49 days. There were no other performance effects.

Problems associated with organic acid

Acidifiers are added to the poultry feed in a solid form which fights mould development in the feed and reduces the pH in the birds' crops. Acids put into the feed tend to be less odorous and less corrosive. This type of acidifier comes in a solid or powder form and can be a severe irritant if inhaled.

The acids may also be injected into the water to kill bacteria, facilitate chlorine in killing bacteria and lowering the pH in the birds' crops. Injecting acids into the water is the most problematic. Adding acid to the water simply compounds that corrosive process. Additionally, acidifiers usually are not a single acid, but a combination of two or more. Sometimes organic acids are sprayed onto the poultry litter. The spray of organic acid to poultry litter attacks the bacteria that facilitate the breakdown of uric acid, limiting the amount of ammonia releases. Another disadvantage to using acidifiers is that they can promote the growth of algae and fungi in the watering system. These too can grow to the point where they clog drinkers and cause them to leak. They also may require additional chemicals to inhibit their growth. Biofilm is also a factor to consider. Biofilm forms when bacteria in water attach to a solid surface. They begin to exude a sticky substance and quickly become an active colony of pathogens embedded in the slime. Research has shown that chlorine, even chlorine boosted by an acidifier, is very ineffective at killing bacteria embedded in a biofilm. Once established, biofilm will grow to the point where it can inhibit drinkers, causing them to leak or preventing the birds from activating them. Portions of the biofilm also can break off and get into the birds during the drinking process.

Microbial adaptation to acids

Bacteria have the capability of becoming resistant to a particular acid. Inducible tolerance (adaptation) to acidic environments is recognized as an important survival strategy for many prokaryotic and eukaryotic microorganisms. It has been known for over 60 years that bacteria possess metabolic capabilities to respond to low external pH (Slonczewski, 1992). A more recent observation is that acid-sensitive bacteria could adapt to acid stress by induction of an acid tolerance response. This response essentially involves growth of the acid sensitive microorganism in a moderately low pH environment that subsequently leads to survival when suddenly exposed to what would normally be considered lethal acidic conditions (Foster, 1995). This adaptation permits the induction of genes involved in an acid-tolerance response and synthesis of a series of acid shock proteins that are protective for extreme acidic conditions (Foster, 1999). Multiple acid-tolerant systems that are physiologically distinct have been identified in food borne pathogens such as Salmonella and E. coli and are classified as log phase and stationary mechanisms (Foster, 1999).

Inducible acid tolerance has been revealed in many gram-negative and gram-positive microorganisms. The acid-tolerance system has been demonstrated to protect Salmonella against lethal effects of organic acids and increase survival in fermented foods (Leyer and Johnson, 1992; Baik *et al.*, 1996). Although it might be anticipated that SCFA at acidic pH levels could induce acid tolerance, it has also been shown that high concentrations of SCFA at neutral pH are capable of inducing acid tolerance in *E. coli* and *S. typhimurium* (Kwon and Ricke, 1998a,b; Arnold *et al.*, 2001). Kwon and Ricke (1998a,b) found that inorganic acid resistance of *S. typhimurium* was increased after exposure to high concentrations of SCFA, and this SCFA-induced acid resistance was further enhanced by acidic pH, anaerobiosis, and prolonged exposure to SCFA. Previous exposure to SCFA also increased *S. typhimurium* survivors against extreme acid (pH 3.0), high osmolarity (2.5 M NaCl), and reactive oxygen (20 mM hydrogen peroxide) and consequently could potentially cross-protect Salmonella from other environmental conditions that exist in food animal and processed food products (Kwon *et al.*, 2000).

Another studies on acid adaptation mechanisms of *Streptococcus mutants* (Quivey *et al.,* 2000) showed that growth at pH 5 resulted in significant changes in membrane fatty acid composition compared with cells grown at pH 7. According to these authors, the shift in the unsaturated/saturated ratio with growth at lower pH suggests that changes in membrane fatty acid composition may be part of the acid adaptive response. This type of shift could affect membrane proton permeability directly by changing the base permeability of the lipid bilayer to protons. Saklani-Jusforgues *et al.* (2000) studied the effect of acid adaptation of *L. monocytogenes* on the survival and translocation after intragastric inoculation in mice. The authors observed a higher survival of acid-adapted *L. monocytogenes* (grown at pH =5.5) in the lumen of caecum and colon of mice compared to non-acid adapted (grown at pH=7.2) *L. monocytogenes*. Moreover, the translocation rate of the acid-adapted bacteria to the mesenteric lymph node was higher than that of the non-adapted bacteria.

Bacteria under *in vitro* conditions are known to adapt to acids, it is not known whether this also occurs in GI-tract, when organic acids are fed to pigs or chickens. The unifying concept is that the microorganism under siege will sense a deteriorating environment and undergo a programmed molecular response by which specific, stress-inducible proteins are synthesized. These proteins presumably act to prevent or repair macromolecular damage caused by the stress. Some stress proteins are induced by a range of stress conditions, whereas others are induced in response to a specific stress (Bearson *et al.,* 1997). Furthermore, different microorganisms have developed different acid survival strategies, even those that are evolutionary close (Lin *et al.,* 1995). According to Bearson *et al.* (1997), there is a correlation between the response of enterobacteria to acid stress and pathogenecity. Kwon and Ricke (2000) suggested that SCFA in the gastrointestinal tract of a host animal or in food materials might contribute to the enhancement of the virulence of *S. typhimurium* by increasing acid tolerance.

The interaction between acid shock and bacteria such as Salmonella and E. coli, however, remains complicated because growth and survival responses to SCFA appear to be dependent on not only the growth phase of the organism but atmospheric conditions (aerobic or anaerobic) used to recover survivors, salt form of the SCFA, and perhaps initial concentration of the SCFA (Kwon *et al.*, 1998; Blankenhorn *et al.*, 1999; Kwon and Ricke, 1999). Future research will require examining growth and survival under more ecologically simulating conditions. In addition, genetic tools that allow for precise dissection of all potential genes directly involved in SCFA and stimulation of acid shock as well as peripheral systems are required. Efforts with transposon mutagenesis and gene array approaches have indicated that these methods may be viable for identifying some of the prospective genes involved in SCFA-induced acid tolerance (Arnold *et al.*, 2001; Kwon *et al.*, 2002).

Limitations of organic acids

- Palatability may be decreased, leading to feed refusal
- Organic acids are corrosive to metallic equipment which may pose handling and equipment issues to the feed manufacturer
- Bacteria are known to develop acid resistance when exposed to acidic environments for over long term
- Presence of other antimicrobial compounds can reduce its efficiency
- Cleanliness of the production environment
- Buffering capacity of dietary Ingredients.

Summary

The increased concern about the potential for antibiotic resistant strains of bacteria has compelled the researchers to utility of other non therapeutic alternatives like probiotics, prebiotics, immune stimulants and organic acids as feed additives in animal production. Due to antimicrobial effect, organic acids result in inhibition of intestinal bacteria leading to the reduced bacterial competition with the host for available nutrients and diminution in the level of toxic bacterial metabolites as a result of lessened bacterial fermentation resulting in the improvement of protein and energy digestibility; thereby ameliorate the performance of bird. The increased villus height in the small intestines induced by organic acids increases the absorptive intestinal surface, facilitates the nutrient absorption and growth performance.

References

Abdel-Fattah, S. A,, El-Sanhoury, M. H., El-Mednay, N M and Abdel-Azeem, F. 2008. Thyroid activity, some blood constituents, organs morphology and performance of broiler chicks fed supplemental organic acids. *International J Poult Sci.* 7(3): 215-222.

Adil, S., Banday, MT., Bhat, GA, Salauddin, M., Mashuq, R and Shanaz, S. 2011. Response of broiler chicken to dietary supplementation of organic acids. *J Central European Agric.* 12(3): 498-508

Alakomi, H.-L., E. Skytta¨, M. Saarela, T. Mattila-Sandholm, K.Latva-Kala, and I. M. Helander. 2000. Lactic acid permeabilizes Gram-negative bacteria by disrupting the outer membrane. *Appl. Environ. Microbiol.* 66:2001–2005.

Alp M, Kocabagli M, Kahraman R and Bostan, K. 1999. Effects of dietary supplementation with organic acids and zinc bacitracin on ileal microflora, pH and performance in broilers. *Turkish J Vet. Anim. Sci..* 23: 451-455.

Arnold, C. N., J. McElhanon, A. Lee, R. Leonhart, and D. A. Siegele. 2001. Global analysis of Escherichia coli gene expression during the acetate-induced tolerance response. *J. Bacteriol.* 183:2178–2186.

Axe, D. D and J. E. Bailey. 1995. Transport of lactate and acetate through the energized cytoplasmic membrane of Escherichia coli. *Biotechnol. Bioeng.* 47:8–19.

Baik, H. S., S. Bearson, S. Dunbar, and J. W. Foster. 1996. The acid tolerance response of Salmonella typhimurium provides protection against organic acids. *Microbiology* 142: 3195–3200.

Bearson, S., Bearson, B. and Foster, J. W. 1997. Acid stress responses in enterobacteria. *FEMS Microbiol. Letters*. 147: 173-180.

Bedford, M. 2000. Removal of antibiotic growth promoters from poultry diets: Implications and strategies to minimise subsequent problems. *World's Poult. Sci. J.* 56:347–365.

Berchieri, A. Jr., and Barrow, P.A. 1996. Reduction in incidence of experimental fowl typhoid by incorporation of a commercial formic acid preparation (Bio-Add™) into poultry feed. *Poult Sci. 75:* 339-341.

Blankenhorn, D., J. Phillips, and J. L. Slonczewski. 1999. Acid and base-induced proteins during aerobic and anaerobic growth of Escherichia coli revealed by two-dimensional gel electrophoresis. *J. Bacteriol.* 181:2209–2216.

Bonem, E., A. Tamm, and M. Hill. 1976. The production of urinary phenols by gut bacteria and their possible role in the causation of large bowel cancer. *J. Clin. Nutr.* 29:1448–1454.

Cherrington, C. A., M. Hinton, G. C. Mead, and I. Chopra. 1991. Organic acids: Chemistry, antibacterial activity and practical applications. *Advances in Medical Biology*. 32:87–108.

Celik, K., Ersoy, I. E., Uzatici, A. and Erturk, M. 2003. The using of organic acids in California turkey chicks and its effect on performance before pasturing. *Int. J. Poult. Sci.,* 2: 446–448.

Christian, L, Nizamettin S, Hasan A and Aylin A. 2004. Acidifier – a modern alternative for anti-biotic free feeding in livestock production, with special focus on broiler production. *Veterinaria Zootechnika*; 27 (49): 91.

Davidson, P. M. 2001. Chemical preservatives and natural antimicrobial compounds. in Food Microbiology- Fundamentals and Frontiers. 2nd ed. M. P. Doyle, L. R. Beuchat, and T. J. Montville, (ed.) American Society for Microbiology, Washington, DC. pp. 593–627

Denli, M., Okan F. and Celik, K. 2003. Effect of dietary probiotic, organic acid and antibiotic supplementation to diets on broiler performance and carcass yield. *Pak. J. Nutr.*, 2: 89-91.

Denyer, S. P., and A. B. Stewart. 1998. Mechanisms of action of disinfectants. *Int. Biodeterior. Biodegradation.* 41:261–268.

Dickson, J. S. 1992. Acetic acid action on beef tissue surfaces contaminated with Salmonella typhimurium. *J. Food S*ci. 57:297–301.

Diebold G, Eidelsburger U 2006. Acidification of diets as an alternative to antibiotic growth promoters. In: D Barug, J de Jong, AK Kies, MWA Verstegen (Eds.) Antimicrobial Growth Promoters. Wageningen Academic Publishers, The Netherlands. pp. 311-327.

Dixon, R. C., and P. B. Hamilton. 1981. Effect of feed ingredients on the antifungal activity of propionic acid. *Poult. Sci.* 60:2407–2411.

Dorsa, W. J. 1997. New and established carcass decontamination procedures commonly used in the beef-processing industry. *J. Food Prot.* 60:1146–1151.

Eckel, B., F. X. Roth, M. Kirchgessner, and U. Eidelsburger. 1992. Zum einfluss von ameisensaure auf die konzentrationen an ammoniak und biogenen aminen im gastrointestinaltrakt. *J. Anim. Physiol. A. Anim. Nutr.* 67:198–205.

Eidelsburger, U., M. Kirchgessner, and F. X. Roth. 1992. Zum einfluss von ameisensaure, calciumformiat und natriumhydrogencarbonat auf pH-wert, trockenmassegehalt, konzentration an carbonsauren und ammoniak in verschiedenen segmenten des gastrointestinaltraktes. *J. Anim. Physiol. A. Anim. Nut*r. 68:20–32.

Engberg, R.M., Hedemann, M.S. and Jensen, B.B. 2002.. The influence of grinding and pelleting of feed on the microbial composition and activity in the digestive tract of broiler chickens. *British Poult Sci.*, 43: 569-579.

Foster, J. W. 1995. Low pH adaptation and the acid tolerance response of Salmonella typhimurium. *Crit. Rev. Microbiol.* 21:215–237.

Foster, J. W. 1999. When protons attack: Microbial strategies of acid adaptation. *Curr. Opin. Microbiol.* 2:170–174.

Franti, C. E., L. M. Julian, H. E. Adler, and A. D. Wiggins. 1972. Antibiotic growth promotion: Effects of zinc bacitracin and oxytetracycline on the digestive, circulatory, and excretory systems of New Hampshire cockerels. *Poult. Sci.* 51:1137–1145.

Fukunaga T M, Sasaki Y, Araki T, Okamoto T, Yasuoka T, Tsujikawa, Fujiyama Y and Bamba T. 2003. Effects of the soluble fibre pectin on intestinal cell proliferation, fecal short chain fatty acid production and microbial population. *Digestion*. 67: 42-49.

Gabert, V. M., and W. C. Sauer. 1995. The effect of fumaric acid and sodium fumarate supplementation to diets for weanling pigs on amino acid digestibility and volatilefatty acid concentrations in ileal digesta. *Anim. Feed Sci. Technol.* 53:243–254.

Garcia V, Catala-Gregori P, Hernandez F, Megias MD and Madrid J. 2007. Effect of Formic Acid and Plant Extracts on Growth, Nutrient Digestibility, Intestine Mucosa Morphology, and Meat Yield of Broilers. *Journal of Applied Poultry Research*. 16: 555–562.

Gauthier R. 2002. Intestinal health, the key to productivity (The case of organic acids) XXVII Convencion ANECA-WPDSA Puerto Vallarta, Jal. Mexico. 30 April 2002.

Gunal M, Yayli G, Kaya O, Karahan N and Sulak O. 2006. The Effects of Antibiotic Growth Promoter, Probiotic or Organic Acid Supplementation on Performance, Intestinal Microflora and Tissue of Broilers. *International* J *Poult. Sci.* 5 (2): 149-155.

Harada, E, Niiyama M, Syuto B. 1986. Comparison of pancreatic exocrine secretion via endogenous secretin by intestinal infusion of hydrochloric acid and monocarb.

Hassan, H. M. A., Mohamed, M. A., Amani, Youssef, W and Hassa, ER. 2010. Effect of Using Organic Acids to Substitute Antibiotic Growth Promoters on Performance and Intestinal Microflora of Broilers, *Asian-Aust. J. Anim. Sci.* 23(10): 1348 – 1353.

Hersey SJ. 1987. Pepsin secretion. In Physiology of the gastrointestinal tract. (Ed. L.R. Johnson). New York: Raven Press. pp. 947-957.

Hinton, A. Jr. , Buhr, R: J. and Ingram, K. D. 2000. Reduction of S*almonella* in the crop of broiler chickens subjected to feed withdrawal. *Poult Sci.*, 79:1566-1570.

Hinton, M., and A. H. Linton. 1988. Control of Salmonella infections in broiler chickens by the acid treatment of their feed. *Vet. Rec.* 123:416–421.

Hume, M. E., Corrier, D. E., Ivie, G. W. and Deloach, J. R. 1993. Metabolism of propionic acid in broiler chicks. *Poult. Sci.* 72:786–793.

Humphrey, T.J. and Lanning, D.G. 1988. The vertical transmission of Salmonella and formic acid treatment of chicken feed. *Epidemiology and Infection*. 100: 43-49.

Izat, A.L., Adams, M.H., Cabel, M.C., Colberg, M., Reiber, M.A., Skinner, J.T. and Waldroup, P.W. 1990b. Effects of formic acid or calcium formate in feed on performance and microbiological characteristics of broilers. *Poult. Sci.* 69: 1876-1882.

Izat, A.L., Tidwell, N.M., Thomas, R.A., Reiber, M.A., Adams, M.H., Colberg, M. and Waldroup, P.W. 1990a. Effects of a buffered propionic acid in diets on the perfrormance of broiler chickens and on microflora in the intestine and carcass. *Poult. Sci.*.69: 818-826.

Kirchgessner, M. and Roth, F.X. 1998. Ergotrope effekte durch organische sauren in der fekelaufzucht und schweinemast. *Ubersichten zur Tiererenährung*. 16: 93-108.

Kwon, Y. M., and S. C. Ricke. 1998a. Induction of acid resistance of Salmonella typhimurium by exposure to short-chain fatty acids. *Appl. Environ. Microbiol.* 64:3458–3463.

Kwon, Y. M., and S. C. Ricke. 1998b. Survival of a Salmonella typhimurium poultry isolate in the presence of propionic acid under aerobic and anaerobic conditions. *Anaerobe* 4:251–256.

Kwon, Y. M., and S. C. Ricke. 1999. Salmonella typhimurium poultry isolate growth response to propionic acid and sodium propionate under aerobic and anaerobic conditions. *Int. Biodeterior. Biodegradation* 43:161–165.

Kwon, Y. M., and S. C. Ricke. 2000. Efficient amplification of multiple transposon-flanking sequences. *J. Microbiol. Methods*. 41:195–199.

Kwon, Y. M., L. F. Kubena, D. J. Nisbet, and S. C. Ricke. 2002. Functional screening of bacterial genome for virulence genes by transposon footprinting. *Methods Enzymol.* 358:141–152.

Kwon, Y. M., S. D. Ha, and S. C. Ricke. 1998. Growth response of Salmonella typhimurium poultry isolate to propionic acid and sodium propionate under aerobic and anaerobic conditions. *J. Food Safety*. 18:139–149.

Kwon, Y. M., S. Y. Park, S. G. Birkhold, and S. C. Ricke. 2000. Induction of resistance of Salmonella typhimurium to environmental stresses by exposure to short-chain fatty acids. *J. Food Sci.* 65:1037–1040.

LeBlay G, Blottiere H M, Ferrier L, LeFoll E C, Bonnet J P, Galmiche and Cherbut C.2000. Short-chain fatty acids induce cytoskeletal and extracellular protein modification associated with modulation of proliferation on primary culture of rat intestinal smooth muscle cells. *Digestive Disease Science*. 45: 1623-1630.

Leyer, G. J., and E. A. Johnson. 1992. Acid adaptation promotes survival of Salmonella spp. in cheese. *Appl. Environ. Microbiol.* 58:2075–2080.

Lin, J., Lee I. S., Frey J., Slonczewski J. L. and Foster J. W. 1995. Comparative analysis of extreme acid survival in Salmonella typhimurium, Shigella flexneri, and *Escherichia coli*. *J Bacteriology*. 177:4097-4104.

Makkink, C. 2001. Acid binding capacity in feed stuffs. *Feed International*. 24: 2729.

Maribo, H., B. B. Jensen and M.S. Hedemann. 2000. Different doses of organic acids to piglets. Danish Bacon and Meat Council, No. 469.

McHan, F. and Shotts, E.B. 1992. Effect of feeding selected short-chain fatty acids on the in vivo attachment of *Salmonella* Typhimurium in chick ceca. *Avian Diseases*, 36, 139-142.

Mead, G.C. 2000. Prospects for competitive exclusion treatment to control *Salmonellas* and other foodborne pathogens in poultry. *Vet. J.*, 159: 111-23.

Moharrery A and Mahzonieh M. 2005. Effect of malic acid on visceral characteristics and coliform counts in small intestine in the broiler and layer chickens. *International J Poult Sci.*. 4 (10): 761-764.

Owens B, Tucker L, Collins M A and McCracken K J . 2008. Effects of different feed additives alone or in combination on broiler performance, gut micro flora and ileal histology. *British Poult Sci.* 49(2): 202-12.

Partanen, K. H. and Z. Mroz. 1999. Organic acids for performance enhancement in pig diets. *Nutrition Res. Reviews* 12: 117-145.

Paster, N. 1979. A commercial scale study of the efficiency of propionic acid and calcium propionate as fungistats in poultry feed. *Poult. Sci.* 58:572–576.

Patten, J. D., and P. W. Waldroup. 1988. Use of organic acids in broiler diets. *Poult. Sci.* 67:1178–1182.

Paul S. K, Halder G, Mondal M. K and Samanta G. 2007. Effect of organic acid salt on the performance and gut health of broiler chicken. *J Poult Sci.*. 44(4): 389-395.

Pirgozliev V, Murphy TC, Owens B, George J and McCann, ME. 2008. Fumaric and sorbic acid as additives in broiler feed. *Research Vet Sci.* 84(3): 387-94.

Rahman, M. S., Howlider1, M. A. R.. Mahiuddin, M and. Rahman, M. M. 2008. effect of supplementation of organic acids on laying performance, body fatness and egg quality of hens. *Bang. J. Anim. Sci.* 37(2): 74 - 81

Ramarao SV, Redddy MR, Raju MVLN and Panda AK. 2004. Growth, nutrient utilisation and immunocompetence in broiler chicken fed probiotic, gut acidifier and antibacterial compounds. *Indian J Poult Sci.*. 39(2): 125-130.

Reddy, V.R. 2004. The Role of Acidifers in Poultry Nutrition. *Avitech Technical Bu*lletin. 1-4.

Ricke, S.C. 2003. Perspectives on the use of organic acids and short chain fatty acids as antimicrobials. *Poult Sci.*. 82:632–639

Runho, R. C., Sakomura, N. K., Kuana, S., Banzatto, D. O. M. Junqueira, and Stringhini, J. H. 1997. Use of an organic acid (fumaric acid) in broiler rations. *R. Bras. Zootec.* 26:1183–1191.

Russell, J. B. 1992. Another explanation for the toxicity of fermentation acids at low pH: Anion accumulation versus uncoupling. *J. Appl. Bacteriol.* 73:363–370.

Salmond, C. V., Kroll, R. G. and Booth, I. R. 1984. The effect of food preservatives on pH homeostasis in Escherichia coli. *J. Gen. Microbiol.* 130:2845–2850.

Samanta, S., Haldar, S and Ghosh, T. K. 2010. Comparative Efficacy of an Organic Acid Blend and Bacitracin Methylene Disalicylate as Growth Promoters in Broiler Chickens: Effects on Performance, Gut Histology, and Small Intestinal Milieu. *Veterinary Medicine International.* pp. 1-8.

Senkoylu, N., Samli, H.E., Kanter, M and Agma, A. 2007. Influence of a combination of formic and propionic acids added to wheat- and barley-based diets on the performance and gut histomorphology of broiler chickens. *Acta Veterinaria Hungarica.* 55(4): 479-90.

Sheikh, A., Banday, T., Bhat, G.A., Mir S., Raquib, M and Shanaz, S. 2011. Response of broiler chicken to dietary Supplementation of organic acids. *J. Central European Agric.* 12(3):498-508.

Skinner. J. T., Izat, A. L and Waldroup, P. W. 1991. Research note: fumaric acid enhances performance of broiler chickens. *Poult. Sci.* 70:1444–1447.

Slonczewski, J. L. 1992. pH-regulated genes in enteric bacteria. *Am. Soc. Microbio*l. *News* 58: 140–144.

Sofos, J.N., Fagerberg, D.J and Quarles, C. L. 1985. Effects of sorbic acid feed fungistat on the intestinal microflora of floor-reared broiler chickens. *Poult Sci.*. 64(5) : 832-40.

Tappenden, K. A and McBurney, M.I. 1998. Systemic short-chain fatty acids rapidly alter gastrointestinal structure, function, and expression of early response genes. *Digestive Diseases and Sciences.* 43: 1526–1536.

Thirumeignanam, D., Swain, R.K., Mohanty, S.P and Pati, P.K. 2006. Effects of dietary supplementation of organic acids on performance of broiler chicken. *Indian J Anim. Nutr.* 23(1): 34-40.

Thompson, J. L., and Hinton, M. 1996. Effect of short-chain fatty acids on the size of enteric bacteria. *Lett. Appl. Microbiol.* 22:408–412.

Thompson, J.L. and Hinton, M. 1997. Antibacterial activity of formic and propionic acids in the diet of hens on *Salmonellas* in the crop. *British Poult Sci.*, *38*, 59-65.

Tung, C. M and Pettigrew, J. E. 2006. Critical review of acidifiers. National Pork Board. http://www.pork.org/Documents/PorkScience/

Vale, M.M., Menten, J. F. M., Morais, S. C. D and Brainer, M.M. D. 2004. Mixture of formic and propionic acid as additives in broiler feeds. *Sci. Agric.* (Piracicaba, Braz.), 61(4): 371-375.

Versteegh, H. A. J and Jongbloed, A. W. 1999. The effect of supplementary lactic acid in diets on the performance of broilers. ID-DLO Rep. No. 99.006. Institute for Animal Science and Health, Branch Runderweg, Lelystad, The Netherlands.

Vieira SL, Oyarzabal OA, Freitas DM, Berres J, Pena JEM, Torres CA and Coneglian JLB. 2008. Performance of Broilers Fed Diets Supplemented with Sanguinarine-Like Alkaloids and Organic Acids. *J. Applied Poult. Res.* 17: 128-133.

Visek, W. J. 1978. The mode of growth promotion by antibiotics. *J. Anim. Sci.* 46:1447–1469.

Vogt, H., and Matthes, S. 1981. Effect of organic acids in rations on the performances of broilers and laying hens. *Arch. Gefluegelkd.* 45:221–232.

Vogt, H., Matthes, S. and Harnisch, S. 1982. Effect of organic acids in rations on the performances of broilers. *Arch. Gefluegelkd.* 46:223–227.

Waldroup A, Kaniawati, S and Mauromoustakos, A. 1995. Performance characteristics and microbiological aspects of broiler fed diets supplemented with organic acids. *J. Food Protection.* 58: 482-489.

8

Significance of Antibiotic Growth Promoters in Animal Feeds

Pankaj Kumar Singh, Kaushal Kumar, Pankaj Kumar and Avinash Kumar

Introduction

After the discovery that low concentrations of antibiotics included in livestock diets had a beneficial effect on production efficiency, the use of antibiotic growth promoters (AGPs) became a common practice, benefiting both the livestock industry and consumers. Antibiotic growth promotion in agricultural and animal production has been practiced for about 50 years (Dibner *et al*., 2007). Orally ingested antibiotics promote growth and efficiency of poultry and other animals. Early indications of a beneficial effect on production efficiency in poultry and swine were reported by Moore *et al*. (1946) and Jukes *et al*. (1950).

Antibiotic Growth Promoter (AGP)

The term "antibiotic growth promoter" is used to describe any medicine that destroys or inhibits bacteria and is administered at a low, subtherapeutic dose. The growth promoting properties of antimicrobial agents in farm animals were discovered in the late 1940s. Trials where fermentation waste from tetracycline production was fed to chickens as a source of vitamin B_{12} revealed that the chickens fed the waste grew more rapidly than did the controls. It was soon found that this effect was not due to the vitamin content of the feed but to residual tetracycline (Stokstad and Jukes, 1949). The practice of feeding sub-therapeutic doses of antimicrobials over long periods of time was readily adopted and soon became an integrated part of the production systems developed in industrialised animal husbandry.

The use of antibiotics for growth promotion has arisen with the intensification of livestock farming. To prevent disease outbreaks and promote growth, low and sub-therapeutic concentrations of antibiotics are often added to the diets of livestock. Antibiotics exert a number of therapeutic effects on the gastrointestinal tract (GIT) of animals, with the majority of these being associated with the microbial population established within. Infectious agents reduce the yield of farmed food animals and to control these infectious agents, the administration of sub-therapeutic antibiotics and antimicrobial agents has been shown to be effective. Apart from increased growth rate and/or increased feed conversion, AGPs at low doses improve egg production in laying hens, and increases litter size in sows.

Mechanism of action of AGP

Following early demonstrations that oral antibiotics do not have growth-promoting effects in germ-free animals (Coates *et al.*, 1955; Coates *et al.*, 1963), studies of the mechanism for growth promotion have focused on interactions between the antibiotic and the gut microbiota. The precise mechanisms of how antibiotics promote growth are not fully understood, but the main effects are focused around the microbiota within the GIT (Gaskins *et al.*, 2002). It has been proven that microbiota does provide real benefits to the monogastric animal such as production of vitamin B complex and protection from pathogenic bacteria through competitive exclusion, but it is often forgotten that these benefits come at a cost. Bedford (2000), Dibner and Richards (2005), and Niewold (2007) proposed that microbiota within the GIT exert their negative effects on animal performance by following mechanism:

- Competing with the host animal for nutrients
- Producing microbial metabolites that suppress growth and increase gut epithelial cell turnover
- Inducing an ongoing immune response which causes a reduction in appetite and an increase in muscle catabolism
- Stimulating inflammatory cells within the GIT
- Causing disease
- Decreasing the ability of the intestine to absorb nutrients.
- By increasing the size of the intestinal tract through the production of certain compounds (such as polyamines and volatile fatty acids) that stimulate its size. The net result is an increases in the energy required to maintain the intestine, thereby leaving less energy available for productive processes, such as muscle growth.

Experiments with germ-free chickens have seemed to indicate that the action of the growth promoters is mediated by their antibacterial effect. Four hypotheses have been proposed to explain their action (Feighner and Dashkevicz, 1987):

i) Nutrients may be protected against bacterial destruction;

ii) Absorption of nutrients may improve because of a thinning of the small intestinal barrier;

iii) The antibiotics may decrease the production of toxins by intestinal bacteria;

iv) There may be a reduction in the incidence of subclinical intestinal infections.

To reduce the competition for nutrients, AGPs are added to prevent excessive growth of pathogenic microbiota; thus, reducing the overall microbial load within the GIT. The best demonstration of this is in the observation that oral antibiotics administered to germ-free animals do not elicit enhanced growth (Coates *et al.*, 1963), while inoculating germ-free animals with GIT bacteria depresses growth (Coates, 1980).

In addition to competition, when pathogenic strains of bacteria increase beyond acceptable concentrations, they can begin to produce toxic compounds that can stimulate an immune response within the host. For example, *Campylobacter jejuni* and *Esherichia coli* have the ability to produce bacterial toxins called 'cytolethal distending toxins', that result in swelling and eventual cell destruction of cultured mammalian cells (Elwell and Dreyfus, 2000; Lara-Tejero and Galan, 2000). Some pathogens also can control nuclear factor kappa B (NF-kB) activation which is a major contributor to the gut inflammatory response (Karin and Ben-Neriah, 2000). Once this immune response is initiated, macrophages and epithelial cells release proinflammatory cytokines and anti-inflammatory cytokines. Upon release of these cytokines, nutrients and energy are shifted away from the growth of the animal towards production of acute phase proteins (Gruys *et al.*, 2006). It has also been shown that a high concentration of ammonia, which is produced from amino acid deamination and urea hydrolysis, depresses growth due to disturbances in the mucosa cycle and increased gut epithelial cell turnover (Pond and Yen, 1987; Visek, 1978). This high rate of cell turnover is accompanied by an increased rate of metabolism and protein synthesis within the gut, which can account for 23 to 36% of the whole body expenditure (Cant *et al.*, 1996).

Thus, the primary reason that AGPs are included in livestock and poultry diets is to reduce pathogen carriage and inhibit subsequent subclinical infections, thus reducing the metabolic costs of immune system activation (Niewold, 2007).

For example, when *Clostridium perfringens*, a bacterium commonly found in the GIT of poultry, grows excessively, a disease called necrotic enteritis can result. The increased growth of this organism leads to toxin production, which, in turn, results in gut mucosal lesions (Parish, 1961). If untreated, the disease leads to intestinal inflammation, diarrhea, increased mortality, growth depression, and liver condemnation at slaughter (Craven, 2000). It can also be associated with other diseases such as gas edema disease, gizzard erosions, and gangrenous dermatitis (Craven, 2000), and manifests into secondary infections during times of coccidiosis infection. Outbreaks of necrotic enteritis and growth of *Clostridium perfringens* are minimized when low concentrations of AGPs are included in the diet, which, in turn, results in unhindered growth of the animal (Collier *et al.*, 2003; George *et al.*, 1982).

Digestion and absorption of nutrients within the GIT also are affected by several physicochemical conditions, including viscosity and pH of the chyme, and these factors may affect the intestinal microbiota and the need for AGPs. Increased intestinal viscosity can contribute to decreased nutrient absorption by the intestine (Jeurissen *et al.*, 2002). When intraluminal viscosity is increased, the thickness of the unstirred water layer covering the mucosa cells increases, thus reducing the rate of nutrient absorption (Johnson and Gee, 1981). Increased digesta viscosity also may interfere with digestion and absorption of nutrients by decreasing their interaction with pancreatic enzymes and bile acids (Edwards *et al.*, 1988), as well as the movement of nutrients towards the mucosal surface (Fengler and Marquardt, 1988). The reduced mixing and passage rate decreases luminal oxygenation, which allows for a more favorable environment for the increased growth of microbial populations due to an increased residence time (Bedford, 1996). Due to the increased viscosity and reduced rate of nutrient absorption, more starch and protein reach the hindgut where they can be fermented and provide energy for microbiota that would normally not have access to them. Increasing the flow of these fermentable substrates provides more opportunities for an increase in pathogenic bacteria. One particular bacterium that thrives in protein-rich environments is *Clostridium perfringens*. The increase in this proteolytic bacterium, as well as the general microbial population, stimulates an increase in mucus production by goblet cells within the GIT. This changing environment can select for mucolytic bacteria and thereby serve as an initiation step in the development of necrotic enteritis (Collier *et al.*, 2003; Jeurissen *et al.*, 2002). This proinflammatory-mediated increase in mucus production combined with an increase in fluid secretion and rate of passage, which characterizes intestinal inflammatory responses, could select for fast-growing pathogenic bacteria (Collier *et al.*, 2003).

The pH of the GIT decreases as the chyme passes from the crop into the proventriculus and gizzard and then becomes less acidic as the digesta progresses down the length of the small intestine (Jeurissen *et al.*, 2002). This variation in pH may be important in terms of nutrient absorption and gut barrier functions, and also in relation to mineral absorption. Langhout (1998) demonstrated an increase in broiler performance and nutrient digestibility due to reduction in gastric pH, and an increase retention time of feed in the gizzard. In addition to the beneficial effect of reduced gastric pH on nutrient digestibility, a reduced pH may also have an antimicrobial effect and limit the growth of microbes in the GIT, particularly pathogenic species.

In order to compensate for the decrease in the ability to absorb nutrients and the increased competition for nutrients with the ever-growing microbial populations, the body increases its production of digestive enzymes, which causes the pancreas to hypertrophy and the intestine to lengthen and increase in weight (Bedford, 2000). Maintaining intestinal health is dependent on the optimal balance between proliferation, maturation, and apoptosis of epithelial cells (Jeurissen *et al.*, 2002); thus, as the intestine increases in length, the rate of enterocyte migration from the intestinal crypts to the villus tip is increased, and the enterocytes are required to become active before they have reached maturity (Silva and Smithard, 1996). The growth of the intestine actually limits enterocyte absorption of nutrients due to the lack of sufficient enzyme activity (Bedford, 2000) and more energy must be diverted away from growth to the replacement of epithelial cells (Tannock, 1997). Also, when the inflammatory response is initiated due to the increasing pathogenic microbial populations, an accumulation of inflammatory cells occurs within the mucosa, leading to a thicker intestinal wall (Niewold, 2007). The addition of AGPs can lead to a reduction in gut weight, including thinner intestinal villi and total gut wall (Coates *et al.*, 1955), which may lead to enhanced nutrient digestibility.

Thus, direct effects of AGP on the microflora can be used to explain decreased competition for nutrients and reduction in microbial metabolites that depress growth (Visek, 1978; Anderson *et al.*, 1999). Additional AGP effects that also occur in germ-free animals include reduction in gut size, including thinner intestinal villi and total gut wall (Coates *et al.*, 1955). This may be due, in part, to the loss of mucosa cell proliferation in the absence of luminal short chain fatty acids derived from microbial fermentation (Frankel *et al.*, 1994). The reduction in gut wall and villus lamina propria has been used to explain the enhanced nutrient digestibility observed with AGP (Jukes *et al.*, 1956; Franti *et al.*, 1972; Anderson *et al.*, 1999). Finally, a reduction in opportunistic pathogens and subclinical infection has also been linked to use of AGP.

Antibiotic Growth Promoters and Animal Production

Antibiotic growth promoters have two major effects:

1. To improve production i.e to improve growth rate and feed conversion efficiency by exerting an antimicrobial affect on commercial bacteria, hence their nutrient sparing effect.
2. As antimicrobials, some AGPs have an effect against pathogenic organisms and can prevent primarily enteric diseases.

According to the National Office of Animal Health (NOAH, 2001), antibiotic growth promoters are used to "help growing animals digest their food more efficiently, get maximum benefit from it and allow them to develop into strong and healthy individuals". It has been estimated that as much as 6 per cent of the net energy in the pig diet could be lost due to microbial fermentation in the intestine (Jensen, 1998). If the microbial population could be better controlled, it is possible that the lost energy could be diverted to growth. There can be no doubt that growth promoters are effective; Prescott & Baggot (1993), however, showed that the effects of growth promoters were much more noticeable in sick animals and those housed in unhygienic conditions.

Several antibiotics have been in use as growth promoters of farm animals ever since. The introduction of these agents coincided with intensive animal rearing. These products improved feed conversion and animal growth and reduced morbidity and mortality due to clinical and subclinical diseases. The average growth improvement was estimated to be between 4 and 8%, and feed utilization was improved by 2 to 5% (Ewing and Cole, 1994). Muramatsu *et al.* (1994) compared the microbial influence on apparent metabolizable energy intake and body weight in germ-free and conventional chicks. They found that when conventional chicks and germ-free chicks were fed a semi-purified diet containing no AGPs, the conventional chicks had a lower body weight (100 vs. 105 g) and had higher energy utilization from the diet (12.4 vs. 11.5 kJ/kg). Miles *et al.* (2006) evaluated the effects of addition of bacitracin methylene disalicylate and virginiamycin on the growth performance and gastrointestinal tract morphology of broiler chickens. Feeding either antibiotics increased body weight and decreased intestinal length and weight as compared with control birds. Dietary addition of virginiamycin resulted in the number of villi per unit length and thinning of muscularis mucosa of ileum. Xia *et al.* (2004) reported increase in villus height and villus height: crypt depth ratio in the small intestine of chicks supplemented with Cu-montmorillonite, probably as a result of fewer *E. coli* and *Clostridium*.

Thus, the use of AGPs affects the animals in many beneficial ways and improves the growth and efficiency of feed utilization by the host animal. Antibiotic growth promoters have been shown (Prescott & Baggot, 1993) to perform best when the animal is in poor health and the living conditions unhygienic. If their local environment is improved, with overcrowding reduced and infection control techniques introduced, then the actual need for growth promoters may be removed.

Impact of antibiotic residue and resistance on human health

However, since the widespread usage of AGPs began, there has been increased concern about the development of antibiotic resistance. The methods that are used in feeding antibiotics to livestock, such as sub-therapeutic dosage, mass treatment during disease outbreaks, and long-term administration, are proposed to elevate the risk for selection of antibiotic-resistant microbes (Mateu and Martin, 2001; Witte, 1998), and this resistance could possibly be transferred to human pathogens (Casewell *et al.*, 2003). Recent consumer concerns about this transmission of antibiotic resistance and political pressures have resulted in the precautionary prohibition of adding AGPs to livestock diets in the European Union. The ban of AGPs in the European Union, has led to a new urgency in the search for possible alternatives, such as new feed additives or modifications (Dibner and Richards, 2005).

Human health can either be affected directly through residues of an antibiotic in meat, which may cause side-effects, or indirectly, through the selection of antibiotic resistance determinants that may spread to a human pathogen. In general, the effect of antibiotic residues in meat is insignificant when compared with the issue of selection and amplification of antibiotic resistant strains of bacteria. A drug that illustrates both potential problems is chloramphenicol. Gassner & Wuethrich (1994) have demonstrated the presence of chloramphenicol metabolites in meat products and have concluded that a link with the presence of these antibiotic residues in meat and the occurrence of aplastic anaemia in humans cannot be ruled out. Over-use of chloramphenicol in animal husbandry is believed to have led to an increase in resistance to the drug in bacteria of the genus *Salmonella*, including *Salmonella typhi*, the causative bacterium of typhoid.

Antibiotic resistance determinants selected in this manner may have various routes by which they may compromise the therapeutic use of antibiotics. Selection may occur in microbes that are pathogenic for humans. Alternatively, resistance may be selected in zoonotic bacteria that subsequently cause human disease. On another level, the resistance determinant may be selected in a bacterium that is a member of the commensal flora of the animal being fed a growth promoter. If such a resistance determinant is mobilisable, it may subsequently transfer to human or animal pathogens. The consequences of selection of resistance can range from prolonged illness and side effects, due to the use of alternative and possibly more toxic drugs to death, following complete treatment failure.

The four types of bacteria most commonly associated with resistance due to use are *Salmonella, Campylobacter*, *Escherichia coli* and the *Enterococci;* these bacteria are likely to be transmitted frequently from animals to humans.

Salmonella sp.

Bacteria of the genus *Salmonella* are responsible for many human diseases. *Salmonella typhi* is the causative agent of the potentially lethal typhoid fever. Other species of *Salmonella* are zoonotic and are usually acquired from contaminated food sources, such as poultry. They commonly cause gastroenteritis. Symptoms can range from mild diarrhoea and nausea to severe vomiting, fever and violent diarrhoea. The most common causative agents of salmonellosis are *Salmonella enterica* var. Enteriditis and *Salmonella enterica* var. Typhimurium and these bacteria are often found as contaminants in poultry and eggs. *Salmonella enterica* var. Typhimurium can usually be treated with fluoroquinolones, chloramphenicol and ampicillin. Spika *et al.* (1987) traced a chloramphenicol-resistant strain of *Salmonella enterica* var. Newport from beef burgers to herds that had been dosed with chloramphenicol. Tacket *et al.* (1985) reported an outbreak of multi-drug resistant *Salmonella enterica* var. Enteriditis following consumption of raw milk.

Using medically important drugs, such as chloramphenicol and tetracycline, as growth promoters would seem to be the most obvious route towards resistant strains that pose a threat to human health, but the selection of resistance is not necessarily that simple. JETACAR (1999) reported a significant correlation between the use of the aminoglycoside apramycin as a growth promoter and the isolation of resistant *Salmonella*, especially *Salmonella enterica* var. Typhimurium DT104, in cattle. Aminoglycoside resistance in these bacteria is due to the acquisition of an acetylating enzyme. This enzyme also confers resistance to gentamicin, an important drug in human medicine.

Campylobacter sp.

Campylobacter, particularly *Campylobacter jejuni* and *C. coli*, is the most common cause of bacterial food poisoning in developed countries, such as the UK and USA. Gastrointestinal disease caused by *Campylobacter* shares many of the clinical symptoms of salmonella infection, including diarrhoea, vomiting and fever. *C. jejuni*, the commonest species to cause of human campylobacter infection, is sensitive to a range of agents, including erythromycin, chloramphenicol, tetracyclines, aminoglycosides and quinolones. Unusually for a Gram-negative bacterium, the agent of choice in infections requiring therapy is the macrolide erythromycin, which is also used as a growth promoter for pigs. Engberg *et al.* (2001) reviewed *in vitro* macrolide and quinolone resistance prevalence and trends in campylobacter isolated from humans, showing a temporal relationship between use of quinolones in food animals and resistant isolates in humans. Endtz *et al.* (1991) reported that the use of fluoroquinolones to treat respiratory diseases in poultry seems to have led to the development of fluoroquinolone-resistant *Campylobacter* in the gut of treated birds.

Escherichia coli

When they are located in the gut, *Escherichia coli* strains are regarded as non-pathogenic, Gram-negative members of the commensal flora of humans and animals. They are, however, the frequent cause of a variety of human infections. Pathogenic strains are most commonly associated with urinary tract infections but strains are also the cause of traveller's diarrhoea. This bacterium is also involved frequently in abdominal infection, such as perforated bowel or appendicitis. It is also one of the most common causes of septicaemia. Some strains like *E. coli* O157, produce Vero cytotoxins and are referred to as Vero-Toxigenic *Escherichia coli* (VTEC) strains. These bacteria may cause haemorrhagic colitis and about 5 per cent of cases progress to the haemolytic uraemic syndrome. The natural reservoir for VTEC strains is the gastrointestinal tract of cattle and possibly other domesticated animals and so these bacteria may be subject to selection pressure from antibiotic growth promoters.

Antibiotic resistance in *E. coli* is widespread globally, with agents such as the penicillins found to be of decreasing efficacy against it (Heritage *et al.,* 2001). LeClerc (1996) warns about the dangers of complacency by reporting high rates of mutation in *E. coli* O157, following observations that they could acquire resistance determinants easily by horizontal gene transfer. It was noted that this was a possible route via which antibiotic resistance, from a pool of environmental pathogens, could be conferred. Even if antibiotic growth-promoters were not directly targeted against the bacteria, it remains possible that strains of this bacterium may acquire resistance from the gut microflora of the food animal. For this reason, and for the sheer volume and severity of disease it causes, it would be sensible not to ignore *E. coli* when considering the risks associated with the use of antibiotic growth-promoters.

Enterococci sp.

The enterococci, such as *Enterococcus faecalis* and *Enterococcus faecium*, are of increasing concern, since they cause illness and death. The enterococci are Gram-positive and so are susceptible to most of the antibiotics used as growth-promoters. Over use of antibiotics in a clinical setting has resulted in the selection of multi-resistant enterococci, including vancomycin-resistant enterococci. Such isolates are resistant to all conventional systemic antimicrobial therapies. Vancomycin-resistant enterococci were first isolated in Europe in the mid-eighties but quickly spread to the USA. Edmond *et al.* (1996) found that patients with blood-borne infections caused by vancomycin-resistant enterococci had over double the mortality of patients who were infected with enterococci susceptible to amoxycillin and vancomycin.

It has been suggested that the use of antibiotic growth-promoters, in particular the drug avoparcin, has contributed to the emergence of vancomycin-resistant enterococci. Both drugs are glycopeptides and the *van* genes found in enterococci confer resistance to both drugs. Use of avoparcin as a growth promoter increases the selective pressure for resistance within the animal. There is thus a risk that, subsequently, resistant bacteria may colonize humans. These resistant bacteria have the potential to cause disease, either in the colonized host or after spreading to another, susceptible host, such as an immunocompromised patient. Alternatively, organisms passing through the gut on meat may be able to transfer resistance genes to the resident microflora, via mobile genetic elements including transposons and plasmids. Khachatourians (1998) reported that avoparcin-resistant bacteria in animals pose a potential threat to humans.

Preventive measures to reduce antibiotic resistance

Regardless of why bacteria become resistant to antibiotics or whether the use of AGPs does, indeed, increase development of resistance, alternative dietary strategies need to be found that can maintain the level of uniform performance seen when using low levels of antibiotics, while limiting the outbreak of disease. It is important that some antibiotics are available for use to treat sick animals. However, we need to limit the ways that antibiotics are used in food production animals. Antibiotics (or similar agents in the same class as antibiotics) that are 'critical' or 'last-line' antibiotics for serious human infection, should not be used in animals or agriculture. There are many serious infections in humans where there are few or, in some cases, no alternate antibiotics that can be used if antibiotic resistance develops to these agents. These can therefore be classified as 'last-resort' or 'critical. There are also many alternatives to these antibiotics that can be used to treat animals successfully that are sick (for example, penicillins, and tetracyclines).

Antibiotic classes that can be classified as 'critical', 'last-resort' or 'reserve'

Class of antibiotic	Examples
Glycopeptides	Vancomycin, Teicoplanin, Avoparcin
3rd and 4th generation cephalosporins	Cefotaxime, Ceftriaxone, Ceftiofur, Cefipime
Anti Pseudomonal Penicillins	Piperacillin, Ticarcillin
Anti tuberculosis drugs	Rifampicin, Isoniazid, Ethambutol, Pyrazinamide
Fluoroquinolones	Ciprofloxacin, Levofloxacin, Enrofloxacin),
Aminoglycosides	Amikacin
Carbapenems	Imipenem, Meropenem
Streptogramins	Synercid, Virginiamycin
Oxazidolones	Linezolid

These 'critical' antibiotics should not be used for therapy or any other purpose in food producing animals. Fluoroquinolones have been approved for use in food production animals in many countries. The use of enrofloxacin has resulted in the development of ciprofloxacin-resistant strains of *Salmonella* spp and *Campylobacter* spp. These resistant bacteria have subsequently caused human infections. When the glycopeptide, avoparcin, was used as a growth promoter in food animals in Europe this resulted in the development and amplification of vancomycin resistant enterococcus (VRE) and subsequent colonisation by a significant percentage of the human population via the food chain (between 2 and 17%). After the ban of avoparcin use in food animals in the EU, the percentage of the general population carrying VRE in their bowel showed a marked reduction (WHO, 2003).

The basic principles we need to follow in order to maintain or facilitate this approach not only now, but also in the future are given below:

- Antibiotics that are 'critical' or 'last-line' antibiotics for serious human infections should not be used in food production animals or agriculture.
- The use of antibiotics for prophylactic purposes in animals should be kept to a minimum. The current usage for this purpose should be significantly reduced. The use of methods (other than antibiotics) to prevent infections should be expanded and developed.

Alternatives to antibiotic growth promoters

The use of AGPs affects the body in many beneficial ways and improves the growth and efficiency of feed utilization by the host animal. The increased public and political pressure to remove AGPs from the diets of livestock creates many new challenges for producers and researchers. Producers will ultimately have to decide upon an alternative(s) that fits well with their nutritional and management practices already in place or that alternative might not be economically viable. Alternatives also need to be found that can maintain the level of uniform performance seen when using low-level antibiotics, while limiting the outbreak of disease. Alternatives of AGPs are probiotic, prebiotic, organic acids etc. (Singh, 2012) In considering phasing out or banning antibiotic growth promoters, the quality of any alternatives, either on the market, that could be developed or that are available illegally, must be assessed. Any replacement for AGP would have to provide an improvement in feed efficiency that is economically viable. Essentially, there are two main ways in which we can reduce our dependence on antibiotic use in animals. An obvious choice is the development of alternatives to antibiotics that work via similar mechanisms, promoting growth whilst enhancing the efficiency of feed conversion. A more difficult route would be to improve animal health.

Future of antibiotics growth promoters

On a global level, a recent joint workshop was held involving the WHO, Food and Agriculture Organization of the United Nations (FAO), and the World Organization for Animal Health (OIE) on nonhuman antimicrobial usage and antimicrobial resistance. The resulting report recommends implementation of the WHO global principles for the containment of antimicrobial resistance in animals intended for food (World Health Organization, 2004). These principles include the withdrawal from food animal production of AGP that are in classes also used to treat human disease unless and until a risk assessment is carried out (World Health Organization, 2000). In addition, the report recommends the implementation on a national level of risk assessment studies and establishment of surveillance programs to monitor AGP use and antimicrobial resistance in bacteria from food animals (WHO, 2004).

Summary

Antibiotic growth promoters improve growth performance and health status in poultry and swine. Antibiotic growth promoters perform best when the animal is in poor health and unhygienic conditions. If their local environment is improved and infection control techniques introduced, then the actual need for growth promoters may be removed. Widespread use of AGPs has resulted in development of antibiotic resistance. Therefore, it is essential that antibiotics should be used wisely and prudently to avoid the development and amplification of resistant bacteria and the genes that encode for this resistance.

References

Anderson, D. B., V. J. McCracken, R. I. Aminov, J. M. Simpson, R. I. Mackie, M. W. A. Vestegen and H. R. Gaskins. 1999. Gut microbiology and growth-promoting antibiotics in swine. *Pig News and Information*. 20:115N-122N.

Bedford, M. R. 1996. Interaction between ingested feed and the digestive system in poultry. *J. Appl. Poult. Res.* 5: 86-95.

Bedford, M. 2000. Removal of antibiotic growth promoters from poultry diets: implications and strategies to minimize subsequent problems. *World Poult. Sci. J.* 56: 347-365.

Casewell, M., C. Friis, E. Marco, P. McMullin, and I. Phillips. 2003. The European ban on growth-promoting antibiotics and emerging consequences for human and animal health. *J Antimicrob. Chemother.* 52: 159-161.

Coates, M. E. 1980. The gut microflora and growth. *In* Growth in Animals. T.L.J. Lawrence (ed). Butterworths, Boston. pp. 175-188.

Coates, M. E., M. K. Davies, and S. K. Kon. 1955. The effect of antibiotics on the intestine of the chick. *Br. J. Nutr.* 9: 110-119.

Coates, M. E., R. Fuller, G.F. Harrison, M. Lev, and S. F. Suffolk. 1963. Comparison of the growth of chicks in the Gustafsson germ-free apparatus and in a conventional environment, with and without dietary supplements of penicillin. *Br. J. Nutr.* 17: 141-151.

Collier, C. T., J. D. van der Klis, B. Deplancke, D. B. Anderson, and H. R. Gaskins. 2003. Effects of tylosin on bacterial mucolysis, *Clostridium perfringens* colonization, and intestinal barrier function in a chick model of necrotic enteritis. *Antimicrob. Agents Chemother.* 47: 3311-3317.

Craven, S. E. 2000. Colonization of the intestinal tract by *Clostridium perfringens* and fecal shedding in diet-stressed and unstressed broiler chickens. *Poult. Sci.* 79: 843-849.

Dibner, J. J., Knight, C., Yi, G.F and Richards, J. D. 2007. Gut Development and Health in the Absence of Antibiotic Growth Promoters. *Asian-Aust. J. Anim. Sci.* 20 (6): 1007-1014.

Dibner, J. J., and J. D. Richards. 2005. Antibiotic growth promoters in agriculture: history and mode of action. *Poult. Sci.* 84: 634-643.

Dibner, J. J., and P. Buttin. 2002. Use of organic acids as a model to study the impact of gut microflora on nutrition and metabolism. *J. Appl. Poult. Res.* 11:453–463.

Dibner, J. 2003. Alimet feed supplement: Value beyond methionine. *Feedstuffs*. 44:12–16.

Edmond, M.B., Ober, J.F., Dawson, J.D., Weinbaum, D.L. and Wenzel, R.P. 1996. Vancomycin-resistant enterococcal bacteraemia: natural history and attributable mortality. *Clinical Infectious Diseases*. 23: 1234-1239.

Edwards, C. A., I. T. Johnson, and N. W. Read. 1988. Do viscous polysaccharides slow absorption by inhibiting diffusion or convection? *Eur. J. Clin. Nutr.* 42: 307-312.

Elwell, C. A., and L. A. Dreyfus. 2000. DNase I homologous residues in CdtB are critical for cytolethal distending toxin-mediated cell cycle arrest. *Mol. Microbiol.* 37: 952-963.

Engberg, J., Aarestrup, F.M., Taylor, D.E., Gerner-Smidt, P., and Nachamkin, I. 2001. Quinolone and macrolide resistance in *Campylobacter jejuni* and *Campylobacter coli*: resistance mechanisms and trends in human isolates. *Emerging Infectious Diseases,* 7: 24.

Fengler, A. I., and R. R. Marquardt. 1988. Water soluble pentosans from rye: II Effects on rate of dialysis and on the retention of nutrients by the chick. *Cereal Chem.* 65: 298-302.

Feighner, S. D., and M. P. Dashkevicz. 1987. Subtherapeutic levels of antibiotics in poultry feeds and their effects on weight gain, feed efficiency, and bacterial cholyltaurine hydrolase activity. *Appl. Environ. Microbiol.* 53:331-336.

Franti, C. E., L. M. Julian, H. E. Adle and A. D. Wiggins. 1972. Antibiotic growth promotion: Effects of zinc bacitracin and oxytetracycline on digestive, circulatory, and excretory systems of New Hampshire cockerels. *Poult. Sci.* 51:1137- 1145.

Frankel, W. L., W. Zhang, A. Singh, D. M. Klurfeld, S. Don, T. Sakata, I. Modlin and J. L. Rombeau. 1994. Mediation of the trophic effects of short-chain fatty acids on the rat jejunum and colon. *Gastroenterol.* 106:375-380.

Gaskins, H. R., C. T. Collier, and D. B. Anderson. 2002. Antibiotics as growth promotants: Mode of action. *Anim. Biotechnol.* 13: 29-42.

Gassner, B. & Wuethrich, A. 1994. Pharmacokinetic and toxicological aspects of the medication of beef-type calves with an oral formulation of chloramphenicol palmitate. *Journal of Veterinary Pharmacology and Therapeutics,* 17: 279-83.

George, B. A., A. M. Ford, D. J. Fagerberg, and C. L. Quarles. 1981. Influence of salinomycin on antimicrobial resistance of coliforms and streptococci from broiler chickens. *Poult. Sci.* 61:1842-1852.

George, B., C. Quarles, and D. Fagerberg. 1982. Virginiamycin effects in controlling necrotic enteritis infections in chickens. *Poult. Sci.* 61: 447-450.

Gruys, E., M. J. Toussaint, T. A. Niewold, S. J. Koopmans, E. van Dijk, and R. H. Meloen. 2006. Monitoring health by values of acute phase proteins. *Acta Histochem.* 108: 229-232.

Heritage, J, Ransome, N., Chambers, P.A. & Wilcox, M.H. 2001. A comparison of culture and PCR to determine the prevalence of ampicillin-resistant bacteria in the faecal flora of general practice patients. *Journal of Antimicrobial Chemotherapy.* 48: 287-289.

JETACAR.1999. The use of antibiotics in food-producing animals: antibiotic-resistant bacteria in animals and humans. Joint Expert Advisory Committee on Antibiotic resistance (JETACAR). Commonwealth Department of Health and Aged Care and the Commonwealth Department of Agriculture, Fisheries and Forestry, Australia.

Jensen, B. B. 1998. The impact of feed additives on the microbial ecology of the gut in young pigs. *J. Anim. Feed Sci.* 7:45–64.

Jeurissen, S. H., F. Lewis, J. D. van der Klis, Z. Mroz, J. M. Rebel, and A. A. ter Huurne. 2002. Parameters and techniques to determine intestinal health of poultry as constituted by immunity, integrity, and functionality. *Intest Microbiol* 3:1-14.

Johnson, I. T., and J. M. Gee. 1981. Effect of gel-forming gums on the intestinal unstirred water layer and sugar transport *in vitro*. *Gut* 22: 332-335.

Jukes, T. H., E. L. R. Stokstad, R. R. Taylor, T. J. Combs, H. M. Edwards and G. B. Meadows. 1950. Growth promoting effect of aureomycin on pigs. *Arch. Biochem.* 26:324-330.

Jukes, T. H., D.C. Hill and H. D. Branion. 1956. Effect of feeding antibiotics on the intestinal tract of the chick. *Poult. Sci.* 35:716-723.

Karin, M., and Y. Ben-Neriah. 2000. Phosphorylation meets ubiquitination: The control of NF-kB activity. *Annu. Rev. Immunol.* 18: 621-663.

Khachatourians, G. 1998. Agricultural use of antibiotics and the evolution and transfer of antibiotic resistant bacteria. *Canadian Medical Association Journal.* 159: 1129-36.

Langhout, D. J. 1998. The role of the intestinal microflora as affected by non-starch polysaccharides in broiler chickens. Ph. D Thesis. Wageningen University, The Netherlands.

Lara-Tejero, M., and J. E. Galan. 2000. A bacterial toxin that controls cell cycle progression as a deoxyribonuclease I-like protein. *Science.* 290: 354-357.

LeClerc, J.E. 1996. High mutation frequencies among *Escherichia coli* and salmonella pathogens. *Science.* 274: 1208-1211.

Mateu, E., and M. Martin. 2001. Why is anti-microbial resistance a veterinary problem as well? *J. Vet. Med. B. Infect. Dis. Vet. Public Health.* 48: 569-581.

Miles, R. D., Butcher, G. D., Henry, P. R., and Littell, R.C. 2006. Effect of Antibiotic Growth Promoters on Broiler Performance, Intestinal Growth Parameters, and Quantitative Morphology. *Poult. Sci.* 85:476–485.

Moore, P. R., A. Evenson, T. D. Luckey, E. McCoy, E. A. Elvehjem, and E. B. Hart. 1946. Use of sulphasuccidine, streptothricin, and streptomycin in nutrition studies with the chick. *J. Biol. Chem.* 165: 437-441.

Muramatsu, T., S. Nakajima, and J. Okamura. 1994. Modification of energy metabolism by the presence of the gut microflora in the chicken. *Br. J. Nutr.* 71: 709-717.

National Office of Animal Health (NOAH). Antibiotics for animals http://www.noah.co.uk/issues/antibiotics.htm .

Niewold, T. A. 2007. The nonantibiotic anti-inflammatory effect of antimicrobial growth promoters, the real mode of action? A hypothesis. *Poult. Sci.* 86: 605-609.

Parish, W. E. 1961. Necrotic enteritis in the fowl. II. Examination of the causal Clostridium welchii. *J.Comp. Pathol*.71: 394-404.

Pond, W. G., and J. T. Yen. 1987. Effect of supplemental carbadox, an antibiotic combination, or clinoptilolite on weight gain and organ weights of growing swine fed maize or rye as the grain sources. *Nutr. Rep. Int.* 35: 801-809.

Prescott, J. F., and J. D. Baggot. 1993. Growth promotion and feed antibiotics, p. 562-568. In J. F. Prescott, and J. D. Baggot (ed.), Antimicrobial therapy in veterinary medicine, 2nd ed. Iowa State University Press, Ames.

Singh, P. K. 2012. Alternatives of Antibiotic Growth Promoters in Poultry Production. LAP Lambert Academic Publishing. AV Akademikerverlag GmbH & Co. KG Deutschland, Germany.

Spika, J.S., Waterman, S.H. and Soo Hoo, G.W. 1987. Chloramphenicol resistant *Salmonella newport* traced through hamburger to dairy farms. *New England Journal of Medicine,* 316: 565-570.

Stokstad, E. L. R. and Jukes, T. H., 1949. 'Further observations on the animal protein factor', *Proceedings of the Society of Biological and Experimental Medicine.* 73: 523–528.

Tannock, G. W. 1997. Influences of the normal microbiota on the animal host. Pages 466-497 in Gastrointestinal Microbiology. 2. Gastrointestinal Microbes and Host Interactions. R. I. Mackie, B. A. White and R. E. Isaacson (ed). Chapman & Hall, New York.

Tacket, C.O., Dominguez, L.B., Fisher, H.J. and Cohen, M.L. 1985. An outbreak of multiple-drug-resistant *Salmonella enteritidis* from raw milk. *Journal of the American Medical Association,* 253: 2058-2060.

Visek, W. J. 1978. Diet and cell growth modulation by ammonia. *Am. J. Clin. Nutr.* 31 (Suppl. 10): S216-S220.

Witte, W. 1998. Medical consequences of antibiotic use in agriculture. *Science.* 279: 996-997.

World Health Organization 2000. 'Global principles for the containment of antimicrobial resistance due to antimicrobial use in animals intended for food', at http://www.who.int/emcdocuments/ zoonoses/ whocdscsraph20004c.html.

World Health Organization. 2003. Impacts of antimicrobial growth promoter termination in Denmark. In: Document No.WHO/CDS/CPE/ZFK/2003.1. Foulum, Denmark. pp. 1-57.

World Health Organization. 2004. Proceedings of the Joint FAO/OIE/WHO expert workshop on non-human antimicrobial usage and antimicrobial resistance: Scientific assessment. Document WHO/CDS/DIP/ZFK/04.20. World Health Organization, Geneva, Switzerland. pp. 1-71.

Xia, M. S., C. H. Hu, and Z. R. Xu. 2004. Effects of copper-bearing montmorillonite on growth performance, digestive enzyme activities, and intestinal microflora and morphology of male broilers. *Poult. Sci.* 83:1868–1875.

9

Uses of Ionophore Antibiotics in Livestock as Feed Additive

S. Shekhar, S.K. Shukla and S.B. Shudhakar

Introduction

An ionophore is a lipid-soluble molecule usually synthesized by microorganisms, which transport ions across cell membranes of susceptible bacteria, dissipating ion gradients and uncoupling energy expenditures from growth to kill these bacteria. The term ionophore was first used in 1967 for their ability of organic molecules to bind metal cations and form lipid soluble complexes that facilitate their transport across cell membrane. The ionophore antibiotics show wide varieties of biological activity ranging from antibacterial, antifungal, antimycoplasma, antiparasitic, antimalarial, antiviral, anti-inflammatory and tumourcell cytotoxic activity (Augustine, *et al.*, 1987; Gumila *et al.*, 1996). Presently seven polyether antibiotics include monensin, lasalocid, salinomycin, narasin, maduramycin, laidomycin and senduramycin are used around the world in various segments of cattle and poultry industries. These ionophores are mainly used as anticoccidial drugs in poultry and growth promoters in the ruminants. Polyether antibiotics are not used in human but recently, it has been shown that some of these compounds are able to selectively kill human cancer stem cells and multidrug resistant cancer cells. Hence, they are recognized as new potential anticancer drugs.

Structure of Polyether ionophores

The polyether ionophores antibiotics came in existence in1951, when two compound nigericin and lasalocid were isolated from different *Streptomyces* species. Since then over fifty microorganisms like, *nocardiopsis, nocardia,* and *actinomadura* etc., were known to produce more than120 naturally occurring ionophores (Dutton *et al.,* 1995; Benno *et al.*, 1988). Monensin previously referred to as monensic acid is produced by *Streptomyces cinnamonensis.* Salinomycin is produced by the fermentation of a *Streptomyces albus*. Narasin is produced by a strain of *Streptomyces aureofaciens.* Lasalocid is produced by a strain of *Streptomyces lasaliensis*. Chemically, the carboxylic acid ionophores appear to be open-chain molecules consisting of an array of heterocyclic ether containing rings. Ionophores molecules are rich in oxygen atoms those are present in various sites in a verity of function groups. They contain one carboxylic group, tetra hydro pyran and tetra hydro furan rings, several hydroxyl groups and ketone group. These groups play an important role in the process of coordination of monovalent and divalent metal cation. However, the involvement of the ionized carboxyl group is not always essential for metal binding. It is required for metal binding by lasalocid, but it is not required by monensin for binding of Na^+ ion. e.g., Lasalocid has a tendency to form dimers, and can form complexes with divalent cations such as Mg^{2+} and Ca^{2+}. Other compound such as Monensin, Salinomycin bind monovalent cations (e.g., Na^+ and K^+).

Classification of Ionophores

Naturally occurring ionophores are divides into four classes, each of which has anti bacterial activity like peptide, cyclic depsipeptide, macrotetrolide, and polyether ionophores. Besides natural products of microorganisms, several chemically modified ionophores prevalent. They belong to a vast group of ionophores, only a subset of which is used as growth promoters or in the prevention of infections in livestock and poultry. On the basis of their transport modes this subset is divided into three major classes:

1. Neutral ionophores
2. Carboxylic ionophores (Poly ether antibiotics)
3. Quasi-ionophores

1. Neutral ionophores

They do not have strong antibacterial activity so they are not used as antibacterial e.g. Valinomycin.

2. Carboxylic ionophores

They have generally incorporated into animal feed as growth promoter. They are sub classified into two groups monovalent and divalent based on their preferential transport of monovalent or divalent cations (Wesley, 1982).

A. Monovalent: e.g. Monensin, Salinomycin, Narasin, and Maduramycin.

B. Divalent: e.g. Lasalocid.

3. Quasi-ionophores

Channel-forming quasi-ionophores form hydrophilic pore in the membrane allowing ion to pass through while avoiding contact with the membrane hydrophobic interiors. e.g., Gramicidin, Polyene, nystatin and amphotericin-B (Watanabe *et al.*, 1981).

Mechanism of action of ionophores

The biological activity of ionophore antibiotics on both prokaryotic and eukaryotic cells is due to their ability to disrupt the flow of ions either into or out of cells. Under normal conditions, cells have a high internal concentration of potassium ions but a low concentration of sodium ions. The concentration of ions in the extracellular medium is just the reverse, high in sodium ions and low in potassium ions, which is essential for normal cell function and is maintained by a specific transport protein (sodium-potassium adenosine tri phosphatase) present in the cell membrane that pumps sodium ions out of the cell in exchange for potassium ions. There are two mechanisms by which ionophores promote the transfer of ions across hydrophobic barriers, ion-ionophore complex formation and ion channel formation. In complex formation, the ion forms a coordination complex with the ionophore in which there is a well-defined ratio (1:1) of ion to ionophore. The ionophore wraps around the ion so that the ion exists in the polar interior of the complex, while the exterior is predominantly hydrophobic in character and as such is soluble in non polar media. The ion is coordinated by oxygen atoms present in the ionophore molecule through ion dipole interactions. The ionophore molecule essentially acts as the solvent for the ion, replacing the aqueous solubility shell that normally surrounds the ion. Some ionophore, valinomycin on move single ion, but ionophores fed to cattle acts as antiporter (Russell and Strobel, 1989). Antiporters' binds protons or metal ions Na^+, K^+ only, uncharged molecules move freely through cell membrane. Metal ion binding is facilitated by the loss of salvation water and the ability of linear molecule to form a donut and shield this charge (Riddell, 2002).Carboxyl group of ionophores monensin, salinomycin, narasin are a monovalent ionophores and Lasalocid is divalent ionophore. Monensin transports Na^+ more efficiently than K^+ (Caffarel-Mendez *et al.*, 1987). Salinomycin and Narasin transports K^+ more efficiently than Na^+. (Droumev, 1983; Caffarel-Mendez *et al.*, 1987). Lasalocid transports bivalent ions such as Ca^{2+} and Mg^{2+} (Watanabe *et al.*, 1981 and Pressman and Fahim, 1982) and it is also an efficient carrier of K^+ ion (Caffarel-Mendez *et al.*, 1987). Monensin present near the surface and its ionization is a pH dependent. Monensin (pKa-7.95) and Lasalocid (pKa-5.8) are more effective when pH is low (Chow and Russell, 1990 and Pressman, 1973). However, if the pH lowers than pKa a large fraction of the ionophore penetrates the cell membrane. Gram positive bacteria appear to be particularly sensitive to the effect of ionophores disrupt normal ion transport. The cell walls of most gram negative bacteria do not permit the penetration of hydrophobic molecules with molecular weights of 600 Dalton and above thus are not susceptible to the action of ionophores (Westley, 1983) and Gram negative microbes have three layers, an outer membrane, a peptidoglycan layer, and a plasma membrane. The outer membrane protects the proton gradient from ionophore activity. Gram positive bacteria lack the protection of an outer membrane, therefore gram positive bacteria do not thrive in an environment containing ionophores. In addition to differences in membrane structure, gram positive and negative bacteria vary in ATP production. Gram negative bacteria depend heavily on fumarate reductase, while gram positive depend on alternative methods.

Ionophores as feed additives

Ionophores are most commonly antimicrobials used in cattle production, yearly sales of ionophores is more than \$150 million. Generally used ionophores are monensin, lasalocid, laidlomycin, salinomycin and narasin to improve feed efficiency and body weight gain (Watanabe *et al.*, 1981). Ionophores not only to increase production of animals, but also make cattle more valuable from a cost of input standpoint. The cost to benefit ratio of ionophore usage has been estimated to save the cattle industry approximately \$1 billion per year. Ionophore specially target on the ruminal bacterial population and alter the rumen microbial environment resulting increase carbon and nitrogen retention by the host animal, increasing production efficiency. Ionophores work in a number of ways to improve growth rate and feed efficiency. Ionophores depress or inhibit the growth of specific rumen microorganisms. This selective inhibition alters rumen fermentation in three major ways:

1. Ionophores improve the efficiency of energy metabolism by changing the types of volatile fatty acids produced in the rumen, increase the VFA propionate, and decrease the VFA acetate (Bergen and Bates, 1984). Propionate is commonly known as the best VFA for cattle to change into

glucose. Propionate has the highest ability to be utilized from feed energy for productive purposes and decreasing energy lost during fermentation of the feed (Russell and Strobel, 1989). Ionophores also reduced methane production and decrease energy loss.

2. Ionophores decrease protein deamination and microbial protein synthesis in the rumen (Morris *et al.,* 1990) and it increases the effectiveness of bypass protein. This has minimal effects on the performance of cattle on high-grain diets, but may have important implications with growing cattle fed high-roughage diets.
3. Ionophores retard abnormal rumen fermentation and reduce the chances of digestive disorders e.g. acidosis, grain bloat, and coccidiosis, reducing these stresses should result in improved animal performance.

Ionophores acts on gram positive bacteria and suppress the growth in rumen but it does not act on gram negative bacteria and promote the gram negative bacteria populations in the rumen (Bergen and Bates, 1984). The gram positive bacteria counts are depleted and gram negative bacteria count increase. Enhancing the proliferation of gram negative bacteria causing an increase presence of fumarate reductase and therefore, increasing the proportion of propionate in the rumen. Propionate is commonly known as the best VFA for cattle to change into glucose and propionate has the highest ability to be utilized feed energy for productive purpose (Russell and Strobel, 1989). Ionophore increase propionate production in 76% and decrease acetate production in 16% and butyrate production in 14% (Perry *et al.,* 1976).

Methane which is eventually eructated, its production causes about 12% losses in feed energy. Supplementation of ionophores in feed cause 30% reduction in energy loss due to methane production (Russell and Strobel, 1989). This reduction is not due to the inhibition of methanogens, but rather a decrease in free protons (Van-Nevel and Demeyer, 1977).

Ionophores decrease protein deamination in the rumen; thereby increase the effectiveness of bypass protein and its production has been increase from 22% to 55%. (Morris *et al.,* 1990; Bergen and Bates, 1984). Ionophore decreases the methanogenesis and therefore decreases in methane production in rumen. Another additional benefit of supplementation of monensin in feed is profound retention of ruminal nitrogen (Russell and Strobel, 1989), that decreases the degradation of amino acid and as a result of this 50% reduction in ammonia accumulation in rumen.This phenomenon has been described as a "protein sparing" effect (Yang and Russell, 1993). Amino acids in the rumen are fermented by several species of ruminal bacteria e.g., *M. elsdenii, Prevotellas* spp. etc, however, the highest specific activities of ammonia production belong to a group of ruminal bacteria i.e. *Peptostreptococcus anerobius, Clostridium aminophilum, Clostridium sticklandi,* these are obligate amino acids fermenting bacteria. These bacteria have capability to deaminating, more than 25% of the protein present in feeds and are very sensitive and their numbers can be reduced tenfold (Callaway, *et al.,* 1997; Krause and Russell, 1996) as a result much of the decrease in ruminal ammonia production.

The extent of improvements observed due to ionophore usage is dependent upon forage quantity and quality available to the animals (Sprott *et al.,* 1988). Supplementation of monensin to cattle on high-roughage diet cause 2 to 10% improvement in live weight gain, 3 to 7% increase in feed conversion efficiency and up to a 6% decrease in food consumption. Others ionophores are generally produce more or less similar effects. Ionophores are readily absorbed from gut, rapidly metabolized from liver, and re-enters the gut from bile. The doses range varies from 6 to 40 ppm in the feed.

Ionophores as methane inhibitor in Ruminants

Ionophore antibiotics such as monensin, typically used to improve efficiency of animal production, are known to decrease methane production (Beauchemin *et al.,* 2008). Ionophores inhibit methanogenesis by lowering the availability of hydrogen and formate; the primary substrates for methanogens bacteria that produce these substrates are sensitive to ionophores, whereas methanogens are more resistant (Chen and Wolin, 1979). Additional evidence for this mechanism is that, in the presence of monensin, methane production by mixed cultures of rumen microbes can be increased by adding hydrogen. Bacteria that reduce succinate to propionate are resistant to ionophores, so propionate production increases. Defaunation by ionophores may also be partly responsible for the effect because protozoa produce hydrogen and are colonized by methanogens (Russell and Strobel, 1989). Similarly rumen fungi which also produce hydrogen, are sensitive to monensin (Marounek and Hodrova, 1989).The effects of ionophores have also been noted in ruminant hindgut fermentation because; approximately 50% of the ingested ionophore passes intact into the hindgut (Davison, 1984, Donoho *et al.,* 1978 and Marounek *et al.,* 1990). Feeding of ionophores changes the site of digestion of dietary carbohydrate fractions. Ruminal digestion of starch may be decreased, but post rumen starch digestion is increased to the extent that total tract digestibility is unchanged (Funk *et al.,* 1986). Fiber digestion is largely unaffected by ionophores (Allen and Harrison, 1979). Increased numbers of ionophore resistant fibrolytic bacteria, such as *F. succinogenes* may offset the reduced numbers of ionophore sensitive ruminococci. Additionally, the somewhat longer rumen retention time caused by ionophores may contribute to maintenance of normal fiber digestion (Lemenager *et al.,* 1978).

Ionophores as anti ruminal acidosis, anti ketosis, anti bloat and anti bovine emphysema

Monensin treatment reduces morbidity and mortality among feed lot animals by reducing the incidence of acute and sub-acute ruminal acidosis, ketosis, bloat, and bovine emphysema (Galyean and Owens, 1988). The rapid increase in milk yield in the transition cow causes a very large increase in energy and nitrogen requirements. These cows are therefore changed from a roughage based diet to a transition diet which is high in concentrates, these contains high content of readily fermentable carbohydrates and resulting decrease in rumen pH due to an increase in the concentration of lactic acid leading to ruminal acidosis (Underwood, 1992a). This condition can be acute or sub acute. Even with this change in diet, transition and high producing dairy cows cause a negative energy balance, this can result in clinical and subclinical ketosis (Sauer *et al.,* 1989). Ionophore has the potential to prevent acidosis and ketosis, the use of monensin in transition dairy cattle could improve health and production of these cows. Monensin inhibits growth of the major lactate producing gram positive bacteria i.e., *Streptococcus* and *Lactobacillus*, but it did not affect any of the major lactate fermenters, i.e., *Anaerovibrio, Megasphera* and *Selenomons*. Monensin inhibits gram positive, but does not affect gram negative bacteria by differences in the dependency on substrate level phosphorylation between these types of bacteria (Bergen and Bates 1984). Normally, the rumen environment is anaerobic with a pH of 6.5, and has a microbial population of protozoa and predominantly gram negative bacteria. Sub acute acidosis has been defined by rumen pH values between 5.2 and 5.6 whereas; values below 5.2 signify acute acidosis (Cooper and Klopfenstein, 1996). These low pH values will reduce fermentation, as the gram-negative cellulolytic and methanogenic bacteria are severely affected when the rumen pH falls below 6.0 (Owens and Goetsch, 1988). The low pH values will also reduce rumen motility and negatively affect the protozoa population in the rumen. Animals respond to this metabolic disorder by reducing dry matter intake, and milk yield (Underwood, 1992a and b). Acidosis can further result in diarrhoea, endotoxin production respiratory collapse and death.

An increased rate of lactic acid production in the rumen leads to an increase in the effective osmotic pressure, resulting in a flow of water from blood to gastro-intestinal tract. Acidosis also causes mucosal damage leading to a deterioration of rumen epithelium tissue, which allows for the systemic invasion of bacteria responsible for liver abscesses (Underwood, 1992a). Ruminal acidosis has been widely reported in feedlot cattle on finishing diets (Cooper and Klopfenstein, 1996), but few reports on the incidence of subclinical and clinical acidosis in dairy cattle. Use of monensin in vitro and in vivo, the relative rates of rumen volatile fatty acid(VFA) production have been consistently altered (Chalupa *et al.*, 1980, Richardson *et al.*, 1976, Prange *et al.*, 1978, Van-Maanen *et al.*, 1978, Rogers and Davis, 1982, and Armentano and Young 1983) as a result of alteration in VFA their will increase in the concentration of propionate and decrease in the concentration of acetate and butyrate (Chen and Wolin, 1979, Bergen and Bates, 1984, Schelling, 1984, Sauer *et al.*, 1989, Weiss and Amiet, 1990). Selection of gram negative bacteria allows enhanced propionate production from succinate because these bacteria have fumarate reductase an enzyme which is necessary to convert fumarate to succinate (Bergen and Bates, 1984).Increase in the net portal flux of propionate with monensin treatment, and after monensin was removed, the net portal flux of propionate and acetate decreased (Harmon *et al.,* 1988).

Bloat is a serious animal production problem that occurs when fermentation gases are retained in the ruminal fluid rather than eructated, often due to an increase in the viscosity of the ruminal fluids, resulting in an accumulation of gas that causes enlargement of the rumen that can asphyxiate the animal (Bartley, et al., 1985; Katz, *et al*., 1986).The anti-bloat effects of monensin are mediated by a direct inhibition of encapsulated (slime producing) bacteria, as well as a decrease in overall ruminal gas production (Galyean and Owens, 1988). Bovine emphysema is another costly problem to the cattle industry that is caused by the eructation of 3-methyl indole (skatole), a by-product of L-tryptophan fermentation, which can be inhaled by the animal, causing asphyxiation. Monensin directly inhibits skatole producing lactobacilli (Honeyfield, *et al.,* 1985).

Ionophore as anticoccidial

Coccidiosis is one of the most common and an important disease that has negative impact on the growth of poultry industry. All the important species of genus *Eimeria* have worldwide distribution. *Eimeria acervulina* and *Eimeria maxima* are the most prevalent, and *Eimeria tenella* is the most common of the highly pathogenic species (Jordan and Pattison 1996). The disease directly influences the production potential of infected chickens due to high mortality, retarded growth and poor feed conversion ratio, causing heavy economic losses up to three billion US dollars annually worldwide (Williams 1999, Dalloul and Lillehoj, 2006). On the other hand, in sub-clinical form, it may render the bird's immune compromised and that paves way to secondary disease conditions (Kabell *et al.,* 2006). Thus management of coccidiosis and maintenance of immune functions for maximum performance, growth and production in poultry industry are primary requirements for profitable farming. Much of the economic loss that is associated with coccidiosis is incurred prior to diagnosis. This makes prevention more important than treatment.The effective use of anticoccidial feed additives over the past 50 years has played a major role in the growth of the poultry industry and has allowed the increased availability of high-quality, affordable poultry products to the consumer. The first drugs used to treat coccidiosis were the sulfonamides, subsequently a wide range of compounds such as clopidol, decoquinate and methylbenzoquate, nicarbazin, toltrazuril and diclazuril, robe-nidine, halofuginone, amprolium and ethopabate have replaced these.

In year 1971, the polyether ionophores antibiotics monensin was first introduces in the USA for the control of coccidiosis in poultry. Today it is most widely used drug in commercial broilers farm for controlling the coccidiosis, beside it is also used to control coccidiosis in game birds, sheep and cattle. Indeed, more animals have been medicated with ionophores for control of disease than any other medicinal agent in the history of veterinary medicine. Since last five decade monensin and other members' ionophores, monensin, narasin, lasalocid, salinomycin, maduramicin and semduramicin have become widely used as anticoccidial drugs particularly in poultry industry. Ionophores affect both extra and intracellular stages of parasite and their principal activity is exerted during the early asexual stage of parasite development. The ionophores can dramatically reduce the mortality, better feed conversion and consumption ratio, more weight gain and lower faecal oocyst count. Monensin is not recommended for layer or chickens over 16 week of age. Combination of monensin with other anticoccidials should be avoided. Monensin activity confined mainly to first generation trophozoites and schizont. The pre slaughter with drawl period is 72 hrs, in chickens. Lasalocid also have similar anticoccidial action like monensin but it transmitted to eggs of hens especially when the treatment continued for one week. The pre slaughter with drawl period is 5 days in chickens. Salinomycin and Narasin are approved mainly for prevention and treatment of coccidiosis in broiler and it has no pre-slaughter withdrawal period. They are active against sporozoites and late asexual stage coccidia in poultry. Maduramicin are more potent than other ionophore antibiotics but it exert adverse effects on growth and feathering when used slightly above than recommended concentration. Semduramicin approved in broiler chicken and it has no pre-slaughter withdrawal period. Among all the coccidiostats developed to date ionophores have proved to be remarkably free from problems of drug resistance, perhaps as a result of their rather non-specific mechanism of action. However, there has been a gradual reduction the sensitivity of coccidia to ionophores indicating that resistant strains may be beginning to emerge and produce residual effect exerting marked cardiovascular effects in increase in coronary flow, indicative of coronary dilatation. It has been estimated that a threshold dose for increased coronary flow in the dog and man, following injection of monensin @ 1.0 mg/ kg (Pressman and Fahim 1983).

Other therapeutic uses of Ionophore

The ionophore antibiotics are active against protozoal infection, including coccidia (Eimeria) and Plasmodium as well as against gram positive organisms and mycoplasmas (Gumila *et al.*, 1997). Monensin controls or prevents swine dysentery caused by *Brachyspira* (formerly known as *Serpulina hyodysenteriae*) and has been proven active against an Enterococcus like pathogen in rainbow trout (Kyriakis, 1989, Carson and Statham, 1993). Lasalocid can be used in the treatment of *Mycoplasma* infections in chickens (Stipkovits *et al.*, 1985). Salinomycin is effective in controlling swine dysentery and porcine intestinal adenomatosis and it can be helpful in controlling *C. perfringens* type A infections in growing pigs (Kyriakis *et al.*, 1995, Kyriakis, 1989). Care should be taken with the dosage of these products increase dose above to recommendation causes impaired growth and performance (Keshavarz and McDougal, 1982). No effect of monensin was observed on the cecal colonization ability of *Salmonella* (Manning *et al.*, 1994) and no resistance selection in coliforms and streptococci could be demonstrated in chickens (George *et al.*, 1981). Monensin inhibits *C. perfringens* (types A and C) in chickens and turkeys, suggesting that it could be used to prevent necrotic enteritis (Stutz and Lawton, 1984). Narasin has is also effective in the treatment and prevention of *C. perfringens* infections in chickens (Elwinger *et. al.,* 1992, Vissiennon *et al.*, 2000). In pigs, salinomycin reduces the lesions and the presence of *Lawsonia intracellularis* (causing proliferative entheropathy) in the intestines in fattening pigs (Kyriakis *et al.*, 1995). Ionophores antibiotics are not used therapeutically in humans.

Ionophore resistance

Several workers found that some ruminal bacteria that are previously sensitive to ionophores are found to be insensitive. *Prevotella ruminicola* is a gram-negative ruminal bacteria that is initially monensin sensitive, but quickly becomes monensin-resistant. Ionophore adapted cells did not bind as much ionophore as did unadapted cells. Ionophore adapted cultures could ferment dipeptides, but lost the ability to ferment tripeptides. Therefore, it was hypothesized that *Prevotella ruminicola* excluded ionophores from the cell membrane by reducing porin size to prevent ionophores reaching the inner membrane (Newbold and Wallace, 1989 and Newbold *et al.,* 1993). *Prevotella bryantii* and other closely related ruminal Prevotella strains can quickly adapt to grow in the presence of ionophores by selecting for a highly resistant subpopulation with altered outer membrane characteristics (Callaway and Russell, 1999). Although gram positive ruminal bacteria are initially quite monensin-sensitive they can also increase their monensin-resistance. Some gram positive ruminal bacteria, *Streptococcus bovis*, *Clostridium aminophilum* and *S. ruminantium* were highly sensitive to monensin upon first exposure, but were able to significantly increase their resistance to monensin following repeated exposures (Callaway, *et al.,* 1999). The resistance acquired by *Clostridium aminophilum* appeared to be mediated by an increase in extracellular polysaccharide that prevented binding of monensin to the cell wall (Rychlik and Russell, 2002). The mechanisms by which resistance is increased in other gram positive species have not been clearly elucidated. Thus it appears that several mechanisms of action can be used by ruminal bacteria to increase their resistance to ionophores. Yet, the observed changes in the rumen appear to be a result of the selection of intrinsically resistant organisms as opposed to the emergence of novel traits. Ionophores are not therapeutic antibiotics, but they are classified as an antibiotic. Antibiotic resistance is an increasing concern in public discourse. However, the increase in ionophore resistant bacteria as a result of its use is not well supported for a number of reasons, ionophores have never been used as antimicrobials for humans, it's have a very different mode of action from therapeutic antibiotics. Ionophore resistance in bacteria seems to be an adaptation rather than a mutation or acquisition of genes (Russell and Houlihan, 2003). It can translocate across cell membranes of animals, which limits their use as therapeutic antibiotics. Ionophore resistance in targeted bacteria shows complexity and a high degree of specificity (Callaway *et al.*, 2003). However, further work is needed to investigate the physiological mechanisms of ionophore resistance.

Ionophore toxicity

Many cases of ionophore poisoning occur in non target species (i.e. species for which ionophore use is not approved). However, poisoning can also occur in the intended species if an animal ingests excessive amounts. Most often this occurs when excessive amounts of ionophores are accidentally added to feeds, or when animals have accidental access to concentrated pre-mix formulations. Horses and rabbits are much more susceptible to ionophore toxicity than are other species. Horses are nearly 20 times more sensitive than cattle and 200 times more sensitive than poultry to monensin toxicity. Ionophore intoxication damages heart and skeletal muscle and has other effects by mechanisms that are not well understood. Probably high concentrations of Ca^{++}

in the cardiac and skeletal muscles are responsible for toxic effect of these drugs. Clinical signs of ionophore poisoning in horses vary depending on the dosage ingested, but can include poor appetite and feed refusal of the grain product, diarrhea, weakness, rapid heart rate, depression, wobbly gait, colic, sweating, recumbency, and sudden death. Animals that recover from sub lethal poisoning can develop chronic heart failure resulting in exercise intolerance, poor performance, and death. However, not all ionophore exposures are necessarily dangerous to horses; risk depends on the amount ingested. If a horse eats a just few mouthfuls of cattle feed containing the approved amount of monensin for cattle (33 ppm) the horse will suffer no adverse effects. Even if the horse eats a small amount of this feed every day for several weeks, the horse might develop only transient anorexia and nothing more serious. However, if a horse ingests several pounds of feed containing very large amounts of monensin, such as 200-300 ppm, this could easily cause death. Monensin @ 1.4 mg per kg of body weight can be fatal to a horse. Minimum toxic dosages for many of the other ionophores have not been well established in horses. Ionophore intoxication is also well known in birds. Not all species of bird are equally sensitive to the toxicity of ionophores. Turkeys, guinea fowl and Japanese quail seem to be more susceptible to monensin intoxication than other birds. Ionophore toxicity causes muscle damage with incordination, leg weakness, diarrhoea, dysponea and reduces feed intake and weight. Birds may have sternal recumbency with neck and limb outstretched.

Ionophore incompatibility

Ionophores are incompatible with several therapeutic antibiotics. It was found that salinomycin and narasin are incompatible with tiamulin, erythromycin, sulfa chlorpyrazine, and sulfa quinoxaline. The effect of incompatibility was shown more markedly with the administration of salinomycin than with narasin. Maduramycin was also shown as incompatible with tiamulin although this interaction is less severe than salinomycin or narasin. It caused a significant loss in body weight gain, without mortality. Ionophores are incompatible with some antioxidants. Salinomycin produce embryo toxicity as they are transported from the laying hen to the egg.

Summary

Ionophores are most commonly used antimicrobial compounds that improve production efficiency and health in cattle by manipulating the composition of the ruminal microbial ecosystem, thereby lessen the incidence of rumen and metabolic related conditions (e.g. rumen acidosis, bloat, bovine emphysema and ketosis). Ionophore antibiotics are also known to decrease methane production by lowering the availability of hydrogen and formate. In poultry, ionophores are most widely used as anticoccidial feed additive to prevent coccidial infection. Therefore, today ionophore appears as integral part of production in dairy, beef and poultry industries due to its potential to increase profitability.

References

Allen, J. D. and Harrison, D. G. 1979. The effect of the dietary addition of monensin upon digestion in the stomachs of sheep. *Proc. Nutr. Soc.* 38:32A.

Armentano, L. E. and Young. J. W. 1983. Production and metabolism of volatile fatty acids, glucose and CO_2 in steers and the effects of monensin on volatile fatty acid kinetics. *J. Nutr.* 113:1265.

Augustine, P. C., Smith II , C. K., Danforth, D. H. and Ruff, D. 1987. Effect of ionophorous anticoccidials on invasion and development of *Eimeria:* comparison of sensitive and resistant isolates and correlation with drug uptake. *Poult. Sci.* 66:960-965.

Bartley, E. E., Nagaraja, T.G. Pressman, E. S., Dayton, A. D., Katz, M. P., and Fina, L.R. 1985. Effects of lasalocid or monensin on legume or grain (feedlot) bloat. *J. Anim. Sci.* 56: 1400-1406.

Beauchemin, K. A., Kreuzer, M., Omara, F. and McAllister, T. A. 2008. Nutritional management for enteric methane abatement: a review *Australian Journal of Experimental Agriculture* 48: 21-27.

Benno, Y., Endo, Shiragami, K. N. and Mitsuoka, T. 1988. Susceptibility of fecal anaerobic bacteria from pigs and chickens to five polyether antibiotics for growth promotion. *J. Vet. Sci.* 50:783-790.

Bergen, W.G. and Bates, D.B. 1984. Ionophores: Their effect on production efficiency and mode of action. *J. Anim. Sci.* 58, 6:1465.

Caffarel-Mendez, S., Demuynck, C. and Jeminet, G. 1987. Etude "in vitro" de quelques antibiotiques ionophores et de certains de leurs dérivés. II. Caractérisation des propriétés ionophores des composés dans un système modèle, pour les ions Na^+ et K^+. *Reprod. Nutr. Dev.* 27: 921-928.

Callaway, T. R., Adams, K. A. and Russell, J. B. 1999. The ability of "low gram-positive" ruminal bacteria to resist monensin and counteract potassium depletion. *Curr. Microbiol.* 39: 226-230.

Callaway, T. R., Edrington ,T. S., Rychilk, J. L., Genovese, K. J., Poole, T. L., Jung, Y. S., ischoff, K. M., Anderson, R. C., and Nisbet, D. J. 2003. "Ionophores: Their Use as Ruminant Growth Promotants and Impact on Food Safety." *Curr. Issues. Intestinal. Microbiol.* 4:43-51.

Carson, J. and Statham, P. 1993. The inhibition by ionophores *in vitro* of an *Enterococcus*-like pathogen of rainbow trout, *Oncorhynchus mykiss*. *Vet. Microbiol.* 36:253-259.

Chalupa, W., Corbett, W. and Brethour, J. R. 1980. Effects of monensin and amicloral on rumen fermentation. *J Anim. Sci.* 51:170.

Chen. M. and Wollin, M. J. 1979. Effect of monensin and lasalocid-sodium on the growth of methanogenic and rumen saccharolytic bacteria. *Appl. Env. Microbiol.* 38:72.

Chow, J. M. and Russell, J. B.1990.Effect of ionophores and pH on the growth of Streptococcus bovis in batch and continuous culture. *Appl. Environ. Microbiol.* 56: 1588-1593.

Cooper, R. and Klopfenstein, T. 1996. Effect of Rumensin- 7 and Feed Intake Variation on Ruminal pH. Update on Rumensin7/Tylan7/Micotil7 for the Professional Feedlot Consultant.

Dalloul, R. A. and Lillehoj, H. S. 2006. Poultry coccidiosis: recent advancements in control measures and vaccine development. *Expert Review of Vaccines,* 5: 143–163.

Davison, K. L. 1984. Monensin absorption and metabolism in calves and chickens. *J. Agric. Food Chem.* 32:1273-1277.

Donoho, A., Manthey, J., Occolowicz, J. and Zornes, L. 1978. Metabolism of monensin in the steer and rat. *J. Agric. Food Chem.* 26:1090-1095.

Droumev, D. 1983. Review of antimicrobial growth promoting agents available. *Vet. Res. Commun.* 7:85-99.

Dutton, C. J., Banks, B. J., and Cooper, C. B. 1995. "Polyether ionophores,"*Natural Product Reports*, vol. 12 (2): 165–181.

Elwinger, K., Schneitz, C., Berndtson, E., Fossum, O., Teglof, B. and Engstrom, B. 1992. Factors affecting the incidence of necrotic enteritis, caecal carriage of *Clostridium perfringens* and bird performance in broiler chickens. *Acta Vet. Scand.* 33:369-378.

Funk, M. A., Galyean, M. L. and Ross, T. T. 1986. Potassium and lasalocid effects on performance and digestion in lambs. *J. Anim. Sci.* 63:685-691.

Galyean, M. L. and Owens, F. N. 1988. Effects of monensin on growth, reproduction, and lactation in ruminants. In: ISI Atlas of Science: *Anim. Plant Sci.* ed. ISI Press, Philadelphia, PA. pp. 71-75.

George, B. A., Ford, A. M.., Fagerberg, D. J. and Quarles, C. L. 1981. Influence of salinomycin on antimicrobial resistance of coliforms and streptococci from broiler chickens. *Poult. Sci.* 61:1842-1852.

Gumila, C., Ancelin, M. L., Delort, A. M., Jeminet, G. and Vial, H. J. 1997. Characterization of the potent in vitro and in vivo anti malarial activities of ionophore compounds. *Antimicrob. Agents. Chemother.* 41:523-529

Gumila, C., Ancelin, M. L., Jeminet, G. A., Delort , M., Miquel, G. and Vial, H. J. 1996. Differential in vitro activities of ionophore compounds against *Plasmodium falciparum* in mammalian cells. *Antimicrob. Agents Chemother.* 40:602-608.

Harmon, D.L., Avery, T.B., Huntington, G. B. and Reynolds, P.J. 1988. Influence of ionophore addition to roughage and high concentrate diets on portal blood flow and net nutrient flux in cattle. *Can. J. Anim. Sci.* 68:419.

Honeyfield, D. C., Carlson, J. R., Nocerini, M. R. and Breeze, R.G. 1985. Duration and inhibition of 3-methylindole production by monensin. *J. Anim. Sci.* 60: 226-231.

Jordan, F.T.W. and Pattison, M. 1996. Poultry Diseases, 4th edition, W.B. Saunders, London. pp: 66–69.

Kabell, S., Handberg, K.J. and Bisgaard, M. 2006. Impact of coccidial infection on vaccine-and IBDV in lymphoid tissues of SPF chickens as detected by RT-PCR. *Acta. Veterinaria. Scandinavica.* 48: 17.

Katz, M. P., Nagaraja, T. G. and Fina, L. R. 1986. Ruminal changes in monensin and lasalocid fed cattle grazing bloat provocative alfalfa pasture. *J. Anim. Sci.* 63: 1246-1257.

Keshavarz, K. and McDouglad, L. B. 1982. Anticoccidial drugs: growth and performance depressing effects in young chickens. *Poult. Sci.* 61:699-705.

Krause, D.O. and Russell, J.B. 1996. An rRNA approach for assessing the role of obligate amino acid fermenting bacteria in ruminal amino acid degradation. *Appl. Environ. Microbiol.* 62: 815-821.

Kyriakis, S.C. 1989. The effect of monensin against swine dysentery. *Br. Vet. J.* 145:373-377.

Kyriakis, S.C., Sarris, K., Kritas, S. K., Saoulidis, K. Tsinas, A. C. and Tsiloyiannis, V. K. 1995. The effect of salinomycine on the control of *Clostridium perfringens* type-A infection in growing pigs. *J. Vet. Med. Ser.* B. 42:355-359.

Lemenager, R. P., Owens, F. N., Shockey, B. J., Lusby, K. S. and Totuscek, R. 1978. Monensin effects on rumen turnover rate, twenty four hour VFA pattern, nitrogen components and cellulose disappearance. *J. Anim. Sci.* 47:255-261.

Manning, J. G., Hargis, B. M., HintonJr., Corrier, A. D. E., Deloach,, J. R. and Creger, C. R.. 1994. Effect of selected antibiotics and anticoccidials on salmonella enteritidis cecal colonization and organ invasion in leghorn chicks. *Avian Dis.* 38:256-261.

Marounek, M., and Hodrova, B. 1989. Susceptibility and resistance of anaerobic rumen fungi to antimicrobial feed additives. *Lett. Appl. Microbiol.* 9:173-175.

Marounek, M., Petr, O. and Machanova, L. 1990. Effect of monensin on in vitro fermentation of maize starch by hind-gut contents of cattle. *J. Agric. Sci.* (Camb.) 115:389-392.

Morris, F. E., Branine, M. E., Galyean, M. L., Hubbert, M. E., Freeman, A.S. and Lofgreen, G.P.1990. Effect of Rotating Monensin Plus Tylosin and Lasalocid on Performance, Ruminal Fermintation, and Site and Extent of Digestion in Feedlot Cattle. *J. Animal Science.* 68: 3069-3078.

Newbold, C. J and Wallace, R. J. 1989. Changes in the rumen bacterium, Bacteroides ruminicola, grown in the presence of the ionophore, tetronasin. Asian Austral. *J. Anim. Sci.* 2:452-453.

Newbold, C. J. Wallace, R. J , and Walker, N. D. 1993. The effect of tetronasin and monensin on fermentation, microbial numbers and the development of ionophore resistant bacteria in the rumen. *J. Appl. Bacteriol.* 75: 129-134.

Owens, F. N. and Goetsch, A. L. 1988. Ruminal fermentation. In: Church, D.C., (Ed.).The ruminant Animal: Digestive Physiology and Nutrition. Prentice-Hall Englewood Cliffs, NJ.

Perry T. W., Beeson, W. M and Mohler, M. T. 1976. Effect of Monensin on Beef Cattle Performanc., *J. Animal Science.* 42: 761-765.

Prange, R. W., Davis, C. L. and Clark, J. H. 1978. Propionate production in the rumen of Holstein steers fed either a control or monensin supplemented diet. *J. Anim. Sci.* 46:1120.

Pressman, B. C. 1973. Properties of ionophores with broad range cation selectivity. *Fed. Proc.* 32: 1689-1703.

Pressman, B. C. and Fahim, M. 1982. Pharmacology and toxicology of the monovalent carboxylic ionophores. *Annu. Rev. Pharmacol. Toxicol.* 22:465-490.

Richardson, L. F., Raun, A. P., Potter, E. L., Cooley , C. O. and Rathmacher, R. P. 1976. Effect of monensin on rumen fermentation in vitro and in vivo. *J. Anim. Sci.* 43:657.

Riddell, F. G. 2002. Structure configuration and mechanism in the membrane transport of alkali metal Ion by ionophoric antibiotics. *Chirality.* 14:121-125.

Rogers, J. A. and Davis, C. L. 1982. Rumen volatile fatty acid production and nutrient utilize-ation in steers fed a diet supplemented with sodium bicarbonate and monensin. *J. Dairy Sci.* 65:944.

Russell, J. B. and Strobel, H. J. 1989. Mini Review: The effect of ionophores on ruminal fermentation. *Appl. Environ. Microbiol.* 55: 1-6.

Russell, J. B. and Houlihan, A. J. 2003. "Ionophore Resistance of Ruminal Bacteria and Its Potential Impact on Human Health." *FEMS Microbiol. Rev.* 27:65-74.

Russell, J. B. and Strobel, H. J. 1989. Effect of ionophores on ruminal fermentation. *Appl. Environ. Microbiol.* 55: 1-6.

Rychlik, J. L. and Russell, J. B. 2002. The adaptation and resistance of Clostridium aminophilum F to the butyrivibriocin-like substance of Butyrivibrio fibrisolvens JL5 and monensin. *FEMS Microbiol. Lett.* 209: 93-98.

Sauer, F. D., Kramer , J. K. G. and Cantwell, W. J. 1989. Antiketogenic effects of monensin in early lactation. *J. Dairy Sci.* 72:436.

Schelling, G. T. 1984. Monensin: Mode of action in the rumen. *J. Anim. Sci.* 58:1518.

Sprott, L. R., Goehring, T. B., Beverly, J. R., and Corah, L. R. 1988. Effects of Ionophores on Cow Herd Production: A Review. *J. Animal Science.* 66: 1340-1346.

Stipkovits, L., Kobulej, T. and Varga, Z. 1985. Efficacy of lasalocid against *Mycoplasma*. *Vet. Bull.* 55: 723.

Stutz, M. W. and Lawton, G. C. 1984. Effects of diet and antimicrobials on growth, feed efficiency, intestinal *Clostridium perfringens*, and ileal weight of broiler chicks. *Poult. Sci.* 63: 2036-2042.

Underwood, J. 1992a. Rumen lactic acidosis. Part I. Epidemiology and pathophysiology. Compend. *Con. Educ. Pract. Vet.* 14 (8): 1127.

Underwood, J. 1992b. Rumen lactic acidosis. Part II. Clinical signs, diagnosis, treatment, and prevention. *Compend. Con. Educ. Pract. Vet.* 14(8):1265.

Van Maanen, R. W., Herbein, J. H., McGilliard, A. D. and Young, J. W. 1978. Effects of monensin on in vivo rumen propionate production and blood glucose kinetics in cattle. *J. Nutr.*108:1002.

Van-Nevel, C. J. and Demeyer, I. 1977. Effect of monensin on rumen metabolism in vitro. *Appl. Environ. Microbiol.* 34:251-257.

Vissiennon, T., Kroger, H., Kohler, T. and Kliche, R. 2000. Effect of avilamycin, tylosin and ionophore anticoccidials on *Clostridium perfringens* enterotoxaemia in chickens. *Berl. Munch. Tierarztl. Wochenschr.* 113:9-13.

Watanabe, K., Watanabe, J., Kuramitsu, S and Maruyama, H. B. 1981. Comparison of the activity of ionophores with other antibacterial agents against anaerobes. *Antimicrob. Agents Chemother*. 19:519-525.

Westley, J. W. 1983.Chemical transformations of polyether antibiotics. *In* J. W. Westley (ed.), Polyether antibiotics: naturally occurring acid ionophores, vol. 2. Chemistry. Marcel Dekker, Inc., New York. pp. 51-87.

Williams, R.B. 1999. A compartmentalized model for the estimation of the cost of coccidiosis to the world's chicken production industry. *International Journal for Parasitology*. 29: 1209–1229.

Yang, M.J. and Russell, J.B. 1993. Effect of monensin on the specific activity of ammonia production by ruminal bacteria and disappearance of amino nitrogen from the rumen. *Appl. Envir. Microbiol.* 59: 3250-3254.

Yokoyama, M.T., Johnson, K.A., Dickerson, P.S. and Bergen, W.G. 1985. Effect of dietary monensin on the cecal fermentation of steers. *J. Anim. Sci.* 61(Suppl. 1): 469 (Abstr.).

10

Immunomodulators as Feed Additives

Indranil Samanta

Introduction

To improve the growth and health status, addition of antibiotics in the poultry feed was a common practice in last decade. Currently, in developed countries it is banned due to fatal consequences like development of antimicrobial resistant organism pool in the gut, persistence of antibiotic residue in the food chain that can reduce the food safety level. Therefore, various alternatives are now a day thought to use as poultry feed additives. Immunomodulaors are presently receiving wide attention as feed additives.

The term 'immunomodulation' is generally used to describe the manipulation of immune system which consists either increase (immunostimulation) or decrease (immunosuppression) in the magnitude of the immune system function. The chemicals or biological substances producing this phenomenon are known as 'immunomodulators'. Specific immunomodulation involves the change in the immune response with a particular antigen (e.g. vaccination). Whereas, non-specific immunomodulation alters the immune response in such a way so that it become sensitive to a wide range of antigens. Generally this kind of non-specific immunomodulators are used as poultry feed additive to increase the resistance against the wide range of infections. The principal components of the immune system targeted for immunomodulation include T cells, B cells, monocytes, macrophages, granulocytes and natural killer cells (Dalloul and Lillehoj, 2005).

Poultry immune system

The immune system is a physiological system, that can discriminate 'self' from 'non-self', thereby protects the host against pathogens. The immune system encompasses heterogeneous subpopulations of immunocompetent cells i.e. T-lymphocytes, B- lymphocytes, natural killer cells, monocytes and antigen presenting cells (dendritic cells) and their products such as immunoglobulin, cytokines, complement proteins which act as effector molecules.

The poultry birds need to survive in an environment full of microbes. After hatching they are exposed to huge numbers of potential pathogens through air, water, feed and contact with other birds. In spite of this exposure the birds rarely suffer from the infection. Innate immunity plays a major role in development of resistance acting as first line of defense after invasion of the pathogen. Innate immune system recognizes the pathogens through the different cellular receptors (pattern recognition receptors, PRR) such as Toll like receptor (TLR) and triggers a cascade of signal transduction to activate phagocytosis and secretion of antimicrobial peptides. The innate immune system is also triggered by Damage-Associated Molecular Patterns (DAMPs), i.e. products of damaged cells and tissues released by necrotic cells and also recognized by PRRs. Recognition causes the liberation of cytokines, which direct the subsequent adaptive immune response. This usually starts at days 4-5 after microbial infections with the early antibody response (IgM), whereas, innate immune reactions offer resistance against the microbial infection from very beginning. Later the innate immunity can also activate the adaptive immune response through the induction of antigen presentation (Kogut, 2009). So, the innate immunity confines the infection in such a level that the birds get sufficient time to develop a specific adaptive immune response.

Importance of immunomodulation

Studies indicate that this innate immune system of poultry can be stimulated by different immunomodulators which is beneficial to them (Genovese *et al.*, 2007). It is important to produce an optimum immune response in birds to resist the variety of infections. It is dependent on environment, nutritional status, age of the bird and the infection status of the bird. Producing maximum immune response may not be optimum. It causes inflammation and metabolic disease syndrome. Whereas, immune response with lesser magnitude may not be able to resist the infection. So, there is a need to stimulate the immune system constantly. For that, there is an expenditure of energy and resources to maintain activated immune cell population and to release antimicrobial peptides in absence of the infection. However, sustained stimulation of innate immune system may produce an unresponsive condition, known as immune-tolerance.

It can prevent the innate immune system to recognize and react the invading pathogens.

Therefore, it is important to know when the immune system of the birds should be modulated to produce the optimum response. During first week of life the birds are most susceptible to the infections due to functional inefficiency of heterophils (bird neutrophils) and macrophages (Van Epps, 2006). In experimental studies it is observed that, feeding of immunomodulators like cytosine-guanosine (He *et al.*, 2005), highly purified β-glucan (Lowry *et al.*, 2005) and antimicrobial peptides (BT) (Kogut *et al.*, 2007) as feed additives to chickens during the first 4 day post hatch can significantly protect the birds against *Salmonella* infection. Chickens fed either the β-glucan or BT-supplemented diet had significantly enhanced heterophil efficacy to phagocytoze and kill the invading *Salmonella.* Another period when modulating the immune response in chickens would be beneficial is during molting. In the commercial egg laying industry molting is required to increase the profitability and longevity of flocks. Different methods of molting in commercial laying hens include feeding different concentrations of minerals (Breeding *et al.*, 1992), or high-fiber and low-energy diets (McReynolds *et al.*, 2006). The standard method is feed deprivation for 12 days. However, feed deprivation negatively affects the innate and acquired immune responses (Holt, 1992). It can be checked by addition of natural products (like Alfalfa) into the diet of layer birds. Alfalfa promotes effective molting without any negative effect on immune system function (Donalson *et al.*, 2005).

Types of Immunomodulators

The categories of immunomodulators used in animals/ poultry are as follows:

I. Vitamin: Vitamin-A, vitamin C, vitamin E etc.

II. Minerals: Zinc, Copper, Iron, Selenium, Magnesium etc.

III. Biological products:

   i) Beta glucan (yeast / mushroom)

   ii) Betaine

   iii) Attenuated avian Poxvirus (PIND-AVI)

   iv) Cytokines : IFN-γ, IL-2, IL-15

IV. Synthetic products: Levamisol, Synthetic Oligonucleotides

V. Herbs:

   i) *Allium sativum* (Lasun, Garlic)

ii) *Aloe vera*

iii) *Andrographis paniculata* (Kalmegh)

iv) *Eclipta alba* (Bhringraj)

v) *Mangifera indica* (Mango)

vi) *Ocinum sanctum* (Tulsi)

vii) *Zingiber officinale* (Ginger)

Role of each immunomodulator in poultry growth and development are discussed as follows:

I. Vitamins

Deficiency of essential nutrients such as vitamins hampers the development of immune system.

(i).Vitamin-A

Vitamin A helps in differentiation of epithelial cells and maintains the integrity of the mucosal surfaces which is essential to prevent the entry of the pathogens (Chew and Park, 2004). Thus it helps in maintaining the function of innate immune system. Vitamin-A deficiency also reduces the population of intraepithelial lymphocytes (IEL), reduction of splenic T-lymphocyte response against mitogen and less secretion of IFN-γ. So, vitamin-A deficiency increases the host susceptibility to different enteric pathogens like *Coccidia* (Dalloul *et al.*, 2002).

(ii). Vitamin-C (Ascorbic acid)

The mammals and birds including the poultry can synthesize ascorbic acid (AA) in the liver or in the kidneys (amphibians, reptiles and birds) using glucose molecules by converting them to L-gulono-lactone and then to ascorbic acid (Padh, 1991). So, generally it is not recommended to supplement ascorbic acid to poultry diets (Mahmoud *et al.*, 2003). However, it is observed that ascorbic acid can improve neutrophil motility to augment the phagocytosis and antimicrobial properties of the neutrophils (Erickson *et al.*, 2000). Ascorbic acid has many other functions like provision of electrons to keep metal ions in their reduced forms, biosynthesis of procollagen and collagen for the formation of extracellular matrix and to act as antioxidant and free radical scavenger to protect cells from oxidative damages (Padh, 1991). Ascorbic acid protects the host cells from inflammatory damages also. As, it inhibits the activation of the oxidant-sensitive transcription factor (NF-κB) which mediates the production of proinflammatory cytokines such as interleukin (IL1) and tumour necrosis factor (TNF-α) (Conner and Grisham, 1996; Erickson *et al.*, 2000). Another beneficial function of ascorbic acid is to protect the tissue detoriation from harmful oxidants generated by activated phagocytes. The tissue damage can be prevented by a process called 'ascorbate recycling'. Ascorbate recycling, induced by both gram-positive and gram-negative bacteria, and pathogenic fungi (*Candida albicans*), occurs when extracellular ascorbate is oxidized, transported as dehydroascorbic acid (DHA) and reduced intracellularly to ascorbate (Wang *et al.*, 1997). This recycling phenomenon does not occur in pathogen (bacteria or fungi) so it may represent a specific eukaryotic defence mechanism against detrimental oxidants. Thus, ascorbic acid has been shown to protect against bacterial and viral diseases and to reduce the impact of detrimental stress in chickens (Lohakare *et al.*, 2005).

(iii) Vitamin-E (α-tocopherol):

Antioxidative property of vitamin-E is related with its immunomodulatory role. If vitamin-E is delivered into the chicken embryo, it can increase the antibody titre in newly hatched chicks (Haq *et al.*, 1996). However, it has no impact on macrophage and lymphocyte population in spleen (Erf *et al*, 1998).

II. Minerals

Several minerals like Zinc, Copper, Iron, Selenium, Magnesium have influence on the immune system.

(i) Zinc

Zinc is a major immunomodulator mineral. It acts as integral part of different enzymes which are essential in cell proliferation and differentiation. So, deficiency of zinc hampers the development of immune cells to produce immunoglobulin, cytokine etc. Studies indicate that, Zinc deficiency causes decreased serum thymulin level ('Thymulin' is a Zinc containing hormone which plays an important role in T-cell maturation, cytokine production and expression of IL-2 receptor on T cells), decreased peripheral T cell count, decreased delayed type of hypersensitivity, decreased natural killer cell activity, decreased macrophage function, decreased phagocytosis and intracellular killing, decreased neutrophil function (oxidative burst) and decreased T-cell dependent antibody production (Goswami *et al.*, 2005). Deficiency of zinc also causes impairment of chemotaxis, phagocytosis and generation of oxidative radicals in neutrophils and monocytes (Keen and Gershwin, 1990). Decreased production of IL-4 is another consequence of Zinc deficiency, hampering the development of Th-2 (T-helper) type of immune response, major antibody mediated immune response against the parasites (Urban *et al.*, 1991).

In poultry, depletion of lymphocytes, degenerative changes in thymus, reduction of lymphoid follicle in bursa of fabricious are commonly observed in Zinc deficiency (Burns, 1983). Enhanced antibody response in the progeny chicks from hen maintained with zinc fortified diet support the immunomodulatory property of this compound (Stahl *et al.*, 1989). Reduction in antibody titre against Newcastle disease and Marek's disease virus, substantially low B cell response against lipopolysaccharide (LPS) and T cell response towards mitogen like Concanavalin-A (Con A) have been reported in zinc deficient chicks (Zhang *et al.*, 1999).

Recently the studies are going on supplementation of zinc in poultry ration from organic sources which are more biologically available to the birds than from the inorganic sources like zinc sulphate and zinc oxide. Addition of inorganic zinc salts in the diet results high level of mineral excretion which is not only wasteful but harmful to the environment. Organic zinc supplementation also improved the immunity in broiler birds (Moghaddam and Jahanian, 2009). The author has personal experience on comparison of supplementation of zinc as zinc proteinate (organic source) and zinc sulfate (inorganic) in broiler ration to measure their immunity level against Newcastle Disease virus. We found better immune response against ND virus when zinc proteinate was added into the broiler diet in comparison to zinc sulfate (Mandal *et al.*, 2011).

(ii) Copper

Addition of copper in optimum level in the poultry ration can stimulate neutrophil killing of the pathogens, phagocytic activity, superoxide dismutase activity and the neutrophil viability (Nockels, 1996). Studies in copper deficient calves showed decreased Ig-M concentration following exposure to Infectious Bovine Rhinotrcaheitis virus (IBR) and *Pasteurella* (a gram negative bacterial poultry pathogen) (Spears, 1988).

(iii) Selenium

Deficiency of selenium is found to produce reduced neutrophil myeloperoxidase activity, required for lysis of the pathogen after phagocytosis (Arthur and Boyne, 1985). In selelnium deficient calves decreased Ig-M titre after primary and secondary inoculation of IBR virus was found (Reffett *et al.*, 1988).

III. Natural Products

(i) Beta Glucan

Beta glucans (β-1, 3) are considered as potent immunomodulator, extracted from the cell wall of yeast, fungi (mushroom), bacteria, algae, oats and barley.

However, *in vitro* and *in vivo* studies in animals and humans show that β-glucans derived from fungi and yeasts only have immunomodulating properties. The appropriate glucanase is not detected in vertebrate body to degrade the beta glucans. So, its half life in the host body is high. It has anti-inflammatory, anti-tumour properties (Tominac *et al.*, 2010).

There are two types of beta glucans i.e. soluble beta glucan (5000-10,000 Daltons) and particulate beta glucan (>10,000 Dalton). Soluble beta glucan acts as beta glucan receptor antagonist which can induce the secretion of IL-6, IL-8, NF-κB. The particulate beta glucan directly stimulates the leukocytes, including their phagocytic, cytotoxic and microbiocidal properties. In addition to that β-glucan stimulates the production of precursor cells in bone marrow with generation of new immunocytes. The increased amount of immunocytes in circulation offers increased protection from potential pathogens. It is important in case of extreme stress (e.g. cancer), when immune system is exhausted by treatments such as irradiation and chemotherapy (Kougias *et al.*, 2001; Hong *et al.*, 2003).

The beta glucan receptor is present in numerous cells like monocytes, macrophages, neutrophils, langerhan cells, NK cells. The receptors include CR3 (Mac-1), scavenger receptor, Dectin-1(Brown and Gordon, 2005). Recently Dectin-1 like receptor is also detected in heterophils of chicken (Nerren and Kogut, 2009).

Several studies show the immunomodulatory roles of beta glucans fed with diet in the broiler chickens (Chen *et al.*, 2008; Revolledo *et al.*, 2009). In a study, purified beta glucan from edible mushroom was fed to one week old broiler chicks @20 mg/kg body weight for 15 days and compared with the control birds without the beta glucan in the diet. Superoxide anion production by neutrophils, lympho-proliferation and IL-2 production was found significantly higher in the treated birds than the control birds. As well as, feeding of beta glucan increased the resistance against New Castle disease in the broiler chicks (Paul *et al.*, 2012). In another study in broilers (Ross 308) we observed the favourable effect of yeast cell wall derivative (glucan) on beneficial bacteria like lactobacilli in the alimentary tract. As a result, colonization of enteric pathogen decreases, feed utilization and feed conversion ratio improves and better live weight of the birds was achieved (Ghosh *et al.*, 2012).

(ii) Betaine

Betaine is a naturally occurring amino acid derivative. It has been investigated as potential immunomodulating agent against coccidiosis in broiler chickens. In *Eimeria* infected chickens, betaine increased the level of duodenal intraepithelial lymphocytes (IEL) and improved the function of the phagocytes (Klasing *et al.*, 2002). Studies indicate that betaine significantly reduced invasion by *Eimeria acervulina* in salinomycin-treated chickens (Allen *et al.*, 1998).

(iii) Attenuated avian Poxvirus (PIND-AVI)

It causes generation of endogenous type I interferon (alpha and beta). Interferons are hormone like proteins having autocrine or paracrine effects in low concentration. Initially it was discovered as antiviral agents (Isaacs and Lindenmann, 1957). Immunomodulation is another major function of interferon detected specially in IFN-α in calf (Abrami *et al.*, 1997).

(iv) Cytokines

The regulatory role of cytokines in directing and controlling immune responses make them an ideal immunomodulator. Initially 'lymphokine cocktails' was used in chickens which increases the number and activity of heterophils and offers protection against challenge (Genovese *et al.*, 1998). This cocktail was generated from the chicken T-cells immunized with *Salmonella enteritidis*. Later, various workers cloned many poultry cytokines in suitable expression vector and used them as immunomodulator, including IL-2 (Tang *et al.*, 2008), interferon-γ (IFN-γ) (Poon *et al.*, 2009), myelomonocytic growth factor (MGF) (Djeraba *et al.*, 2002) and IL-15. Among them, chicken IL-2 and IL-15 are considered as potent immune stimulatory molecules which involve activation, proliferation and differentiation of T cells (CD8+) and Natural killer cells (Poon *et al.*, 2009). Whereas, antiviral and macrophage-activating factor (MAF) activity was detected in recombinant chicken IFN- γ (chIFN- γ). It strongly up-regulates the expression of the genes for MHC class II, guanylate-binding proteins and inducible nitric oxide synthase (iNOS) in appropriate cell systems (Schultz *et al.*, 2004). In addition to that IFN- γ also offered protection from infection after challenge and reduced the weight loss in chickens (Hilton *et al.*, 2002). Chicken myelomonocytic growth factor enhanced the number and functional activity of blood monocytes. However, its resistance property against the challenge is not established.

(IV) Synthetic products

(i) Levamisole

Levamisole was originally synthesized as anti-helmenthic drug (Amery, 1979). It is the first chemically defined compound to show immunoenhancing activity (Renoux and Renoux, 1979). It has thymomimetic properties which cause increased T cell recruitment and activation. Macrophages and granulocytes are also activated by it (Giroud, 1980). Its mode of action depends on activation of cyclic guanosine mono-phosphate (GMP) and release of a thymic hormone-like factor. Studies in India found that that levamisole significantly improved the cell mediated immune response during the early stages of Marek's disease infection in poultry (Narang *et al.*, 2005).

(ii) Short Oligonucleotides

Short oligonucleotides containing unmethylated cytosine-guanosine (CpG) motif can enhance the immune response in domestic animals including chicken (Dalloul *et al.*, 2004). It can activate T-cells with enhanced production of cytokines. Studies are going on with detection of increasing numbers of oligonucleotides having immunomodulatory properties. One of them (CpG2006) is found to increase pro-inflammatory cytokine (IL-6) secretion, enhanced nitric oxide release and most importantly to increase intracellular bacterial lysis (Xie *et al*, 2003). *In ovo* administration of CpG oligonucleotides also enhanced the immunity against *Eimeria* challenge (Dalloul and Lillehoj, 2005).

(V) Herbs

There are several plants which can produce immunostimulation in animals and human. They are safe without side effects, cheap and much more effective. Traditional Indian system of medicines like Ayurveda and Siddha has mentioned several plants having such kind of immunostimulatory role (Atal *et al.*, 1986). We will discuss some of such plants or their parts having immunostimulatory properties.

(i) *Allium sativum* (Lasun, Garlic)

The garlic may augment macrophage and T-lymphocyte activities and enhanced the production of IL-2. It can also protect from ultraviolet induced suppression of contact hypersensitivity (Reeve *et al.*, 1993). We have conducted a study on Yaks regarding the therapeutic application of methanolic extract of Garlic in naturally occurring infectious kerato conjunctivitis, caused by *Moraxella bovis*. We found that at a concentration of 7.5mg/ml methanolic extract of Garlic caused a marked decline (80%) of the viablility of multi drug resistant isolates of *Moraxella bovis* (Bandyopadhyay *et al.*, 2010).

(ii) *Aloe vera*

It is most commonly used medicinal plant specially in arthritis, gout, dermatitis, ulcer and burns (Grindlay and Reynolds, 1986). 'Acemannan' is the major sugar present in the plant extract responsible for increasing respiratory burst, phagocytosis of the macrophages with increased secretion of IL-1 (Duke, 1992).

(iii) *Andrographis paniculata* (Kalmegh)

It can stimulate both the antigen specific and non-specific immune response. Studies indicated significant enhancement in antibody production and delayed type of hypersensitivity response to sheep-RBC in mice (Puri *et al.*, 1993).

(iv) *Eclipta alba* (Bhringraj)

It was found to increase the level of phagocytic index and antibody tire significantly (Upadhyay *et al.*, 2001).

(v) *Mangifera indica* (Mango)

The alcoholioc extract of stem bark stimulated T and B cells to enhance the immunity (Makare *et al.*, 2001).

(vi) *Ocinum sanctum* (Tulsi)

It has potent immunostimulatory properties. A steam distilled extract of the leaves enhanced Ig-E antibody titer against sheep-RBC. Probably this immunostimulatory property is mediated by gamma aminobutyric acid (GABAergic) pathways (Mediratta *et al*, 2002).

(vii) *Zingiber officinale* (Ginger)

Ginger root is widely used as a spice (Larsen *et al.*, 1999) and medical treatment for certain diseases (Tapsell *et al.*, 2006). Ginger contains several active principles such as gingerol, gingerdiol, and gingerdione that possess strong antioxidant activity (Kikuzaki and Nakatani, 1996). The ginger rhizome had antiviral activity (Lee *et al.*, 2008). As well as, the gingerol has been reported as active inhibitor of *M. avium* and *M. tuberculosis in vitro* (Hiserodt *et al.*, 1998). The garlic-derived volatile allyl methyl sulfide (AMS) as a lead compound of volatile garlic metabolites was shown to exhibit an antibacterial effect against the pig pathogen *Actinobacillus pleuropneumoniae (Becker et al.,* 2012). Studies indicated that supplementation of ginger powder at the level of 5 g/kg to diet tended to increase growth rate of broilers and increased carcass yield without affecting feed intake or feed conversion rate. Inclusion of ginger in the diet at this level also enhanced oxidative stability, increased total protein (TP), but lowered cholesterol concentrations in the serum of broilers (Zhang *et al.*, 2009).

References

Abrami, S., Amadori, M., Archetti, I.L and Faccoli, S. 1997. Riduzione dell'impiego di antibiotici nell'allevamento di vitelli a carne bianca mediante la somministrazione di Interferone alfa (IFN-α) umano. *Atti Soc Ital Buiatria*. 29:215-22.

Allen, P. C., H. D. Danforth, and Augustine, P. C. 1998. Dietary modulation of avian coccidiosis. *Int. J. Parasitol.* 28:1131–1140.

Amery, W.K. 1979. Time in chemo-immunotherapy models: An absolute or relative variable? *Int. J. Iinmtaiophannacol.* 1: 65-68.

Arthur, J.R. and Boyne, R. 1985. Superoxide dismutase and glutathione peroxidise activities in neutrophils from selenium and copper deficient cattle. *Life Sci.* 36: 1569-1575.

Atal, C.K., Sharma, M.L., Kaul, A., Khajuria, A. 1986. Immunomodulating agents of plant origin. I : Preliminary screening. *J. Ethonopharmacol.* 18(2): 133-141.

Bandyopadhyay, S., Biswas, T.K., Sasmal, D., Samanta, I., Ghosh, M.K. 2010. Evaluation of methanolic extract of *Allium sativum* and *Saussurea costus* in yaks with infectious keratoconjunctivitis. *Ind. J. Anim. Sci.* 80(3): 199-202.

Becker, P.M., van Wikselaar, P.G., Mul, M.F., Pol, A., Engel, B., Wijdenes, J.W., van der Peet-Schwering, C.M.C., Wisselink, H.J and Zurwieden, N.S. 2012. *Actinobacillus pleuropneumoniae* is impaired by the garlic volatile allyl methyl sulfide (AMS) *in vitro* and in-feed garlic alleviates pleuropneumonia in a pig model. *Vet. Microbiol.* 154 (3-4): 316-324.

Brown G. D. and Gordon S. 2005. Immune recognition of fungal β-glucans. *Cell Microbiol* 7: 471-479.

Burns, R. B. 1983. Antibody production suppressed in the domestic fowl by zinc deficiency. *Avian. Pathol.* 12:141-146.

Chen, K.L., Weng, B.C., Chang, M.T., Liao, Y.H., Chen, T.T and Chu, C. 2008. Direct enhancement of the phagocytic and bactericidal capability of abdominal macrophage of chicks by β-1,3-1,6- Glucan. *Poult. Sci.* 87: 2242-2249.

Chew, B. P and Park, J. S. 2004. Carotenoid action on the immune response. *J. Nutr.* 134:257S – 261S.

Conner, E.M. and Grisham, M.B. 1996. Inflammation, free radicals, and antioxidants. *Nutrition.* 12: 274- 277.

Dalloul, R. A., Lillehoj, H. S., Okamura, M., Xie, H., Min, W., Ding, X and Heckert, R. A. 2004. In vivo effects of CpG oligodeoxynucleotide on Eimeria infection in chickens. *Avian Dis.* 48:783–790.

Dalloul, R. A., Lillehoj, H. S., Shellem, T. A and Doerr, J. A. 2002. Effect of vitamin A deficiency on host intestinal immune response to Eimeria acervulina in broiler chickens. *Poult. Sci.* 81:1509–1515.

Dalloul, R.A. and Lillehoj, H. S. 2005. Recent Advances in immunomodulation and vaccination strategies against coccidiosis. *Avian Dis.* 49: 1-8.

Donalson, L. M., Kim, W. K., Woodward, C. L., Herrera, P., Kubena, L. F., Nisbet, D. J and Ricke, S. C. 2005. Utilizing different ratios of alfalfa and layer ration for molt induction and performance in commercial laying hens. *Poult. Sci.* 84:362–369.

Duke, J.A. 1992. Handbook of biologically active phytochemicals and their activities. Boca Raton, CRC Press Inc.

Erf, G.F., Bottje, W.G., Bersi, T.K., Headrick, M.D and Fritts, C.A. 1998. Effects of dietary vitamin E on the immune system in broilers: altered proportions of CD4 T cells in the thymus and spleen. *Poult. Sci.* 77 (4): 529-537.

Erickson, K.L., Medina, E.A and Hubbard, N.E. 2000. Micronutrients and innate immunity. *J. Infect. Dis.* 182: S5-S10.

Genovese, K.J., Lowry, V.K., Stanker, L.H and Kogut, M. H. 1998. Administration of Salmonella enteritidis-immune lymphokine to day old turkeys by subcutaneous, oral and nasal routes: a comparison of effects of Salmonella enteritidis liver invasion, peripheral blood heterophilia and heterophil action. *Avian Pathol.* 27:597-604.

Genovese, K. J., H. Haiqi, V. K. Lowry, D. J. Nisbet, and Kogut, M. H. 2007. Dynamics of the avian inflammatory response to *Salmonella* following administration of the Toll-like receptor 5 agonist, flagellin. *FEMS Immunol. Med. Microbiol.* 51:112–117.

Ghosh, T.K., Haldar, S., Bedford, M.R., Muthusamy. N and Samanta, I. 2012. Assessment of yeast cell wall as replacements for antibiotic growth promoters in broiler diets: effects on performance, intestinal histo-morphology and humoral immune responses. *J.Anim. Physiol.Anim. Nutri.*(Berl). 96(2): 275-84.

Giroud, J.P., Roch-Arveiller, M., Muntaner, O and Bradshaw, D. 1980. Action comparee du levamisole du muramyldipeptide et de la tuftsine sur le chimiotactisme des polynucleaires du rat. Nouv. *Rev.Fr. Hematol.* 22: 69-76.

Goswami, T. K., Bhar, R., Jadhav, S. E., Joardar, S. N and Ram, G. C. 2005. Role of Dietary Zinc as a Nutritional Immunomodulator. *Asian-Aust. J. Anim. Sci.* 18(3): 439-452.

Grindlay, D. and Reynolds, T. 1986. The *Aloe vera* phenomenon: review of the properties and modern uses of the leaf parenchyma gel. *J. Ethonopharmacol.* 16(2-3): 115-151.

Haq, A., Bailey, C.A and Chinnah, A. 1996. Effect of β-Carotene, Canthaxanthin, Lutein, and Vitamin E on Neonatal Immunity of Chicks When Supplemented in the Broiler Breeder Diets. *Poult. Sci.* 75(9):1092-1097.

He, H., V. K. Lowry, C. L. Swaggerty, P. Ferro, and Kogut, M. H. 2005. In vitro activation of chicken leukocytes and in vivo protection against *Salmonella* Enteritidis organ invasion and peritoneal *S.* Enteritidis infection-induced mortality in neonatal chickens by immunomodulatory CpG oligpdeoxynucleotide. *FEMS Immunol. Med. Microbiol.* 43:81-89.

Hilton, L.S., Bean, A.G, Kimpton, W.G and Lowenthal, J.W. 2002. Interleukin-2 directly induces activation and proliferation of chicken T cells in vivo. *J Interferon Cytokine Res.* 22(7):755-63.

Hiserodt, R. D., Franzblau, S. G and Rosen, R. T. 1998. Isolation of 6-, 8-, and 10-Gingerol from Ginger Rhizome by HPLC and Preliminary Evaluation of Inhibition of *Mycobacterium avium* and *Mycobacterium tuberculosis. Agric. Food Chem.* 46: 2504-2508.

Holt, P. S. 1992. Effect of induced molting on B cell and CT4 and CTS T cell numbers in spleens and peripheral blood of White Leghorn hens. *Poult. Sci.* 71:2027–2034.

Hong, F., Hansen, R. D., Yan, J., Allendorf, D. J., Baran, J. T., Ostroff, G. R and Ross G. D. 2003. Glucan functions as an adjuvant for monoclonal antibody immunotherapy by recruiting tumoricidal granulocytes as killer cells. *Cancer Res.* 63: 9023–9031.

Isaacs, A. and Lindenmann, I. 1957.Virus interference. The interferon. *Proc R. Soc (London) Ser B*; 147:258-73.

Keen, C. and Gershwin, M. 1990. Zinc deficiency and immune function. Ann. Rev. Nutr. 10: 415-431.

Kikuzaki, H., and Nakatani, N. 1996. Cyclic diarylheptanoids from rhizomes of *Zingiber officinale. Phytochemistry.* 43:273–277.

Klasing, K. C., Adler, K. L., Remus, J. C and Calvert, C. C. 2002. Dietary betaine increases intraepithelial lymphocytes in the duodenum of coccidiainfected chicks and increases functional properties of phagocytes. *J. Nutr.* 132: 2274–2282.

Kogut, M. H., Genovese, K. J., He, H., Li, M. A and Jiang, Y. W. 2007. The effects of the BT/TAMUS 2032 cationic peptides on innate immunity and susceptibility of young chickens to extraintestinal *Salmonella enterica* serovar Enteritidis infection. *Int. Immunopharmacol.* 7:912–919.

Kogut, M.H. 2009. Impact of nutrition on the innate immune response to infection in poultry. *J. Appl. Poult. Res.* 18 :111–124.

Kougias P., Wei D., Rice P. J., Ensley H. E., Kalbfl eisch , J. H., Williams D. L and Browder I. W. 2001. Normal human fi broblasts express pattern recognition receptors for fungal (1→3)-β-D glucans. *Infect Immun.* 69: 3933-3938.

Lee, H. S., Lim, S.S., Lim, G J., Lee, J.S., Kim, E.J and Hong, K.J. 2008. Antiviral effect of ingenol and gingerol during HIV-1 replication in MT4 Human T lymphocytes. *Antiviral Res.* 12:34-37.

Lohakare, J.D., Ryu, M.H., Hahn, T.W., Lee, J.K and Chae, B.J. 2005. Effects of supplemental ascorbic acid on the performance and immunity of commercial broilers. *J. Appl. Poult. Res.* 14: 10-19.

Lowry, V. K., Farnell, M. B., Ferro, P. J., Swaggerty, C. L., Bahl, A and Kogut, M. H. 2005. Purified β-glucan as an abiotic feed additive up-regulates the innate immune response in immature chickens against *Salmonella enterica* serovar Enteritidis. *Int. J. Food Microbiol.* 98:309–318.

Mahmoud, K.Z., Edens, F.W., Eisen, E.J and Havenstein, GB. 2003. Effect of ascorbic acid and acute heat exposure on heat shock protein 70 expression by young white Leghorn chickens. *Comp. Biochem. Physiol.*136: 329-335.

Makare, N., Bodhankar, S and Rangari, V. 2001. Immunomodulatory activity of alcoholic extract of *Mangifera indica* L. in mice. *J. Ethonopharmacol.* 78(2-3): 133-137.

Mandal, G.P., Roy, A., Samanta, I and Biswas, P. 2011. Influence of dietary zinc and its sources on growth, body zinc deposition and immunity in broiler chicks. *Ind. J. Anim. Nutr.* 28(4): 433-436.

McReynolds, J. L., Moore, R. W., Kubena, L. F., Byrd, J. A., Woodward, C. L., Nisbet, D. J and Ricke, S.C. 2006. Effect of various combinations of alfalfa and standard layer diet on susceptibility of laying hens to *Salmonella* Enteriditis during forced molt. *Poult. Sci.* 85:1123–1128.

Mediratta, P.K., Sharma, K.K and Singh, S. 2002. Evaluation of immunomodualtory potential of *Ocimum sanctum* seed oil and its possible mechanism of action. *J. Ethonopharmacol.* 80(1): 15-20.

Moghaddam, H.N and Jahanian, R. 2009. Immunological response of broilrr chicks can be modulated by dietary supplementation of Zinc-methionine in place of inorganic Zinc sources. *Asian-Australasian. J. Anim. Sci.* 22: 396-403.

Narang, G., and Pruthi, A. K. 2005. Effect of levamisole along with HVT vaccine on cell mediated immunity against Marek's disease. *Haryana Veterinarian* 44: 17-20.

Nerren, J.R and Kogut, M. H. 2009. The selective Dectin-1 agonist, curdlan, induces an oxidative burst response in chicken heterophils and peripheral blood mononuclear cells. *Vet. Immunol. Immunopathol.* 127:162-166.

Nockels, C.F. 1996. Antioxidants improve cattle immunity following stress. *Anim. Feed Sci. Tech.* 62: 59-68.

Padh, H. 1991. Vitamin C: Newer insights into its biochemical functions. *Nutr. Rev.* 49: 65-70.

Paul, I., Isore, D.P., Joardar, S.N., Samanta, I., Biswas, U., Maiti, T.K., Ganguly, S and Mukhopadhay, S. K. 2012. Orally administered beta glucan of edible mushroom origin upregulates innate immune response in broiler. *Ind. J. Anim. Sci.* 82(7): 745-748.

Puri, A., Saxena, R., Saxena, R.P., Saxena, K.C., Srivastava, V and Tandom, J.S. 1993. Immunostimulant agents from *Andrographis paniculata*. *J. Nat. Prod.* 56(7): 995-999.

Reeve, V.E., Bosnic, M., Rozinova, E and Boehm-Wilcox, C. 1993. A garlic extract protects from ultraviolet B (280-320 nm) radiation induced suppression of contact hypersensitivity. *Photochem Photobiol.* 58(6): 813-817.

Reffett, J. K., Spears, J.W and Brown, T.T. 1988. Effect of dietary selenium on the primary and secondary immune response in calves challenged with IBR virus. *J. Nutr.* 118: 229-235.

Renoux, G and Renoux, M. 1971. Effect immunostimulant d'un imidothiazole sur l'immunisation des souris par Brucella abortus. *C.R. Acad. Sci. (Paris)* 269D: 1467-1469.

Renoux, G and Renoux, M. 1981. Immunologic activity of DTC Potential for cancer therapy. In: *AtrgnientingAgents in Cancer Therapy,* EM Hersh, MA Chirigos, and MJ Mastrangelo (eds). Raven Press, New York, pp. 427-440.

Revolledo, L., Ferreira, C.S.A and Ferreira, A. J. P. 2009. Prevention of *Salmonella* Typhimurium colonization and organ invasion by combination treatment in broiler chicks. *Poult. Sci.* 88:734-743.

Schultz, U., Kaspers, B and Staeheli, P. 2004. The interferon system of non-mammalian vertebrates. *Dev Comp Immunol.* 28(5):499-508.

Spears, J.W., Harvey, R.W and Brown, T.T. 1988. Trace minerals and the immune system. Proc of Symposium on Trace minerals in beef cattle nutrition. Texas A & M Research and Extension Centre, Amarillo, TX, p.1-6.

Stahl, J. L., Cook, M. E., Sunde, M. L and Greeger, J. L. 1989. Enhanced humoral immunity in progeny chicks from hens fed practical diets supplemented with zinc. *Appl. Agric. Res.* 4(2):86-89.

Tang, M., Wang, H., Zhou, S., Tian, G. 2008. Enhancement of the immunogenicity of an infectious bronchitis virus DNA vaccine by a bicistronic plasmid encoding nucleocapsid protein and interleukin-2. *J Virol Methods*. 149(1):42-8.

Tapsell, L. C., Hemphill, I., Cobiac, L., Patch, C. S., Sullivan, D. R., Fenech, Roodenrys, M., S., Keogh, J. B., Clifton, P. M., Williams, P. G., Fazio, V. A and Inge, K. E. 2006. Health benefits of herbs and spices: The past, the present, the future. *Med. J. Aust.* 185:4-24.

Tominac, V.P., Krpan, V.Z., Grba, S., Sreèec, S., Krbavèiæ, I.P and Vidoviæ, L. 2010. Biological effects of Yeast β-Glucans. *Agri. Conspectus Scientificus*. 75 (4): 149-158.

Upadhyay, R.K., Pandey, M.B., Jha, R.N and Pandey, V.B. 2001. Eclalbatin, a triterpene saponin from *Eclipta alba*. *J. Asian Nat. Prod. Res*. 3(3): 213-217.

Urban, J. F., Katona, I. M., Paul, W. E and Frinkelman, F. D. 1991. Interleukin-4 is important in protective immunity in gastrointestinal nematodes of mice. *Proc. Natl. Acad. Sci. USA*. 88:3513-3517.

Van Epps, H. L. 2006. Ignoring endotoxin. *J. Exp. Med.* 203:1137.

Wang, Y., Russo, T.A., Kwon, O., Chanock, S., Rumsey, S.C and Levine, M. 1997. Ascorbate recycling in human neutrophils: Induction by bacteria. *Proc. Natl. Acad. Sci.* U.S.A. 94: 13816-13819.

Xie, H., Raybourne, R. B., Babu, U. S., Lillehoj, H. S and Heckert, R. A. 2003. CpG-induced immunomodulation and intracellular bacterial killing in a chicken macrophage cell line. *Dev. Comp. Immunol*. 27: 823– 834.

Zhang, R., Zhou, Y., Huang, Y and Yang, H. 1999. The modulation effects of zinc on immune organs development and function in broiler. *Acta Veterinaria Zootechnia Sinica*. 30: 504-512.

Zhang , G. F., Yang , Z. B., Wang, Y., Yang , W. R., Jiang, S. Z. and Gai, G. S. 2009. Effects of ginger root (*Zingiber officinale*) processed to different particle sizes on growth performance, antioxidant status, and serum metabolites of broiler chickens. *Poul. Sci.* 88: 2159–2166.

11

Hen Egg Antibody as a Feed Additive for Oral Immunotherapy

Rajni Kumari, Sanjay Kumar, Shanker Dayal, S.V. Lal, Kaushalendra Kumar and A.K Srivastava

Introduction

There is concern that antibiotic-resistance in bacteria may make commonly used antibiotics less effective. Oral immunotherapy (passive immunization) with specific antibodies is a strategy that has been actively pursued in laboratory and clinical studies for the last two decades. Feeding of egg yolk antibodies to neutralize specific pathogens, especially enteric microorganisms, is a potential alternative to antibiotics. Oral administration of these antibodies has met with some degree of success in prevention of viral and bacterial enteric infections in humans, piglets, calves, fish and rabbits. It is proposed that the egg yolk antibodies may act against enteric pathogens by binding, immobilising and consequently reducing or inhibiting the growth, replication, or colony forming abilities of these pathogens. Three immunoglobulin classes (IgA, IgM and IgY) have been shown to exist in chickens. Chicken immunoglobulin G (IgG) has been designated as immunoglobulin Y(IgY) because it varies in several aspects from mammalian IgG. IgY is the main serum immunoglobulin in chickens. It is transported from the hen to the embryo via the egg yolk. The egg yolk thus contains high concentrations of IgY. Other immunoglobulin (Ig) classes are present only in negligible amounts in the egg yolk. It has been shown that the presence of immunoglobulins in eggs is an example of passive immunity. This is because these antibodies originated from the hen and are used to protect the offspring from various infectious diseases after hatch.

The acquisition of passive immunity in birds was first noted in 1893 when Klemperer demonstrated the transfer of immunity to tetanus toxin from hen to chick. The amounts of IgG in yolk have been reported to be 20-25 mg/ml in the hen's egg. A laying hen can produce approximately 300 eggs annually and the volume of one egg yolk is approximately 15 ml. This could supply close to 100g of antibody per hen per year. It has been demonstrated that egg yolk weight and the percent of hen-day production in laying hens may affect efficiency of IgY production. The IgY concentration in the egg yolk is an important parameter for commercial production of antibodies. It has been shown that the IgY concentration varies among different lines and among individual hens. This suggests the possibility of increasing IgY production by choosing high-producing lines and by genetic selection within the lines. Egg yolk antibodies are stable over time; even when storing IgY for 5-10 yr at 4°C there is no significant loss in antibody activity. These antibodies also retain their activity after 6 months at room temperature or 1 month at 37°C.

Production of immunoglobulin Y (IgY)

In order to produce these antibodies, hens are exposed (usually injected) to particular antigens which induce immune responses, including the production of antibodies. Normally, these antibodies are then transferred to the egg yolk. Booster immunization (second exposure) is usually given at a later time to ensure continued transfer of antibodies from hen to the egg yolk. These antibodies are then extracted from the egg yolk and processed to be administered directly to the animal or included in the feed. In the last few years, several commercial sources of egg yolk antibodies have become available. The production of large amounts of IgY in a cost-effective manner is the key to its successful use for passive immunization. To this end, several aspects of hen immunization have been studied in order to improve IgY production and yolk deposition, including immunization route, vaccine adjuvant, and type of antigen (Levesque *et al.,*2007).

The most common injection route is the intramuscular route. Chang *et al.* (1999) demonstrated that intramuscular immunization resulted in higher levels of specific IgY when compared to antigen injected subcutaneously. Oil-based adjuvants, such as Freund's complete adjuvant (FCA) remain the most effective adjuvants for antibody production: however, FCA has been associated with potentially severe injection site reactions (Wanke *et al.,* 1996). Alternatively, the addition of immunostimulating components to vaccine formulations can markedly increase antibody levels in the yolk and utilize less antigen, thereby, being more cost-effective (Herath *et al.,* 2010, Levesque *et al.,* 2007, Trott *et al.,* 2008). Adding whole-killed *Streptococcus suis* or *Staphylococcus aureus* was found to increase IgY levels against phospholipase A2 (PLA2) by 51% or 62%, respectively (Trott *et al.,* 2008). Moreover, Levesque *et al.,* (2007) found that adding synthetic oligonucleotides containing unmethylated CpG dinucleotides, characteristic of bacterial DNA, to Freund's incomplete adjuvant (FIA) increased specific IgY production by up to 480%. The use of DNA vaccines has also been described for IgY production (Cova 2005). This process involves immunizing birds with plasmid DNA encoding the antigen of interest and climinates the potentially tedious and costly process of purifying antigens.

Although, the amount of IgY deposited into the yolk varies depending on several factors, including the age, breed of chicken, and antigen used, IgY yields have been reported to range from 60 to 150 mg IgY per egg (Cook and Trott 2010; Pauly *et al.,* 2009). Given that a typical hen can lay approximately 325 eggs per year, this can result in a potential yield of around 20-40 g of IgY per year (Pauly *et al.,* 2009), of which 2% to 10% is antigen specific (Schade *et al.,* 1991; Tini *et al.,* 2002). Pouly *et al.* (2009) monitored IgY levels in chickens over a two-year period and found that although the laying capacity decreased in the second year, the was compensated for by a greater although the laying capacity decreased in the second year, this was compensated for by a greater amount of total IgY per egg. Similar results were observed by Trott *et al.,* (2009), who found that older hens had higher IgY titers compared with younger hens. One of the major challenges in IgY purification is separating the water soluble IgY from the yolk lipoproteins (Hattra *et al.,* 2008), and a number of methods have been reported that result in different yields and purities. This typically involves isolating the IgY containing water soluble fraction, followed by additional purification steps. Dilution of the yolk with water, which results in the aggregation of yolk lipoproteins at low ionic strength followed by centrifugation or ultrafiltration, has been reported (Akita & Nakai 1992), Kim and Nakai, 1998). Likewise, freezing and thawing of diluted yolk, producing lipid aggregates that are large enough to be removed by conventional low speed centrifugation, have also been used (Jensenius and Koch 1993), resulting in a purity of approximately 70%. For dilution methods, pH and extent of dilution are very important for optimal IgY recovery, and Nakai *et al.,* (1994) found that the best results were obtained using a six-fold water dilution, at pH 5.0.

Other methods of removing lipoproteins prior to IgY purification include organic solvent delipidation (Horikoshi *et al.,* 1993; Kawan *et al.,* 1991; Polson 1990) and use of lipoprotein coagulating agents, such as polyethylene glycol (Akita & Nakai 1993, Svendsen *et al.,* 1995) ordextran sulfate (Jensenius *et al.,* 1981); however, application of these methods for large scale production of IgY for passive immunization are limited by problems related to safety as well as cost constraints (Hatta *et al.,* 2008).

Properties of IgY

Despite sharing a similar biological function, there are some striking differences in the structures and immunoreactivities of IgY and IgG. IgY is composed of two identical heavy (H) and two identical light (L) chain linked by a disulfide bridge and has molecular mass of ~180 kDa, larger than mammalian IgG (~ 150 kDa). The light chain of both antibodies consists of one variable domain (VL) and one constant (CL) domain. The heavy chain of IgY consists of one variable domain (Vh) and four constant domains (CH1, CH2, CH3 and CH4), in contrast to IgG, which has only three constant domains, and IgY has a shorter and less flexible hing region (Warr *et al.,* 1995). Both antibodies contain Asn-linked oligosaccharides; however, IgY contains high-mannose-type oligosaccharides, which differ from those of IgG (Taylor *et al.,* 2009). In addition, the b-sheet content of IgY has been reported to be lower, suggesting that the conformation of IgY is more disordered and therefore less stable than IgG (Shimizu *et al.,* 1992). The isoelectric point of IgY is in the range of 5.7 to 7.6 and is lower than that of IgG and it has been suggested that IgY may be more hydrophobic, which wound correspond to the lipid rich environment of the yolk (Davalos-Pantoja *et al.,* 2000).

The stability of IgY under various processing and physiological conditions is an important consideration for passive immunotherapy applications. The activity of IgY has been shown to decrease with increasing temperature and time. Minimal loss of activity was observed following heating between 60^{0}C and 65^{0}C but was decreased markedly by heating for 15 min at 70^{0}C or higher (Shimizu *et al.,* 1988). However, addition of carbohydrates has been found to improve its heat resistance (Shimizu *et al.,* 1994). Likewise, it has been shown that the addition of disaccharides or complex carbohydrates to IgY prior to freeze drying can improve stability during storage (Nilsson and Larsson 2007).

IgY is relatively stable between pH 4 and pH 11 but displays a rapid reduction in activity above pH 12. At pH 3.5, IgY activity decreases and is almost completely lost at pH 3 because of rapid conformational changes (Shimizu *et al.,* 1988,). However, pH stability may be improved by the addition of stabilizers, such as sugars, complex carbohydrates, and polyols (Change *et al.,* 2000). As with any protein, antibodies are susceptible to proteolytic digestion. The sensitivity of IgY to pepsin digestion is highly dependent on the pH and enzyme to substrate ratio used. Shimizn *et al.,* (1988) found that antibody activity was completely lost following pepsin digestion below pH 4.5 at an enzyme to substrate ratio of 1:20. Using similar conditions, Hatta *et al.* (1993) found that 63% of IgY activity remained even after 4h when pepsin was used at a enzyme to substrate ratio of 1:200; however, activity was rapidly lost using this ratio at pH 2. Following trypsin and chymotrypsin digestion, IgY retained 39% and 41% of its activity, respectively, after 8 h of digestion, despite the apparent production of peptides (Shimize *et al.,* 1988).

Gastric pH, enzyme levels, and gastric transit times can vary depending on a number of factors, including age and health status, thus inactivation of IgY during digestive processes may be a major concern for oral immunotherapy applications of IgY. Because gastric content and diet can also affect these parameters, it should be noted that the matrix in which IgY is administered (e.g., purified antibody varsus whole egg yolk) may also affect antibody stability (Cook and Trott, 2010; Jaradat and Marquardt, 2000). A number of studies have looked at the survival of orally administered antibodies from different species (i.e., bovine and avian sources) following passage through the gastrointestinal tract in both humars and animals, with varying results, Roos *et al.* (1995) found that 19% of orally ingested bovine IgG was still active following digestion in humans, whereas other studies have reported that up to 50%, of orally administered antibodies could be recovered in adult or infant stool (reviewed in Reilly *et al.,* 1997).

In order to prevent inactivation of IgY following oral administration, a number of encapsulation techniques have been examined. Multiple emulsions (Shimizu and Nakane 1995) and microencapsulation of IgY using liposmes (Shimizu *et al.,* 1993) have previously been described; however, these have shown limited effectiveness. The macro encapsulation of IgY using enteric-coated gelatin capsules has also been examined and was found to significantly improve antibody stability (Akita and Nakai, 2000). Recently, Xun *et al.* (2010) described the use of the cytoprotective agent sucralfate and found that it cold improve IgY resistance to low pH and pepsin treatment, as well as enhance storage stability, in a dose-dependent manner. Finding an effective method of protecting IgY from degradation would not only ensure the delivery of a consistent antibody dose but could also allow the effective dose to be lowered, thereby reducing cost (Bogstedt *et al.,* 1997).

Use of IgY in Veterinary Sciences

Antimicrobials are used extensively in agriculture to both prevent disease and promote growth in livestock. However, the use of antibiotics, especially for growth enhancement, has come under security, as it has been shown to contribute to the increased prevalence of anitibiotic-resistant bacteria (Mathew *et al.,* 2007). As such, IgY has been examined as a feed additive for livestock to both target specific pathogens and improve growth and feed efficiency (Cook and Trott, 2010) (Table 1).

IgY has been tested against a number of enteric pathogens. IgY produced against the porcine enterotoxigenic *Escherichia coli* (ETEC) fimbrial antigens K88, K99, and 987 P was found to inhibit (Ikemori *et al.,* 1992) and procine intestinal mucus (Jin *et al.,* 1998) when given orally to piglets, these antibodies dose-dependently protected against E. coli infection (Yokoyama *et al.,* 1992). More recently, Li *et al.,* (2009) found that when anti-K88+ ETEC IgY was encapsulated in chitosan-alginate microparticles, it exerted its anti-diarrheal effect much faster (24 h versus 72 h postinfection in pigs given nonencapsulated IgY) and led to increased weight gain when compared to pigs fed nonencapsulated antibodies. The passive protective effect of anti- *E. coli* IgY in cattle has also been shown. Neonatal calves fed milk containing anti–ETEC IgY had transient diarrhea, 100% survival, and improved body weight gain (Ikemori *et al.,* 1992).

Bovine rotavirus (BRV) is an important cause of diarrhea in newborn calves, and local passive immunity is the most efficient protective strategy to control the disease (Vega *et al.,* 2011). Kuroki *et al.* (1994) had previously found that orally administered anti-BRV IgY could protect against infection in mice and calves. More recently, it was shown that anti BRV IgY containing yolk provided up to 80% protection against BRV-induced diarrhea in neonatal claves when compared with calves given nonimmunized egg yolk (Vega *et al.,* 2011), suggesting that supplementing newborn calves diets for the first 14 days of life with BRV-specific IgY may be a promising strategy to prevent BRV-related mortality. Other uses of IgY in cattle include the prevention of respiratory diseases and mastitis. Dahlen *et al.* (2008) found that IgY preparation against several bovine pathogens administered intranasally reduced morbidity and mortality in calves. IgY against bovine mastitis-causing E. coli (0111) Zhen *et al.,* 2008) and *S. aureus* (Wange *et al.,* 2011; Zhen *et al.,* 2009) have both shown promise by inhibiting bacterial growth and internalization (Wang *et al.,* 2011) and enhancing phagocytic activity of bovine macrophages (Zhen *et al.,* 2008).

Table 1: Uses of IgY for passive immunization in animals and aquaculture

Pathogen/ antigen	Target species	Effects of IgY	Reference
Escherichia coli	Pigs	Protected against infection by K88+K99+, and 987 P+ E. coli	Yokoyama *et al.*, 1992
		Encapsulated anti-K88+ E.coli IgY enhanced protection and led to improved weight gain	Li *et al.*, 2009a
	Cattle	Protected against K99+ E.coli infection in calves	Ikemori *et al.*, 1992
	Cattle	Reduced 0157:H7 fecal shedding in feedlot steer	Di Lorenzo *et al.*, 2008
	Chickens	Improved intestinal health and immune responses in broilers challenged with 078:K80	Mahdavi *et al.*, 2010
	Cattle	Inhibited growth and internalization of O111 and enhanced uptake by macrophages	Zhen *et al.*, 2008
Salmonella spp.	Chicken	Reduced rate of Salmonella-contaminated eggs in Salmonella Enteritidis (SE)- infected chickens	Gurtler *et al.*, 2004
	Chickens	Reduced fecal shedding and cecal colonization in SE-infected broilers	Rahimi *et al.*, 2007
Bovine rotavirus	Cattle	Protected neonatal calves from BRV –induced diarrhea	Kuroki *et al.*, 1994; Vega *et al.*, 2011
Infectious bursal disease virus	Chickens	Protected chicks from IBDV infection	Yousif *et al.*, 2006
Porcine epidemic diarrhea virus	Pigs	Protected piglets against PEDV infection	Kweon *et al.*, 2000
Eimeria spp.	Chickens	Protected chicks against avian coccidiosis	Lee *et al.*, 2009
Canine parvovirus	Dogs	Protected dogs against CPV2–induced disease symptoms	Van Nguyen *et al.*, 2006
Phospholipase A2	Chickens	Improved growth and feed efficiency is chickens	Cook 2004
Influenza	Chickens/ Humans	Protected mice from lethal H5N1, H5N2, and H1N1 challenge	Nguyen *et al.*, 2010
Staphylococcus aureus	Cattle	Reduced symptoms of clinical and experimental mastitis	Zhen *et al.*, 2009

Contd.

Pathogen/ antigen	Target species	Effects of IgY	Reference
Edwardsiella tarda	Fish	Prevented E. tarda-induced mortality in eels	Hatta *et al.*, 1994
Yersinia ruckeri	Fish	Reduced mortality and infection rates in rainbow trout	Lee *et al.*, 2000
Vibrio anguil-larum	Fish	Proteced rainbow trout against vibriosis	Arasteh *et al.*, 2004
White spot syndrome virus	Shrimp	Protected shrimp and crayfish from WSSV infection	Kumaran *et al.*, 2010

IgY has also been used for the passive protection of chicks against infectious bursal disease virus (Yousif *et al.,* 2006) and avian coccidiosis caused by Eimeria spp. (Lee *et al.,* 2009), for the protection of piglets against porcine epidemic diarrhea virus (Kweon *et al.,* 2000), and for the protection dogs against canine parvovirus -2 (Van Nguyen *et al.,* 2006). *Salmonella enteritidis* (SE) is the main causes of outbreaks in infections in chickens (Lee *et al.,* 2002). Chalghoumi *et al.* (2009) found that IgY against the outer membrane proteins of SE and ST reduced Salmonella sp. adhesion to intestinal epithelial cells in vitro, suggesting that passive immunization with Salmonella-specific IgY could be useful to prevent Salmonella colonization in broiler chickens. Moreover, feeding chickens egg powder containing SE –specific antibodies was found to reduce fecal shedding, cecal colonization, and the rate of Sabmonella – contaminated eggs in experimentally infected chickens (Gurtler *et al.,* 2004; Rahimi *et al.,* 2007).

Along with preventing disease, antibiotics can also improve growth performance and feed efficiency in livestock by reducing the number of immune-stimulating bacteria in the gut and minimizing the host antimicrobial inflammatory response (Cook and Trott, 2010). To this end, the use of IgY targeting specific molecules involved in inflammation and animal's growth has been described. Antibodies directed against the gut neuropeptides cholcystokinin and neuropeptide Y, believed to be involved in regulating appetite, as well as the enzyme PLA2. Which plays a role in the production of the inflammatory mediators prostaglandins and leukotrienes, have both been shown to improve growth and feed efficiency which fed to chickens (Cook *et al.,* 2004). Similarly, Mahdavi *et al.,* (2010) found that feeding yolk powder containing IgY against E.coli O78: K80 to chickens improved intestinal health and feed conversion ratio, and enhanced intestinal health and feed conversion ratio, and enhanced intestinal responses upon oral challenge. IgY feeding has also been found to have beneficial effects on growth in cattle. Di Lorenzo *et al.* (2008) found that feeding egg yolk preparations containing IgY against *Fusobacterium necrophorum* and *Streptococcus bovis* in place of antibiotics reduced ruminal counts of the target bacteria and improved feed efficiency and growth performance. Recently proteomic analysis carried out on pigs infected with *E. coli* and *S. Typhimurium* to reduce the stress of microbial infections (Park *et al.,* 2011).

Use of IgY in Aquaculture

Passive immunization has also been applied in the aquaculture industry, where infectious disease can result in significant economic loss (Table. 1). *Edwardsiella tarda* is a fish pathogen spread by infection through the intestinal mucosa, and Edwardsiellosis in Japanese eels constitutes a serous problem for the ell- farming industry, especially with the appearance of antibiotic-resistant strain (Hatta *et al.,* 1994). Eels challenged with E. tarda followed by administration of anti-E. tarda IgY survived without any symptoms of infection, in contrast to control eels that died within 15 days Hatta *et al.* (1994).

Enteric redmouth disease, caused by *Yersimia ruckeri*, is a systemic bacterial septicemia of salmonid fish, and the persistence of *Y. ruckeri* in carrier fish and shedding of bacteria in feces can present a continuing source of infection. Feeding anti-Y. ruckeri IgY before or after bacterial challenge resulted in lower mortality and reduced infection rates in rainbow trout (Lee *et al.,* 2000). Similarly, IgY against *Vibrio anguillarum* protectd rainbow trout against vibriosis for at least 14 days when given by intraperitoneal injection, oral intubation, or feeding (Arasteh *et al.,* 2004). White spot syndrome virus (WSSV) causes high mortality in cultured shrimp. IgY produced against WSSV was shown to passively protect shrimp (Lu *et al.,* 2008) and crayfish (Lu *et al.,* 2009) from WSSV infection when used as an immersion solution or incorporated into feed (Kumaran *et al.,* 2010).

Commercially available IgY preparations

There are a number of IgY, or hyperimmune egg products for sale to treat specific diseases or for the promotion of overall health in humans, livestock, and companion animals. Although this list is by no means exhaustive, it highlights some of the IgY preparations that are currently commercially available. I 26- Companion has been developed to improve immune function in cats and dogs. Arkion has also developed a hyperimmune feed supplement containing IgY against bovine and porcine enteric pathogens (Protimax), designed to improve performance of weanling pigs and calves. EW Nutrition, along with Ghen Corporation in Japan, Produces IgY supplements for both livestock and companion animals (Globigen) as well as a full line or products for human health (Ovalgen) containing IgY against a number of the pathogens described above targeting oral, stomach, intestinal, and skin and mucosa care. Similarly, AD Bootech Co. Ltd. and Dan Biotech Inc. produce several IgY-containing products (including Ig-Guard and Ig-Lock, respectively) that target gastrointestinal pathogens in livestock and companion animals as well as viral diseases in the aquaculture industry. Finally, Aova Technilogies sells a line of products (BIG for pigs, cows, poultry, and the aquaculture industry that contains IgY against the enzyme PLS2, and a number of animal trials have demonstrated significant improvement in feed efficiency, growth rate, carcass yield, and general health in livestock.

Challenges and Future Prospects

With the push to reduce antibiotic use in livestock and the emergence of antibiotic-resistant pathogens, the role of passive immunotherapy is more important than ever. The production of specific antibodies in egg yolk continues to attract the attention of the scientific community, as evidenced by the significant body of IgY related literature (Schade *et al.,* 2005). Indeed, the immunization of hens is an excellent method to efficiently generate large quantities of antibodies because antibody production is nonstressful and noninvasive, and the isolation and purification of IgY are relatively simple and high yielding (Zhang 2003). However, the production cost of high-quality IgY for large-scale applications still remains higher than that of other drug therapies, such as routine antibiotics (Casadevall *et al.,* 2004), suggesting that the cost of egg production may still be a significant barrier for the expanded use of IgY in some commercial applications (e.g., animal feed supplements). Trott *et al.,* (2009) found that aged or spent hens, although no longer valued for egg production for human consumption, were still a valuable source of high titer antigen-specific IgY and provided a more cost-effective source of eggs, thereby potentially reducing IgY production costs. Continuing research into IgY technology, improving chicken immunization protocols and adjuvants, as well as developments in extraction techniques and increasing antibody yield, will no doubt lead to new applications for IgY immunotherapy to improve animal health.

Summary

Hens deposit large amounts of an antibody called IgY into the egg yolk to transfer passive immunity to the developing chick. This process can be harnessed for large scale antibody production by immunizing hens with the antigen of interest, then collecting and isolating the resulting antigen- specific antibodies from the egg yolk. IgY production using chickens is more cost effective, convenient, and less invasive than antibody production in mammals, making it a promising alternative for passive immunization applications. Antigen-specific IgY has been shown to be effective at treating and preventing a number of different human and animal diseases, and commercially available IgY preparations have been shown to promote gut health and immunity in humans and improve feed efficiency and growth rates in livestock.

References

Akita, EM and Nakai S. 1992, Immunoglobulins from egg yolk: isolation and purification. *J. Food Sci.* 57:629-34.

Akita, EM and Nakai, S. 1993. Comparison of four purification methods for the production of immunoglobion from eggs laid by hens immunized with an enterotoxigenic E. coli strain. *J. Immunol. Methods*. 160: 207-14.

Akita, E.M and Nakai, S. 2000. Preparation of enteric-coated gelatin capsules of IgY with cellulose acetate phthalate. In: *Egg Nutrition and Biotechnology* (Ed. JS Sim, S Naki and W Guenter) . CAB International. New York. pp. 301-10.

Arasteh, N., Aminirissehei, A., Yousif, A., Albright, L and Durance, T. 2004. Passive immunization of ranbow trout (*Oncorbyncbus mykiss*) with chicken egg yolk immunoglobins (IgY). *Aquaculture*. 231:23-26.

Bogstedt, A.K., Hammarstrom, L and Robertson, A. K. 1997. Survival of immunoglobulins from different species through the gastrointestinal tract in healthy adult volunteers: implications for human therapy. *Antimicrob. Agents Chemother.* 41 : 2320.

Casadevall, A., Dadachova, E and Pirofski, L. A. 2004. Passive antibody therapy for infectious diseases. *Nat. Rec. Microbiol.* 2: 695-703.

Chalghoumi, R., Thewis, A., Beckers, Y., Marcq, C., Portetelle, D and Schneider, Y.J. 2009. Adbesion and growth inhibitory effect of chicken egg yolk antibody (IgY) on Salmonella enteric serovars Enteritidis and Typhimurium in vitro. *Foodborne Pathog. Dis.* 6:593-604.

Chang, H.M., Lu, T.C., Chen, C.C., Tu, Y.Y and Hwang, J.Y. 2000. Isolation of immunoglobulin from egg yolk by anionic polysaccharides. *J. Agric. Food Chem.* 48:995-99.

Chang, H.M., Ou-Yang, R.F., Chen, Y.T and Chen, C. C. 1999. Productivity and some properties of immunoglobulin specific against Streptococcu mutans serotype c in chicken egg yolk (IgY). *J. Agric. Food Chem.* 47: 61-66.

Cook, M.E. 2004. Antibodies : alternatives to antibiotics in improving growth and feed efficiency. *F. Appl. Poult. Res.* 13 : 106-9.

Cook, M.E and Trott, D. L. 2010. IgY: immune components of eggs as a source of passive immuniry for animals and humans. *World Poult. Sci. J.* 66: 215-25.

Cova L. 2005. DNA-designed avian IgY antibodies : novel tools for research, diagnostics and therapy. *J. Chin. Virol.* 34 (Suppl. 1) : 70-74.

Dahlen, C., Di., Lorenzo, N and DiCostanzo, A. 2008. Efficacy of a po9lyclonal antibody preparation against respiratory disease pathogens on cattle morbidity and performance during the step-up period. *J. Anim. Sci.* 86 (Suppl. 2): 195.

Davalos-Pantoja, L., Ortega-Vinuesa, J.L., Bastos-Gonzalez, D and Hidalgo-Alvarez, R. 2000. A comparative study between the absorption of IgY and IgG on latex particles. *J.Biomater. Sci. Polym. Ed.* 11: 657-73.

DiLorenzo, N., Dahlen, C and DiCostanzo, A. 2008. Effects of feeding a polyclonal antibody preparation against Escbericbia coli O157:H7 on performance, carcass characteristics and *E. coli* O157:H7 fecal shedding of feedlog steers. *J. Anim. Sci.* 86 (Suppl. 2): 281.

Gurtler, M., Methner, U., Kobilke, H and Fehlhaber, K. 2004. Effect of orally administered egg yolk antibodies on Sabmonella enteritidis contamination of hen's eggs. *J. Vet. Med. B. Infect. Dis. Vet. Public Health* 51: 129-34.

Hatta, H., Kapoor, M.P and Juneja, L.R. 2008. Bioactive components in egg yolk. In *Egg Bioscience and Biotechnolog* (Ed. Y, Mine), John Wiley Sons, Inc. Hoboken, NJ. pp. 185-237.

Hatta, H., Mabe, K., Kim, M., Yamamoto, T., Gutierrez, M. A and Miyazaki, T. 1994. Prevention of fish disease using egg yolk antibody. In *Egg Uses and Processing Technologies, New Development*, (Ed. J.S Sim and S. Nakai). CAB International, Oxon, UK. pp. 241-49.

Hatta, H., Sim, J.S and Nakai, S. 1988. Separation of phospholipids from egg yolk and recovery of water-soluble proteins. *J. Food Sci.* 53: 425-27.

Hatta, H., Tsuda, K,, Akachi, S., Kim, M., Yamomoto, T and Ebina T. 1993. Oral passive immunization effect of antihuman rotavirus IgY and its hehavior against proteolytic enzymes. *Biosci. Biotechnol. Biochem.* 57:1077-81.

Herath, C., Kumar, P., Singh, M., Kumar, D and Ramakrishnan, S. 2010. Experimental iron-inactivated *Pasteurilla multocida* A: 1 vaccine adjuvanted with bacterial DNA is safe and protects chickens from fowl cholera. *Vaccine.* 28 :2284-89.

Horikoshi, T., Hiraoka, J., Saito, M and Hamada, S. 1993. IgG antibody from hen egg yolk : purification by ethanol fractionation. *J. Food Sci.* 58: 739-42.

Ikemori, Y., Kuroki, M., Peralta, R.C., Yokoyama, H and Kodama Y. 1992. Protection of neonatal calves against fatal enteric colibacillosis by administration of egg yolk powder from hens immunized with K99-piliated enterotoxigenic *Escbericbia coli. Am. J. Vet. Res.* 53: 2005-8.

Jensenius, J. C., Andersen, I., Hau, J., Crone, M and Koch, C. 1981. Eggs : convenientlypackaged antibodies. Methods for purification of yolk IgG. *J. Immunol. Methods.* 46 : 63-68.

Jensenius, J.C and Koch, C. 1993. On the purification of IgG from egg yolk. *J. Immunol. Methods* 164 : 141-42.

Jin, L.Z., Baidoo, S.K., Marquardt, R.R and Frohlich, A.A. 1998. In vitro inhibition of adhesion of enterotoxigenic Escbericbia coli K88 to piglet intestinal mucus by egg-yolk antibodies. *FEMS Immunol. Med. Microbid.* 21:313-21.

Kim, H and Nakai, S. 1998. Simple separation of immunoglobulin from egg yolk by ultra filtration. *J. Food Sci.* 63:485-90.

Kumaran, T., Michaelbabu, A., Selvaraj, T., Albindhas, S and Citarasu, T. 2010. Production of anti WSSV IgY edible antibody using herbal immunoadjuvant asparagus recemosus and its immunological influence against WSSV infection in penaeus monodon. *F. Aquaculture Feed Sci. Nutr.* 2: 1-5.

Kuroki, M., Ohta, M., Ikemori, Y., Peralra, R.C., Yokoyama, H and Kodama, Y. 1994. Passive protection against bovine rotavirus in calves by specific immunoglobulins from chicken egg yolk. *Arch. Virol.* 138: 143-48.

Kwan, L., Li-Chan, E.C., Helbig, N and Nakai, S. 1991. Fractionation of water-soluble and insoluble components from egg yolk with minimum use of organic solvents. *J. Food Sci.* 56 : 1537-41.

Kweon, C.H., Kwon, B.J., Woo, S.R., Kim, J.M and Woo, G.H. 2000. Immunoprophylactic effect of chicken egg yolk immunoglobulin (IgY) against porcine epidemic diarrhea virus (PEDV) in piglets. *J. Vet. Med. Sci.* 62: 961-64.

Lee, F.N., Sunwoo, H.H., Menninen, K and Sim J.S. 2002. In vitro studies of chicken egg yolk antibody (IgY) against Salmonella enteritidis and Salmanella tybimurium. *Poult. Sci.* 81: 632-41

Lee, S.B., Mine, Y and Stevenson, R. M. 2000. Effects of hen egg yolk immunoglobulin in passive protection of rainbow trout against Yersinia ruckeri. *J. Agric. Food Chem.* 48 : 110-115.

Lee, S.H., Lillehoj, H.S., Park, D.W., Jang, S.I and Morales, A. 2009. Induction of passive immunity in broiler chickens against Eimeria acervulina by hyperimmune egg yolk immunoglobulin *Y. Poult. Sci.* 88: 562-66.

Levesque, S., Martinez, G and Fairbrother, J. M. 2007. Improvement of adjuvant systems to obtain a cost-effective production of high levels of specific IgY. *Poult. Sci.* 86: 630-35.

Li, X.Y., Jin, Lj, Lu, Y.N., Zhen, Y.H and Li, S.Y. 2009. Chitosan-alginate microcapsules for oral delivery of egg yolk immunoglobulin (IgY): effects of chitosan concentration. *Appl. Biochem. Biotechnol.* 159-778-87.

Lu, Y., Liu, J., Jin, L., Li, X and Zhen, Y. 2009 Passive immunization of crayfish (Procambarus clarkia) with chicken egg yolk immunoglobulin (IgY) against white spot syndrome virus (WSSV). Appl. *Biochem. Biotechnol.* 159: 750-58.

Lu, Y., Liu, J., Jin, L., Li, X and Zhen, Y. 2008. Passive protection of shrimp against white spot syndrome virus (WSSV) using specific antibody from egg yolk of chickens immunized with inactivated virus or a WSSV-DNA vaccine. *Fish Shellfish Immunol.* 25: 604-10.

Mahadavi, A. H., Rahmani, H. R., Nili, N., Samie, A. H., Soleimanian –Zad, S and Jahanian, R. 2010. Effects of dietary egg yolk antibody powder on growth performance, intestinal Escbericbia coli colonization, and immunocompetence of ehallenged broiler chicks. *Poult. Sci.* 89 : 484-94.

Mathew, A.G., Cissell, R and Liamthong, S. 2007. Antibiotic resistance in bacteria associated with food animals: a United States perspective of livestock production. *Foodbornc Patbog. Dis.* 4:115-33.

Nakai S, Li-Chan EC, Lo KV, 1994. Separation of immunoglobulin from egg yolk. In Egg Uses and Processing Technologies: New Developments (Ed. J.S Sim and S. Nakai), CAB Int. Wallingford, UK. pp. 94-105.

Nguyen, H.H., Turnpey, T.M., Park, H.J., Byun, Y.H and Tran, L.D. 2010. Prophylactic and therapeutic efficacy of avian antibodies against influenza virus H5N1 and H1N1 in mice. *PLoS ONE 5:e* 10152.

Nilsson, K and Larsson, A. 2007. Stability of chicken IgY antibodies freeze-dried in the presence of lactose, sucrose and threalose. *J. Poult. Sci.* 44:58-62.

Park, H.S., Park, K.I., Nagappan, A., Lee, D.H and Kang, S.R. 2011. Proteomic analysis of effects on natural herb additive containing immunoglobulin yolk sac (IgY) in pigs. *Am. J. Chin. Med.* 39 : 477-88.

Pauly, D., Dorner, M., Zhang, X., Hlinak, A., Dorner, B and Schade, R. 2009. Monitoring of laying capacity, immunoglobulin Y concentration, and antibody titer development in chickens immuniezed with ricin and botulinum toxins over a two-year period. *Poult. Sci.* 88:281-90.

Polson, A. 1990. Isolation of IgY from the yolks of eggs by a chloroform polyethylene glycol procedure. *Immunol. Investig.* 19: 253-58.

Rahimi, S., Moghadam Shiraz, Z., Zahraci Salehi, T., Karimi Torshizi, M.A and Grimes, J.I. 2007. Prevention of Salmonella infection in poultry by specific egg-derived antibody. *Int. J. Poult. Sci.* 6: 230-35.

Reilly, R., Domingo, R and Sandhu, J. 1997. Oral delivery of antibodies: future pharmacokinetic trends. *Clin. Pharmacokinet.* 4: 313-23.

Roos, N., Mahe, S., Benamouzig, R., Sick, H., Rautureau, J and Tome, D. 1995. 15N-labeled immunoglobulins from bovine colostrums are partially resistant to digestion in human intestine. *J. Nutr.* 125: 1238-44.

Schade, R., Calzado, E.G., Sarmiento, R., Chaema, P.A,. Porankiewiez-Asplund, J and Terzolo, H.R. 2005. Chicken egg yolk antibodies (IgY-technology). A review of progress in production and use in reaearch and human and veterinary medicine. Altern. *Lab. Anim.* 33: 129-54.

Schade, R., Pfister, C., Halastsch, R and Henklein, P. 1991. Polyclonal IgY antibodies from chicken egg-an alternative to the production of mammalian IgG type antibodies in rabbits. *ATLA* 19: 403-19.

Shimizu, M., Fitzsimmons, R.C. and Nakai, S. 1988. Anti-E.coli immunoglobulin Y isolated from egg yolk of immunized chickens as a potential food ingredient. *J. Food Sci.* 53: 1360-68.

Shimizu, M., Miwa, Y., Hashimoto, K and Goto, A. 1993. Encapsulation of chicken egg yolk immunoglobulin G (IgY) by liposomes. Biosci. *Biotechnol. Biochem.* 57: 445-49.

Shimizu, M., Nagashima, H., Hashimoto, K and Suzuki, T. 1994. Egg yolk antibody (IgY) stability in aqueous solution with high sugar concentration. *J. Food Sci.* 59: 763-72.

Shimizu, M., Nagashima, H., Sano, K., Hashimoto, K and Ozeki, M. 1992. Molecular stability of chicken and rabit immunoglobulin *G. Biosci. Biotechnol. Biochem.* 56 : 270-74.

Shimizu, M and Nakane, Y. 1995. Encapsulation of biologically active proteins in a multiple emulsion. *Biosci. Biotechnol. Biochem.* 59: 492-96.

Svendsen, I., Crowley, A., Ostergaard, L.H., Stodulski, G and Hau, J. 1995. Development and comparison of purification strategies for chicken antibodies from egg yolk. *Lab. Anim. Sci.* 45: 89-93.

Taylor, A. I., Fabiane, S. M., Sutton, B.J and Calvert, R.A. 2009. The crystal structure of an avian IgY –Fc fragment reveals conservation with both mammalian IgG and IgE, *Biochemistry.* 45: 558-62.

Tini, M., Jewell, U. R., Camenisch, G., Chilov, D and Gassmann, M. 2002. Generation and application of chicken egg-yolk antibodies. *Camp. Biochem. Physiol. A. Mol. Integr. Physiol.* 131: 569-74.

Trott, D.L., Hellestad, E.M., Yang, M and Cook, M.E. 2008. Additions of killed whole cell bacteria preparations to Freund complete adjuvant alter laying hen antibody response to soluble protein antigen. *Poul. Sci.* 87: 912-17.

Trott, D.L., Yang, M., Utterback, P.L., Utterback, C.W., Koelkeback, K and Cook, M.E. 2009. Utility of spent single comb white leghorn hens for production of polyclonal egg yolk antibody. *J. Appl. Poult. Res* 18: 679-89.

Van Nguyen, S., Umeda, K., Yokoyama, H., Tohya, Y and Kodama, Y. 2006. Passive protection of dogs against clinical disease due to canine parvovirus-2 by specific antibody from chicken egg yolk. *Can. J. Vet. Res.* 70: 62-64.

Vega, C., Bok, M., Chacana, P., Saif, L., Fernandez, F and Parreno, V. 2011. Egg yolk IgY : Protection against rotavirus induced diarrhea and modulatory effect on the systemic and mucosal antibody responses in newborn calves. *Vet. Immunol. Immunopatbol.* 142: 156-69.

Wang, L.H., Li, X.Y., Jin, L.J., You, J.S and Zhou, Y. 2011. Characterization of chicken egg yolk immunoglobulins (IgYs) specific for the most prevalent capsular serotypes of mastitis-causing *Staphylococcus aurous. Vet. Microbiol.* 149: 415-21.

Wanke, R., Schmidt, P., Erhard, M.H., Sprick-Sanjosc, A and Stangassinger, M. 1996. Freund's complete adjuvant in the chicken efficient immuno stimulation with severe local inflammatory reaction. *Zentralbl. Veterinarmed.* 43 : 243-53.

Warr, G.W., Magor, K.E and Higgins, D.A. 1995. IgY: clues to the origins of modern antibodies. *Immunol. Today* 16: 392-98.

Xun, Z., Li-Yuan, G., Zhihang, Y and Xiaoping, C. 2010. Protective effects of sucralfate on anti-H. Pylori VacA IgY in vivo and in vitro. *Afr. J. Microbiol. Res.* 4 : 1091-99.

Yokoyama, H., Peralta, R. C,. Diaz, R., Sendo, S., Ikemori, Y., Kodama, Y. 1992. Passive protective effect of chicken egg yolk immunoglobulins against experimental enterotoxigenic *Escherichia coli* infection in neonatal piglets. *Infect. Immun.* 60 : 998-1007.

Yousif, A.A., Mohammad, W.A., Khodeir, M.H., Zeid, A.Z., el –Sanousi, A. A. 2006. Oral administration of hyperimmune IgY: an immunoecological approach to curbing acute infections bursal disease virus infections. *Egypt. J. Immunol.* 13: 85-94.

Zhang, W.W. 2003. The use of gene-specific IgY antibodies for drug target discovery. *Drug Discov. Today.* 8: 364-71.

Zhen, Y.H., Jin, L.J., Guo, J., Li, X.Y and Lu, Y.N. 2008. Characterization of specific egg yolk immunoglobulin (IgY) against mastitis-causing *Eschericbia coli. Vet. Microbiol.* 130: 126-33.

Zhen, Y. H., Jin, L. J., Li, X.Y., Guo, J and Li, Z. 2009. Efficacy of specific egg yolk immunoglobulin (IgY) to bovine mastitis caused by *Staphylococcus aureus. Vet. Microbiol.* 133: 317-22.

12

Use of Coccidiostats for Sustainable Poultry Production

Rinesh Kumar and Suman Kumar

Introduction

Coccidiosis is the most economically harmful disease encountered in commercial poultry production. It is caused by single celled protozoan parasites of the genus Eimeria which are commonly referred to as coccidia. Williams (1998) reported that the annual worldwide loss is about $ 800 million. The total loss due to coccidiosis is found to be Rs. 1.14 billion in India. The important finding is that almost 70 percent of the estimated cost is due to subclinical coccidiosis, by impact on weight gain and feed conversion rate. One of the reasons for these remarkable finding is probably the difficult diagnosis of subclinical coccidiosis.

Chickens are susceptible to at least seven species of Eimeria. The most common species are *Eimeria tenella*, which causes the caecal coccidiosis, *E. necatrix* & *E. maxima*, which causes mid intestinal coccidiosis, *E. acervulina* which causes anterior intestinal coccidiosis and *E. brunetti* which causes posterior intestinal or rectal coccidiosis. *E. mitis* and *E. praecox* are generally mildly pathogenic and commonly occurring as co-infection/mixed infection in with other species. Coccidiosis is particularly difficult to combat because several different species of *Eimeria* exist in the field. Poultry may become infected with different species because the immunity that develops after infection is specific only to one species. *Eimeria* has direct and complex life cycle that involves many developmental stages within the host cells. Each *Eimeria* species has characteristic intestinal lesions and type of diarrhoea (with or without blood).

The bird develops reduced ability to absorb nutrients, which results in weight loss and eventually death. Sub clinically, it is manifested by poor performance, impaired feed conversion, poor flock uniformity and poor growth. Coccidia can also damage the immune system and leave poultry more vulnerable to pathogens like *Clostridium, Salmonella* and *E. coli.*

Coccidiostats

There are basically two means of prevention of coccidiosis: Chemoprophylaxis and vaccination. Despite the introduction of vaccine, in most countries prophylactic chemotherapy is still the prefer method for the control of coccidiosis (Chapman, 2000). Chemoprophylaxis compound used against coccidia are known as anticoccidial product or anticoccidials and most popularly as coccidiostat. Coccidiostat are given in feed to prevent coccidiosis and its prophylactic use is preferred because most of the damage occurs before sign become apparent, and because drugs cannot completely stop an outbreak. The effects of anticoccidial drugs may be coccidiostatic, in which growth of intracellular coccidia is arrested but development may continue after drug withdrawal, or coccidiocidal, in which coccidia are killed during their development. Some anticoccidial drugs may be coccidiostatic when given short-term but coccidiocidal when given long term. The coccidiostat can be classified as:

1. Synthetic anti-coocidial (coccidiostat) or chemicals:- These compounds have specific mode of action against parasite metabolism., such as Amprolium, Clopidol and quinolines, Halofuginone etc.
2. Polyether ionophores such as monensin, salinomycin, lasalocid, narasin, maduramicin etc, which act through general mechanism of altering ion transport and disrupting osmotic balance. These compounds are now the most widely used drugs for coccidiosis prevention in poultry.

History of coccidiostat

Control of coccidiosis in poultry was attempted of prior to 1940s by using various kinds of alchemic recipes, skimmed milk, vinegar, and flowers of sulphur. Discovery of sulphaquinoxaline as potent anticoccidial activity in early 1948 transformed the treatment pattern of poultry coccidiosis. In 1948, sulphaquinoxaline was the first drug administered in the feed continuously and at lower doses (Chapman, 2003, McDougald, 2003). Other chemicals followed in the years allowing the expansion of the poultry industry to what it is today. A major breakthrough in the prevention of coccidiosis through feed medication occurred, however, in 1972, with the launch of the first polyether ionophore anticoccidials, monensin, since then this type of anticoccidial agents have been commonly referred to as ionophores, a term derived from their general structure, and have become the most widely used drugs for coccidiosis prevention in poultry. With the remarkable growth of poultry industry there is a steady procession of chemical agents used to combat coccidiosis.

Classification of coccidiostat

Anticoccidial drugs are generally divided into several classes on the basis of mechanism of action and chemical structures as follows:

i. Thiamine(Vitamin B_1) antagonists e.g., amprolium

i. Ionophores/Polyether antibiotics. e.g.,monensin, lasalocid, salinomycin, narasin, maduramicin and semiduramicine.

iii. Folic acid antagonists

- Sulphonamides; e.g. sulphaguinoxaline, sulphaguinidine, sulphadimidin, sulphadiazine, sulphadimethoxine, sulphamethoxazole, sulphachlorpyrazine.
- Pyrimidines; e.g. trimethoprim, pyrimethamine, diaveridine and ormetoprim.
- Substituted benzoic acid e.g. ethopabate.
- Potentiated Sulphonamides e.g. trimethoprim-sulphadimethoxime, trimethoprim-sulphadiazine, ormetoprim-sulphadimethoxime, pyrimethamine-sulphadiazine.
- Quinolones; e.g. decoquinate, methylbenzoquate and buquinonate.
- Pyridinoles e.g clopidol.
- Guanidines; e.g robenidine.
- Nitrobenzamides e.g. dinitolmide and alkomide.
- Carbanilides; e.g. nicarbazin.
- Quinazolines; e.g. halofuginone.
- Benzeneacetonitriles; diclazuril and clazuril.
- Triazinones e.g. toltrazuril.
- Benzyl purines e.g. arprinocid.
- Miscellaneous drugs.

- Nitrofurans e.g. furazolidone.
- Tetracyclines e.g. oxytetracyclines and chlortetracyclines.

Table 1: History of coccidiostat

Anticoccidial compound	First approved by FDA	Drug withdrawal period
Sulfaquinoxaline	1948	5
Nitrofurazone	1948	5
Arsenilic acid	1949	5
Butynorate	1954	28
Nicarbazine	1955	4
Furazolidone	1957	5
Mitromide+Sufanitran+ Roxarasone	1958	5
Oxytetracyclin	1959	3
Amprolium	1960	0
Chrortetracyclin	1960	-
Zoalene	1960	5
Amprolium + ethopabate	1963	0
Buquinolate	1967	0
Sodium sulfachlorpyrazine monohydrate	1967	4
Sulfadimethoxine	1968	5
Clopidol	1968	5
Decoquinate	1970	0
Sulfadimethoxine+ Ormetoprin	1970	5
Monensin	1971	0
Robenidin	1972	5
Lasalocid	1976	3
Salinomycin	1983	0
Halofuginone	1987	5
Narasin	1988	0
Maduramicin	1989	5
Narasine+nicarbazine	1989	5
Toltrazuril	1985	-
Diclazuril	1998	-

I. Thiamine (Vitamin B_1) antagonists

Amprolium

The coccidiostatic activity of amprolium is related to its mimicry of thiamine and competition for absorption of thiamine by the parasite. The activity occurs because of the structural similarity between thiamine and amprolium (Looker *et al.*, 1986; United States Pharmacopeia, 1998). The anticoccidial effect may be reversed by feeding of excess thiamine. Amprolium is most effective against the first generation schizont stage and thus is more effective as a preventative than as a treatment. Amprolium (Amprol) is fed in poultry ration or drinking water to prevent or treat coccidiosis. Amprolium is given in drinking water for 2 weeks at 0.0125% (0.025% for severe outbreaks), then given at 0.006% for another 2 weeks.

II. Important Ionophores/ Polyether antibiotics

Monensin

Monensin is an antibiotic produced as fermentation product of *Streptomyces cinnamonensis*. It is used in cattle, goats, poultry, and quail for its coccidiostatic activities. It forms ionophores with sodium and potassium in the host and in the parasite. When the parasite mitochondrial membrane is affected, it is rendered permeable to potassium and sodium ions. Monensin (Coban) is used in broilers and pullets to prevent coccidiosis caused by *Eimeria nacatrix, Eimeria tenella, Eimeria acervulina, Eimeria brunetti, Eimeria mivati,* and *Eimeria maxima.* It is fed at the rate of 90 to 110 g/ton of complete feed. It is also approved for use in turkey to prevent infection with *Eimeria adenoeides, Eimeria meleagrimitis* and *Eimeria gallopavonis* when fed at 54 to 90 g/ton. Bobwhite quail can be fed monensin at 73g/ton to prevent coccidiosis caused by *Eimeria dispersa* and *Eimeria lettyae.*

Lasalocid

Lasalocid, an ionophore closely related to monensin, is produced by a *Streptomyces lasaliensis*. Like other ionophores, it forms complexes with sodium and potassium ions. This action renders the parasite membranes permeable to ions, and mitochondrial functions are inhibited. The trophozoite stage is most susceptible to lasalocid (Guyonnet *et al*., 1990). Lasalocid is the least toxic of the ionophores. It is approved for use in cattle, sheep, rabbits, and poultry for the control of coccidia and improvement of feed efficiency. Lasalocid is approved in broilers, turkeys, and chukar partridges to prevent coccidiosis caused by *Eimeria tenella, Eimeria nacatrix, Eimeria acervulina, Eimeria brunetti, Eimeria mivati, Eimeria maxima, Eimeria meleagrimitis, Eimeria gallopavonis, Eimeria adenoeides, and Eimeria legionensis.* The product (Avatec) is mixed into a complete ration for broilers and turkeys at a rate 68 to 113 g/ton. Chukar partridges should be fed lasalocid at a dose rate of 113g/ton. Withdraw product at least 3 days before slaughter.

Narasin

Narasin is an ionophore coccidiostat produced by *Streptomyces aureofaciens*. Similar in structure to salinomycin (Lindsay and Blagburn, 2001), it is available as a feed additive (Monteban) for use only in broilers. The product is fed at a rate of 54 to 72 g/ton of feed for prevention of coccidiosis caused by *Eimeria nacatrix, Eimeria tenella, Eimeria acervulina, Eimeria brunetti, Eimeria mivati,* and *Eimeria maxima.* No withdrawal period is required before slaughter. It should not be fed to laying hens as it affects egg production negatively. Ingestion by adult turkey may be fatal.

Salinomycin

Salinomycin was the third ionophore coccidiostat to enter the market in the United States. A fermentation product of *Streptomyces albus*, it is most active against the sporozoite stage. Salinomycin is available in a feed additive (Bio-Cox) for use in broilers, pullets, and quail. It is fed at 40 to 60 g/ton (50g/ton for quail) for prevention of coccidiosis caused by *Eimeria tenella, Eimeria nacatrix, Eimeria acervulina, Eimeria maxima, Eimeria brunetti,* and *Eimeria mivati* in chickens and coccidiosis caused by *Eimeria dispersa* and *Eimeria lettyae* in quail. Do not feed salinomycin to laying hens. No withdrawal period is required before slaughter. The product may be fatal if fed to adult turkeys.

Maduramicin

Maduramicin is a monovalent monoglycoside polyether ionophore antibiotic produced by *Actinomadura yumaense.* It is more potent than other ionophore antibiotics and is approved for prevention of coccidiosis in broilers. Maduramicin has some adverse effects on growth and feathering when administered at slightly above the recommended concentrations. The pre-slaughter withdrawal period of maduramicin is 5 days. For prophylaxis of coccidiosis dose rate of maduramycin in broilers is 5g/ tonne feed.

Semduramicin

Semduramicin is an ionophore coccidiostat produced by *Actinomadura roseorufa*. It is available as a feed additive (Aviax) for use in broiler chickens only. The product is fed at 22.7g/ton for prevention of coccidiosis caused by *Eimeria tenella, Eimeria acervulina, Eimeria maxima, Eimeria brunetti, Eimeria nacatrix,* and *Eimeria mivati.* It should not be fed to broilers within 5 days of slaughter.

III. Important Folic acid antagonists

Sulfaquinoxaline

Sulfaquinoxaline is a sulfonamide approved for use in chickens, turkeys for control and treatment of coccidiosis. It is not well absorbed from the gastrointestinal tract. Sulfaquinoxaline is available as a water medication (Sul-Q-Nox). Chickens should receive 10 to 45 mg/pound/day. Turkeys should be given 3.5 to 55 mg/pound/day. Treatment should be administered for 2 to 3 day in chickens and turkeys. Make a fresh solution every day.

Substituted benzoic acid

Ethopabate is quite effective against *Eimeria acervulina* and some stain of *Eimeria maxima* and *Eimeria brunetti.* It has been used only in combination with amprolium, first @ 4ppm, and later @ 40ppm.

Decoquinate

Decoquinate is an approved coccidiostatic drug for the control of coccidial (*Eimeria*) infection in chickens, cattle, sheep, and goats. This quinolone product kills the sporozoite stage of life cycle. It disrupts the electron transport in the mitochondrial cytochrome system of the parasite (Plumb, 2005). Decoquinate is primarily indicated for prevention rather than treatment of coccidiosis.

Decoquinate is indicated for prevention of coccidiosis caused by *Eimeria bovis* and *Eimeria zuernii* in ruminating calves and older cattle. It is fed (Deccox) at 0.5 mg/kg body weight per day for at least 28 days during period of exposure to infective oocysts. In young sheep and goats, decoquinate is used at the same rate for the prevention of infections caused by *Eimeria* species.

Clopidol

Clopidol is a pyridinol coccidiostat that has some activity against ionophore-resistant strain of coccidia. It acts against the sporozoite stage, allowing host cell penetration without development. Long (1993) suggested that the mode of action of clopidol was similar to that of quinolone anticoccidials because of similar structure and biological activity of the agents. However, cross resistance between clopidol and quinolone anticoccidials does not occur (Long, 1993). It is insoluble in water but is available as feed additive.The product is fed to chickens at 0.0125% or 0.025%. It should not be fed to laying hens, to chickens older than 16 week of age, or within 5 days of slaughter (Lindsay and Blagburn, 2001).

Robenidine

Robenidine is a synthetic coccidiostat chemically similar to guanidine. It is an older drug with a history of developing resistant strains of coccidian but is now used to treat ionophore-resistant strains. Robenidine is available as a feed additive (Robenz) for use in broilers only. The product is fed at 30g/ton of feed for prevention of coccidiosis caused by *Eimeria mivati, Eimeria brunetti, Eimeria tenella, Eimeria acervulina, Eimeria maxima,* and *Eimeria nacatrix.* It should not be fed to laying hens or within 5 days of slaughter, as studies have shown that robenidine is transmitted to the eggs (Long *et al.*, 1981). Meat and eggs from treated birds have an unpleasant taste if the withdrawal period is not followed (Lindsay and Blagburn, 2001).

Aklomide and Dinitolmide

Aklomide and Dinitolmide act primarily on the first generation schizonts and Dinitolmide inhibits sporulation oocysts (Mathis and McDougald, 1981). These are fodder additive for poultry, used to prevent coccidiosis. It also known under trade names such as Coccidine A, Coccidot, Zoalene and Zoamix. Zoalene is usually added to feed in dose of 125 ppm (preventive) or 250 ppm (curative). It is a broad spectrum anticoccidial drug, preventing seven main species of *Eimeria.*

Nitromide

Nitromide is effective against *Eimeria tenella, Eimeria nacatrix* and *Eimeria acervulina*. It is given @ 0.1 per cent in the feed.

Nicarbazin

Nicarbazin is a synthetic coccidiostate effective in preventing ceacal coccidiosis and intestinal coccidiosis caused by *Eimeria tenella, Eimeria acervulina, Eimeria maxima, Eimeria nacatrix* and *Eimeria brunetti.* The mechanism of action is unknown (Lindsay and Blagburn, 2001). It is available as a 25% feed additive (Nicarb) and is approved for use at 0.0125% in feed of broilers. The product should not be fed within 4 days of slaughter or to laying hens. It is not effective for treatment of coccidiosis, and it may depress growth in young birds (Lindsay and Blagburn, 2001).

Halofuginone

Halofuginone is a coccidiostat used in veterinary medicine. It is a synthetic halogenated derivative of febrifugine, a natural quinazolinone alkaloid which can be found in the Chinese herb *Dichroa febrifuge*. It is effective against all asexual stages of *Eimeria,* but probably most effective against developing first generation schizonts. Its anticoccidial mode action is not known. Halofuginone is transmitted to eggs from hens that have been fed halofuginone for one week(Long *et al*., 1981), but no adverse effect on egg production or egg quality have been associated with administration (At a feed concentration 3 ppm, the drug is effective against pathogenic *Eimeria* species in chicken.

Diclazuril

This drug has anticoccidial activity when fed at low level in the feed. The exact mode action of diclazuril in not known. Diclazuril is active against various species of coccidia, but its anticoccidial activity may vary with species of cocidia. Diclazuril is used for prophylaxis of coccidiosis in chicken, turkey and rabbite. It is a safe compound and is compatible other anticoccidials and feed additives. Diclazuril is used mostly for prevention at 1ppm in the feed.

Clazuril

It is structurally related to diclazuril. It is highly effective against different species of pigeon coccidian.

Toltrazuril

Toltrazuril (Baycox) was first produced by Bayer in the 80s, and is a chemical substance that belongs to the class of symmetric triazinones. It was introduced to the market in 1986, and due to the fact that it affects all intracellular stages of *Eimeria*, it has been used with excellent results to combat coccidiosis problems in various animal species since its launch.

Toltrazuril (Baycox) damages all intracellular developmental stages of *Eimeria.* Toltrazuril affects schizonts, micro and macrogametes, but not the tissue cells of the host animals, as was shown in light and electron microscopic studies. These findings suggest that toltrazuril interferes with the division of the nucleus and with the activity of the mitochondria, which is responsible for the respiratory mechanism of *Eimeria*. In the macrogametes, toltrazuril damages the so- called wall forming bodies. In all intracellular developmental stages, severe vacuolisation occurs due to inflation of the endoplasmic reticulum. Toltrazuril thus has a coccidiocidal mode of action. The special mode of action results in the following advantage:

- Toltrazuril acts on all intracellular developmental stages,
- Toltrazuril does not interfere with development of immunury,
- Follow-up treatment usually is not necessary,

- Even an advanced infection (after 3-5 days; gametogony) can still be treated successfully,
- The efficacy of toltrazuril is independent of the severity of the infection. Dose for treatment is 7mg/kg, PO, once daily by addition to drinking water for 2 days.

Miscellaneous drugs

Nitrofurazone

It is used for therapeutic purposes @ 0.222 per cent but, if this level is continued for more than ten days, toxic effects like nervous signs may be seen. For preventive medication, it is recommended @ 0.0055-0.0056 per cent in feed.

Furazolidone

It is of some value against *Eimeria tenella* @ 0.011 per cent in the feed. However, it is generally used in combination with nitrofurazone at the dose rate mentioned above.

Tetracyclines

These are broad spectrum antibiotics. In addition to antibacterial properties, these also have anticoccidial action.

Oxytetracycline

Oxytetracycline is used as curative @ 0.022% in feed +0.25% calcium for a maximum period of 5 days; as preventive it may be @ 0.020 % in feed.

Chlortetracyclines

Chlortetracyclines is used as curative @ 0.022% in feed + 0.8% calcium for a maximum period of 3 weeks.

Roxarsone

Roxarsone is an organic arsenical compound. It has significant activity against *Eimeria tenella* and is used in combination with ionophores to improve control of that species. A withdrawal period is required.

Glycarbylamide

The drug is coccidiostatic and active against early asexual stages of *Eimeria tenelle, Eimeria necatrix and Eimeria acervulina.* This drug is safe @ 0.003 per cent in feed. It is more potent than nicarbazin.

Resistance against coccidiostat

The development of drug resistance in Eimeria is common because of extensive use of anticoccidial drug for the control of avian coccidiosis. There is practice of incorporating a coccidiostats at low dose level in the feed by the firms dealing with them. This continual low level dosing of a coccidiostat must be permitting the species of Eimeria to develop resistance to it. More and more coccidiostat become useless due to drug resistance throughout the world (Ryley, 1980).It is worth mentioned that this has not been the cases for all chemical anticoccidials, for example nicarbazin was first introduced in the market place in 1955 and it remains today a regularly used and effective medication for the prevention and control of coccidiosis in broiler chuicken.. it is intriguing to ask if the successful life and longevity of nicarbazin id due to its unique mode of action coupled with its seasonal use (typically used during cool weather) and limited exposure (typically only used in the starter feed).

Development of resistance to anticoccidials is a known phenomenon. With some compound (eg. Glycoamide) drug resistance in coccidia is not as precise as for bacteria and this often leads to controversies among farmers, feed manufacturers and pharmaceutical companies. Laboratory methods for diagnosing drug resistance involves, in vivo testing of field isolates against different anticoccidials. However these require elaborate laboratory facilities. In future PCR technique with DNA markers could provide easy tools. Under field conditions (i) Down sliding flock performance (High FCR, low weight gain/production) (ii) Increased oocysts output in droppings and (iii) Higher lesion scores could be considered to indicate developing resistance against the drug in use. But a variety of local factor on the farm *viz* managemental and nutritional faults; poor biosecurity; immunosuppression; stress; and intercurrent diseases may also be responsible for the above situation. It is therefore necessary that fairly well managed farms, where local contributing factor are minimum, should only be monitored for detection of anticoccidial resistance.

Presently, although coccidiosis control programmes are based mainly on use of anticoccidials, there is increasing concern on the wide spread use of these drugs because the following reasons:-

- Rising levels of drug resistance.
- Falling success rates in discovery of new compound.
- High cost involved in development of new anticoccidials (estimated to be around 5 to 10 million dollars).
- Consumer pressure to phase out drug additives in poultry feed.

- Realisation from the experience of last 50 years that was against coccidia cannot be won with use of only anticoccidials.

In general drug resistance is more common to chemical anticoccidials and is rare to nil in ionophores. These drugs are in use for the last 25 years and although resistance to ionophores is being reported from some studies, it has been a slow process, not been absolute and has been overcome by slight increase in dose (Bafundo, 1999). During last decades, search for new anticoccidial compounds has slowed down and no new drugs are in pipeline. However various programmes are being followed to maintain the efficacy of present anticoccidials in the field.

Steps to avoid resistance

Typically coccidiosis prevention is achieved through the use of anticoccial agent added to the feed of poultry. Continuous use of anticoccidial drugs often leads to ineffective treatment due to development of drug resistance in the coccidial population. Resistance to anticoccidial drug is a serious problem and has markedly reduced the efficacy of many agents. Various strategies are employed in the poultry industry to minimise this problem. These include shuttle programme, rotation programme, and use of alternative drug mixtures.

1. ***Shuttle Programme***: In this programme, drugs of different classes are used in single grow-out i.e., different drugs are used in starter and grower rations. Shuttle programme is a common practice in many countries.
2. ***Rotation (Switch) programme***: In this programme, different anticoccidials drugs are used between two or more grow out i.e., rotation of drugs is done after one or more crops of birds.
3. ***Drug mixtures***: In this programme, combination of two or more drugs is used. This practice of using drugs mixture is applicable only for a few agents because many drugs are not compatible with other agents.

References

Bafundo, K.W. 1999. Drug resistance development in poultry coccidia. *World poultry suppl*: 20-21.

Bowman, D.D. 2009. Antiparasitic Drugs. In: Georgis' Parasitology for Veterinarians 9th ed. Elsevier Press.

Chapman, H.D. 2007. Rotation programmes for coccidiosis control. *International Poultry Production*. 15: 7-9.

Chapman, H. D. 2003. Origins of coccidiosis research in the fowl-The first fifty years. *Avian Diseases*. 47: 1-20.

Gumila, C., Ancelin, M., Jeminet, G., Delort,M., Miguel,G. and Vial, H. 1996. Differential in vitro activities of ionophore compounds against Plasmodium falciparum and mammalian cells. *Antimicrobial Agent and Chemotherapy*. 40: 602-608.

Guyonnet, V., Johnson, J. K., Long, P. L. 1990. Studies on the stage of action of lasalocid against *Eimeria tenella* and *Eimeria acervulina* in the chicken. *Veterinary Parasitology*. 37: 93.

Lindsay, D. S. and Blagburn, B.L. 2001. Antiprotozoan drugs. *In* Adams HR, ed: Veterinary Pharmacology and Therapeutics, Iowa State University Press, Ames.

Long, P.L., Sheridan, K. and McDougald, L.R. 1981. Maternal transfer of some anticoccidial drugs in the chicken. *Poultry Science*. 60: 2342-2345.

Long,P.L.1993. Avian coccidiosis. *In* J.P.Kreier, ed., Parasitic protozoa, vol. 4, 2nd ed., Academic Press, San Diego. pp.1-88.

Looker ,D.L., Marr, J.J. and Stotish, R.L 1986. Modes of action of anti protozoal agents. In W. L. Campbell and R. S. Rew, eds., Chemotherapy of Parasitic Diseases. Plenum Press, New York. pp. 193-207.

Mathis, G.F. and Mcdougald, L.R. 1981. Experimental development of resistance to amprolium or dinitolmide in Eimeria acervulina and its effect on inhibition of sporulation of oocysts. *J. Parasitol*, 67: 956- 957.

McDougald, L.R. 2003. Coccidiosis.Disease of poultry (11th edn) Iowa State University press, Ames, IA, USA.

Plumb, D.C 2005. Plumb's Veterinary Drug Handbook, Blackwell, Ames, Iowa

Pressman, B.C. 1976. Biological applications of ionophores. *Annual Review of Biochemistry*. 45: 501-530.

Ryley, J, F. 1980. Drug resistance in coccidia. *Ad. Vety.Sc. & Com. Med.* 24: 99-120.

United State Pharmacopeia. 1998. USP Drug information update (September), United States Pharmacopeial Convention, Rockville, Maryland.

Weppelman, R.M., Olson,G., Smith,D.A., Tamas,T., Vaniderstine, A. 1977. Comparison of anticoccidial efficacy, resistance and tolerance of narasin, monensin and lasalocid in chicken battery trials. *Poultry Science*. 56(5): 1550- 1569.

Wesley, J. W. 1982. Polyethers antibiotics; naturally occurring acid ionophores. *In* J.W. Westley (ed.), Biology, vol. 1.Marcel Dekker, Inc., New York.

13

Bioactive Phytochemicals as Modifiers of Rumen Fermentation

Amlan K. Patra

Introduction

For the past few decades, a number of chemical feed additives such as antibiotics, ionophores, methane inhibitors and defaunating agents have been introduced into ruminant nutrition in order to improve rumen fermentation, and increase growth and milk production by improving intake and feed conversion efficiency. However, most of these supplements are not used routinely because of toxicity problems to the host animals and microbial adaptation to these additives. Most importantly, growing awareness from public health aspects has centered on the safety of tissue residues of these chemicals and also bacterial resistance to antibiotics as a result of increased use in the food chain. These supplements have been criticized by the consumers' organizations on the ground of product safety and quality. Antibiotic feed additives are banned in European Union. The consumers' demands have drived the search for natural alternatives to chemical feed additives for eco-friendly animal production. Plants are part of herbivore diets, and these plants contain bioactive compounds with antimicrobial properties can be of interest in animal nutrition. Therefore, recent research has been greatly focused to exploit bioactive plant secondary compounds as natural feed additives to improve rumen fermentation such as enhancing protein metabolism, decreasing methane production, reducing nutritional stress like bloat, and improving animal health and productivity (Wallace *et al.,* 2002; Patra and Saxena, 2009a, b; 2010; 2011), which may be useful in eco-friendly livestock farming (Patra, 2007). This review discusses the effects of different plant metabolites on rumen fermentation, rumen microbial ecosystems and ruminant performance.

Rumen Manipulations to Improve Rumen Fermentation

Improving the fibre degradation in the rumen

Diets of ruminants comprises of low quality roughages in developing countries where the animals usually have low production potential. These feedstuffs are degraded in the rumen by the synergistic activities of the bacteria, protozoa and fungi, with bacteria and fungi contributing approximately 80% of the degradative activity, and the protozoa only 20% (Dijkstra and Tamminga, 1995). The most active fibrolytic bacteria *Fibrobacter succinogenes, Ruminococcus albus* and *Ruminococcus flavefaciens* are generally considered as the primary organisms responsible for degradation of plant cell wall in the rumen while *Butyrivibrio fibrisolvens, Clostridium locheadii* and *Clostridium longisporum* are some of the secondary fibrolytic bacteria. Ruminal fungi produce a broad array of enzymes and generally degrade a wider range of substrates than do rumen bacteria. Furthermore, ruminal fungi are able to degrade the most resistant plant cell wall polymers and the cellulases and xylanases produced by them are among the most active fibrolytic enzymes. The ruminal protozoa also contribute to the degradation of plant cell wall polymers, but their contribution in fibre degradation is considered not as important as that of the bacteria and fungi. This may be due to the fact that in the absence of ciliate protozoa (in defaunated animals) the increased number of bacteria and fungi compensate for the protozoal fibrolytic ability of the rumen (Williams and Coleman, 1997).

Defaunation

The presence of protozoa in the rumen has been shown to stabilize rumen pH and decrease the redox potential of rumen digesta, which should indirectly stimulate the cellulolytic bacterial activity. Furthermore, approximately one-quarter to one-third of fibre degradation and fibre degrading enzymes in the rumen are of protozoal origin (Dijkstra and Tamminga, 1995). However, the engulfment and digestion of bacteria in the rumen by protozoa is the most important mode of microbial protein turnover in the rumen, decreasing the efficiency of protein utilization in ruminants. The removal of protozoa (defaunation) can also reduce methane production substantially and improve feed utilization efficiency depending upon diet composition. Despite the protozoal contribution to ruminal cellulolysis, there is evidence demonstrating the increase in digestibility of fibrous feeds and the potential for improvement in ruminant productivity maintaining the defaunated state (Williams and

Coleman, 1997). Many of the negative effects of defaunation may be transitory and disappear as the bacterial and the fungal population increases and occupy the niche previously filled by the protozoa (Williams and Coleman, 1997). In fact, the effect of defaunation depends on the balance between energy and protein needed for animals and nutrient supplied through diets. Defaunation is useful in case of diet low in true protein, but not limited in energy for animals with high protein requirements (Patra and Saxena, 2009b). Therefore, removal of protozoa has been a target of rumen microbiologist for rumen manipulation for more than three decades.

Inhibition of ruminal methanogenesis

Methane produced during anaerobic fermentation in the rumen represents a feed energy loss and contributes to greenhouse effect in the environment. Therefore, decreasing methane emission has been an important objective for ensuring the sustainability of ruminant production. Methane is produced normally during fermentation of feeds in the rumen by methanogenic archaea. The removal of protozoa can also reduce methane production as some population of methanogens remains attached with protozoa as commensal (Patra and Saxena, 2010).

Improving the protein metabolism in the rumen

The quantity of protein flowing from the rumen is a major factor determining the productivity of ruminant livestock. The protein reaching the abomasum consists of a mixture of dietary and microbial protein. Therefore, increased flow of protein from the rumen depends on decreased proteolysis by the rumen microorganisms and the growth of beneficial microbes and increased efficiency of microbial protein synthesis. When ruminant are fed on a high quality fresh forages containing high concentration of N (25 to 35 g/kg DM), degradation of forage N is excessive, resulting in surplus levels of ammonia (20 to 35%) which is absorbed from the rumen and excreted in urine (Ulyatt *et al.,* 1975). Therefore, a reduction of protein degradation in the rumen will increase the quantity of protein digested in the small intestine.

Inhibition of ruminal biohydrogenation

Increasing conjugated linoleic acid levels in food derived from ruminants has been of recent interests in many researches due to its beneficial effects on human health. Conjugated linoleic acid levels in ruminant-derived foods can be increased by nutritional and management practices that facilitate higher forestomach output of conjugated linoleic acid and vaccenic acid. Many rumen bacteria species are associated in ruminal biohydrogenation, including species of the genera *Butyrivibrio, Ruminococcus, Treponema-Borrelia, Micrococcus, Megasphaera, Eubacterium, Fusocillus* and *Clostridium* (Maia *et al.*, 2007). *Butyrivibrio* group are most active species, where all bacteria form conjugated linoleic acid from linoleic acid, while only *Clostridium proteoclasticum* converts vaccenic acid to stearic acid (Maia *et al.*, 2007). Therefore, selective inhibition of *C. proteoclasticum* without affecting *B. fibrisolvens* may provide more unsaturated acids, including vaccenic acids and conjugated linoleic acid, escaping the rumen for absorption and incorporated into animal tissues (Patra and Saxena, 2011).

Plant Secondary Metabolites

The term plant secondary metabolite is a generic term used to describe a group of chemical compounds in plants that are not primarily involved in the biochemical processes such as plant growth, development and reproduction. These secondary metabolites in plants function as defense mechanisms those ensure survival of their structure and reproductive elements protecting against insect predation or by restricting grazing by herbivores. More than 200,000 different plant secondary metabolites of various classes have been reported in various plants. Some of these metabolites, which have specific to broad range of antimicrobial properties (Patra, 2012), may modify rumen microbial ecosystem.

Effects of saponins on rumen fermentation and rumen microbes

Chemistry of saponins

Saponins are chemical complex compounds found in many plant genera. Chemically, saponins are high molecular weight glycosides in which sugars are linked to a triterpene or steroidal aglycone moiety. The number of sugars, the type of sugars and the stereochemistry of aglycone moieties may vary, producing a diverse array of metabolites in this compound class. Structures of triterpenoid and steroidal saponins are presented elsewhere (Wallace *et al.*, 2002).

Effects of saponins on rumen protozoa

Numerous studies have demonstrated that saponins and saponin-containing plants have toxic effects on protozoa. One of the earliest observation that sarsaponin from *Yucca schidigera* decreased the protozoal population but not bacterial population in *in vitro* continuous culture systems (Valdez *et al.*, 1986). Saponins from *Y. schidigera* have been shown to suppress the growth of rumen ciliates *in vitro* (Wallace *et al.*, 1994) and with rumen simulation technique

(RUSITEC) without affecting dry matter disappearance and other rumen microbial populations (Wang *et al.*, 1998). Butanol extraction of *Y. schidigera* caused defaunation (Makkar *et al.*, 1998; Wang *et al.*, 1998) and the butanol extraction of the plant showed all the anti-protozoal activity, the active component being proved to be saponins. Alfalfa saponin isolated by ethanol extraction and partial acid hydrolysis decreased protozoal number in the rumen of sheep by 34 and 66% at the levels of 2 and 4% of dietary dry matter, respectively (Lu and Jorgensen, 1987). Saponin like material in the organic phase of a butanol-water mixture of *Sesbania sesban* foliage, a leguminous tree, was found to be inhibitory to protozoal activity (Newbold *et al.*, 1997). However, protozoal activity of *S. sesban* varied with the different accessions having different types of saponins, but there had been found to be adaptation to these plants in rumen (Teferedegne, 2000; Wallace *et al.*, 2002). Protozoal activity, as measured by the breakdown of [^{14}C] leucine-labelled *Selenomonas ruminantium* in rumen fluid incubated *in vitro,* was inhibited by the addition of 1% *Y. schidigera* extract. Similarly, saponins from *Quillaja saponaria* and *Acacia auriculoformis* (Makkar *et al.*, 1998) and soapnut (Kamra *et al.*, 2000) have also shown anti-protozoal activity *in vitro*. Hess *et al.* (2003) studied three saponin rich tropical fruits - *Sapindus saponaria* (12% saponins), *Enterolobium cyclocarpum* (1.9% saponins) and *Pithecellobium saman* (1.7% saponins) supplemented to forage based diets in RUSITEC using rumen fluid from faunated and defaunated cow. *S. saponaria* reduced protozoal count (54%) and daily methane release (20%) without affecting the methanogens in faunated rumen fluid. Patra *et al.* (2006a) noted that water, methanol and ethanol extracts of *Acacia concinna* inhibited growth of protozoa *in vitro*, which may be attributed to presence of saponins in it. Agarwal *et al.* (2006) noted that methanol, ethanol and water extracts of berries of *Sapindus mucorossi* inhibited protozoal number by 70-90% in *in vitro* gas production system.

A numer of *in vivo* studies have also confirmed the anitiprozoal effects of saponins or saponin-containing plants. For example, *Y. schidigera* decreased rumen protozoal numbers in heifer (Hristov *et al.*, 1999). Similarly, a decrease in protozoal numbers was reported in the rumen of sheep infused with pure alfalfa saponins (Lu and Jorgensen, 1987) or fed saponin-containing plants, including *S. sesban* (Newbold *et al.*, 1997) and *Enterolobium cyclocarpum* (Navas-Camacho *et al.*, 1993).

The sensitivity of ciliate protozoa towards saponins may be explained by the presence of sterols in protozoa, but not bacterial membranes (Williams and Coleman, 1997). Thus, sterol-binding capability of saponins (Hostettmann and Marston, 1995) most likely causes the destruction of protozoal cell membranes.

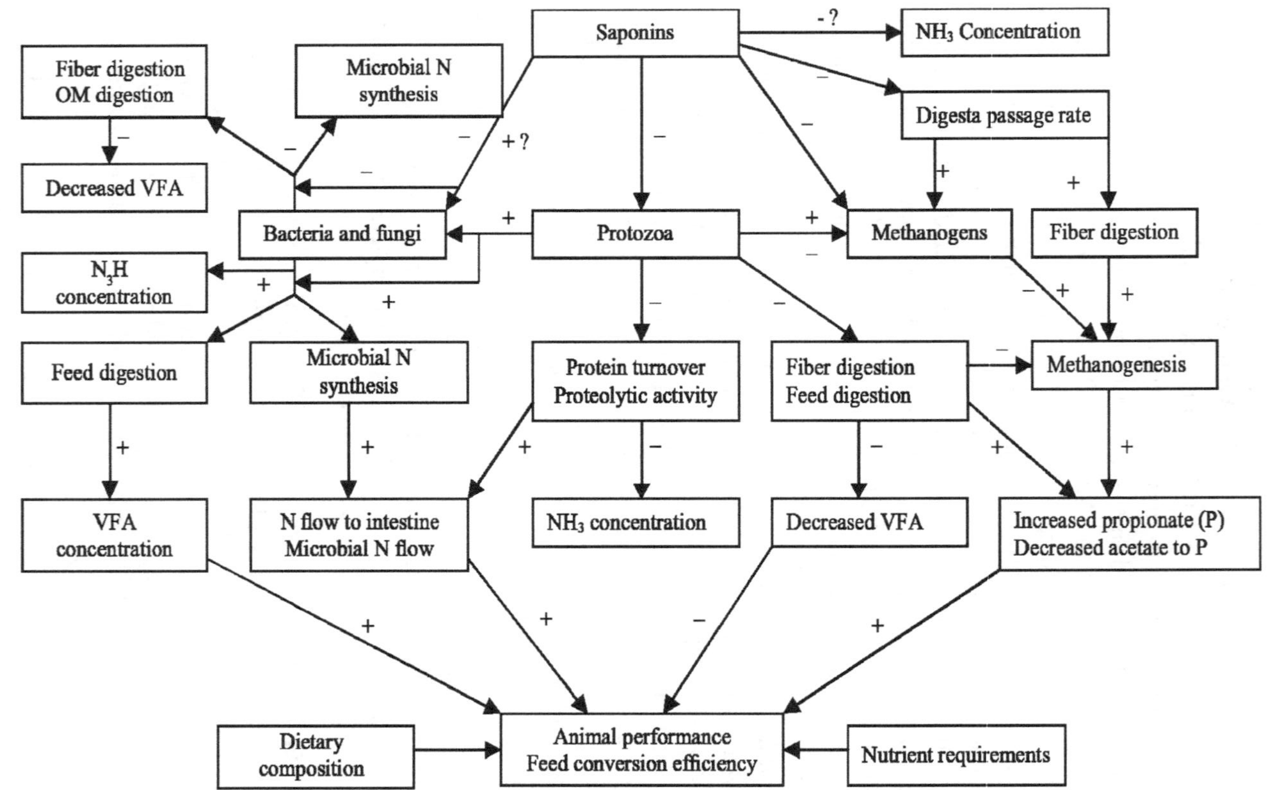

Fig. 1: A schematic presentation of proposed effects of saponins on rumen microbes and fermentation. Primary effects modify the composition of rumen microbes and secondary effects modify the rumen fermentation. The signs '+' and '-' indicate an increase and decrease, respectively, VFA, volatile fatty acids; NH_3, ammonia (Adapted from Patra and Saxena, 2009b).

Effects of saponins on rumen bacteria and fungi

The bacterial growth in the rumen is also affected by the presence of saponins. Studies on pure cultures showed that *Y. schidigera* extract (1% v/v) inhibited the growth of *Butyrivibrio fibrisolvens* and *Streptococcus bovis*, while the growth of *Prevotella ruminicola* was stimulated and the growth of *Selenomonas ruminantium* Z108 was unaffected (Wallace *et al.*, 1994). Wang *et al.* (2000a) also observed that growth of some bacteria, *Streptococcus bovis, Prevotella bryantii* and *Ruminobacter amylophilus* was reduced as a result of alteration of the cell wall of the bacteria, while the growth of *S. ruminantium* was enhanced in the presence of steroidal saponins. Steroidal saponins from *Y. schidigera* had no effect on bacterial count in RUSITEC; however, inoculating fluid from the fermenter into the medium containing saponins decreased the viable count (Wang *et al.*, 1998). Another type of saponin, theasaponin (oleanene-type triterpene) reduced the number of microbes in a culture solution of rumen fluid; however, quantity of gas and propionate produced were increased significantly, while methane and isobutyrate production were decreased (Ye *et al.*, 2001). However, some studies with saponin-containing plants showed increased bacterial numbers, presumbly as a consequence of inhibition of protozoal number (Newbold *et al.*, 1997). Ruminal fungi, *Neocallimastix frontalis* and *Piromyces rhizinflata*, were also sensitive to *Y. schidigera* saponins (Wang *et al.*, 2000a). However, Wina *et al.* (2006) noted that at low levels of *S. rarak* saponins had a positive effect on *Chytridiomycetes*, rumen fungus, during the long-term feeding, while no effect was observed with higher concentration. Muetzel *et al.* (2003) found that the addition of saponin containing *Sesbania pachycarpa* leaves did not inhibit the growth of *F. succinogenes* and *R. flavefaciens,* but negatively affected *R. albus* in *in vitro*. Wina *et al.* (2005) investigated the effect of saponins from a methanol extract of *Sapindus rarak* at concentrations of 0, 0.25, 0.5, 1.0, 2.0 and 4.0 mg/ml to an elephant grass and wheat bran (7:3 w/w) diet *in vitro*. Microbial biomass gradually increased with increasing extract from 102.8 (control) to 147.3 at 4 mg/ml and protozoal count was decreased. Membrane hybridization studies also showed no eukaryotic RNA in the presence of extract higher than 1 mg/ml. Bacterial RNA concentration increased only at 1 mg/ml compared to control and methanogen RNA concentration was lower at the highest concentration (4 mg/ml). There was no effect on *Fibrobacter* spp., however, there was a negative effect on *Ruminococcus albus* and *Ruminococcus flavefaciens* at 2 and 4 mg/ml extract. *Chytridiomycetes* was reduced at the highest level. The xylanase activity and neutral detergent fiber digestibility decreased at higher levels of extract. The *in vivo* trial with sheep fed an extract of *Sapindus rarak* up to 0.72 g/kg body weight and goats fed 0.60 g/kg body weight also confirmed that *Fibrobacter* spp. were not affected by the S. *rarak* extract, while *R. albus, R. flavefaciens* and *Chytridiomycetes* (rumen fungus) showed a short-term response to the application of saponins, but were not affected when this saponins extract was fed longer period (105) days (Wina *et al.*, 2006). In experiment with sheep, Hess *et al.* (2004) reported that supplementation with fruits of *S. saponaria* (250 g/kg concentrate) increased total bacteria count, and decreased total ciliate protozoa count in lambs fed basal diets of grass hay or grass and legume mixture (1:2 or 2: 1) along with concentrate (1:3). Different types of saponins (for example, yucca vs. quillaja saponins) exert different influences on the rumen bacterial and archaeal populations and diversity (Patra *et al.*, 2012a).

Microbial adaptation and metabolism of saponins

One of the problems of using saponins or saponin-containing plants is that anti-protozoal activity was found to be transient (Patra and Saxena, 2009a,b). Teferedgne (2000) noted a time dependence detoxification process of the saponins in ruminal fluid *in vitro*. However, some of the studies suggested that protozoa did not become resistant to these anti-protozoal compounds (Newbold *et al.*, 1997). Therefore, it is possible that bacterial population of the rumen degraded the saponins or saponin-containing plants (Newbold *et al.*, 1997). Wina *et al.* (2006) also reported that there was an adaptation of *R. albus, R. flavefaciens and Chytridiomycetes* (fungi) to saponin when fed over a period of 105 days. Odenyo *et al.* (1997) confirmed that dietary *S. sesban* had no effects in Ethiopian sheep, whereas, other plants were more effective, suggesting that different saponins have different efficacy against protozoa.

Saponins are degraded by rumen microbes to sapogenins, the aglycone remaining after the removal of sugar moiety. Thereafter, the resultant sapogenins are degraded more slowly (Wang *et al.*, 1998). Since the saponins, not the aglycone sapogenins, are toxic to the protozoa, it is likely that cleavage of the sugar moieties of saponins by rumen microbes renders the saponins ineffective to protozoa (Patra and Saxena, 2009b). Wang *et al.* (2000a) suggested that *Fibrobacter succinogenes* apparently deglycosylated the saponins from *Y. shidigera*. The observation made by Odenyo *et al.* (1997) that *S. sesban* introduced directly into the rumen remained toxic to protozoa, but not dietary *S. sesban* implies that chewing caused detoxification, perhaps by salivary amylase, or that the larger particle size protected saponins from degradation. However, the reasons that the anti-protozoal effects of *E. cyclocarpum* were much more persistent than *S. sesban* are not clearly understood (Teferedegne, 2000).

Alfalfa saponins were degraded by 96-98% in the total digestive tract of sheep (Mathison *et al.*, 1999), but mowrin, extracted sapogenins from mahua (*Bassia latifolia*), degraded 46% after 24 h of incubation in *in vitro* (Chahal and Sharma,

1991). Makkar and Becker (1997) observed that quillaja saponin was degraded 100% in *in vitro* medium containing rumen fluid of cow at 24 h of incubation, while no degradation was observed up to 6 h of incubation. Meagher *et al.* (2001) investigated the metabolism of diosgenin-derived saponins in the gastrointestinal tract of sheep. In the rumen, saponins underwent rapid hydrolysis and reduction into epismilagenin, smilagenone, smilagenin and tigogenin. In the duodenum, jejunum and ilium, these products were absorbed and then secreted through bile as conjugated and free sapogenins. The epimerisation of sapogenins was continued in the caecum and colon.

Effects of saponins on rumen fermentation

The effects of saponins due to its anti-protozoal properties are expected to have mixed observations on ruminant productivity depending on the diets and the saponins involved. Ciliate protozoa are primarily responsible for the substantial turnover of bacterial protein (Wallace *et al.*, 2002). Therefore, nitrogen retention may be improved by defaunation. The argument in favour of defaunation also depends on other factors as the fibre degradation decreases due to defaunation. It is usually agreed that removing or suppressing protozoa would result in increased ruminant performance, particularly on low-protein diet. This antiprotozoal effect could explain the reduction of rumen ammonia N concentration observed in different studies conducted with saponins *in vivo* (Pen *et al.*, 2007) and *in vitro* (Busquet *et al.*, 2006). Hess *et al.* (2004) reported that ammonia concentration in ruminal fluid was not significantly affected by the inclusion of *S. saponaria* in the concentrate diets, although it reduced protozoal numbers.

There are varying reports on the effects of saponins on ruminal volatile fatty acids (VFA) concentrations. No difference in total VFA concentrations with supplementation of saponin (Pen *et al.*, 2007) had been reported. However, Lu and Jorgensen (1987) noted a reduction in VFA concentrations when saponin was administered at a level of 4% of DM intake. Some studies reported that saponins supplementation increased propionate and decreased acetate proportion (Devant *et al.*, 2007). Gas and total VFA productions from barley grain were increased by *Y. schidigera*; whereas, those were reduced from alfalfa hay (Wang *et al.*, 2000b). Patra *et al.* (2006a) reported that extracts of *A. concinna* containing saponins that inhibited protozoa number to the extent of 80% did not affect total VFA productions, but increased propionate production, and there was a decrease in acetate to propionate ratio. Similarly, an increased molar proportion of propionate level in the rumen by the *Biophytum* extract that contains saponins had been reported (Santoso *et al.*, 2007). In a study by Hess *et al.*, (2003), *S. saponaria* in RUSITEC did not affect organic matter degradation despite 54% reduction in protozoa counts, and there was 20% reduction in methane production. In another studies, Hess *et al.* (2004) observed that daily methane release was reduced by *S. saponaria* supplementation in lambs fed basal diets of grass hay or grass and legume mixture (1:2 or 2:1) along with concentrate (Hess *et al.*, 2004). Agarwal *et al.* (2006) reported that the depression in methane production was 96% and 39.4%, 20% with ethanol, water and methanol extracts of *S. murkossi*, respectively, compared with control. The digestibility of organic matter and neutral-detergent fibre also reduced in both studies. However, saponins from *A. concinna* extracts did not affect methane production and dry mater digestibility, although it reduced protozoa numbers (Patra *et al.*, 2006a). Pen *et al.*, (2007) reported an increase in neutral detergent fiber digestibility by sheep supplemented with quillaja saponins. It has been observed that effect of *S. saponaria* on methane was more pronounced in defaunated (29%) than faunated (14%) rumen fluid indicating that reduced methane production was not entirely due to associated depression in protozoal count (Hess *et al.*, 2003). Gunjan *et al.* (2007) indicated that methane inhibition effect of *S. sesban* and fenugreek (*Trigonella foenum-graecum* L.) was pronounced in concentrate-based diets compared to roughage-based diets.

In animal feeding trials, with sheep receiving alfalfa saponins and cattle receiving *Y. schidigera* saponins, a decrease in the efficiency of microbial protein synthesis was observed, probably due to the suppression bacterial and protozoal growth (Lu and Jorgensen, 1987; Goetsch and Owen, 1985). Santosa *et al.* (2006) reported that daily supplementation of a basal diet consisting of timothy silage and a concentrate (85:15 on a dry matter (DM) basis) with 240 ppm (DM) *Y. schidigera* reduced rumen ammonia N concentrations compared to the control diet in sheep. The molar proportion of acetate was lower in sheep receiving the *Yucca* diets compared to the control diet, and molar proportions of butyrate and *iso*-acids were higher. Microbial N supply and efficiency of microbial N synthesis were greater in sheep fed the *Yucca* diet *versus* those fed control diets. In another experiment, Santosa *et al.* (2007) reported that saponins from *Biophytum petersianum* decreased ruminal ammonia N concentrations, urinary N and total N excretion, and increased retained N as a proportion of N digested and efficiency of microbial N synthesis when an extract of saponins was infused orally.

Effects of saponins on ruminant performance

It seems that the effect of saponins is diet dependent. Supplementation of 150 mg/day of sarsaponin and 1% urea in corn silage improved the body mass gain of steers during the first 28 days of an experiment (Mader and Brumm, 1987), but by the end of the trial (62 days), there was no significant treatment effect.

However, supplementation of low quality roughage diet (90% concentrate) with 150 mg/kg *Yucca* powder did not improve the weight gain of male lambs (Gorgulu *et al.*, 2004). Mirza *et al.* (2002) reported an improvement in feed efficiency and economics when grazing zebu cattle were fed with *Yucca* extract along with urea molasses mineral block. Few milk yield data from moderate or high yielding cows (20-30 kg/day) indicated that *Yucca* extract in diets containing 10 to 20% crude protein or less than 10% crude protein did not improve milk yield and milk composition (Wilson *et al.*, 1998). The effects of sarsaponins and saponins on feed consumption are contradictory. Some studies have reported increases (Zinn *et al.*, 1998), whereas others (Hussain and Cheeke, 1995) have reported a decrease from 2 to 3%, and others (Hristov *et al.*, 1999) did not find any effect of *Y. schidigera* (rich in sarsaponins and saponins) supplementation on feed consumption. Devant *et al.* (2007) reported a numerical improvement in average daily gain due to the supplementation of a blend of plant extracts (cynarin, gingsen and fenugreek). This has been attributed to the effect on rumen fermentation similar to that of monensin due to the gingseng and fenugreek saponin content. However, water washings of mahua (*Bassia latifolium*), which almost completely removed saponins, in the diet of crossbred calves did not affect the growth and nutrient utilization (Joshi *et al.*, 1989). Even the addition of 250 mg /kg of *Yucca* extract to a mixed diet (45% hay and 50% rolled barley, 5% soybean meal) did not enhance the growth of steers (Hussain and Cheek, 1995). Addition of 30 mg/kg DM of sarsaponin (*Y. schidigera* extract) to a diet of hay: concentrate (1:1) did not increase the daily gain of sheep (Sliwinski *et al.*, 2002). It appears that various saponins have different responses. Therefore, there is need to identify the saponins that would be beneficial for ruminal manipulation and hence ruminant production.

Future research needs to focus on the influence of saponin structure, stabilization of saponins breakdown, the influence of different bacterial species, and the role of the animal's own digestive system in order to maximize their usefulness in ruminant production.

Effect of tannins on rumen fermentation, microbial ecology and ruminant production

Chemistry of tannins

Tannins are water-soluble polyphenolic compounds of high molecular weight having the ability to precipitate protein. Tannins are widely distributed in nutritionally important forage trees, shrubs and legumes, cereals and grains that often limit their utilization as feedstuffs. Tannins are primarly of two distinct types;

Hydrolysable tannins (HT): hydrolysable tannins are derivatives of gallic acid (3, 4, 5-trihydroxy benzoic acid) that is partially or totally esterified to a core polyol such as glucose, glucitol, quinic acids, quercitol and shikimic acid (for structures of tannins, see Haslam, 1989). The resultant galloyl groups may further be esterified or oxidatively crosslinked to yield more complex hydrolysable tannins. Hydrolysable tannins are susceptible to hydrolysis by acids, bases or esterases to yield polyol and the constituent phenolic acids (Haslam, 1989).

Condensed tannins (CT): they are divided into (a) proanthocyanidines that are di-, tri, or polymers of the anthocyanidins and/or catechin flavan-3-ol units and (b) leucoanthocyanidins that are dimers of the flavan-3, 4-diol flavonoids (Haslam, 1989). The flavan-3-ol monomer units are linked by C4-C9 or C4-C6 interflavonoid linkages that give the shape of CT polymer chain (Patra and Saxena, 2011). The number of monomeric units can vary and this determines the degree of polymerisation from di-, tri- and tetraflavonoids to higher oligomers. These can then produce an infinite variety of chemical structures, which in turn can produce different biological properties. Condensed tannins are degraded to form monomeric anthocyanidins (e.g., cyanidin, delphinidin) pigments upon heating in strong acid (Haslam, 1989; Patra and Saxena, 2011). Proanthocyanidins can react by hydrogen bonding with plant protein at near neutral pH to form proanthocyanidin-protein complexes, which are stable and insoluble at pH 3.5-7.0, but dissociate and release protein pH < 3.5 (Jones and Mangan, 1977).

These polyphenolic polymers are of relatively high molecular weight and have capacity to form complexes with carbohydrates and proteins. Low molecular weight CT oligomers are more reactive and have higher protein precipitating capacities than high molecular weight polymeric tannins (Patra and Saxena, 2011).

Effects of tannins on rumen bacteria and fungi

Tannins are generally regarded as inhibitory to the growth of microorganism, which may be due to complexing ability of tannins with the cell wall of bacteria causing morphological changes of the cell wall and the extracellular enzymes secreted (Jones *et al.*, 1994). Bae *et al.* (1993) reported that cell associated and extracellular endoglucanase activity of *Fibrobacter succinogenes* were inhibited by CT (100-400 mg/ml) from *Lotus corniculatus* under *in vitro* condition. Jones *et al.* (1994) studied the effects of CT on growth and proteolysis by four strains of ruminal bacteria. They observed that growth of proteolytic bacteria (*Butyrivibrio fibrisolvens, Ruminobacter amylophilus* and *Streptococcus bovis*) was reduced by CT but a strain of *Prevotella ruminicola* was tolerant to CT from sainfoin (*Onobrychris viciifolia*). Similarly, CT in *L. corniculatus* reduced the populations of *Clostridium proteoclasticum*, *B. fibrisolvens, Eubacterium* spp. and *S. bovis* in sheep (Min *et al.*, 2002). Tagari *et al.* (1965) also observed the inhibition of growth of cellulolytic and proteolytic bacteria by carob tannins in artificial rumen. This indicated that different sources of tannins might have different effect as microbial agents. It has been further suggested that the toxicity of tannins towards microorganisms correlates with their molecular weight. McAllister *et al.* (1994) reported that *Lotus corniculatus* CT caused a considerable detachment of *Fibrobacter succinogenes* S85 from colonised filter paper after a 30 min exposure. Inclusion of 30% *Calliandra* leaves in the diet significantly reduced rumen cellulolytic bacteria including *F. succinogenes* and *Ruminococcus* spp. without affecting the total protozoal population, fungi and proteolytic bacteria (McSweeney *et al.*, 2001b). Although, tannins affected some bacteria, rumen microbial population appeared to adapt and efficiency of microbial protein synthesis (g microbial protein per kg organic matter apparently digested in the rumen) was unaffected. Paul *et al.* (2003) reported that fungus could grow at tannic acid concentration up to 20 g/L and the growth was not appreciably affected up to 10g/L concentration acid. However, fibre-degrading ability of rumen fungi may be less sensitive to the inhibitory effects of CT compared to cellulolytic bacteria (Patra and Saxena, 2009a, 2011; Patra *et al.*, 2012b).

Several species of tannins tolerating bacteria have been identified including *Streptococcus gallolyticus*, strain related to *S. bovis* (Patra *et al.*, 2012b). An anaerobic diplococcoid bacterium, isolated from the rumen fluid of goat fed on *Desmodium ovalifolium* (contains 17% CT), was able to grow in the presence of up to 30 g of tannic acid/litre and this bacterium degraded tannic acid to pyrogallol (Nelson *et al.*, 1995). These bacteria were able to hydrolyze tannic acid or gallic acid. Similarly, Odenyo and Osuji (1998) isolated three strains of tannin-tolerant bacteria related to *Selenomonas* spp. which were able to grow in media containing tannin extract 50 g/litre and tannic acid 50-70 g/litre from sheep, goat and antelope. The tannin degrading microorganisms have also been isolated from several species of ruminants including sheep, goat, deer and elk that have access to tannin rich forages. Tolerance of some bacteria to tannins has been explained due to exo-polysaccharide secreted by these microbes that form a protective layer around the cells (Krause *et al.*, 2005). Brooker *et al.* (1995) also isolated a tannin resistant *S. caprinus* from feral goats browsing *Acacia aneura* (11-14% tannins). *S. caprinus* which was absent in domestic goats and sheep was able to grow in the medium containing 2.5% (w/v) tannic acid. Trans-inoculation of these bacteria of rumen fluid from feral

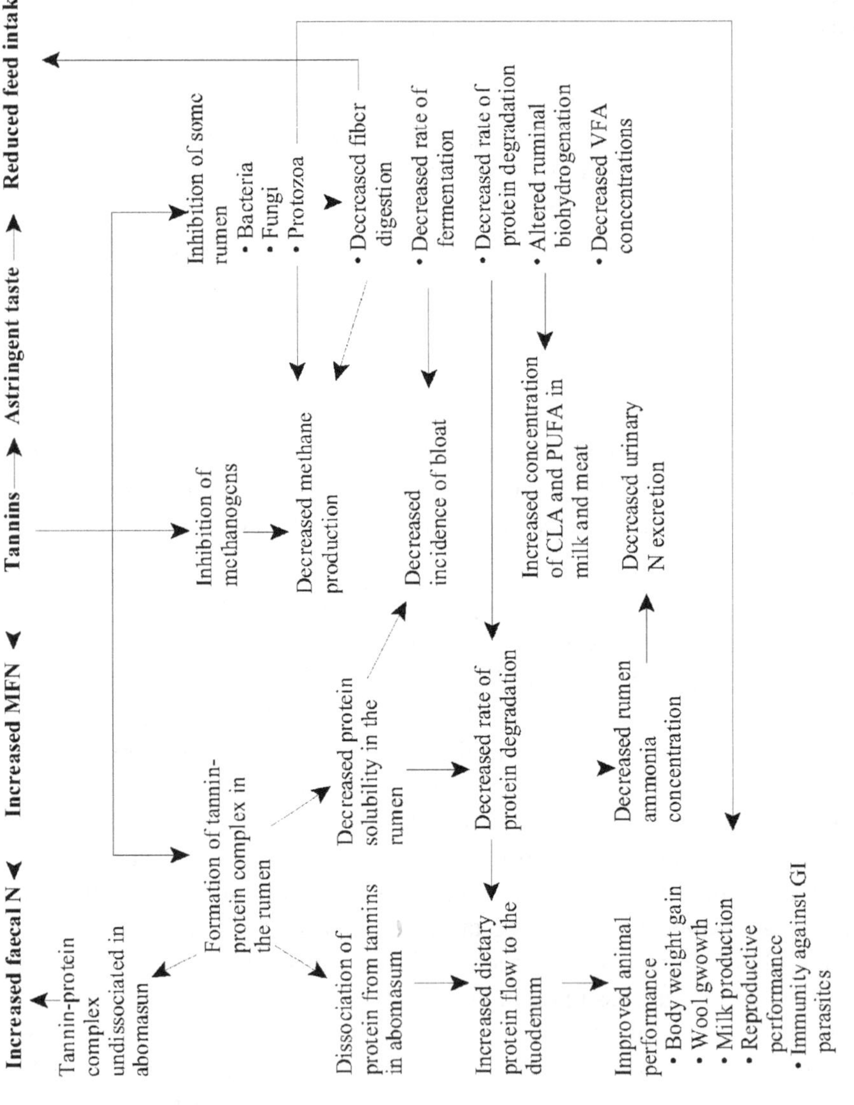

Fig. 2: Schematic representation of the effect of tannins on ruminal metabolism and ruminant performance (Adapted from Patra and Saxena, 2011)

goats to sheep fed on *Acacia* improved dry matter intake and nitrogen digestibility in sheep and the population was maintained for as long as the animal was fed the *Acacia* diet.

Effects of tannins on rumen protozoa

Effects of tannins on rumen protozoa are variable (Patra and Saxena, 2009a; 2011). Newbold *et al.* (1997) established that tannins were not responsible for the anti-protozoal activity of *S. sesban.* However, CT in birdsfoot treefoil (*L. corniculatus*) and sulla (*Hedysarum coronarium*) increased protozoal numbers in the rumen of sheep (Terrill *et al.*, 1992). In contrast, Makkar *et al.* (1995a) reported that quebracho tannin (condensed tannin) significantly reduced the number of total protozoa, entodiniomorphs and holotrichs, the effect being higher on holotrichs, which may increase the efficiency of microbial protein synthesis in the rumen.

Metabolism of tannins in rumen

Rumen microbes are able to degrade simple plant phenolic compounds to some extent to counter anti-nutritive effects of forage tannins. Phenolic glycosides such as rutin and naringin and flavonoid ring system of the common plant flavonoids (e.g. quercitrin) are readily degraded in the rumen by hydrolysis of the glycoside and cleavage of the heterocyclic ring to acetate, butyrate, di- and monohydroxyphenolics, and phloroglucinol (Patra and Saxena, 2011). Tannin degrading bacteria isolated from animals exposed to tannins can reduce tannic acid and CT from the medium (Patra and Saxena, 2011, Patra *et al.*, 2012b). This apparent loss could be due to the degradation of monomeric, dimeric and oligomeric tannins. However, there is no evidence of cleavage of carbon-carbon bonds of heterocyclic ring system of flavan-3-ols (e.g. catechin, epicatechin) that are the monomeric units of CT and these observations need to bc confirmed by definitive experiments on microbial degradation of model CT (McSweeney *et al.*, 2001a). In contrast to CT, the ester bonds and depside bonds of hydrolysable tannins are both probably cleaved in the rumen by organisms such as *Selenomonas ruminantium, Lonepinella koalarum* and *Streptococcus* spp. which produce esterase, tannin acylhydrolase (Nelson *et al.*, 1998) to form metabolites, gallic acid and ellagic acid. Gallic acid is then decarboxylated in the rumen to pyrrogallol and converted to resorcinol and phloroglucinol before cleavage of the phloroglucinol ring to acetate and butyrate by rumen microorganisms (Murdiati *et al.*, 1992).

Effects of tannins on nitrogen metabolism

In ruminant fed fresh forages, most proteins are rapidly solubilized releasing 56 to 65% of the protein concentration in the rumen as soluble protein during mastication; consequently, a large amount of soluble protein is degraded by rumen micro-organisms and is lost as ammonia, absorbed from the rumen (Ulyatt *et al.*, 1975). Condensed tannins in *L. pedunculatus* and *L. corniculatus* forages slowed down the rates of both solubilization and degradation of forage protein to ammonia (Min *et al.*, 2000). Min and Hart (2003) reported that CT progressively increased the duodenal non-ammonia N; whereas, rumen microbial outflow was little affected. Therefore, tannins may be advantageous by protecting dietary protein from degradation in the rumen and thus increasing total supply of protein for absorption from intestine. The absorption process depends upon the reversibility of protein-tannin complex post ruminally as tannins differ in their ability to bind protein at rumen pH (Perez-Maldonado *et al.*, 1995) and occasionally the tannin-protein complexes may not dissociate fully depending upon the postruminal pH and tannin-protein complex that again could be attributed to differences in structures of tannins and proteins.

CT from *L. corniculatus* reduced the rate of proteolysis and inhibited the growth of proteolytic rumen microorganisms (Min and Hart, 2003). Puchala *et al.* (2005) also reported that ruminal ammonia N concentration (3.7 and 9.9 mg/dL) were lower in goats fed CT containing pasture *Sericea lespedeza* than crabgrass/tall fescue that might have been due to reduced protein breakdown in the rumen. Aerts *et al.*, (1999) reviewed that forages containing moderate concentrations of CT (2-4% of DM) can exert beneficial effects on protein metabolism; however, high dietary concentrations (6-12% of DM) may depress voluntary feed intake, digestive efficiency and animal productivity. The *in vitro* studies suggested that the minimal concentration of CT (g/g of protein) needed to reduce proteolysis is 1:10 (wt/wt; Tanner *et al.* 1994) or 1:12 (wt/wt; Jones and Mangan 1977). In contrast, Komolong *et al.* (2001) did not observe any improvement on protein metabolism such as protein degradation and post ruminal protein supply in sheep fed low to moderate levels (0 to 6% of DM) of quabracho tannins as liquid drench.

The effects of CT on plant protein degradation depend upon the chemical structures and molecular weight of tannins (Patra *et al.*, 2012b). For example, Molan *et al.* (2000) showed that trimers and tetramers were more effective in decreasing the rate of proteolysis of soluble leaf protein extracted from white clover than dimmers in *in vitro* incubation media of mixed rumen microbes, and pentamers were more effective than dimers, trimers and tetramers of procyanidins that were extracted and purified from apple skin. At similar concentration of CT (0.25 to 1.75 mg/mg of total soluble protein), CT of *L. pedunculatus* was more effective than *L. corniculatus* to protect plant protein from degradation (Aerts *et al.*, 1999). Proteins also differ greatly in their affinity for phenolics. The proteins having high molecular weight, open and flexible tertiary structure bind strongly with phenolics (Patra and Saxena, 2011). It is assumed that CT form a complex with protein in rumen pH (6 to 7) thus protects the protein being degraded in the rumen. Both hydrolysable and CT form complex compounds with protein by forming hydrogen bonds between the phenolic subunits of the polymer and aliphatic and aromatic side chains (carbonyl groups of peptides) of the protein. However, the condensed tannin-protein complex is not dissociated by the rumen microbes (McSweeney *et al.*, 1999), where as hydrolysable tannin-protein complex is degraded as a result of depolymerization of tannins polymers by cleaving the ester linkages between glucose and the phenolic subunits by the enzymes (tannin acylhydrolases and esterases) secreted by the rumen microbes (Skene and Brooker 1995; McSweeney *et al.*, 2001a). The tannin-protein complex is dissociated in the abomasum at pH less than 3.5 (Barry *et al.*, 1999). However, tannins differ in their ability to bind protein at rumen pH and thus reversibility of the process post ruminally may differ. Tannins may also interact with the digestive enzymes and epithelium lining the digestive tracts if significant amount of tannins reach the duodenum, and therefore reduce amino acids absorption from the intestine (Patra and Saxena, 2011). The reduction in proteolysis might be attributed to direct effects of CT on microbial proteolytic enzyme activity or to indirect effects on rumen metabolite concentration that can regulate proteolytic activity in some bacteria. The CT has shown to inhibit endogenous enzyme activity (Patra and Saxena, 2001). It is suggested that at higher tannins/protein ratio, the inhibition of proteolysis is likely due to polyphenolic compounds covering the protein surface (McManus *et al.*, 1981) leading to interference with the interaction of enzyme and substrate.

Effects of tannins on fiber utilization

Tannins could reduce fibre digestion by complexing with lignocellulose and preventing microbial digestion or by directly inhibiting cellulolytic microorganisms or both (McSweeney *et al.*, 2001a; Patra and Saxena, 2011; Patra *et al.*, 2012b). Although the presence of *Callindra* tannins in the diet (2-3% tannins) reduced the population of fiber degrading bacteria, the efficiency of microbial protein synthesis was not affected. Makkar *et al.* (1995a) studied different sources and levels of tannins on the rate of fermentation of hay in *in vitro*. The decrease in rate of fermentation of dry matter was 3 and 17% for *Q. incana*, 6 and 21% for *D. cinerea* and 7 and 27% for *A. barteri* at 0.47 and 0.93 mg/ml tannins in the medium, respectively. These results suggested that different tannins exhibit different effects on dry matter digestibility at the same inclusion level. Tannins and saponins interaction was found to be additive (Makkar *et al.*, 1995b), which probably suggests that tannins and saponins affect different genera or strains of rumen microbes. Tannins present in *T. chebula* inhibited DM degradability at high concentration *in vitro* (Patra *et al.*, 2006a).

Effects of tannins on ruminal methanogenesis

Many types of forage known to contain CT have been shown to decrease methane production both *in vivo* and *in vitro*. A reduction in methane production was observed in RUSITEC as the proportions of sainfoin (*Onobrychis viciifolia*) incubated were increased (McMohon *et al.*, 1999). Woodward *et al.* (2002) investigated the feeding of sulla on methane emission and milk yield in Friesian and Jersey dairy cows. Cows grazing on sulla had higher daily dry matter intake (13.1 vs. 10.7 kg DM) and daily milk solid production (1.07 vs. 0.81 kg) than grazing perennial ryegrass pasture. Total daily methane emission was similar (253.9 vs. 260 g). However, cows fed sulla produced less methane per kg DM intake (19.5 vs. 24.6 g) and per kg milk solid yield (243.3 vs. 327.8 g). Similar trends in methane emission and milk production have been observed in sheep fed lotus silage (Woodward *et al.*, 2001). There was also a 16% reduction in methane production in lambs fed on *L. pedunculatus* (lotus), which is due to the presence of CT present in this forage (Waghorn *et al.*, 2002). Another CT containing forage *Sericea lespedeza* (17.7% CT) decreased methane emission (7.4 vs. 10.6 g/d and 6.9 vs. 16.2 g/kg DMI for *Sericea lespedeza* and crabgrass/tall fescue, respectively) in Angora goats (Puchala *et al.*, 2005). Dry matter intake (1.11 vs. 0.67 kg/d) and digestible DM intake (0.71 vs. 0.51 kg/d) were greater for *S. lespedeza* than for crabgrass/tall fescue in this study. Animut *et al.* (2007) also observed that feeding of different levels of *Kobe lespedeza* decreased methane production linearly in goats and it has been attributed to the presence of CT (Animut *et al.*, 2008). Smilarly, supplementation of tannin containing extract of *Acacia mearnsii* to forage fed sheep decreased methanogenesis (Carulla *et al.*, 2005). Methane production was also inhibited by inclusion of methanol extract of pericarp of *T. chebula* (a tropical fruit) in *in vitro* up to 90% (Patra *et al.*, 2006a). Ground materials of the pericarp of *T. chebula* also inhibited methane production *in vitro* (Patra *et al.*, 2006b) and in sheep fed 10 g/kg of dry matter intake (Patra *et al.*, 2011). This is assumed to be due to presence of tannins especially hydrolysable tannins in this fruits. Tavendale *et al.* (2005) reported that methane production (ml/g DM) at 12 h for *M. sativa* (25 ml) was higher than *L. pedunculatus* (17.6 ml) that contains 0.10% CT and after addition of polyethelene glycol increased methane production for *L. pedunculatus* (17%), but not for *M. sativa*. They also studied the mechanism of inhibitory effects of extractable condensed tannin fractions from *L. pedunculatus* on the common rumen methanogens *Methanobrevibacter ruminantium*, strains YLM-1 and DSM1093. The oligomeric CT fractions were inactive against both strains, as determined by methane production measurements. The polymeric fraction completely inhibited methane production. Inhibitory effects in broth culture for strain YLM-1 were bacteriostatic, while strain DSM1093 did not recover growth, as indicated by methane production, even upon prolonged incubation. In a plate assay, the zone of inhibition with the polymeric fraction remained after a further week of incubation. The CT in *S. lespedeza* markedly decreased N digestibility and elicited a moderate decline in ruminal methane emission. Supplementation with PEG alleviated the effect of CT on N digestibility but not ruminal methane emission, presumably because of different modes of action (Puchala *et al.*, 2012a). It has been noted that the influence of CT containing *S. lespedeza* on ruminal methane emission was immediate and short-lived, and the effect appeared attributable to activity of methanogenic bacteria and possibly ciliate protozoa (Puchala *et al.*, 2012b). These results indicated that the action of CT on methanogenesis may be attributed to the indirect effect via reduced hydrogen production (hence a reduced supply of substarte for methane formation by methanogens) presumably due to reduced forage digestibility, decreasing protozoal numbers and via direct inhibitory effects on methanogens.

Effects of tannins on ruminant performance

The increased animal performance is observed when the diet contains low levels of tannins, which has generally been attributed to the protection of feed protein from degradation in the rumen, leading to an increase in the flux of essential amino acids to small intestine and an increase in the absorption of essential amino acid to blood (Waghorn and Shelton, 1997). Wang *et al.*, (1996a) reported that lambs grazing *L. corniculatus* had better wool growth and carcass gain than grazing lucerne, which was attributed to the presence of CT (3.4%) in the diets. Similarly, *L. corniculatus* fed to lactating ewes increased milk yield, lactose and protein by 21, 12 and 14%, respectively, during mid and late lactation (Wang *et al.*, 1996b). Similar experiment with dairy cows were conducted by Woodward *et al.* (1999) and they reported that cows fed on *L. corniculaus* had 42% higher milk yields than those fed on *L. corniculatus* + polyethylene glycol (binds with tannins), ryegrass, ryegrass + polyethylene glycol, indicating that CT in *L. corniculatus* caused an increase in milk yield. It was also responsible for 57% increase in milk protein of cows fed on *L. corniculatus*. Feeding of 7.5% tamarind seed husk, a tannin rich by-product to cross-bred dairy cows also resulted in an increased body weight gain and milk protein content in mid lactation (Bhatta *et al.*, 2000). Ramirez-Restrepo *et al.*

(2005) reported higher body growth, reduced parasites burden, and wool production in lambs grazing *L. corniculatus* compared perennial ryegrass (*Lolium perenne*)/white clover (*Trifolium repens*) pasture. One of the reasons for these effects can be possibly due to the increased metabolizable protein supply, from the protein binding action of CT. Besides, potential beneficial biological effects of CT in forages may also be mediated through the control of gastrointestinal parasites. The CT also has the potential to decrease bloat by altering ruminal gas production and soluble protein digestibility (Min *et al.*, 2005).

Effects of flavonoids on rumen fermentation and rumen microbial ecology

Chemistry of flavonoids

Flavonoids are a large class of chemicals constructed basically with A and C rings of benzo-1-pyran-4-quinone and B ring. It has been again classified into (for structure see IUPAC, 1997) following categories;

1 flavones (basic structures): luteolin, diosmetin, chrysoeriol, tangeretin, sinensetin, gardenin, apigenin, vitexin, baicalein, etc.

2. flavonols (having a hydroxyl group at the 3-position): galangin, datiscetin, kaempferol, morin, quercetin, robinetin, isorhamnetin, tamarixetin, quercetagetin, myricetin, etc.

3. flavonones (2-3 bond saturated): hesperetin, taxifolin, eriodictyol, naringenin, etc.

Flavonoids have been suspected to have anti-nutritional characteristics for ruminants and also antibacterial effects, which was found to be more effective against gram-positive bacterial strains than on gram-negative (Patra, 2012).

Effects on rumen fermentation and rumen microbes

Broudiscou *et al.* (2000) studied the effect of thirteen dry plant extracts, selected for their high flavonoid content on rumen fermentation and methanogenesis in continuous cultures of rumen microbes. *Lavandula officinalis* and *Solidago virga-aurea* stimulated fermentation, whereas methanogenesis was decreased with *Equisetum arvense* and *Salvia officinalis*. In another experiment, Broudiscou and Lassalas (2000) studied the effects of *L. officinalis* and *E. arvense* dry extracts, and of isoqercitrin, a flavonoid present in *Equisetum arvense* on fermentation of diets varying in forage contents by rumen microbes in batch culture. The amounts of acetate and propionate produced from 100%

hay diet were increased by the plant extracts, strongly by *L. officinalis* (60% and 37%, respectively) and *E. arvense* (59% and 40%, respectively), to a lesser degree by isoquercitrin (29% and 15%, respectively). In dual outflow fermenters receiving a 50:50 hay-barley diet for seven days, the addition of *L. officinalis* appeared to increase the amount of fermented organic matter, while *E. arvense* tended to inhibit methanogenesis (Broudiscou and Lassalas 2000). *Achillea millefolium* appeared to induce extensive stimulation of microbial metabolism as it increased both degradabilities of crude protein and cell wall constituents and the efficiency and yield of biomass production (Broudiscou *et al.*, 2002).

Unlike flavan-3-ols (monomeric units of CT), the flavonoids of common plants are readily degraded in the rumen by hydrolysis of the glycosides and cleavage of the heterocyclic ring (Patra and Saxena, 2011). The products of flavonoid degradation in the rumen include acetate, butyrate, di- and monohydroxyphenolics, and phloroglucinol. The di- and monohydroxyphenolics are probably released from the B ring while phloroglucinol would be derived from the A ring of the flavonoids (Cheng *et al.*, 1969). The degradation products of flavonoids could modify the microbial metabolism in the rumen (Broudiscou *et al.*, 2002). The bacterial ring fission of aglycone flavonoids leads to the production of 3,4-dihydroxyphenylacetic acid from isoquercutrin and quercitrin, phenylacetic acid from naringenin (Schneider *et al.*, 1999). Some of these simple phenolic compounds may interact with the biosynthesis of aromatic amino acids. Phenylpropanoic acid and phenylacetic acid have been reported to enhance cellulose-degrading bacteria in the rumen (Stack *et al.*, 1986).

Effects of essential oils on rumen microbial ecology and rumen fermentation

Essential oils as feed additives in animal nutrition have been reviewed recently (Wallace *et al.*, 2002). Essential oils were examined many years ago in ruminal bacteria from the point of view of the oils contributing to poor palatability in some plant species (Oh *et al.*, 1968). Estell *et al.* (1998) studied the effect of some volatile compounds (camphor, limonene, cis-jasmone, beta-caryophyllene, borneol and alfa-pinene) on the consumption of alfalfa pellets by sheep. Camphor, alfa-pinene and borneol depressed the consumption of alfalfa pellets, whereas, other three compounds had no discernable effect on consumption.

Chemistry of essential oils

Essential oils are steam-volatile or organic-solvent extracts of plants. They are commonly derived from herbs and spices, and present to some degree in many plants for their protective role against bacterial, fungal or insect attack. They are mainly cyclic hydrocarbons and their alcohol, aldehyde or ester derivatives (Wallace *et al.*, 2002).

Effects of essential oil on rumen bacteria

Essential oils inhibit the hyper-ammonial producing bacteria (HAP) in the rumen, which results in decreased amino acids deamination. McIntosh *et al.* (2003) observed that a mixture of essential oils inhibited growth of some HAP bacteria (i.e., *Clostridium sticklandii* and *Peptostreptococcus anaerobius*), but other HAP bacteria (e.g., *Clostridium aminophilus*) were less sensitive. HAP are present in low numbers in the rumen (<0.01% of the rumen bacterial population), but they possess a very high deamination activity (Wallace *et al.*, 2002). Wallace *et al.* (2002) investigated the rate of degradation of different protein meals and colonisation of feedstuffs incubated in nylon bags by attached enzyme activity in the presence of essential oils. Essential oil had significant effect only on the breakdown of pea meal, the most rapidly degraded meal of the protein meals tested. Bacterial proteinase and amylase associated with plant (pea, rapeseed, etc.) protein supplement tended to be lower in animals receiving essential oils, while corresponding activities associated with fishmeal were unaffected. Glutamate dehydrogenase activity, a measure of total microbial colonization, associated with grass hay suspended in the bovine rumen was decreased by essential oils, while colonization of the less degradable fibrous substrates such as grass silage and barley straw was unaffected. Carboxymethylcellulase activity, a measure of fibrolytic bacterial population was similar. This indicated that essential oils might suppress the colonization and/or digestion of readily degradable substrates by amylolytic and proteolytic bacteria without affecting fibre digestion (Wallace *et al.*, 2002). However, it had been noted that activities of carboxymethcellulase and xylanase were reduced by extracts of clove and fennel (Patra *et al.*, 2010). Wallace (2004) reported that the number of HAP bacteria was reduced by 77% in sheep receiving a low protein diet supplemented with essential oils at 100 mg/day, but that essential oils had no effect on HAP bacteria when sheep were fed a high-protein diet. Individual essential oils had different effects on mixed ruminal bacteria. Monoterpene hydrocarbons were less toxic and sometimes stimulatory to microbial activity compared to the corresponding oxygenated compounds, the monoterpene alcohols and aldehyde (Oh *et al.*, 1968). Total viable count of bacteria was unaffected by essential oil, but HAP that include *Cl. sticklandi* and *Prevotella anaerobius* were decreased by 77% in sheep receiving low protein diet without affecting protozoal numbers or activity (Wallace *et al.*, 2002). HAP bacteria have a high capability to generate ammonia from amino acids (Wallace *et al.*, 2002). Therefore, the inhibition of HAP bacteria by essential oil generally decreases rumen ammonia concentration. However, Castillejos *et al.* (2005) observed that a blend of essential oil added at 1.5 mg/liter increased total VFAs without affecting nitrogen metabolism in dual flow continuous culture fermenters. Therefore, doses and types of the essential oils and type of diet and adaptation to essential oils by rumen microorganism are important to get benefits from it. It is known that HAP comprises only around 1% of the rumen bacterial population. Nevertheless, even a small decrease in the rate of ammonia production may be beneficial nutritionally (Wallace *et al.*, 2002).

Effects of essential oils on rumen protozoa

There are mixed reports on the effects on rumen protozoa. McIntosh *et al.* (2003) observed that the bacteriolytic activity of rumen ciliate protozoa was unaffected in dairy cows supplemented with 1 g/day of mixed essential oil. Similarly, Newbold *et al.* (2004) and Benchaar *et al.* (2007) reported that ruminal protozoa counts were not affected when sheep and dairy cows were fed 110 and 750 mg/day of mixture of essential oil, respectively. Supplementation of dairy cows diets with 0.5 g of cinnamaldehyde per liter of rumen fluid had also no effect on the number of ciliate protozoa (Fraser *et al.*, 2007). The extract of fennel seeds containing essential oil had no affect on protozoa (Patra *et al.*, 2010). In contrast, Ando *et al.* (2003) reported that feeding 200 g/day (i.e. 30 g/kg of total dietary DM) of peppermint (*Mentha piperita* L.) to Holstein steers decreased the total number of protozoa, and the numbers of *Entodinum*, *Isotrica*, and *Diplodium*, which is attributed to essential oils. It had also been observed that clove extract probably containing essential oils decreased numbers of protozoa, small entodiniomomphs and holotrich protozoa, but did not affect large entodiniomorphs (Patra *et al.*, 2010). However, extracts of ginger, garlic and onion increased total protozoal number *in vitro* (Patra *et al.*, 2010), although it was not possible to confirm whether it was due to essential oils or not from that experiment. Likewise, Cardozo *et al.* (2006) observed that addition of a mixture of cinnamaldehyde (180 mg/day) and eugenol (90 mg/day) to the diets of beef heifers increased numbers of holotrichs, and had no effect on entodiniomorphs, but there was no effect on numbers of those protozoal species when the mixture contained higher concentrations of cinnamaldehyde (600 mg/day) and eugenol (300 mg/day). In contrast, feeding 2 g/day of anise extract containing 100 g/kg of anethol to beef heifers' decreased counts of holotrichs and entodiniomorphs (Cardozo *et al.*, 2006). Overall, essential oils and their components might not have marked effect on numbers and/or activity of ruminal ciliate protozoa.

Effects of essential oils on rumen fermentation

The main effects of essential in the rumen involve reduction of protein and starch degradation and an inhibition of amino acid degradation, due to selective action on certain rumen microorganisms, specifically some bacteria (Patra, 2011). One mode of action suggested for essential oils have an effect on the pattern of bacterial colonisation in particular starch rich substrates as they enter the rumen. A second possible mode of action is their inhibition of HAP involved in amino acid deamination. However, the effect of essential oils depends on the chemical structures of the essential oils used.

Yang *et al.* (2007) reported that ruminal digestibilities of dry matter and organic matter were higher (13%) for garlic oil (5 g/d) and juniper oil (2 g/d) than for the control diet consisting of 40% forage and 60% barley-based concentrate in Holstein cows. However, total tract digestibilities of dry matter, organic matter, fiber, and starch were not affected by experimental treatments. They suggested that increased ruminal digestibility was due to the increased ruminal digestion of dietary protein by 11% as compared with the control. The dry matter intake, milk production, ruminal microbial protein synthesis, ruminal pH, and ruminal concentrations of volatile fatty acids and ammonia N were not affected by garlic and juniper oils. There is also concerns that residue of essential oil in milk and meat may affect the quality of these products.

Evans and Martin (2000) observed that thymol (400 mg/l), a main component of essential oil derived from *Thymus* and *Origanum* plants, was a strong inhibitor of methane *in vitro*, but acetate and propionate concentrations also decreased. Methanol and ethanol extracts of fennel, garlic and clove inhibited *in vitro* methane production, which was also accompanied by reduction of degradability of organic matter by cloves, whereas as extracts of garlic and fennel had no effects on degradability of organic matter (Patra *et al.*, 2010). Acetate to propionate ratio increased with cloves extract, while it was decreased by garlic extracts in this experiment. Evaluating the effects of garlic oil and four of its main components (diallyl sulfide, diallyl disulfide, allyl mercaptan, and allicin), Busquet *et al.* (2005) observed, in batch culture that garlic oil and diallyl disulfide (300 mg/l of ruminal fluid) reduced methane production by 74 and 69%, respectively, without altering digestibility of nutrients. Busquet *et al.* (2005) suggested that garlic oil and diallyl disulfide inhibited methane production due to the direct inhibition of rumen methanogenic archaea. In experiment with sheep fed on wheat straw and concentrate (1:1), inclusion of garlic at 10 g/kg of dry matter intake reduced methane inhibition per unit organic matter digested and increased digestibility of fiber (Patra *et al.*, 2010). Increased digestibility due to feeding of garlic might have been due to increased number of fiber degradaing microbial population as observed in fistulated buffaloes fed on same diet (Patra *et al.*, 2008a). Patra and Yu (2012) tested five different types of essential oils, and noted that they have different efficacy in inhibiting methanogenesis, rumen fermentation, rumen bacterial and archeal populations *in vitro*.

Other bioactive plant secondary compounds

Many other plant secondary compounds have been known to affect rumen microorganisms. Allyl isothiocyanate, a plant metabolites present in horseradish oil, decreased methane production significantly (19%) when horseradish oil was fed at 2% of feed DM to steers without affecting number of protozoa in ruminal fluid (Mohammed *et al.*, 2004). Lila *et al.* (2003) also noted that synthetic allyl isothiocyanate inhibited *in vitro* methane production and ruminal methanogenic bacteria. Neem seed kernel cake caused a decrease in protozoal populations both *in vitro* (Patra *et al.*, 2006a) and *in vivo* (Mondal and Garg, 2002), which is possibly due to bitter principles like nimbin, nimbidin, azadirachtin present in it, but it did not decrease methane production *in vitro* (Patra *et al.*, 2006a). There was a decrease in acetate to propionate ratio, enzyme activities and dry matter degradability due to addition of the extracts of *A. indica* in *in vitro* incubation medium (Patra *et al.*, 2006a). However, Agarwal *et al.* (1991) reported that water extract of neem seed kernel cake stimulated fibre degrading enzymes of rumen when tested in *in vitro*.

Recently, Bodas *et al.* (2008) screened several plants for their possible antimethanogenic effects. Out of 450 plants, 35 plants decreased methane production by more than 15%, and 6 of these plant additives decreased methane production by more than 25%, with no adverse effects on digestibility, total gas and VFA production. *Rheum nobile* and *Carduus pycnocephalus* consistently decreased methane production without adversely affecting other parameters of the rumen fermentation. Garcia-Gonzalez *et al.* (2008) reported that rhizomes and roots of *Rheum officinale* (rhubarb) and bark of *Frangula alnus* (frangula or alder buckthorn) might contain active secondary compounds against methanogens after screening of 158 plant materials. The inhibition of methanogenesis was also noted with the extracts of leaves of *Psidium guajava* (guava), *Populus deltoides* (poplar) and *Magnifera indica* (mango) (Kamra *et al.*, 2008; Patra *et al.*, 2008b). It has been observed that *P. deltoides* inhibited methane production, decreased acetate to propionate ration with out affecting dry matter degradability and protozoal counts (Patra *et al.*, 2008b). Goel *et al.*, (2007) reported that leaves of *Carduus pycnocephalus* decreased methane production by 21% and suggested that active compound present in this leaves was non-tannin and non-saponins in nature.

The increased proportions of conjugated linoleic acids in milk and intramuscular fat of ruminants fed botanically diverse forages have been associated with increased forestomach outflow of vaccenic acid, which may be attributed to the presence of plant bioactives in herbs of botanically diverse forages. Feeding of red and white clover increases forestomach outflow of linoleic acids. It has been suggested that polyphenol oxidase particularly active in red clover inhibits rumen lipolysis (Lourenco *et al.*, 2007). Durmic *et al.* (2007) screened 91 Australian plants for their effect on bacteria responsible for biohydrogenation of lipids. Plant secondary compounds extracted with ethanol and essential oils from a wide range of plants had selective inhibitory effect towards *Clostridium proteoclasticum* (which forms stearate from linoleic acid) without affecting *Butyrivibrio fibrisolvens* (which forms conjugated linoleic acids and vaccenic acid, but not stearate). However, only a few plants, including *Acacia iteaphylla* and *Kennedia eximia*, inhibited linoleic acid metabolism or stearate formation in the mixed bacterial community *in vitro*. However, it needs *in vivo* evaluations whether these results can be nutritionally significant to increase conjugated linoleic acid conten in ruminant derived food products.

Summary

Plants or their extracts with high concentration of plant secondary metabolites such as saponins, tannins, flavonoids, essential oils and many other metabolites seem to have potential to act as natural rumen manipulators. The possible mechanisms and effects of many plant secondary metabolites on rumen fermentation are not well defined. The chemical structures and molecular weight of the plant secondary metabolites, chemical composition of diets, and rumen environmental conditions depending upon the different feeding regimes and geographical distribution may explain some of the discrepancy of research findings. These phytochemicals such as tannins, saponins, and essential oils influence differently on the rumen fermentation characteristics. Nevertheless, these natural products can be used for increasing the effectiveness of environment friendly organic livestock farming in near future. They are also generally regarded as safe and may bypass many of the regulatory hurdles that are required before the introduction of a new synthetic compound into feedstuffs. Identifying the potential of phytoadditives for rumen manipulation and inclusion of these plants or their extracts containing high content of secondary metabolites into feeding regimes of animals may be economically feasible, easily available and acceptable to the farmers for sustainable animal production. Again, some plant secondary metabolites could be useful against pathogenic bacteria and gastro-intestinal parasites. Enormous opportunities will emerge to improve livestock production using plant bioactives as feed additives after banning of chemical feed additive. The beneficial effects of these metabolites can also be used intensively if more active metabolites can be isolated and identified from plants or biotechnological tools can be employed to produce target bioactive principles. There is need for systematic and comprehensive research for evaluation of plant compounds on their effect on rumen fermentation and microbial interactions based on chemical structures of these plant secondary compounds to obtain consistent beneficial effects on ruminant production.

References

Aerts, R.J., T.N. Barry, and W. C. McNabb. 1999. Polyphenols and agriculture: beneficial effects of proanthocyanidins in forages. *Agric. Ecosys. Environ.* 75:1-12.

Agarwal, N., N. Kewalramani, D. N. Kamra, D. K. Agrawal, and K. Nath. 1991. Effects of water extracts of neem (*Azadirachta indica*) cake on the activity of hydrolytic enzymes of mixed rumen bacteria from buffalo. *J. Sci. Food Agric.* 57:147-150.

Agarwal, N., D. N. Kamra, L. C. Chaudhary, and A. K. Patra. 2006. Effect of *Sapindus mukorossi* extracts on in vitro methanogenesis and fermentation characteristics in buffalo rumen liquor. *J. Appl. Anim. Res.* 30:1-4.

Ando, S., T. Nishida, M. Ishida, K. Hosoda, and E. Bayaru. 2003. Effect of peppermint feeding on the digestibility, ruminal fermentation and protozoa. *Lives. Prod. Sci.* 82:245-248.

Animut, G., A. L. Goetsch, R. Puchala, A. K. Patra, T. Sahlu, V. H. Varel, and J. Wells. 2008. Methane emission by goats consuming different sources of condensed tannins. *Anim. Feed Sci. Technol.* 144:228-241.

Animut, G., A. L. Goetsch, R. Puchala, A. K. Patra, T. Sahlu, V. H. Varel, and J. Wells. 2007. Methane emission by goats consuming diets with different levels of condensed tannins from lespedeza. *Anim. Feed Sci. Technol.* 144:212-227.

Bae, H. D., T. A. McAllister, L. J. Yanke, K. J. Cheng, and A. D. Muir. 1993. Effect of condensed tannins on endoglucanase activity and filter paper digestion by *Fibrobacter succinogenes* S85. *Appl.Environ. Microbiol.* 59:2132-2138.

Barry, T. N. and W. C. McNabb. 1999. The implications of condensed tannins on the nutritive value of temperate forages fed to ruminants. *Br. J. Nutr.* 81:263-272.

Benchaar, C., H. V. Petit, R. Berthiaume, D. R. Ouellet, J. Chiquette, and P. Y. Chouinard. 2007. Effects of essential oils on digestion, ruminal fermentation, rumen microbial populations, milk production, and milk composition in dairy cows fed alfalfa silage or corn silage. *J. Dairy Sci.* 90:886-897.

Bhatta, R., U. Krishnamurty, and F. Mohammed. 2000. Effect of feeding tamarind (*Tamarindus indica*) seed husk as a source of tannin on dry matter intake, digestibility of nutrients and production performance of cross-bred dairy cows in mid laction. *Anim. Feed Sci. Technol.* 83:67-74.

Bodas, R., S. Lopez, M. Fernandez, R. Garcia-Gonzalez, A. B. Rodriguez, R. J. Wallace, and J. S. Gonzalez. 2008. *In vitro* screening of the potential of numerous plant species as antimethanogenic feed additives for ruminants. *Anim. Feed Sci. Technol.* 145:245-258.

Brooker, J. D., D. K. Lum, S. Miller, I. Skene, and L. O'Donovan. 1995. Rumen microorganisms as providers of high quality protein. *Livst. Res. Rural Develop.* 6, http://www.cipav.org.co/lrrd/lrrd6/3/1.htm

Broudiscou, L. P., Y. Papon, and A. F. Broudiscou. 2002. Effects of dry plant extract on feed degradation and proportion of rumen microbial biomass in a dual flow ferments. *Anim. Feed Sci. Technol.* 101:183-189.

Broudiscou, L. P., and B. Lassalas. 2000. Effects of *Lavandula officinalis* and *Equisetum arvense* dry extracts and isoquercitrin on the fermentation of diets varying in forage contents by rumen microorganisms in batch culture. *Reprod. Nutr. Develop.* 40:431-440.

Broudiscou, L.P., Y. Papon, and A. F. Broudiscou. 2000. Effect of dry plant extract on fermentation and methanogenesis in continuous culture of rumen microbes. *Anim. Feed Sci. Technol.* 87:263-277.

Busquet, M., S. Calsamiglia, A. Ferret, M. D. Carro, and C. Kamel. 2005. Effect of garlic oil and four of its compounds on rumen microbial fermentation. *J. Dairy Sci.* 88:4393-4404.

Busquet, M., S. Calsamiglia, A. Ferret, and C. Kamel. 2006. Plant extracts affect in vitro rumen microbial fermentation. *J. Dairy Sci.* 89:761-771.

Cardozo, P. W., S. Calsamiglia, A. Ferret, and C. Kamel. 2006. Effects of natural plant extracts on ruminal protein degradation and fermentation profiles in continuous culture. *J. Anim. Sci.* 82:3230-3236.

Carulla, J. E., M. Kreuzer, A. Machmuller, and H. D. Hess. 2005. Supplementation of *Acacia mearnsii* tannins decreases methanogenesis and urinary nitrogen in forage-fed sheep. *Aust. J. Agric. Res.* 56:961-970.

Castillejos, L., S. Calsamiglia, A. Ferret, and R. Losa. 2005. Effects of a specific blend of essential oil compounds and the type of diet on rumen microbial fermentation and nutrient flow from a continuous culture system. *Anim. Feed Sci. Technol.* 119:29-41.

Chahal, S. M., and D. D. Sharma. 1991. In vitro ruminal degradation of mowrin-an incriminating factor of mahua (*Bassia latifolia*) seed cake. *Indian J. Anim. Nutr.* 8:143-144.

Cheng, K. J., G. A. Jones, F. J. Simpson, and M. P. Bryant. 1969. Isolation and identification of rumen bacteria capable of anaerobic rutin degradation. *Can. J. Microbiol.* 15:1365-1371

Devant, M., A. Anglada, and A. Bacha. 2007. Effects of plant extract supplementation on rumen fermentation and metabolism in young Holstein bulls consuming high levels of concentrate. *Anim. Feed Sci. Technol.* 137:46-57.

Dijkstra, J., and S. Tamminga. 1995. Simulation of the effects of diet on the contribution of rumen protozoa to degradation of fibre in the rumen. *Br. J. Nutr.* 74:617-634.

Durmic, Z., C. S. McSweeney, G. W. Kemp, P. Hutton, R. J. Wallace, and P. E. Vercoe. 2007. Australian plants with potential to inhibit bacteria and processes involved in ruminal biohydrogenation of fatty acids. *Anim. Feed Sci. Technol.* 145:271-284.

Estell, R. E., E. L. Fredrikson, M. R. Tellez, K. M. Havstad, W. L. Shupe, D. M. Anderson, and M. D. Remmenga. 1998. Effects of volatile compounds on consumption of alfalfa pellets by sheep. *J. Anim. Sci.* 76:228-232.

Evans, J. D., and S. A. Martin. 2000. Effects of thymol on ruminal microorganisms. *Curr. Microbiol.* 41:33-340

Fraser, G. R., A. V. Chaves, Y. Wang, T. A. McAllister, K. A. Beauchemin, and C. Benchaar. 2007. Assessment of the effects of cinnamon leaf oil on rumen microbial fermentation using two continuous culture systems. *J. Dairy Sci.* 90:2315-2328.

Garcia-Gonzalez, R., S. Lopez, M. Fernandez, R. Bodas, and J. S. Gonzalez. 2008. Screening the activity of plants and spices for decreasing ruminal methane production in vitro. *Anim. Feed Sci. Technol.* 147:36-52.

Goel, G., H. P. S. Makkar, and K. Becker. 2007. Effect of *Sesbania sesban* and *Carduus pycnocephalus* leaves and fenugreek (*Trigonella foenum-graecum* L.) seeds and their extracts on partitioning of nutrient from roughage and concentrate based feeds to methane. *Anim. Feed Sci. Technol.* 147:72-89.

Goetsch, A. L., and F. N. Owen. 1985. Effects of sarsaponin on digestion and passage rates in cattle fed medium to low concentrates. *J. Dairy Sci.* 68:2377-2384.

Gorgulu, M., S. Yurtseven, I. Unsal, H. R. Kutle. 2004. Effect of *Yucca* powder on fattening performance male lambs. *J. Appl. Anim. Res.* 25:33-36.

Haslam, E. 1989. *Plant polyphenols*. Cambridge University Press, Cambridge, UK

Hess, H. D., R. R. Beuret, M. Lotscher, I. K. Hindrichsen, A. Machmüller, J. E. Carulla, C. E. Lascano, and M. Kreuzer. 2004. Ruminal fermentation, methanogenesis and nitrogen utilization of sheep receiving tropical grass hay-concentrate diets offered with *Sapindus saponaria* fruits and *Cratylia argentea* foliage. *Anim. Sci.* 79:177-189.

Hess, H. D., L. M. Monsalve, C. E. Lascano, J. E. Carulla, T. E. Diaz, and M. Kreuzer. 2003. Supplementation of a tropical grass diet with forage legumes and *Sapindus saponaria* fruits: effects on in vitro ruminal nitrogen turnover and methanogenesis. *Aust. J. Agric. Res.* 54:703-713.

Hostettmann, K, and A. Marston. 1995. *Saponins*. Cambridge University Press, Cambridge, UK

Hristov, A. N., T. A. McAllister, F. H. van Herk, C. J. Newbold, and K. J. Cheng. 1999. Effect of *Yucca schidigera* on ruminal fermentation and nutrient digestion in heifer. *J. Anim. Sci.* 77:2554-2563.

Hussain, I, and P. R. Cheeke. 1995. Effect of dietary *Yucca schidigera* extract on rumen and blood profiles of steers fed concentrate- or roughage-based diets. *Anim. Feed Sci. Technol.* 51:231-242.

IUPAC. 1997. IUPAC compendium of chemical terminology, second edition, International Union of Pure and Applied Chemistry, NY, USA.

Jones, G A., T. A. McAllister, A. D. Muir, and K. J. Cheng. 1994. Effects of sainfoin (*Onobrychis viciifolia* Scop.) condensed tannins on growth and proteolysis by four strains of ruminal bacterium. *Appl. Environ. Microbiol.* 60:1374-1378.

Jones, W. T., and J. L. Mangan. 1977. Complexes of the condensed tannins of sainfoin (*Onobrychis viciifolia* Scop.) with fraction. 1 Leaf protein and with submaxillary mucoprotein and their reversal by polyethelene glycol and pH. *J. Sci. Food Agric.* 28:126-136.

Joshi, D. C., R. C. Katiyar, M. Y. Khan, R. Banerji, G Misra, and S. K. Nigam. 1989. Studies on mahua (*Bassia latifolia*) seed cake saponin (mowrin) in cattle. *Indian J. Anim. Nutr.* 6:13-17.

Kamra, D. N., R. Singh, N. Agarwal, and N. N. Pathak. 2000. Soapnut (reetha) as natural defaunating agent - its effect on rumen fermentation and *in sacco* degradability of jowar hay in buffaloes. *Buffalo J.* 16:99-104.

Kamra, D. N., A. K. Patra, P. N. Chatterjee, R. Kumar, N. Agarwal, and L. C. Chaudhary. 2008. Effect of plant extract on methanogenesis and microbial profile of the rumen of buffalo: a brief overview. *Aust. J. Exp. Agric.* 48:175-178.

Komolong, M. K., D. G Barber, and D. M. McNeill. 2001. Post-ruminal protein supply and N retention of weaner sheep fed on a basal diet of lucerne hay (*Medicago sativa*) with increasing levels of quebracho tannins. *Anim. Feed Sci. Technol.* 92:59-72.

Krause, D. O., W. J. M. Smith, J. D. Brooker, and C. S. McSweeney. 2005. Tolerance mechanisms of *Streptococci* to hydrolysable and condensed tannins. *Anim. Feed Sci. Technol.* 121: 59-75.

Lila, Z. A., N. Mohammed, S. Kanda, T. Kamada, and H. Itabashi. 2003. Effect of á-cyclodextrin allyl isothiocyanate on ruminal microbial methane production in vitro. *Anim. Sci. J.* 74: 321-326.

Lourenco,[a] M., G van Ranst,[b] B. Vlaeminck,[a] S. D. Smet,[a] and V. Fievez. 2007. Influence of different dietary forages on the fatty acid composition of rumen digesta as well as ruminant meat and milk. *Anim. Feed Sci. Technol.* 145:418-437.

Lu, C. D., and N. A. Jorgensen. 1987. Alfalfa saponins affect site and extent of nutrient digestion in ruminants. *J. Nutr.* 117:919-927.

Mader, T. L., and M. C. Brumm. 1987. Effect of feeding sarsaponin in cattle and swine diets. *J. Anim. Sci.* 65:9-15.

Maia, M. R., L. C. Chaudhary, L. Figueres, and R. J. Wallace. 2007. Metabolism of polyunsaturated fatty acids and their toxicity to the microflora of the rumen. *Antonie Van Leeuwenhok.* 91:303-314.

Makkar, H. P. S., M. Blummel, and K. Becker. 1995a. In vitro effects and interaction between tannins and saponins and fate of tannins in the rumen. *J. Sci. Food and Agric.* 69: 481-493.

Makkar, H. P. S., K. Becker., H. Abel, and C. Szegletti. 1995b. Degradation of condensed tannins by rumen microbes exposed to querbracho tannin (QT) in rumen simulation technique (RUSITEC) and effects of QT on fermentative processes in the RUSITEC. *J. Sci. Food and Agric.*69:495-500.

Makkar, H. P. S., and K. Becker. 1997. Degradation of *Quillaja* saponins by mixed culture of rumen microbes. *Lett. Appl. Microbiol.* 25:243-245.

Makkar, H. P. S., S. Sen, M. Blummel, and K. Becker. 1998. Effects of fractions containing saponins from *Yucca schidigera, Quillaja saponaria* and *Acacia auriculoformis* on rumen fermentation. *J. Agric. Food Chem.* 46:4324-4328.

Mathison, G. W., S. R. Soofi, P. T. Klita, E. K. Okine, and G. Sedgwick. 1999. Degradability of alfalfa saponins in the digestive tract of sheep and their rate of accumulation in rumen fluid. Can. *J. Anim. Sci.*79, 315-319.

McAllister, T. A., H. D. Bae, G. A. Jones, and K. J. Cheng. 1994. Microbial attachment and feed digestion in the rumen. *J. Anim. Sci.* 72:3004-3018.

McIntosh, F. M., P. Williams, R. Losa, R. J. Wallace, D. A. Beever, and C. J. Newbold. 2003. Effects of essential oils on ruminal microorganisms and their protein metabolism. *Appl. Environ. Microbiol.* 69:5011-5014.

McManus, J. P., K. G. Davis, T. H. Lilley, and E. Haslam. 1981. The association of protein with polyphenols. *J. Chem. Soc. Chem. Commun.* 7:309-311.

McSweeney, C. S., B. Palmer, R. Bunch, and D.O. Krause. 1999. Isolation and characterization of proteolytic ruminal bacteria from sheep and goats fed the tannin-containing shrub legume *Calliandra calothyrsus. J. Appl. Environ. Mircobiol.* 65:3075-3083.

McSweeney, C. S., B. Palmer, D. M. McNeill, and D. O. Krause. 2001a. Microbial interactions with tannins: nutritional consequences for ruminants. *Anim. Feed Sci. Technol.* 91: 83-93.

McSweeney, C. S., B. Palmer, R. Bunch, and D. O. Krause. 2001b. Effect of tropical forage *Callindra* on microbial protein synthesis and ecology in the rumen. *J. Appl. Microbiol.* 90: 78-88.

Meagher, L. P., B. L. Smith, and A. L. Wilkins. 2001. Metabolism of diosgenin derived saponins: implications for hepatogenous photosensitization diseases in ruminants. *Anim. Feed Sci. Technol.* 91:157-170.

Min, B. R., McNabb, W. C., T. N. Barry, and J. S. Peters. 2000. Solubilization and degradation of ribulose-1,5-*bis*-phosphate carboxylase/oxygenase (*EC* 4.1.1.39; Rubisco) protein from white clover (*Trifolium repens*) and *Lotus corniculatus* by rumen microorganisms and the effect of condensed tannins on these processes. *J. Agric. Sci.* 134:305-317.

Min, B.R., G.T. Attwood, K. Reilly, W. Sun, J.S. Peters, T.N. Barry, and W.C. McNabb. 2002. *Lotus corniculus* condensed tannins decrease in vivo populations of proteolytic bacteria and affect nitrogen metabolism in the rumen of sheep. *Can. J. Anim. Sci.*48:911-921.

Min, B. R. and S. P. Hart. 2003. Tannins for suppression of intestinal parasites. *J. Anim. Sci.* 81:(E Suppl. 2), E102-E109.

Min, B. R., W. E. Pinchak, J. D. Fulford, and R. Puchala. 2005. Wheat pasture bloat dynamics, in vitro ruminal gas production, and potential bloat mitigation with condensed tannins. *J. Anim. Sci.* 83:1322-1331.

Mirza, I. H., A. G. Khan, A. Azim, and M. A. Mirza. 2002. Effect of supplementing grazing cattle calves with urea-molasses blocks, with and without *Yucca schidigera* extract, on performance and carcass traits. Asian-Aust. *J. Anim. Sci.*15:1300-1306.

Mohammed, N., N. Ajisaka, Z. A. Lila, K. Hara, K. Mikuni, K. Hara, S. Kanda, and H. Itabashi. 2004. Effect of Japanese horseradish oil on methane production and ruminal fermentation in vitro and in steers. *J. Anim. Sci.* 82:1839-1846.

Molan, A.L., L.Y. Foo, and W.C. McNabb. 2000. The effect of different molecular weight procyanidins on in vitro protein degradation. *Asian-Aust. J. Anim. Sci.*13 (Suppl), 215-218.

Mondal, D.P., and A.K. Garg. 2002. Effect of feeding untreated and water washed neem (*Azadiracta indica* a juss) seed kernel cake on rumen enzyme profile and fermentation pattern in crossbred calves. *Anim. Nutr. Feed Technol.* 2:27-37.

Muetzel, S., E. M. Hoffmann, and K. Becker. 2003. Supplementation of barley straw with *Sesbania pachycarpa* leaves in vitro, effects on fermentation variables and rumen microbial concentration structure quantified by ribosomal RNA-targeted probes. *Br. J. Nutr.* 89: 445-453.

Murdiati, T. B., C. S. McSweeney, and J. B. Lowry. 1992. Metabolism in sheep of gallic acid, tannic acid and hydrolysable tannin from *Terminalia oblongata. Aust. J. Agric. Res.* 43:1307-1319.

Nelson, K. A., P. Schofield, and S. Zinder, S. 1995. Isolation and characterization of an anaerobic bacterium capable of degrading hydrolysable tannins. *Appl. Environ. Microbiol.* 61:3293-3298.

Nelson, K. A., M. L. Thonney, T. K. Woolston, S. H. Zinder, and A. N. Pell. 1998. Phenotypic and phylogenic characterization of ruminal tannin-tolerant bacteria. *Appl. Environ. Microbiol.* 64:3824-3830.

Newbold, C. J., S. M. E. Hassan, J. Wang, M. E. Ortega, and R. J. Wallace. 1997. Influence of foilage from African multipurpose tree on activity of rumen protozoa and bacteria. *Br. J. Nutr.* 78:237-249.

Newbold, C. J., F. M. McIntosh, P. Williams, R. Losa, and R. J. Wallace. 2004. Effects of a specific blend of essential oil compounds on rumen fermentation. *Anim. Feed Sci. Technol.* 114:105-112.

Odenyo, A. A., Osuji, P. O. and Karanfil, O. 1997. Effect of multipurpose tree (MPT) supplements on ruminal ciliate protozoa. *Anim. Feed Sci. Technol.* 67, 169-180.

Odenyo, A. A., and P. O. Osuji. 1998. Tannin tolerant ruminal bacteria from East African ruminants. *Can. J. Microbiol.* 44:905-909.

Oh, H. K., M. B. Jones, and W. M. Longhurst. 1968. Comparison of rumen microbial inhibition resulting from various essential oils isolated from relatively unpalatable plant species. *Appl. Microbiol.* 16:39-44.

Patra, A. K., D. N. Kamra, and N. Agarwal. 2006a. Effect of plant extracts on in vitro methanogenesis, enzyme activities and fermentation of feed in rumen liquor of buffalo. *Anim. Feed Sci. Technol.* 128:276-291.

Patra, A. K., D. N. Kamra, and N. Agarwal. 2006b. Effect of plants containing secondary metabolites on in vitro methanogenesis, enzyme profile and fermentation of feed with rumen liquor of buffalo. *Anim. Nutr. Feed Technol.* 6:203-213.

Patra, A. K. 2007. Nutritional management in organic livestock farming for improved ruminant health and production - an overview. *Livest. Res. Rural Develop.* 19, http://www.cipav.org.co/lrrd/lrrd19/3/patr19041.htm

Patra, A. K., D. N. Kamra, N. Agarwal, P. N. Chatterjee. 2008a. Effect of *Terminalia chebula* and *Allium sativum* on rumen fermentation, enzyme activities and microbial counts in buffalo. *Indian J. Anim. Nutr.* 24:251-255.

Patra, A. K., D. N. Kamra, and N. Agarwal. 2008b. Effect of extracts of leaves on rumen methanogenesis, enzyme activities and fermentation in in vitro gas production test. *Indian J. Anim. Sci.* 78:91-96.

Patra, A. K., and J. Saxena 2009a. Dietary phytochemicals as rumen modifiers: a review of the effects on microbial populations. *Antonie van Leeuwenhoek*, 96:363-375.

Patra, A. K., and J. Saxena, 2009b. The effect and mode of action of saponins on the microbial populations and fermentation in the rumen and ruminant production. *Nutr. Res. Rev.* 22:204-219.

Patra, A. K., D. N. Kamra, and N. Agarwal. 2010. Effects of extracts of spices on rumen methanogenesis, enzyme activities and fermentation of feeds in vitro. *J. Sci. Food Agric.* 90:511-520.

Patra, A. K. 2010. Meta-analyses of effects of phytochemicals on digestibility and rumen fermentation characteristics associated with methanogenesis. *J. Sci. Food Agric.* 90:2700-2708.

Patra, A. K., and J. Saxena, J. 2010. A new perspective on the use of plant secondary metabolites to inhibit methanogenesis in the rumen. *Phytochemistry,* 71:1198-1222.

Patra, A. K., D. N. Kamra, R. Bhar, R. Kumar, V. B. Chaturvedi, and N. Agarwal. 2011. Effect of *Terminalia chebula* and *Allium sativum* on in vivo methane emission by sheep. *J. Anim. Physiol. Anim. Nutr.* 95:187-191.

Patra, A. K., and J. Saxena. 2011. Exploitation of dietary tannins to improve rumen metabolism and ruminant nutrition. *J. Sci. Food Agric.* 91:24-37.

Patra, A. K. 2011. Effects of essential oils on rumen fermentation, microbial ecology and ruminant production. *Asian J. Anim. Vet. Adv.* 6: 416-428.

Patra, A. K., and Z. Yu. 2012. Effects of essential oils on methane production and fermentation by, and abundance and diversity of, rumen microbial populations. *Appl. Environ. Microbiol.* 78:4271-4280.

Patra, A. K., J. Stiverson, ad Z. Yu. 2012a. Effects of quillaja and yucca saponins on communities and select populations of rumen bacteria and archaea, and fermentation in vitro. *J. Appl. Microbiol.* 113:1329-1340.

Patra, A. K. 2012. An overview of antimicrobial properties of different classes of phytochemicals. In: Dietary phytochemicals and microbes, Patra, A.K. (ed.), Springer, Dordrecht, Netherlands, pp.1-20.

Patra, A. K., B. R. Min, J. Saxena. 2012b. Dietary tannins on microbial ecology of the gastrointestinal tract in ruminants. In: Dietary phytochemicals and microbes, Patra, A.K. (ed.), Springer, Dordrecht, Netherlands. pp. 312-325.

Paul, S. S., D. N. Kamra, V. R. B. Sastry, N. P. Sahu, A. Kumar. 2003. Effect of phenolic monomers on biomasss and hydrolytic enzyme activities of an anaerobic fungus isolated from wild nil gai (*Baselophus tragocamelus*). *Lett. Appl. Microbiol.* 36:377-381.

Pen, B., K. Takaura, S. Yamaguchia, R. Asa, and J. Takahashi. 2007. Effects of *Yucca schidigera* and *Quillaja saponaria* with or without â-1,4 galacto-oligosaccharides on ruminal fermentation, methane production and nitrogen utilization in sheep. *Anim. Feed Sci. Technol.* 138, 75-88.

Perez-Maldonado, R. A., B. W. Norton, and G. L. Kerven. 1995. Factors affecting in vitro formation of tannin-protein complexes. *J. Sci. Food Agric.* 69:291-298.

Puchala, R., B. R. Min, A. L. Goetsch, and T. Sahlu. 2005. The effect of a condensed tannin-containing forage on methane emission by goats. *J. Anim. Sci.* 83:182-186.

Puchala R, G. Animut, A. K. Patra, G. D. Detweiler, J. E. Wells, V. H. Varel, T. Sahlu, and A. L. Goetsch. 2012a. Effects of different fresh-cut forages and their hays on feed intake, digestibility, heat production, and ruminal methane emission by Boer x Spanish goats. *J. Anim. Sci.* 90:2754-2762.

Puchala R, G. Animut, A. K. Patra, G. D. Detweiler, J. E. Wells, V. H. Varel, T. Sahlu, and A. L. Goetsch. 2012b. Methane emissions by goats consuming *Sericea lespedeza* at different feeding frequencies. *Anim. Feed Sci. Technol.* 175:76-84.

Ramirez-restrepo, C. A., T. N. Barry, W. E. Pomroy, N. Lopez-Villalobos, W. C. McNabb, and P. D. Kemp. 2005. Use of *Lotus corniculatus* containing condensed tannins to increase summer lamb growth under commercial dryland farming conditions with minimal anthelmintic drench input. *Anim. Feed Sci. Technol.* 122:197-217.

Santoso, B., A. Kilmaskossub, and P. Sambodo. 2007. Effects of saponins from *Biophytum petersianum* Klotzsch on ruminal fermentation, microbial protein synthesis and nitrogen utilization in goats. *Anim. Feed Sci. Technol.* 137:58-68.

Schneider, H., A. Schwiertz, M.D. Collins, and M. Blaut. 1999. Anaerobic transformation of quercetin-3-glucoside by bacteria from the human intestinal tract. *Arch. Microbiol.* 171: 81-91.

Skene, I. K., and J. D. Brooker. 1995. Characterization of tannin acylhydrolase activity in the ruminal bacterium *Selenomonas ruminantium*. *Anaerobes* 1:321-327.

Sliwinski, B. J., M. Kreuzer, H. R. Wettstein, and A. Machmuller. 2002. Rumen fermentation and nitrogen balance of lambs fed diets containing plant extracts rich in tannins and saponins and associated emissions of nitrogen and methane. *Arch. Anim. Nutr.* 56: 379-392.

Stack, R. J. and M. A. Cotta. 1986. Effect of phenylpropanoic acid on growth of, and cellulose utilization by, cellulolytic ruminal bacteria. *Appl. Environ. Microbiol.* 52:209-210.

Tagari, H., Y. Heins, M. Tamir and R. Volcani. 1965. Effect of carob pod extract on cellulolysis, proteolysis, deamination, and protein biosynthesis in an artificial rumen. *Appl. Microbiol.* 13:437-442.

Tanner, G. J., A. E. Moore, and P. J. Larkin. 1994. Proanthocyanidins inhibit hydrolysis of leaf proteins by rumen microflora *in vitro*. *Br. J. Nutr.* 71, 947-958.

Tavendale, M. H., L. P. Meagher, D. Pacheco, N. Walker, G. T. Attwood, and S. Sivakumaran. 2005. Methane production from *in vitro* rumen incubations with *Lotus pedunculatus* and *Medicago sativa*, and effects of extractable condensed tannin fractions on methanogenesis. *Anim. Feed Sci. Technol.* 123-124, 403-419.

Teferedegne, B. 2000. New perspective on the use of tropical plants to improve ruminant nutrition. *Proc. Nutr. Soc.* 59:209-214.

Terrill, T. H., G. B. Douglas, A. G. Foote, R. W. Purchas, G. F. Wilson, and T. N. Barry. 1992. Effect of condensed tannins upon body growth and rumen metabolism in sheep grazing sulla (*Hedysarum coronarium*) and perennial pasture. *J. Agric. Sci.* 119: 265-273.

Ulyatt, M. J., J. C. McRae, T. J. Clarke, and P. D. Pearce. 1975. Quantitative digestion of fresh forages by sheep. 4. Protein synthesis in the stomach. *J. Agric. Sci.* 84:453-458.

Valdez, F. R., L. J. Bush, A. L. Goetsch, and F. N. Owens. 1986. Effect of steroidal sapogenins on rumen fermentation and on production of lactating dairy cows. *J. Dairy Sci.* 69: 1568-1575.

Waghorn, G. C., and I. D. Shelton. 1997. Effect of condensed tannins in *Lotus corniculatus* on the nutritive value of pasture for sheep. *J. Agric. Sci.* 128 : 365-372.

Waghorn, G. C., M. H. Tavendale, and D. R. Woodfield. 2002. Methanogenesis in forages fed to sheep. *Proc. N. Z. Grassl. Assoc. Sixty-fourth Conf.* 64: 167-171.

Wallace, R. J., L. Arthaud, and C. J. Newbold. 1994. Influence of *Yucca schidigera* extract on ruminal ammonia concentration and ruminal microorganisms. *Appl. Environ. Microbiol.* 60:1762-1767.

Wallace, R. J., N. R. McEwan, N. R., F. M. McIntosh, B. Teferedegne, and C. J. Newbold. 2002. Natural products as manipulators of rumen fermentation. *Asian-Austr. J. Anim. Sci.* 15:1458-1468.

Wallace, R. J. 2004. Antimicrobial properties of plant secondary metabolites. *Proc. Nutr. Soc.* 63: 621-629.

Wang, Y., G. B. Douglas,G. C. Waghorn, T. N. Barry, A. G. Foote, and R. W. Purchas. 1996a. Effect of condensed tannins upon the performance of lambs grazing *Lotus corniculatus* and lucern (*Medicago sativa*). *J. Agric. Sci.* 126:87-98.

Wang, Y., G. B. Douglas, G. C. Waghorn, T. N. Barry, A. G. Foote, and R. W. Purchas. 1996b. Effect of condnsed tannins in *Lotus corniculatus* upon lactation performance in ewes. *J. Agric. Sci.* 126:353-362.

Wang, Y., T. A. McAllister, C. J. Newbold, L. M. Rode, P. R. Cheeke, and Cheng, K. J. 1998. Effect of *Yucca schidigera* extract on fermentation and degradation of steriodal saponins in the rumen simulation technique (RUSITEC). *Anim. Feed Sci. Technol.* 74:143-153.

Wang, Y., T. A. McAllister, L. J. Yanke, and P. R. Cheeker. 2000a. Effect of steroidal saponin from *Yucca schidigera* extract on ruminal microbes. *J. Appl. Microbiol.* 88:887-896.

Wang, Y., T. A. Mcallister, L. J. Yanke, Z. J. Xu, P. R. Cheeke, K. Cheng, Y. X. Wang, and K. J. Cheng. 2000b. *In vitro* effects of steroidal saponins from *Yucca schidigera* extract on rumen microbial protein synthesis and ruminal fermentation. *J. Sci. Food Agric.* 80: 2114-2122.

Williams, A. G., and G. S. Coleman. 1997. *The Rumen Protozoa*. In *The rumen microbial ecosystem* (Eds P. N. Hobson and C. S. Stewart). Academic Professional, Blackie, London, pp. 72-120.

Wilson, R. C., T. R. Overton, and J. H. Clark. 1998. Effects of *Yucca schidigera* extract and soluble protein on performance of cows and concentrations of urea nitrogen in plasma and milk. *J. Dairy Sci.* 81:1022-1027.

Wina, E., S. Muetzel, E. Hoffmann, H. P. S. Makkar, and K. Becker. 2005. Saponins containing methanol extract of *Sapindus rarak* affect microbial fermentation, microbial activity and microbial community structure in vitro. *Anim. Feed Sci. Technol.* 121:159-174.

Wina, E., S. Muetzel, and K. Becker. 2006. The dynamics of major fibrolytic microbes and enzyme activity in the rumen in response to short- and long-term feeding of *Sapindus rarak* saponins. *J. Appl. Microbiol.* 100:114-122.

Woodward, S. L., M. J. Auldist, P. J. Laboyrie, and E. B. L. Jansen. 1999. Effect of *Lotus corniculatus* and condensed tannins on milk yield and milk composition of dairy cows. *Proc. N. Z. Soc. Anim. Prod.* 59:152-155.

Woodward, S. L., G. C. Waghorn, M. J. Ulyatt, and K. R. Lassey. 2001. Early indication that feeding lotus will reduce methane emission from ruminants. *Proc. N. Z. Soc. Anim. Prod.* 61:23-26.

Woodward, S. L., G. C. Waghorn, K. R. Lassey, and P. G. Laboyrie. 2002. Does feeding sulla (*Hedysarum coronarium*) reduce methane emission from dairy cows? *Proc. N. Z. Soc. Anim. Prod.* 62:227-230.

Yang, W. Z., C. Benchaar, B. N. Ametaj, A. V. Chaves, M. L. He, and T. A. McAllister. 2007. Effects of garlic and juniper berry essential oils on ruminal fermentation and on the site and extent of digestion in lactating cows. *J. Dairy Sci.* 90:5671-5681.

Ye, J. A., J. Liu, Z. Q. Shi, L. S. Jiang, and J. X. Liu. 2001. Effect of theasaponin on fermentation of cultured rumen fluid. Chinese *J. Anim. Sci.* 37:29-30.

Zinn, R. A., E. G. Alvarez, M. Montano, A. Plascencia and J.E. Ramirez. 1998. Influence of tempering on the feeding value of rolled corn in finishing diets for feedlot cattle. *J. Anim. Sci.* 76:2239-2246.

14

Methane Inhibitors: Trends in Manipulation of Rumen Fermentation

Kaushalendra Kumar, Sanjay Kumar and Rajni Kumari

Introduction

There is growing worldwide interest in reducing methane emissions from domestic ruminants. Methane is a potent greenhouse gas and its release into the atmosphere is directly linked with animal agriculture, particularly ruminant production. Methane emitted from ruminant livestock is regarded as a loss of feed energy and also a contributor to global warming. Methane is synthesized in the rumen as one of the hydrogen sink products that are unavoidable for efficient succession of anaerobic microbial fermentation. Various attempts have been made to reduce methane emission, mainly through rumen microbial manipulation, by the use of agents including chemicals, antibiotics and natural products such as oils, fatty acids and plant extracts. A newer approach is the development of vaccines against methanogenic bacteria. While ionophore antibiotics have been widely used due to their efficacy and affordable prices, the use of alternative natural materials is becoming more attractive due to health concerns regarding antibiotics. An important feature of a natural material that constitutes a possible alternative methane inhibitor is that the material does not reduce feed intake or digestibility but does enhance propionate that is the major hydrogen sink alternative to methane. Since methane contains energy, its emission during rumen fermentation is considered to be a loss of feed energy that is equivalent to 2-12% of the gross energy of animal feed (Johnson and Johnson, 1995). Some implications of these approaches, as well as an introduction to antibiotic-alternative natural materials and novel approaches, are provided.

Hydrogen sinks in the rumen

The rumen is an anaerobic fermentation chamber, in which diverse and dense microbial populations have symbiotic relationships in which metabolites are exchanged that promote or compensate each others growth, a process which is termed "cross feeding" (Wolin *et al.*, 1997). Methane synthesis is regarded as one such cross feeding between hydrogen-producing microbes and hydrogen consuming methanogens. Since the hydrogen-producing microbes include fibrolytic fungi and bacteria, their co-association with methanogens allows efficient removal of hydrogen, which facilitates continuous fibre degradation.

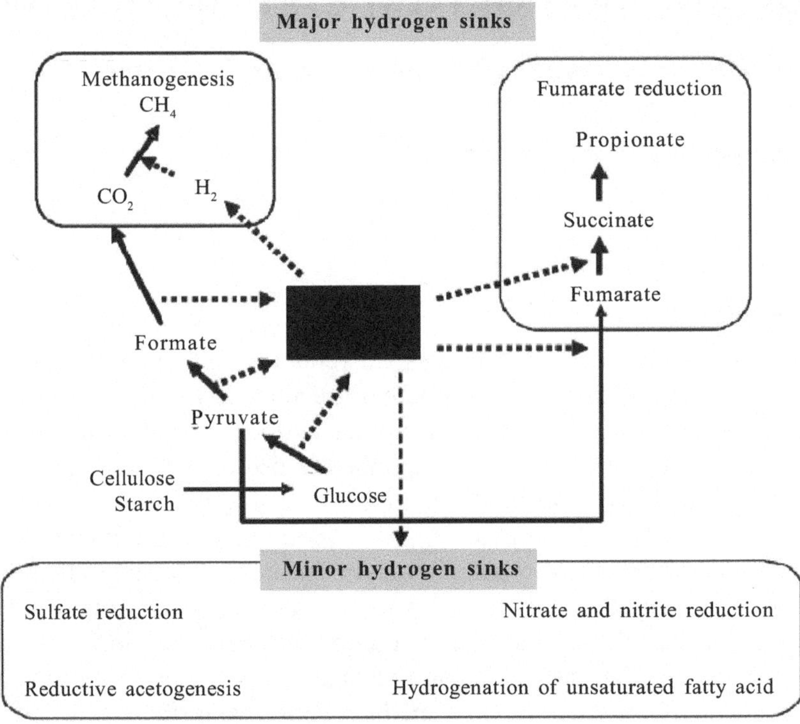

Fig. 1: Hydrogen-consuming pathways recognized in the rumen.

When methane reduction is attempted, it is therefore necessary to consider alternative hydrogen sinks to methanogenesis. Figure 1 shows the hydrogen-consuming pathways currently known to exist in the rumen. Methanogenesis is the primary pathway followed by propionate production (fumerate reduction) (Mitsumori and Sung, 2008). Other reactions (nitrate- and nitrite reduction, reductive acetogenesis, biohydrogenation of unsaturated fatty acid) play a relatively minor role in hydrogen consumption within the rumen. Thus, it is reasonable that a strategy for methane mitigation is developed concomitantly with a strategy to enhance propionate production. Otherwise, rumen fermentation could be hindered by hydrogen accumulation caused by the lack of hydrogen removal by methanogenesis. Accordingly, propionate enhancement could be a good indicator of simultaneous methane reduction in the rumen. A number of studies on methane reduction have been performed along these lines and, indeed, rumen microbial numbers and their metabolic activities do change with methane reduction. However, the manner of these changes varies depending on the manipulation procedure, *i.e.* whether chemicals and vaccines directly active against methanogens, or antibiotics and plant-derived antimicrobial compounds that indirectly affect methanogenesis, are used.

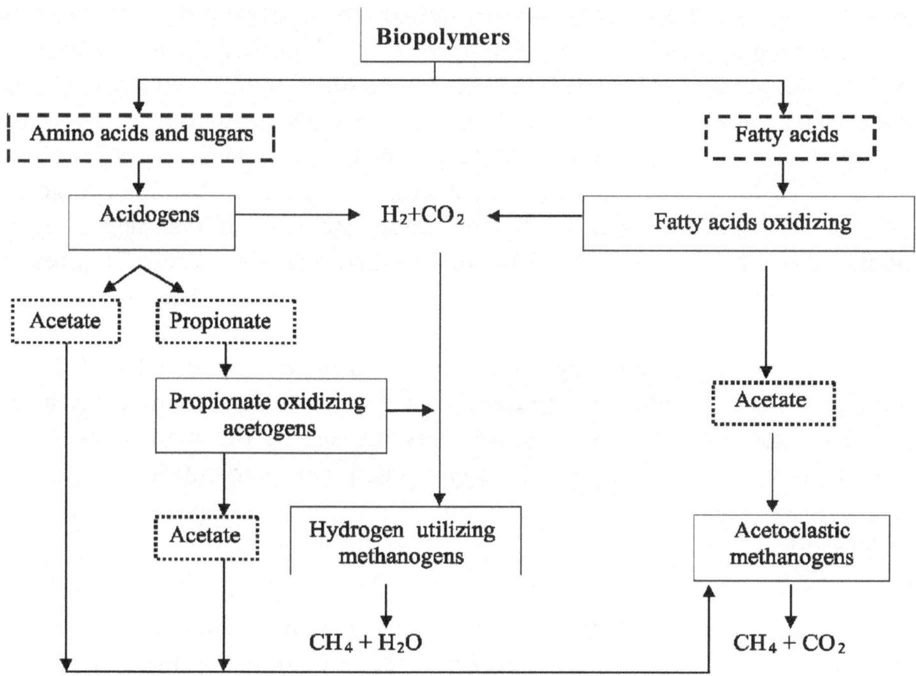

Fig. 2: Flow chart of methane production.

Halogens and other methane-inhibiting chemicals

Several anti-methanogenic compounds have been documented since the 1970s (Chalupa *et al.*, 1977). Although all these compounds are effective in the reduction of methane production, individual compounds have specific disadvantages which inhibit their current practical use. Thus some compounds are regarded as ozone disruptive agents, while others are expensive or have safety concerns. However, these compounds are good models for the study of shifts in rumen fermentation and microbiota that occur when methanogens and methane production are drastically inhibited.

Ungerfeld *et al.* (2004) evaluated the sensitivity of representative rumen methanogens to potent methanogen inhibitors including 2-bromoethanesulphonate (BES), 3-bromopropanesulphonate (BPS), limazine, propynoic acid and ethyl 2-butynoate. All of these chemicals, with the exception of BPS, inhibited methane production from *Methanobrevibacter ruminantium*, *Methanosarcina mazei* and *Methanomicrobium mobile*. The sensitivity of the methanogens to each chemical was species-dependent, suggesting that sensitive methanogens can be replaced by resistant methanogens following administration of the chemical over a certain period of time.

Bromochloromethane (BCM) inhibits cobamide dependent methanogenesis in which the majority of rumen methanogens are involved. When BCM was fed to cattle, the total methane emission was reduced by 30% with a resultant increase in propionate and branched chain fatty acids as alternative hydrogen sinks (Denman *et al.*, 2007). These changes were accompanied by an average decrease of 34% in the number of methanogens quantified by *mcrA*-targeted real-time PCR. BCM feeding led to diversification of the methanogen population even though the total population size was decreased. This result implies that alternative methanogens (*Methanomicrobium*, *Methanosarcina*, *Methanococcus* and unknown methanogens) are developed following the suppression of major methanogens such as *Methanobacterium* by BCM. Thus, a change in the methanogen population diversity in response to an inhibitor should be considered as a possible consequence of the manipulation of methanogens via the mitigation strategy of chemical intervention.

Probiotics

There is very little information on the effects of probiotics on CH_4 production in dairy cattle. The effects of the most widely used microbial feed additives, *Saccharomyces cerevisiae* and *Aspergillus oryzae*, on rumen fermentation were earlier studied in vitro (Mutsvangwa *et al.*, 1992). *Aspergillus oryzae* was shown to reduce CH_4 by 50% as a result of a reduction in the protozoal population (Frumholtz *et al.*, 1989). The addition of *Saccharomyces cerevisiae* reduced CH_4 by 10% *in vitro*, but was not sustain over a long period (Mutsvangwa *et al.*, 1992). It has been proposed that probiotics provide nutrients, including metabolic intermediates and vitamins that stimulate the growth of ruminal bacteria, resulting in increased bacterial population (Newbold *et al.*, 1996). However, the specific mode of action is still unknown.

Another theory indicates that probiotics stimulate lactic acid utilizing bacteria, resulting in a reduction of lactic acid and a more stable ruminal environment. Eun *et al.* (2003) reported that brewer's yeast culture enhanced the activity of bacteria that convert H_2 to acetate and decreased CH_4 output by 25% in a continuous culture of ruminal microorganisms. Although microbial preparations are commercially available as ruminant feed additives, there is a need for further research to establish the potential of probiotics for reducing CH_4 production *in vivo*. Producers are skeptical about the benefits of probiotics and there is a need to identify the dietary and management situations in which probiotics can give consistent production benefits as well as the added effect of reducing CH_4 emissions (Moss *et al.*, 2000).

Reductive acetogenesis

A technology that may hold some promise in the long term of diverting electrons from methanogens is the production of acetic acid by acetogens (Joblin, 1999). In the gut of termites and rodents, acetogens convert excess H_2 to acetic acid, which is then utilized by the host (Joblin, 1999). However, in the rumen, acetogens are few and cannot compete effectively with methanogens for H_2 ions, because they have a lower affinity for H_2 than methanogens (Nollet *et al.*, 1998). Carbon flux studies in the rumen of sheep revealed that rumen acetogenesis occurs in the first 24 h after birth, but is subsequently displaced by methanogenesis (Morvan *et al.*, 1994); methanogens easily out compete the acetogens for the low concentration of H_2 normally encountered in the rumen (Joblin, 1999). Thus, methanogens have to be inhibited to allow H_2 pressure to rise before acetogenesis can be significant as an alternate H_2 sink in the rumen. Increasing the populations of acetogens through exogenous inoculations into the rumen could be useful for competing against methanogens (Joblin, 1999).

Methane oxidizers

Methane oxidizing bacteria have been isolated from different environments, including the rumen (Moss *et al.*, 2000). *In vitro* studies with stable carbon isotopes suggest that the extent of CH_4 oxidation to CO_2 is quantitatively minor (0.3 to 8%) in the rumen (Kajikawa and Newbold, 2000). Valdez *et al.* (1996) isolated a CH4 oxidizing bacterium from the gut of young pigs, which decreased CH4 accumulation when added to rumen fluid *in vitro*. However, this approach has not been validated *in vivo*. In the long-term, CH4 oxidizers from gut sources could be screened for their activity in the rumen to reduce the proportion of ruminal gas in the form of CH_4.

Propionate enhancers

As a result of the growing awareness of the threat of microbial resistance to antibiotics, there is an increasing interest in alternatives to antibiotics as growth promoters (Moss *et al.*, 2000). Dicarboxylic acids such as fumaric and malic acids have been studied in vitro as feed additives in ruminant diets (Asanuma *et al.*, 1999). Fumaric acid is an intermediate in the propionic acid pathway, in which it is reduced to succinic acid. In this reaction, H_2 ions are needed and therefore reducing fumaric acid may provide an alternative electron sink for H_2. Asanuma *et al.* (1999) showed that the addition of 20 mM of fumarate to cultures that were fermenting hay powder and concentrate incubated for 6 h significantly decreased CH_4 production by 5% and increased propionate production by 56%, while with the addition of 30 mM of fumarate, CH_4 declined by 11%, and propionate production increased by 58% compared to the control. Their data suggested that most of the fumarate consumed was metabolized to propionate with little production of acetate and succinate, whereas a much larger amount of succinate accumulated with the addition of 30 mM of fumarate. However, when incubation time was prolonged to 12 h, most of the succinate was metabolized to propionate.

The effects of salinomycin (15 ppm) plus fumaric acid (2%) supplemented to diets of Holstein steers increased the molar proportion of propionic acid and decreased CH_4 production by 16% and had no effect on DM digestibility (Itabashi *et al.*, 2000). Bayaru *et al.* (2001) found that CH_4 production was reduced by 23% when fumaric acid added to sorghum silage was fed to Holstein steers. The authors observed that the addition of fumaric acid increased propionic acid formation and had no effect on DM digestibility. Fumaric acid was also shown to increase concentration of plasma glucose and milk protein synthesis in dairy cows due to an increase in propionic acid production (Itabashi, 2001). The authors concluded that fumaric acid may be put to practical use for ruminant diets since it has the dual benefit of decreasing CH_4 production and increasing net energy retention. Malate, which is converted to propionate via fumarate, also increased propionate production and inhibited CH_4 production in vitro (Martin *et al.*, 1999). However, malate failed to increase ruminal propionate concentrations in feedlot cattle and did not affect CH_4 production (Montano *et al.*, 1999). There is a need for further testing and evaluation of these enhancers *in vivo* to assess their potential as feed additives in the industry.

Bacteriocins

Naturally occurring anti-bacterial agents, bacteriocins, originating from rumen bacteria have been reviewed by Teather and Forster (1998) who pointed out their possible use as modifiers of rumen fermentation. These bacteriocins may also be useful for the prevention of animal metabolic disorders such as lactic acidosis and bloat and may even prove useful for food storage.

Bovicin, a bacteriocin produced by *Streptococcus bovis*, has been reported as a possible methane-mitigating agent in the rumen (Lee *et al.*, 2002).

Supplementation of a mixed rumen bacterial culture with bovicin inhibited methane production by as much as 53%. When the culture was transferred successively (50% v/v) with bovicin, it lost the ability to produce methane after only 5 transfers. Moreover, the restriction pattern of amplified 16S archaeal rDNA was not different between cultures with and without bovicin indicating that the effect of bovicin on rumen methanogens might not be selective. Activity of bovicin against other rumen bacteria or an effect of bovicin on the fermentation pattern has not been reported and remains to be investigated.

The first described bacteriocin, nicin, that is produced by *Lactococcus lactis*, also has a methane-mitigating ability that was observed in a monensin-supplemented *in vitro* culture (20% inhibition without a negative effect on volatile fatty acid (VFA) production) (Callaway *et al.*, 1997). Although no mechanism was proposed to explain its effect on rumen bacteria, nicin does potentiate propionate production and possibly shows selective activity against Gram-positive rumen bacteria. The above bacteriocins are known to be as potent as monensin and are also active even at low pH. However, further investigation is necessary before they can be considered as candidate additives for ruminant livestock, in particular for beef cattle fed a high grain diet.

Ionophore antibiotics

Ionophore antibiotics, represented by monensin, have been widely used all over the world as feed additives for ruminant livestock since the mid 1970s. Monensin is considered as a growth promoter due to its favorable effects on rumen fermentation including methane reduction, propionate enhancement and ammonia reduction, together with its preventive effects on coccidiasis, bloat and lactic acidosis. These effects are attributed to a selective antimicrobial action of monensin on rumen microbes. Monensin is inhibitory for protozoa, Gram-positive bacteria including *Ruminococci*, *Streptococci* and *Lactobacilli* but not for Gram negative bacteria, and therefore leads to rumen microbiota that produce more propionate and less acetate, butyrate, formate and hydrogen (Russell and Strobel, 1989). Partial inhibition of hydrogen- and formate-producing microbes contributes to methane reduction, the extent of which varies between reports. Review papers indicate that methane reduction by monensin ranges from 4 to 31% (Rumpler *et al.*, 1986). A recent report indicated that long term administration of monensin to dairy cattle stably reduced methane by 7% and that this reduction persisted for 6 months with no adverse effect on milk yield (Odongo *et al.*, 2007). However, beef steers that had been given monensin only showed methane reduction during the first 4-6 weeks of administration. Nevertheless, it should be noted that propionate enhancement persisted throughout the 14 wk experiment (Guan *et al.*, 2006).

The number of rumen protozoa is decreased by ionophores and this decrease causes a reduction in methane, because rumen protozoa accommodate methanogens on their cell surface and within the cell (Tokura *et al.*, 1999). Therefore the number and/or activity of methanogens are believed to be indirectly reduced by ionophores. This is part of the reason why methane reduction by ionophores occurs only at the early stage of feeding since rumen protozoan populations that are depressed by ionophores tend to restore their numbers when ionophores are administered for a long time (Kobayashi *et al.*, 1988).

However, long term feeding of monensin does not affect the number of methanogens (Hook *et al.*, 2009). These data suggest that reduction of methane by monensin feeding is not due to a reduction in the population size of methanogens but is more likely due to the development of an alternative hydrogen consuming pathway such as propionate enhancement by stimulation of the proliferation of propionate and succinate producing bacteria such as *Selenomonas* and *Megasphaera* (Russell and Strobel, 1989).

Such rumen bacterial selection by monensin appears to be maintained even months following administration which probably explains the persistence of propionate enhancement. Although the effect of monensin on rumen fiber digestion is inconsistent, one of the most dominant fibrolytic bacteria, *Fibrobacter succinogenes*, appears to be insensitive to monensin as its abundance within the rumen, monitored by a DNA probing method, was not affected by monensin (Stahl *et al.*, 1988).

Oils and fatty acids

Plant oils rich in medium chain fatty acids are known to inhibit rumen methanogenesis. One such oil, coconut oil, is particularly effective (Dohme *et al.*, 2000). The major component of coconut oil is lauric acid (C14:0) which is more potent in the reduction of methane in a semi-continuous fermenter that simulates the rumen (RUSITEC) than other fatty acids including palmitic (C16:0), stearic (C18:0) and linoleic (C18:2) acids (Dohme *et al.*, 2001). A similar reduction in methane was observed in batch cultures, in which coconut oil and lauric acid were directly compared, and which showed that lauric acid inhibited methanogenesis to a greater extent (Yabuuchi *et al.*, 2006, 2007).

Lauric acid is inhibitory for Gram-positive rumen bacteria including cellulolytic ruminococci. Therefore, addition of lauric acid to feed might decrease feed digestibility of a high roughage diet. However, a decrease in feed digestibility would be negligible with the high concentrate diet that is fed to beef cattle. Lauric acid was also shown to depress the metabolic activity of the sacchalolytic bacterium *Streptococcus bovis* without affecting its maximal growth. The decreased lactate production by *S. bovis* in the presence of lauric acid may explain the preventive and curing effects of lauric acid on rumen lactic acidosis. These data suggest that lauric acid may not alter the size of a specific bacterial population but may modulate metabolic activity when it is fed over a long period of time. Indeed, the abundance of other rumen bacterial species was not altered following lauric acid feeding (Yabuuchi *et al.*, 2007).

Most of the oils and fatty acids that reduce methanogenesis reduce the ruminal level of protozoa that are known to be co-symbionts of methanogens as mentioned in the section on ionophores. Therefore, a reduction in protozoan numbers is partly responsible for the decreased methane production induced by oils and fatty acids.

Plants and their extracts

Many candidate feed additives originating from plant materials have been screened for their potential ability to reduce rumen methanogenesis. One such compound is saponin. Although the inhibitory effect of saponin and sarsaponin on methanogenesis varies with the plant source, inhibition ranging from approximately 5-60%, accompanied by enhanced propionate, has been reported (Wina *et al.*, 2005). Saponins have a detergent action that disrupts microbial cell membranes by formation of a complex with membrane sterols. Rumen protozoa are particularly sensitive to saponins which reduce their level in the rumen, resulting in the depression of methanogens associated with protozoa. Guo *et al.* (2008) have suggested that a decrease in methanogens associated with protozoa as exo- and endo-symbionts could be the main mechanism by which saponin feeding reduces methanogenesis.

The antimicrobial activity of essential oils has prompted interest in whether these compounds could be used to inhibit methanogenesis in the rumen. The challenge is to identify essential oils that reduce methane production without a concomitant reduction in feed digestion. Some essential oils that possess antibacterial activity are commercially available. The main components of essential oils that exert antibacterial activity are considered to be a variety of compounds that are mainly classified as terpenoids or phenylpropanoids. Their antibacterial spectra are relatively broad and their mechanism of action involves interaction of the antibacterial compound with the bacterial cell membrane which destabilizes the membrane. Although favorable depressive effects of essential oils on rumen proteolysis and deamination have been demonstrated, reports of their potency for the reduction of rumen methanogenesis are inconsistent (Calsamiglia *et al.*, 2007).

The effects of essential oils on colonisation were, in contrast, very clear. The effect may stem from a decreased adhesion to readily digested solids; alternatively, the rate of development of solids-associated bacteria - once attachment has already occurred-may be inhibited. The exact cause is as yet unclear. Thus, essential oils may have several independent actions, depending perhaps on individual oils within the mixture, and they may be related to each other in their biochemical consequences (Figure 3).

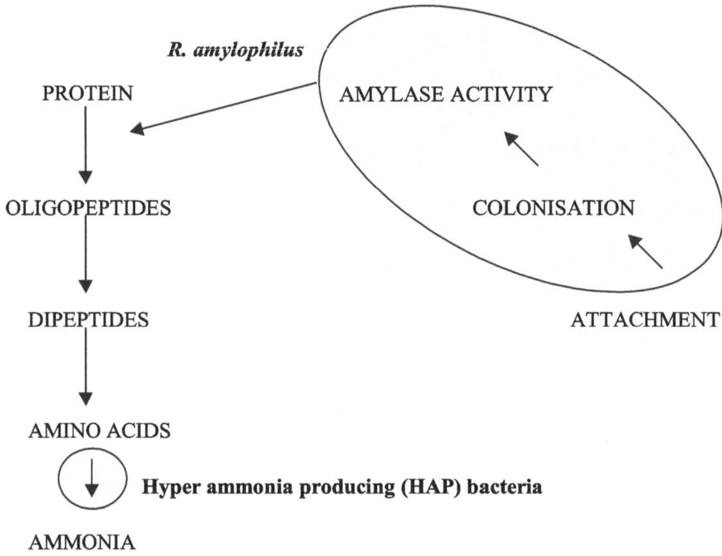

Fig. 3: A proposed mode of action for essential oils in the rumen.

Evans and Martin (2000) observed that thymol (400 mg/l), a main component of essential oil derived from Thymus and Origanum plants, was a strong inhibitor of methane *in vitro*, but acetate and propionate concentrations also decreased. Kamra *et al.* (2006) investigated methanol and ethanol extracts of various spices, including fennel, clove, garlic, onion, and ginger for effects on methane production *in vitro*. Among the extracts tested, methanol extract of garlic was the most effective suppressant of methane, with a 64% reduction *in vitro* and no adverse effects on feed digestibility. Evaluating the effects of garlic oil and four of its main components (diallyl sulfide, diallyl disulfide, allyl mercaptan, and allicin), reduced methane production by 69 percent, without altering digestibility. Patra *et al.* (2004) reported that ethanol and methanol extracts of cloves and the methanol extract of fennel also inhibited methane production *in vitro*, but digestibility of the feed was also reduced. The ground root from Rhubarb (*Rheum officinale*, 1.6 g/l) was reported to reduce methane production in vitro by 20% without affecting digestibility.

Tannin-containing forages and tannin extracts have been demonstrated to decrease methane production both *in vivo* and *in vitro*. Puchala *et al.* (2005) worked on CH4 emission relative to dry matter intake by goats fed *Lespedeza cuneata* (177 g CT/kg DM) 0.55 of that by goats consuming a mixture of grasses (*Digitaria ischaemum* and *Festuca arundinacea*; 5 g CT/kg DM) and found that relative reduction in methane production. Carulla *et al.* (2005) suggested that the inhibition of methanogens by condensed tannin was primarily the result of suppressed fibre degradation that limits hydrogen derived from synthesis of acetate. The reason of depressed fibre degradation could be due to a reduced number of cellulolytic bacteria, formation of tannin-cellulose complexes and impaired bacterial adhesion to substrate and fibrolytic activity of rumen microbes (Bento *et al.*, 2005a).

Goetsch *et al.* (2008) reported that condensed tannins from different sources had disparate influence on nitrogen metabolism, but similar effects on rumen microbial CH4 emission by goats, possibly by altering the activity of methanogens, though change in actions of other bacteria and/or protozoa may also be involved. Thus, various tannins sources could be used in future strategies to reduce ruminal CH4 emission. Patra *et al.* (2008) suggested that the condensed tannins containing forage decreased CH4 emission by goats regardless of its level and the effect per unit of CT increased with decreasing CH4. However, the impact of condensed tannins on CH4 emission appeared attributable to changes in archea activity, which might also involve alterations of protozoal activity. This suggests that relatively low dietary levels of condensed tannins could be employed to decrease methane emission without any marked detrimental effects on other conditions such as total tract nitrogen metabolism.

There may be potential to select essential oils compounds that reduce methane by selectively inhibiting protozoal numbers, which would be expected to decrease methane production because ruminal protozoa provide a habitat for methanogens that live on and within them. However, the anti-protozoal effects of essential oils have been inconsistent and variable among essential oils and essential oils active components. McIntosh *et al.* (2003) reported that growth of the methanogen *Methanobrevibacter smithii* was inhibited, but only when the concentration of the product was 33 fold higher than that fed in vivo, as reported by Beauchemin and McGinn (2006), a feeding rate that is impractical due to potentially deleterious effects on diet digestibility. Using another high supplementation rate (i.e., 20 g/kg of DM intake) of encapsulated horseradish, Mohammed *et al.* (2004) observed a 19% decrease in methane production in steers that was not accompanied by a reduction in protozoal numbers or feed digestibility. It seems clear that there is potential to select essential oil compounds that selectively reduce methane when used at levels that do not depress feed utilization, but further research is necessary to evaluate these compounds *in vivo*.

European scientists have been collaborating in an exploration of plants that might be useful as alternatives to antibiotics for inhibition of methanogenesis in ruminant livestock. Seven potential candidates were ultimately selected from 500 different plant species based on their ability to inhibit methane production by 15-27% without a detrimental effect on total VFA production or feed digestibility. The plant species selected were the Italian plumeless thistle (*Carduus pycnocephalus*, 30% inhibition), the Chinese peony (*Paeonia lactiflora*, 8-53%), the European aspen (*Populus tremula*, 25%), the sweet cherry (*Prunus avium*, 20%), goat willow (*Salix caprea*, 30%), English oak (*Quercus pedunculata*, 25%) and Sikkim rhubarb (*Rheum nobile*, 25%). From these species, a final 2 species (*Carduus* and *Rheum*) were evaluated as to their potency in a RUSITEC analysis. On a high forage diet 16 and 22% inhibition of methanogenesis respectively was noted, while less inhibition (5 and 15% respectively, not significant) was observed on a high concentrate diet. Methane reduction was not accompanied by propionate enhancement or other favorable fermentation changes. No clear dose response was observed and solvent extraction diminished the inhibitory effect. Therefore, details of the inhibition, such as the identity of the effective compound and its mechanism of action, remain to be clarified.

Archaeal Viruses

Another possible method of biological control of methanogens is the use of archaeal viruses (bacteriophages) (Klieve and Hegarty 1999). Bacteriophages are obligate pathogens that can infect and lyse bacteria and methanogens. They are highly host-specific. Although the presence of bacteriophages in the rumen is well known (Mackie and White 1990), knowledge of archaeal viruses is still limited. No viruses have been recorded from the rumen archaea and few from methanogens in general (Klieve and Hegarty 1999). Klieve and Hegarty (1999) noted that without a considerable increase in knowledge of the genetic diversity and viral susceptibility of methanogens and the host range of archaeal viruses, it is difficult to assess the potential of archaeal viruses as a biological control agent.

Immunization

A unique attempt to reduce rumen methane is ongoing in Australia. This approach does not involve dietary manipulation by the inclusion of additives but involves vaccination of the animal against methanogenic bacteria in the rumen. Wright *et al.* (2004) reported that a significant 7.7% reduction in rumen methane production, corrected for dry matter intake level, was achieved by this immunization strategy. They estimated that less than 20% of the methanogens were targeted by the vaccine that was prepared using 3 *Methanobrevibacter* strains. A vaccine of broader range is being developed to induce a greater extent of methane reduction. One such vaccine (targeting >52% of different species/strains of methanogens that were tested based on a survey of sheep prior to vaccination) was applied to 32 sheep. Although the animals showed specific IgG titers in plasma, saliva and rumen fluid, neither methane output nor the number of rumen methanogens was significantly changed (Williams *et al.*, 2009). The development of alternative methanogens after immunization is a possible reason for this failure and a much more broad spectrum approach together with a more comprehensive understanding of the rumen methanogen population is surely required for the vaccination approach to be successful.

Genetic selection

Robertson and Waghorn (2002) observed that Dutch/US cross Holstein cows produced 8–11% less CH_4 (% of GEI) than New Zealand Friesian cows for about 150 d post calving, either when grazing or receiving a TMR. Hegarty (2001a) noted that the natural variation among animals in the quantity of feed eaten per unit of live weight gain can be exploited to breed animals that consume less feed than the unselected population while achieving a desired rate of growth. Accordingly, to exploit such traits, the concept of Residual (Net) Feed Intake (RFI) was developed and used (Basarab *et al.*, 2003). Okine *et al.* (2002) calculated annual CH_4 emissions from Canadian high NFE steers to be 21% lower than that for low NFE steers. Selecting animals for a faster passage rate of feed from the rumen would reduce CH_4 emissions per unit of food ingested. Faster passage rate of feed also affects propionate and microbial yield; thus, selection of animals for this would also have major production benefits. Selecting animals with high NFE offers an opportunity to reduce daily CH_4 emissions without reducing livestock numbers.

New materials

Recent research in Japan has revealed two potential natural materials for the reduction of rumen methanogenesis; plant-derived liquid (PDL) and yeast derived surfactant (YDS). Both of these materials have induced a dramatic reduction in methane production in batch cultures (>95%) and in RUSITEC (>70%) without any adverse effect on feed digestibility or total VFA production. The extent of inhibition induced by these new materials is much greater than that induced by the materials proposed in above studies.

PDL contains anacardic acid, a salicylic acid derivative with an alkyl group that inhibits Gram-positive bacteria including *bacilli* and *staphylococci* (Kubo *et al.*, 1993). Therefore, PDL is expected to selectively inhibit Gram positive rumen bacteria. Anacardic acid was suggested to be a propionate enhancer in early studies (Van Nevel *et al.*, 1971), although this fact has not been highlighted for a long time. The surfactant YDS disrupts bacterial cell walls in a selective manner depending on the structure of the bacterial surface. Gram-negative bacteria possess an outer membrane that minimizes bacterial cell damage from such a surfactant. Thus, YDS might also selectively inhibit Gram positive rumen bacteria. Indeed, YDS and PDL showed a similar antibacterial spectrum when tested against representative rumen bacterial species. Propionate and succinate producers such as *Selenomonas ruminantium*, *Megasphaera elsedenii*, and *Succinivibrio dextrinosolvens* were tolerant to these two materials, while hydrogen and formate producers such as *Ruminococcus flavefaciens*, *Ruminococcus albus*, *Butyrivibrio fibrisolvens*, and *Eubacterium ruminantium* were sensitive. Therefore, both of these materials are believed to shift rumen fermentation toward more propionate and less methane production through selective anti-bacterial activities.

The anti-bacterial selectivity of these materials that was indicated in the above pure culture study was fairly well reflected when the bacteria were analyzed in RUSITEC, in which the DGGE banding pattern was apparently changed following supplementation with either material. Thus, the abundance of *M. elsedenii* and *S. dextrinosolvens* increased as estimated by real-time PCR assays and as judged by their detection frequency in clone library analyses. A more favorable fermentation pattern was observed following addition of the materials that was shifted, as expected, towards more propionate and less methane and there were no changes in total VFA production or feed digestibility. Sheep that were fed a diet supplemented with PDL or YDS showed a fermentation pattern that was similar to that observed in RUSITEC and was accompanied by similar bacterial population shifts. Further evaluation of these materials as additives needs to be investigated *in vivo*.

Summary

Methane is a potent greenhouse gas and its release into the atmosphere is directly linked with animal agriculture, particularly ruminant production. Methane emitted from ruminant livestock is regarded as a loss of feed energy and also a contributor to global warming. Methane is synthesized in the rumen as one of the hydrogen sink products that are unavoidable for efficient succession of anaerobic microbial fermentation. Various attempts have been made to reduce methane emission, mainly through rumen microbial manipulation, by the use of agents including chemicals, antibiotics and natural products such as oils, fatty acids and plant extracts. Stimulation of propionate production could be the best alternative hydrogen sink to methanogenesis in the rumen. Therefore, a strategy for abatement of methane production should be considered concurrently with a strategy to enhance propionate production. Although various feed additive candidates are now available to achieve this aim, the choice of additive must depend on the potency, safety, and expense of the candidate additive. Since our understanding of rumen microbes is still incomplete, elucidation of microbial diversity and microbial interrelationships is absolutely essential for the successful manipulation of rumen fermentation towards a significant reduction in ruminant methane emission. Attainment of such knowledge would permit the realization of abatement of rumen methane production in a more successful manner than heretofore. New approaches for methane reduction such as vaccination of ruminants against methanogens and the application of wallaby foregut microbiota that produce much less methane than the microbiota of cattle rumen (Morrison *et al.*, 2008), are still at a fundamental stage of development. However, the backbone of the future success of these approaches is also a comprehensive analysis of microbiota and a systematic understanding of their biological function.

References

Asanuma, N., M. Iwamoto and T. Hino .1999. Effect of the addition of fumarate on methane production by ruminal microorganisms *in vitro*. *J. Dairy Sci.* 82: 780-787.

Basarab, J. A., M. A. Price, J. L. Aalhus, E. K. Okine, W. M. Snelling and K. L. Lyle. 2003. Residual feed intake and body composition in young growing cattle. *Can. J. Anim. Sci.* 83: 189-204.

Bayaru, E., Kanda, S., Toshihiko, K., Hisao, I., Andoh, S., Nishida, T., Ishida, M., Itoh, T., Nagara, K. and Isobe, Y. 2001. Effect of fumaric acid on methane production, rumen fermentation and digestibility of cattle fed roughages alone. *Anim. Sci. J.* 72: 139 - 146.

Beauchemin, K. A. and S. M. McGinn. 2006. Methane emissions from beef cattle: effects of fumaric acid, essential oil, and canola oil. *J. Anim. Sci.* 84(6): 1489-1496.

Bento, B.H.L., T. Acamovic and H.P.S. Makkar. 2005a. The influence of tannin, pectin, and polyethylene glycol on attachment of 15N-labeled rumen microorganisms to cellulose. *Anim. Feed Sci. Technol.* 122: 41-57.

Callaway, T. R., A. M. S. Cameiro De Melo and J. B. Russell. 1997. The effect of nicin and monensin on ruminal fermentations *in vitro*. 35:90-96.

Calsamiglia, S., M. Busquet, P. W. Cardozo, L. Castillejos and A. Ferret. 2007. Essential oils as modifiers of rumen microbial fermentation *Curr. Microbiol.*. J. Dairy Sci. 90: 2580-2595.

Carulla, J. E., M. Kreuzer, A. Machmuller and H. D. Hess. 2005 Supplementation of *Acacia mearnsii* tannins decreases methanogenesis and urinary nitrogen in forage fed sheep. *Aust. J. Agri. Res.* 56(9): 961-970.

Chalupa, W. 1977. Manipulating rumen fermentation. *J. Anim. Sci.* 46:585-599.

Denman, S. E., N. W. Tomkins and C. S. McSweeney. 2007. Quantitation and diversity analysis of ruminal methanogenic populations in response to the antimethanogenic compound bromochloromethane. *FEMS Microbiol.* Ecol. 62:313-322.

Dohme, F., A. Machmuller, A. Wasserfallen and M. Kreuzer. 2000. Comparative efficiency of various fats rich in mediumchain fatty acids to suppress ruminal methanogenesis as measured with Rusitec. *Can. J. Anim. Sci.* 80:473-482.

Dohme, F., A. Machmuller, A. Wasserfallen and M. Kreuzer. 2001. Ruminal methanogenesis as infuenced by individual fatty acids supplemented to complete ruminant diets. *Lett. Appl. Microbiol.* 32:47-51.

Eun, J.S., V. Fellner, L.W. Whitlow and B. A. Hopkins. 2003. Influence of yeast culture on fermentation by ruminal microorganisms in continuous culture. Department of Animal Science Bulletin, North Carolina State University, Raleigh, NC.

Evan, J.D. and S.A. Martin. 2000. Effects of thymol on ruminal microorganisms. *Curr. Microbiol.,* 41: 336 – 340.

Frumholtz, P. P., C.J. Newbold and R.J. Wallace. 1989. Influence of Aspergillus oryzae fermentation extract on the fermentation of a basal ration in the rumen simulation technique (Rusitec). *J. Agric. Sci. (Camb.)* 113: 169-172.

Goetsch, A.L., G Animut, R. Puchala, A.K. Patra, T. Sahlu, V.H. Varel and J. Wells. 2008. Methane emission by goats consuming different sources of condensed tannins. *Anim. Feed Sci. Technol.* 144: 228-241.

Guan, H., K.M. Wittenberg, K. H. Ominski and D. O. Krause. 2006. Efficacy of ionophores in cattle diets for mitigation of enteric methane. *J. Anim. Sci.* 84:1896-1906.

Guo, Y. Q., J. X. Liu, Y. Lu, W. Y. Zhu, S. E. Denman and C. S. McSweeney. 2008. Effect of tea saponin on methanogenesis, microbial community structure and expression of *mcrA* gene, in cultures of rumen micro-organisms. *Lett. Appl. Microbiol.* 47:421-426.

Hegarty, R. S. 2001a. Strategies for mitigating methane emissions from livestock- Australian options and opportunities. Pages 31–34 in Proc. 1st International Conference on Greenhouse Gases and Animal Agriculture. Obihiro, Hokkaido, Japan.

Hook, S. E., K. S. Northwood, A. D. G Wright and B. W. McBride. 2009. Long-term monensin supplementation does not significantly affect the quantity or diversity of methanogens in the rumen of the lactating dairy cow. *Appl. Environ. Microbiol.* 75:374-380.

Itabashi, H. 2001. Reducing ruminal methane production by chemical and biological manipulation. Pages 106 - 111 in Proc. 1st International Conference on Greenhouse Gases and Animal Agriculture. Obihiro, Hokkaido, Japan.

Itabashi, H., E. Bayaru, S. Kanda, T. Nishida, S. Ando, M. Ishida, T. Itoh, Y. Isobe, K. Nagara and K. Takei. 2000. Effect of salinomycin (SL) plus fumaric acid on rumen fermentation and methane production in cattle. *Asian Aust. J. Anim. Sci.* 13 (Suppl.): 287 (Abstr.).

Joblin, K. N. 1999. Ruminal acetogens and their potential to lower ruminant methane emissions. *Aust. J. Agric. Res.* 50: 1307-1313.

Johnson, K. A. and D. E. Johson. 1995. Methane emissions from cattle. *J. Anim. Sci.* 73:2483-2492.

Kajikawa, H. and C. J. Newbold. 2000. Methane oxidation in the rumen. *J. Anim. Sci.* 78 (Suppl.): 291 (Abstr.)

Kamra, D. N., N. Agarwal and L.C. Chaudhary. 2006. Inhibition of ruminal methanogenesis by tropical plants containing secondary compounds. *International Congress Series* 1293: 156-163.

Klieve, A. V. and R. S. Hegarty. 1999. Opportunities for biological control of methanogenesis. *Aust. J. Agric. Res.* 50: 1315-1319.

Kobayashi, Y., M. Wakita and S. Hoshino. 1988. Persistency of salinomycin effect on ruminal fermentation in wethers. *Nutr. Rep. Int.* 38: 987-999.

Kubo, I., H. Muroi and M. Himejima.1993. Structure-antibacterial activity relationships of anacardic acids. *J. Agric. Food Chem.* 41:1016-1019.

Lee, S. S., J. T. Hsu, H. C. Mantovani and J. B. Russell. 2002. The effect of bovicin HC5, a bacteriocin from *Streptococcus bovis* HC5, on ruminal methane production *in vitro*. *FEMS Microbiol. Lett.* 217:51-55.

Mackie, R. I. and B. A. White. 1990. Recent advances in rumen microbial ecology and metabolism: Potential impact on nutrient output. J. *Dairy Sci.* 73: 2971-2995.

Martin, S. A., M. N. Streeter, D. J. Nisbet, G. M. Hill and S. E. Williams. 1999. Effects of DL-malate on ruminal metabolism and performance of cattle fed a high-concentrate diet. *J. Anim. Sci.* 77: 1008 - 1015.

McIntosh, F.M., P. Williams, R. Losa, R.J. Wallace, D.A. Beever and C.J. Newbold. 2003. Effects of essential oils on ruminal metabolism and their protein metabolism. *Appl. Environ. Microbiol.* 69: 5011-5014.

Mitsumori, M., and W. Sun. 2008. Control of rumen microbial fermentation for mitigating methane emissions from the rumen. *Asian-Aust. J. Anim. Sci.* 21:144-154.

Mohammed, N., N. Ajsaka, Z.A. Lila, K. Mikuni, K. Hara, S. Kanda and H. Itabashi. 2004. Effect of Japanese horseradish oil on methane production and ruminal fermentation *in vitro* and in steers. *J. Anim. Sci.* 82: 1839-1846.

Montano, M. F., W. Chai, T. E. Zinn Ware and R. A. Zinn. 1999. Influence of malic acid supplementation on ruminal pH, lactic acid utilization, and digestive function in steers fed high- concentrate finishing diets. *J. Anim Sci.* 77: 780-784.

Morrison, M. 2008. The ecophysiology of plant biomass conversion in vertebrate herbivores: new insights from metagenomics. Proc. Mie Bioforum 2008 (Ed. K. Sakka). CDROM.

Morvan, B., J. Dore, F. Rieu-Lesme, L. Foucat, G. Fonty and P. Gouet. 1994. Establishment of hydrogen-utilizing bacteria in the rumen of new born lambs. *FEMS Microbiol. Lett.* 117: 249-256.

Moss, A. R., J. P. Jouany and J. Newbold. 2000. Methane production by ruminants: its contribution to global warming. *Ann. Zootech.* 49: 231-253.

Mutsvangwa, T., I. E. Edwards, J. H. Topps and G. F. M. Paterson. 1992. The effects of dietary inclusion of yeast culture (Yea- Sacc) on patterns of rumen fermentation, food intake and growth of intensive fed bulls. *Anim. Prod.* 55: 35-40.

Newbold, C. J., R. J. Wallace and F. M. McIntosh. 1996. Mode of action of the yeast *Sacchararomyces cerevisiae* as a feed additive for ruminants. *Br. J. Nutr.*76: 249 - 261.

Nollet, L., L. Mbanzamihigo, D. Demeyer and W. Verstrete. 1998. Effect of the addition of *Peptostreptococcus productus* ATCC 35244 on reductive acetogenesis in the ruminal ecosystem after inhibition of methanogenesis by cell-free supernatant *Lactobacillus plantarum* 80. *Anim. Feed Sci. Technol.* 71: 49-66.

Odongo, N. E., R. Bagg, G. Vessie, P. Dick, M. M. Or-Rashid, S. E. Hook, J. T. Gray, E. Kebreab, J. France and B. W. McBride. 2007. Long-term effects of feeding monensin on methane production in lactating dairy cows. *J. Dairy Sci.* 90:1781-1788.

Okine, E. K., J. A. Basarab, V. Baron and M. A. Price. 2002. Methane and manure production in cattle with different net feed intake. *J. Anim. Sci.* 80 (Suppl. 1): 206 (Abstr.).

Patra, A.K. 2004. Studies on inhibition of ciliate protozoa and stimulation of fibre degrading microbes in the rumen of buffalo by plant secondary metabolites. Ph.D. Thesis, submitted to IVRI, Izatnagar, India.

Patra, A.K., D.N. Kamra and N. Agrawal. 2008. Effect of leave extract on *in vitro* fermentation of feed and methanogenesis with rumen liquor of buffalo. *Indian J. Anim. Sci.* 78: 91-96.

Puchala, R., B .R. Min, A.L. Goetsch and T. Sahlu. 2005. The effect of a condensed tannin containing forage on methane emission by goats. *J. Anim. Sci.* 83(1): 182-186.

Robertson, L. J. and G. C. Waghorn. 2002. Dairy industry perspectives on methane emissions and production from cattle fed pasture or total mixed rations in New Zealand. *Proc. N. Z. Soc. Anim. Prod.* 62: 213-218.

Rumpler, W. V., D. E. Johnson and D. B. Bates. 1986. The effect of high dietary cation concentrations of methanogenesis by steers fed with or without ionophores. *J. Anim. Sci.* 62:1737-1741.

Russell, J. B. and H. J. Strobel. 1989. Effect of ionophores on ruminal fermentation. *Appl. Environ. Microbiol.* 55:1-6.

Stahl, D. A., B. Flesher, H. R. Mansfield and L. Montgomery. 1988. Use of phylogenetically based hybridization probes for studies of ruminal microbial ecology. *Appl. Environ. Microbiol.* 54:1079-1084.

Teather, R. M. and R. J. Forster. 1998. Manupulating the rumen microflora with bacteriocins to improve ruminant production. *Can. J. Anim. Sci.* 78:57-69.

Tokura, M., I. Changan, K. Ushida and Y. Kojima. 1999. Phylogenetic study of methanogens associated with rumen ciliates. *Curr. Microbiol.* 39:123-128.

Ungerfeld, E. M., S. R. Rust, D. R. Boone and Y. Liu. 2004. Effects of several inhibitors on pure cultures of ruminal methanogens. *J. Appl. Microbiol.* 97:520-526.

Van Nevel, C. J., D. I. Demeyer and H. K. Henderickx. 1971. Effect of fatty acid derivatives on rumen methane and propionate *in vitro*. *Appl. Environ. Microbiol.* 21:365-366ÿ

Williams, Y. J., S. Popovski, S. M. Rea, L. C. Skillman, A. F. Toovey, K. S. Northwood and A-D. G. Wright. 2009. A vaccine against rumen methanogens can alter the composition of archaeal populations. *Appl. Environ. Microbiol.* 75:1860-1866.

Wina, E., S. Muetzel and K. Becker. 2005. The impact of saponins or saponin-containing plant materials on ruminant production - a review. *J. Agric. Food Chem.* 53:8093-8105.

Wolin, M. J., T. L. Miller and C. S. Stewart. 1997. Microbemicrobe interactions. In: The Rumen Microbial Ecosystem. 2nd ed. (Ed. P. J. Hobson and C. S. Stewart), Blackie Acad. Press. London. pp. 467-491.

Wright, A. D. G., P. Kennedy, C. J. O'Neill, A. F. Toovey, S. Popovski, S. M. Rea, C. L. Pimm and L. Klein. 2004. Reducing methane emissions in sheep by immunization against rumen methanogens. *Vaccine.* 22: 3976-3985.

Yabuuchi, Y., M. Tani, Y. Matsushita, H. Otsuka and Y. Kobayashi. 2007. Effects of lauric acid on physical, chemical and microbial characteristics in the rumen of steers on a high grain diet. *Anim. Sci. J.* 78:387-394.

Yabuuchi, Y., Y. Matsushita, K. Otsuka, K. Fukamachi and Y. Kobayashi. 2006. Effects of supplemental lauric acid-rich oils in high-grain diet on *in vitro* rumen fermentation. *Anim. Sci. J.* 77:300-307.

15

Probiotics for Ruminants

Ravindra Kumar

Introduction

Ruminants are provided with a microbial ecosystem in the gastro-intestinal tract which helps in the bioconversion of lignocellulosic feeds into volatile fatty acids which are utilized by the animal as a source of energy. This microbial biomass is also helpful in synthesis of microbial protein and B-complex vitamins for the host animals. Any disturbance in the microbial balance may affect the performance of the animal adversely and any improvement in the microbial environment may lead to the improvement in productivity of the animal. A number of chemical feed additive like hormone, antibiotics have been used to improve the performance of the animals. But due to residual left over in the animal products and risk of creating disease, resistance against antibiotics, the use of these chemical feed additives have been discouraged and even banned in some countries. As an alternative microbial feed additive seems to be more natural; as it is not creating any new thing in GIT but it is merely restoring the normal and ideal flora in the digestive tract to its full capacity.

The concept of probiotics is not new, however. Approximately 100 years ago, Elie Metchnikoff, the father of immunology, investigated intestinal microbes as causative agents in aging, a process he called "autointoxication." He believed that lactic acid–producing bacteria (such as those found in yogurt) would suppress the growth of more proteolytic, autointoxicating bacteria. He viewed that the consumption of yoghurt by Bulgarian peasants as conferring a long span of life. The term "Probiotic" was first of all used by Parker in 1974 and he described it as the organisms and substances that contribute to intestinal microbial balance. The term probiotic means "for life" and has a contrast with the term antibiotic means "against life". Later Fuller (1989) redefined probiotics as live microbial feed supplements which beneficially affect the host animal by improving its intestinal microbial balance. Probiotic products are available in the form of oral pastes, water dispersible powders or liquids or directly fed feed additives and include microbial cells, microbial cultures and microbial metabolites. The US Food and Drug Administration used the term direct fed microbials (DFM) instead of probiotic and the manufacturers were directed to write DFM on their products.

Microorganisms used as probiotics in ruminants

1. Bacteria: *Lactobacillus acidophilus, L. bifidus, L. bulgaricus, L. casei, L. fermentum, L. lactis etc. Bacteroides* spp., *Leuconostoc* spp., *Pediococcus* spp., *Streptococcus spp.*
2. Fungi: *Aspergillus oryzae, Saccharomyces cerevisiae, Candida pintolopesii*

Characteristics of a good probiotics

1. The culture should exert a positive effect on the host. It should be gram positive, acid resistant, bile resistant and contain a minimum 30 x 10^6 CFU (colony forming unit) per gram.
2. The culture should possess high survival rate and multiply faster in the digestive tract. It should be strain specific.
3. The culture microorganisms should neither be pathogenic nor toxic to the host.
4. The adhesive capability of microorganisms must be firm and faster.
5. Be durable enough to withstand the duress of commercial manufacturing, processing and distribution so that product can be delivered alive to the intestine.

Mechanism of action of probiotics

The mechanisms by which probiotics exert their effects are largely unknown, but may involve modifying gut pH, antagonizing pathogens through production of antimicrobial and antibacterial compounds, competing for pathogen binding and receptor sites as well as for available nutrients and growth factors, stimulating immunomodulatory cells, and producing lactase. The mode of action may be specific for a micro-organism or generally common to the microorganisms used as probiotics. In ruminants, generally lactic acid bacteria and yeast culture are used as probiotics. These microorganisms have different mode of action to produce beneficial effect on the animals.

Mode of Action of lactic acid bacteria

Several mechanisms of lactic acid bacteria have been proposed like adhesion to the digestive tract to prevent colonization of pathogens, neutralization of toxins, bactericidal activity, prevention of amine synthesis and enhanced immuno-competence which are described below.

1. Competitive exclusion

Competitive exclusion refers to physical blocking of pathogenic bacteria colonization by probiotic bacteria from their favourite site such as intestinal villus, goblet cells and colonic crypts. The competition for space to adhere between indigenous bacteria and exogenous pathogens result in the competitive exclusion of pathogenic bacteria. Probiotic bacteria alter the physical environment of the intestines in such a way that pathogenic bacteria cannot survive. Probiotic bacteria exclude the opportunistic bacteria in two ways. First, the probiotic bacteria compete with pathogenic bacteria for nutrients and energy source thus, preventing them from acquiring energy required for growth and proliferation of pathogenic bacteria in the gut environment. Second, probiotics produce several organic acid and volatile fatty acids (VFA) as a result of their metabolism and fermentation. Consequently, the pH of the gut is lowered below that essential for survival of pathogenic bacteria such as *E. coli* and Salmonella. Probiotic bacteria also eject the colonization of pathogenic bacteria by attaching themselves to the surface of the gut thus preventing the adhesion of the pathogenic bacteria to gastrointestinal epithelium.

Antimicrobial activity

Lactic acid bacteria have the potential in limiting the growth of pathogenic microorganisms in calves. The protective effect of indigenous gut flora has been confirmed against *Salmonella, E. coli*, *Campylobactor foetus* sub sp. *Jejuni*, *Clostridium botulinum,* and *Mersinia enterocolitica. Streptococcus thermophilus* suppressed the growth of pathogenic bacterial, *Clostridium perfrigens* and *Salmonella entridides* by 73.3 and 51.8 per cent, respectively (Sikes and Hilton, 1987). *Lactobacilli* has been reported to produce various metabolites which have inhibitory effect on pathogenic bacteria (Vila *et al.*, 2010). *Lactobacillus acidophilus* has been reported to produce acidophilin, lactocidin and acidolin and *L. plantarum* produces lactolin (Vila *et al.*, 2010).

Following possible mechanisms have been suggested to explain the anti-microbial activity of lactobacilli and streptococci.

Mode of action	Reference
Adhesion to the intestinal wall preventing colonization by pathogens	Fuller (1989)
Lowering of intestinal pH (due to production of H_2O_2 or organic acid)	Renner (1991)
Competition of nutrients	Wilson and Perini (1988)
Production of antibacterial substances	Renner (1991)
Increase in phagocytic activity and immunoglobulin levels	Saito *et al.* (1981)

Immunomodulation

The gut is often referred as the largest immune organ of the body as more lymphocytes reside in the gut than any other organ of the body (Chichlowski *et al.*, 2007). The enterocytes of intestines provide such a barrier which prevents the passive loss of nutrients on one hand and on the other prevent the access of pathogens into the body. The lamina properia of the intestines is enriched with lymphocytes, macrophages, heterophils and dendritic cells all of them fighting against the pathogens. Probiotics have the ability to enhance the capacity of the host immune system against the pathogens and ultimately improve their health.

Probiotic bacteria are able to influence the inflammatory response elicited by pathogens through specific signalling pathway (Yurong *et al.*, 2005; Chichlowski *et al.*, 2007). Probiotic bacteria may exert its beneficial effects and modulate the immune system of the host against potentially harmful antigens via activation of lymphocytes and antibody production (Ng *et al.*, 2009). For example, *L. rhamnosus* admistration resulted in enhanced non-specific humoral response reflected by an increase production of IgG, IgA and IgM from circulating lymphocytes. Similar results were recorded by feeding yogurts containing *L. acidophilus, L. bulgaricus, S. thermophilus, B. bifidum and B. infantis* probiotic bacteria (Tejada-Simon *et al.*, 1999). Accumulated body of evidence have shown that the protective effect of probiotics is associated with elevated humoral and cellular immune response which is achieved through increased production of T lymphocytes and antibody secreting cells, expression of pro and anti-inflammatory cytokines, interleukins, IFN-γ, natural killer cells, antibody production, respiratory burst of macrophages and delayed type hypersensitivity reaction.

Lactobacillus as probiotics

Neonates of all species can suffer from a number of adverse environmental conditions soon after birth, including dampness, chilling, drought, inadequate milk and periods of separation from their dams. These conditions favour for non-beneficial bacteria to colonize the digestive tract and result in increased neonatal morbidity and mortality. In neonatal calves, probiotics aid in the prompt establishment of an animal-microbes relationship beneficial to the newborn's intestinal tract. These studies have demonstrated improved animal health and performance. Since Metchnikoff's early study data in several species have shown the ability of lactobacilli to suppress coliform growth. Viable *lactobacilli* feeding were shown to reduce the incidence of diarrhoea in young dairy calves. These findings contrast with those of other workers who have not observed benefits from feeding with *L. acidophilus* (Hatch *et al.*, 1973). No mortality of kids of Barbari breed (2-13 weeks of age) was recorded under probiotics (*Lactobacillus sporogens*) supplemented milk replacers.

Feeding of bacterial preparations increased feed conversion efficiency and live weight gain which may be due to improved digestion performance or indirectly due to the suppression of gut pathogens. It is feasible that extra-cellular enzymes of gut micro-flora supplement endogenous secretions, particularly in neonates when the digestive system is immature. Supplementation of lactic acid bacteria increased the bioavailability of calcium, phosphorus, magnesium and zinc from all diets (Schaafsma *et al.*, 1988). Consumption of lactic acid bacteria resulted into increased bone calcium and improved bone formation (Kaup, 1988).

Mode of action of yeast

Maintenance of ruminal pH

Yeast culture stimulates digestion of forage by stabilization of pH and the metabolism of readily fermentable carbohydrate (RFC). Cereal grains in ruminant rations provide substrates for rapidly growing ruminal bacteria (*Streptococcus bovis*) that produce large quantities of lactate. Rumen pH is mainly associated with concentration of lactic acid in the rumen. Diets having RFC depressed ruminal pH which lead to reduction in number of cellulolytic bacteria, impaired forage degradation. The effect of yeast supplementation on pH stabilization seems to be strongly dependent on the type of diet tested. Generally higher rumen pH has been observed with yeast supplementation when control pH tended to be below 6. The effect of yeast culture on the rumen pH was more pronounced in sheep fed with a maize silage/cereal based concentrate diet (high sugars/starch content) than with grass hay and sugar beet pulp-based concentrate. pH stabilization is generally associated with decreased levels of lactic acid in rumen. Nisbet and Martin (1991) have demonstrated that soluble components in *A. oryzae* and *S. cerevisiae* culture filtrate stimulated lactate uptake by *Selenomonas ruminantium* and *Megasphaera elsdenii*. For this reason in concentrate diet there is more cellulolytic activity due to addition of yeast culture. Mathieu *et al.* (1996) have found an increase of the pH with yeast only in faunated sheep and not in defaunated sheep suggesting that protozoa are involved in the effect of *saccharomyces cerevisiae* on the increase of rumen pH.

Production of growth stimulating factors

Yeast cells in the rumen may supply a chemical growth factor to the cellulolytic microorganisms. *S. cerevisiae* produces specific factors like B-vitamins or branched chain fatty acids to stimulate the growth of cellulolytic bacteria. Yeast culture provided soluble growth factor (i.e., organic acids, B-vitamins and amino acids) that stimulated growth of ruminal bacteria that utilize lactate and digest cellulose.

Modification in the ruminal environment

Production potential of yeast culture can be associated to their effect on the increased microbial population in the rumen. Addition of yeast culture resulted into increase in concentration of total anaerobic bacteria, but the increase was associated with fibre digesting and lactic acid utilising bacteria. Yeast culture supplementation stimulates the growth of cellulolytic bacteria in the rumen. It is particularly pronounced when forage containing diets are feed. Combined culture of yeast (*S. cerevisiae*), lactobacilli and enterococci supplementation to the diet of steers enhanced the concentration of cellulolytic bacteria in rumen (5 to 40 times). Enhanced growth of rumen bacteria could result in enhanced microbial protein synthesis and increased contribution of the microbial population to the digestive process. Cell yields were increased by 7 to 63% when cultures of *Ruminobactor amylophilus*-1024 and *Fibrobactor succinogenes*-S-85 were treated with yeast. Total bacterial counts under *in vitro* condition were increased significantly ($P<0.01$) in viable culture supplementation groups (*S. Cerevisiae-NCDC-47* and *S. cerevisiae-NCDC-47 + L. plantarum-NCDC-25 + E. faecium-NCDC-124)* when the substrate had high or medium concentrate. But, in high roughage groups no difference was recorded among treatment groups (Dutta *et al.*, 2001). It has been suggested that an increased bacterial flora in animals fed *S. cerevisiae* is central to the action of yeast in the rumen.

Increase in the number of total culturable bacteria in the rumen appears to be one of the most consistently reported responses to yeast supplementation. Increased levels of rumen protozoa following *Saccharomyces cerevisiae* ingestion were also reported. *Aspergillus* fermentation extracts and yeast cultures have been shown to directly stimulate rumen fungi, which may improve fiber digestion. *Saccharomyces cerevisiae* supplementation was associated with an increased flow of microbial protein leaving the rumen and enhanced supply of amino acids entering the small intestine.

One of the common observations associated with the addition of yeast culture to ruminants and in rumen simulating fermentors has been the reduction of rumen ammonia concentration (Sohn and Song, 1996). The concentration of ammonia was decreased by 10 to 35 per cent *in vitro* (Carro *et al.*, 1992). Similar results have been reported by others worker *in vivo*, suggesting an improved microbial capture of ammonia. Whereas, Moloney and Drennan (1994) reported that rumen ammonia concentration was not affected when yeast was included in low concentrate diet of steer, but was reduced when yeast was added to the diet having high concentrate. Reduced ammonia levels have not been associated with decreased protein degradation or deamination (Williams and Newbold, 1990). Incorporation of ammonia into microbial protein was enhanced due to supplementation of yeast (Carro *et al.*, 1992), which was confirmed by greater microbial yield and microbial true protein reaching the duodenum (Erasmus, 1991). VFA production also gets improved due to addition of yeast culture in ruminants. Yeast culture may alter the pattern of VFA production. In lamb, yeast culture addition along with sugar-beet pulp and rolled barley (0.8:0.2) and 60g soybean meal increased total VFA and acetate. Whereas in other study, yeast culture addition increased TVFA non-significantly and acetate: propionate ratio was decreased from 5.01 to 3.81 mainly due to increase in propionate in sheep. In lactating cow, some workers observed that there was higher production of propionate and reduction of acetate to propionate ratio due to supplementation of yeast culture. Similarly, enhanced production of propionate and reduction of acetate to propionate was reported by Moloney and Drennan (1994). Supplementation of commercial yeast to the high concentrate diet of steer reduced acetate and acetate to propionate ratio; molar proportion of butyrate and iso-acids were higher. It was reported that neither the yeast culture supplement nor the mixed microbial (yeast, *lactobacilli* and *enterococci*) supplement consistently alter the patterns of VFA production in the continuous cultures or in the rumen of steers.

Yeast culture supplementation resulted into reduction of methane production in steers by 28% (Williams, 1989). The addition of yeast culture resulted in a lower methane production for medium concentrate diet, but in a higher for the high concentrate diet (Carro *et al.*, 1992). Many workers have reported improvement in degradability patterns of nutrients either *in vitro* or *in sacco* due to supplementation of yeast culture. Supplementing the high concentrate diet with yeast culture resulted in significantly higher DM and NDF degradabilities (Carro *et al.*, 1992). Moreover, filter paper (cellulose) degradability (24h) tended to be higher for the vessels supplemented with yeast culture than for the control (33% Vs 27%). These results might suggest that yeast culture increased the population of fibre--degrading bacteria and/ or their activity. However, yeast culture had no significant effect on DM and NDF degradabilities with medium or low concentrate diet (Carro *et al.*, 1992). In *vitro* digestibility of NDF was increased by yeast or yeast + white rot fungi (*Armillaria heimii*) treatment, with the yeast + fungi combination showed highest values in hay based ration. Strain variability of *S. cerevisiae* was observed on nylon bag degradability of straw and hay.

Feed Intake and nutrient utilization as influenced by *S. cerevisiae*

Nutrient supply to the animals has been improved due to yeast culture supplement at a fixed intake, but in farm trials, its effect on intake appears the most important cause of improved performance. In several studies it has been observed that yeast culture addition increased the feed intake in calves. However, some studies indicated no added advantage of yeast culture supplementation to lactating cows on DMI. Differences in response to added yeast might have been due to interactions among yeast, diet and stage of lactation.

A number of studies provided evidence indicating that live yeast cells enhances the digestive process in the gastrointestinal tract. Supplementation of *S. cerevisiae* increased the digestibility of protein, cellulose, fibre, NDF and ADF. Similarly, supplementation of yeast culture increased hemi cellulose and CP digestibility in ruminants. DM digestion in the rumen of dairy cows was increased receiving yeast culture. Carro *et al.* (1992) reported that effect on digestibility is dependent on the forage to concentrate ratio, supplementation of yeast culture with high concentrate diet resulted in significantly higher DM and NDF digestibilities in rumen (Rusitech). However, on high forage diet yeast culture had no effect on DM, NDF and cellulose digestibility.

Effect of probiotics supplementation on the productive performance of animals

Growth performances

In ruminants, higher growth is positively related with the higher rumen fermentation pattern and nutrient utilization pattern. Probiotics, either lactic acid bacteria or yeast culture, have beneficial effect on higher nutrient utilization. Significantly ($P<0.05$) higher weight gain and average daily gain (ADG) were observed in lambs fed complete rations supplemented with Yea-Sacc[1026], followed by *Lactobacillus acidophilus* and control (without probiotics) (Rao *et al.*, 2003). Earlier similar observations were recorded by Rouzbehan *et al.* (1994) in lambs fed Yea-Sacc supplemented barley. Some workers also observed higher growth rates in lambs fed probiotics. The higher growth rate in Yea-Sacc supplemented ration might be due to increased flow of microbial protein leaving the rumen and an enhanced supply of amino acids entering the small intestine.

Milk production

Several workers have reported that supplementation of yeast culture to the lactating cows improved milk yield or FCM yield (Williams *et al.*, 1991). Teh *et al.* (1987) tested the effect of feeding yeast culture with or without added $NaHCO_3$ on lactating goats and found that both yeast culture and $NaHCO_3$ increased milk yield without an increase in DM intake. In the high producing cows, there was a 6% increase in milk production due to supplementation of viable mixed microbial feed additives (*S. cerevisiae, Streptococcus faecium* and *Lactobacillus acidophilus*), whereas there was no treatment effect in the low productive cows. Williams *et al.* (1991) have shown that there was interaction between diet composition and yeast culture regarding the DMI and milk yield, effects of yeast culture were greatest in diets containing 60:40 (concentrate: forage) ratio.

Some workers reported that milk yield was unaffected by the addition of yeast culture to the diet of cows (Erdman and Sharma, 1989). Dawson (1993) stated that yeast culture supplements did not have equal beneficial effects with all types of diets, and it is currently not possible to define the dietary conditions that provide optimum response to yeast culture supplementation. Teh *et al.* (1987) found that both yeast culture and $NaHCO_3$ increased milk fat per cent in early lactation goats. Other authors have also observed significant improvement in milk fat per cent of lactating cows due to supplementation of yeast culture. Williams *et al.* (1991) have noticed that milk fat per cent slightly improved only in high concentrate diet (60%). Whereas, other authors did not record any difference in milk fat per centage due to yeast culture supplementation (Erdman and Sharma, 1989).

Summary

Probiotics or direct fed microbial have great potential to enhance growth and immunity status of ruminants. Lactic acid bacteria have specific function in digestive system and immunity of ruminants. Whereas, yeast culture trigger the beneficial effect on rumen fermentation and intestinal digestion.

References

Carro, M. D., Libzien, P. and Rohr, K. 1992. Influence of yeast culture on the *in vitro* fermentation (Rusitech) of diets containing variable portions of concentrates. *Anim. Feed Sci. Technol.* 37:209-220.

Dawson, K.A. 1993. Current and future role of yeast culture in animal production: a review of research over the last six years. Proc. Alltech's 9th Annu. Symp. Alltech Tech. Publ., Nicholasville, KY. pp 269.

Dutta, T. K., Kundu, S. S. and Sharma, D. D. 2001. Potential of probiotics supplementation on *in vitro* rumen fermentation and ^{35}S incorporation in microbial protein. *Indian J. Anim. Nutr*. 18:227-234.

Erasmus, L.J. 1991. The importance of the duodenal amino acid profile for dairy cows and the impact of changes in these profiles following the use of Yea-sacc[1026]. *Feed Compounder*. 11:24-29.

Erdman, R.A. and Sharma, B.K. 1989. Effect of yeast culture and sodium bicarbonate on milk yield and composition in dairy cows. *J. Dairy Sci.* 72:1929-1932.

Fuller, R. 1989. Probiotics in man and animals. *Appl. Bacteriol.*66:365-378.

Hatch, R.C., Thomas, R.O., Thayne, W.V. 1973. Effect of adding *Bacillus acidophilus* to milk fed to baby calves. *J. Dairy Sci*. 56:682 (abstr.)

Kaup, S.M. 1988. Bioavailability of calcium on yoghurt and its relationship to the hypocholesterolemic properties of yoghurt. *Disease Abstr. Int*. B. 48:1959.

Mathieu, F., Jouany, J.P., Senaud, J., Bohatier, J., Bertin, G. and Mercier, M. 1996. The effect of accharomyces cerevisiae and Aspergillus oryzae on fermentations in the rumen of faunated and defaunated sheep; protozoal and probiotic interactions. *Reprod. Nutr. Dev*. 36: 271-287.

Metchnikoff, E. 1907. *Prolongation of life*. London, Heinemann. Miller, B.G., Phillips, A., Newby, T.J., Stokes, C.R. and Bourne, F.J. 1985. A transient hypersensitivity to dietary antigens in the early weaned pig:a factor in the aetiology of post weaned diarrhoea. *Proc. 3rd Int. Seminar on Digestive Physiology of the Pig*, Copenhagen, May, *Nat. Inst. Anim. Sci*., Denmark, pp. 65-66.

Moloney, A.P. and Drennan, M.J. 1994. The influence of the basal diet on the effects of yeast culture on ruminal fermentation and digestibility in steers. *Anim. Feed Sci. and Technol.* 50:55-73.

Nisbet, D.J. and Martin, S.A. 1991. Effect of a *Saccharomyces cerevisiae* culture on lactate utilization by the rumen bacterium *Selenomonas ruminantium*. *J. Anim. Sci.* 69: 4628-4633.

Parker, R.B. 1974. Probiotics, the other half of the antibiotics story. *Anim. Nutri. Health*. 29:4-8.

Rao, T.N., Rao, Z.P., Rama Prasad, J. and Prasad, P.E. 2003. Supplementation of probiotics on growth performance in sheep. *Indian J. Anim. Nutr*. 20:224-226.

Renner, E. 1991. *Cultured dairy products in human nutrition*. Bull. IDF, 255:1-24.

Rouzbehan, Y., Galbraith, H., Rooke, J.A. and Perrott, J.G. 1994. A note on the effects of dietary inclusion of lambs given diets containing unground pelleted molassed dried sugar-beet pulp and barley in various proportions. *Anim. Prod.* 59:147-150.

Saito, H., Tomioka, H.and Sato, K. 1981. Enhanced resistance of *Lactobacillus* against *Listeria* infection in mice. *Med. and Biol.* 102:273-277.

Schaafsma, G., Dekker, P.R. and deWaard, H. (1988). Nutritional aspects of yoghurt. II. Bioavailability of essential minerals and trace elements. *Netherland Milk Dairy J.* 42:135-146.

Sikes, A. and Hilton, T. 1987. Sensitivity of food brone bacteria (spoilage & pathogenic) to a methanol-acetone extract of milk fermented by *Streptococcus thermophilus*. *J. Food Prot.* 50:812-814.

Sohn, H.J. and Song, M.K. 1996. Effect of feeding yeast diets on the digestibility by sheep. *Korean J. Anim. Sci.* 38:578-588.

Teh, T.H., Sahlu, T., Escobar, E.N and Cushaw, J.L. 1987. Effects of live yeast culture and sodium bicarbonate on lactating goats. *J. Dairy Sci.* 70 (Suppl.1) :200 (Abstr.).

Williams, P.E.V. 1989. The mode of action of yeast culture in ruminant diets: A review of the effect on rumen fermentation patterns. Biotechnology in the feed industry. Proceedings of Alltech's fifth Annual Symposium (Ed. Lyons,T.P.). Alltech Technical Publications. Nicholasville, USA. pp 65-84.

Williams, P.E.V. and Newbold, C.J. 1990. Rumen probiosis: The effects of novel microorganisms on rumen fermentation and ruminant productivity. In: W.Haresing and D.J.A.Cole (Ed.). *Recent Advances in Animal Nutrition*. Butterworths, London, pp. 211-227.

Williams, P.E.V., Tait, C.A.G., Innes, G.M. and Newbold, C.J. 1991. Effects of the inclusion of yeast culture (*Saccharomyces cerevisiae* plus growth medium) in the diet of dairy cows on milk yield and forage degradation and fermentation patterns in the rumen of steers. *J. Anim. Sci.* 69:3016-3026.

Wilson, K.H. and Perini, F. 1988. Role of competition for nutrients in suppression of *Clostridium difficile* by the colonic microflora. *Infection and Immunity.* 56:2610-2614.

Chichlowski, M.,J. Croom,B.W. McBride,G.B. Havenestein and Koci,M.D. 2007.Metabolic and physiological impact of probiotics or diret-fed-microbials on poultry: A brief review of current knowledge. *Int.J. Poult.Sci.,*6:694-704.

Ng, S.C., A.L. Hart, M.A. Kamm, A.J. Stagg and Knight S.C. 2009. Mechanisms of action of probiotics: Recent advances. *Inflammation Bowel Dis.*, 15:300-310.

Tejada-Simon,M.V.,Z. Ustunol and Pestka, J. J. 1999. Ex vivo effects of lactobacilli,streptococci and bifidobacteria ingestion on cytokine and nitric oxide production in a murine model. *J. Food Prot.,* 62:162-169.

Vila, B., E. Esteve-Garcia and Brufau, J. 2010. Probiotic micro-organisms: 100 years of innovation and efficiency; modes of action. *World's Poult. Sci.* J. 65: 369-380.

Yurong, Y., S. Ruiping, Z. Shimin and Yibao, J. 2005. Effect of probiotics on intestinal mucosal immunity and ultrastructure of cecal tonsils of chickens. *Arch. Anim.Nutr.*, 59:237-246.

16

Scope of Propionic Acid Producing Bacteria as Feed Additive in Ruminants

Haidar Ali Ahmed, S. K. Sirohi and A. M. Ganai

Introduction

The original concept of feeding bacterial direct fed microbial (DFM) to livestock was based primarily on potential post–ruminal effects, including improved establishment of beneficial gut microflora (Fuller, 1999). Calves fed *Lactobacillus acidophilus* have been reported to have reduced incidence of diarrhoea (Beecham *et al.*, 1977) and reduced intestinal coli form count (Bruce *et al.*, 1979). But now it is known that certain bacterial DFM also have beneficial effects in the rumen and one of the important groups of this type of bacteria is propionic acid producing bacteria (Elsden, 1945; Johns, 1951a; Langa, 2010). This group of bacteria includes *Propionibacterium* spp., *Veillonella* spp., *Megasphaera* spp., *Selenomonas* spp. and *Succiniclasticum* spp. (Table 1). Propionic acid producing bacteria (PAPB) are secondary bacteria which enhance propionate production by utilizing the end products of primary fibrolytic or starch utilizing bacteria which produce succinic acid and lactic acid (Swenson and Reece, 1996). Inoculation of *in vitro* fermentation with the lactate utilizing ruminal bacterium *Megasphaera elsdenii* prevented lactate accumulation and produced propionic acid when a highly fermentable substrate was used (Kung and Hession, 1995). Propionibacteria are also rumen active bacteria which can produce propionic acid by utilizing glucose, lactic acid and succinic acid (Elsden, 1945; Grinstead and Barefoot, 1992; Hugenholtz *et al.*, 2002). Ghorbani *et al.* (2002) also reported that Propionibacteria convert lactate and glucose to acetate, propionate and CO_2. They also reduce nitrate to nitrite and reduce nitrite load by 40-60% and minimize changes in blood nitrite and met hemoglobin (Rehberger *et al.,* 1993). Succinic acid is an important end–product, but in practice this is converted into propionic acid by other bacteria such as *Selenomonas ruminantium* (Swenson and Reece, 1996; McDonald *et al.*, 2002; Langa, 2010); such interactions between microorganisms are an important feature of rumen fermentation. *Veillonella gazogenes* are also active in the rumen and produce propionic acid by fermenting succinic acid (Johns, 1951a). Moreover in low quality forage-based diet *Fibrobacter succinogenes* are predominant fibrolytic bacteria which digest fibre faster and produce sufficient amounts of succinic acid in the rumen (Cheng *et al.*, 1984; Miron and Ben-Ghedalia., 1992) and synergistic increases in propionate production and fiber digestion was found in co-cultures of *Fibrobacter succinogenes* and *Selenomonas ruminantium* (Sawanon *et al,* 2003), and of *Ruminococcus flavefaciens* and *Selenomonas ruminantium* (Sawanon *et al.*, 2006). These findings suggest that positive interactions occur between bacteria, mostly via metabolite exchange.

Propionibacterium sp. and *Veillonella gazogenes* can produce propionic acid in the rumen either through decarboxylic acid pathway (Succinate pathway) by decarboxylation of succinic acid ((Sijpesteijn and Elsden, 1951) or through acrylate pathway from lactate. *Megasphaera elsdenii* produce propionic acid from lactate through acrylate pathway (Langa, 2010). *Selenomonas ruminantium* and *Succiniclasticum ruminis* utilize succinate and convert them to propionate in the rumen (Swenson and Reece, 1996; McDonald *et al.*, 2002 and Gylswyk, 1995). *Enterococcus* spp. also increased propionate production indirectly by providing lactic acid to lactate utilizing bacteria which converts lactate to propionate (Melo, 2007).

Table 1: Different species of propionic acid producing bacteria

Attributes	***Propionibacterium* spp**	***Veillonella* spp**	***Megas-pha-era* spp**	***Selenomonas* spp**	***Succiniclasticum* spp**
Species	*P. acidipropionici, P. jensenii, P. theoni, P. freudenreichii*	*P. freudenreichii P. freudenreichii*	*M. elsdenii*	*S. ruminantium*	*S. ruminis*
Population	103-104/ml	6x106/ml	-	3.6x107/ml	108/ml
Reference	Davidson et al, 1995	Sijpesteijn and Elsden, 1951	Rogossa, 1971	Prins, 1971	Gylswyk, 1995

Propionibacterium sp.

They are natural inhabitant of the rumen and comprise 1.4 to 4.3 % of the total microbial population in the rumen (Mead and Jones., 1981; Oshio *et al.*, 1987). In dairy cows, the population of propionic bacteria range from 10^3 to 10^4 cfu / ml (Davidson, 1998). These bacteria are naturally found in high numbers in the rumen of animals fed forage and medium concentrate diets. E. Freudenreich and O. Orla-Jensen isolated the first propionic acid bacterium from emmental cheese in 1906 (Jan *et al.*, 2007) and the genus *Propionibacterium* was first proposed by Orla-Jensen (1909) for these bacteria. In early studies the propionic acid bacteria were exclusively isolated from cheese, milk and dairy products (Lind, 2010) and known as dairy propionibacteria or classic propionibacteria as they were the first to be included in the genus propionibacteria (Lind ,2010). Another group of propionibacteria is "Cutaneous propionibacteria " present in faeces, in mouth and on moist part of human and animal skin (Cummin and Johnson, 1992). Refined isolation techniques have made it possible to isolate dairy propionibacteria from soil and silage too (Rossi *et al.*, 1999). Now it is known that their natural habitat can also be the digestive tract of ruminants (Rinta-Koski *et al.*, 2001; Jarvis *el al.*, 1989). Elsden (1945) also observed that propionic acid was present normally in the rumen, and that it was produced during *in vitro* fermentation of cellulose, glucose and lactic acid, particularly the last mentioned, suggested at once that members of the genus *Propionibacterium* were playing a part in the fermentations in the rumen. Organisms of the genus *Propionibacterium* comprise a small proportion of the ruminal microflora, are slow growing and therefore are difficult to isolate from ruminal fluid (Davidson *et al.*, 1995). A strain of *Propionibacterium acnes* (a cutaneous propionibacteria) was isolated successfully from healthy cow rumen fluid (Wu *et al*, 2009) in China. The cutaneous propionibacteria, included in the genus *Propionibacterium* in the 1970s, are distinguished from dairy propionibacteria (Vorobjeva, 1999) as presented in Table 2.

Table 2: Difference between dairy and cutaneous Propionibacteria (Lind, 2010)

Characteristic	Dairy Propionibacteria	Cutaneous Propionibacteria
Respiration	Microaerophilic or facultative	Anaerobic
Pathogenicity	-	+
Colony Colour	Cream, yellow, orange, red, brown	Pinkish
DNA (G+C mol %)	65-68	58-64
Esculin hydrolysis	-	+
Gelatin liquidation	+	-
Growth optimum (°C)	30-32	36-37

Propionibacterium acnes is a non lactate utilizing rumen bacteria whereas *Propionibacterium cyclohexanicum* is lactate producing bacteria.

Veillonella gazogenes

Johns (1951a) reports that the rumen contains at least $6x10^6$ *V. gazogenes* /ml of rumen liquor. This bacterium is mainly found in sheep. They are also present in cattle, but population is less. *Veillonella gazogenes* is more active in decarboxylation of succinic acid to propionic acid than *Propionibacterium sp.* (Sijpesteijn and Elsden, 1951). *Veillonella parvula* also utilize lactate and convert them to propionate (Langa, 2010).

Megasphaera elsdenii

Elsden and Lewis (1953) described a large, VFA-producing, strictly anaerobic, non-33motile cocci that they isolated from a sheep. Elsden *et al.* (1956) isolated another similar appearing organism approximately 2.4 × 2.6 ìm in diameter, which occurred in pairs and in chains up to 20 cells in length. A similar organism was discovered in the rumens of bloating cattle by Gutierrez *et al.* (1959) and they proposed the name *Peptostreptococcus elsdenii*. In 1971, Rogosa proposed the transfer of *P. elsdenii* to a new genus, *Megasphaera elsdenii.* This bacterium was first discovered in sheep, but it is also an important lactate fermenter in cattle and produce propionic acid through acrylate pathway (Sun *et al.*, 2009). It has been shown that ruminal populations of *Megasphaera elsdenii are* diet dependent, and increases are normal with the introduction of grain to the diet (Mackie and Gilchrist, 1979). It is a filamentous bacteria and found in close association with *Streptococcus bovis* which ferments starch to lactate, whatever lactate is produced is taken by it. Its fermentation end products are acetic, propionic, butyric, valeric, caproic acid and CO_2.

Selenomonas ruminantium

Scheifinger and Wolin (1973) first demonstrated that growth of the non-fibrolytic *Selenomonas ruminantium* occurs on cellulose in co-culture with *F. succinogenes*. Russell (1985) subsequently showed that cellodextrins (primarily cellotetraose and cellopentaose) support growth of the non-fibrolytic *S. ruminantium* and *P. ruminicola*. Cross feeding was also demonstrated in succinate propionate metabolism. *F. succinogenes* and *R. flavefaciens* produce succinate during fiber digestion. However, succinate does not accumulate in the rumen, since it is rapidly converted into propionate. For this conversion, succinate-decarboxylating bacteria such as *S. ruminantium* are considered to play a central role in the rumen (Scheifinger and Wolin, 1973; Wolin *et al.*, 1997; Sawanon and Kobayashi, 2006). *S. ruminantium* has been detected as a member of the fiber-associated bacterial community in the rumen (Koike *et al.,* 2003b), and its population size on plant fiber was estimated to be 9% (Koike *et al.*, 2007). This ecological information suggests that *S. ruminantium* may be involved in fiber digestion. Sawanon *et al.* (2003) isolated *S. ruminantium* strains belonging to a phylogenetically novel group from sheep rumen. Subsequently, they reported that fiber digestion in a co-culture of *F. succinogenes* and the novel *S. ruminantium* strains exceeded the value achieved by *F. succinogenes* alone. Furthermore, propionate production and growth of *S. ruminantium* was notable in co-cultures, while succinate accumulated in monocultures of *F. succinogenes*. These results indicate that *F. succinogenes* provides fiber hydrolysis products to *S. ruminantium* as growth substrates, while *S. ruminantium* may activate *F. succinogenes* by rapidly consuming the products. The occurrence of similar events was confirmed in co-culture of *S. ruminantium* with *R. flavefaciens* (Sawanon and Kobayashi, 2006). Such cross feeding between fibrolytic and non-fibrolytic bacteria could enhance fiber digestion in the rumen.

Succiniclasticum ruminis

Succiniclasticum ruminis isolated and identified by Gylswyk (1995) from rumen of cattle fed on production diet (grass silage and equal amount of concentrate on dry matter basis) and in the rumen of cows on pasture. Their population in the rumen is 10^8/g of ingesta and convert succinate quantitatively to propionate and CO_2 as the sole energy yielding mechanism.

Characteristics of propionic acid producing bacteria

Propionibacteria are Gram-positive, irregular, none spore forming, nonemotile, catalase- positive pleomorphic rods ("coryneform") with characteristic V or Y configurations (Davidson *et al.*, 1995; Vorobjeva, 1999). Depending on stage of growth and environment, their size can vary distinctly (Jan *et al.*, 2007). The main metabolites are propionic acid ,acetic acid and carbon dioxide, but succinic acid, formic acid vitamin B_{12} are also formed (Vorobjeva, 1999).They can ferment both lactic acid with the exception of *Propionibacterium acnes* (Douglas and Gunter, 1946) and decarboxylate succinic acid (Delwiche,1948) to produce propionic acid. They can utilise a variety of carbone sources such as glucose, lactose, xylose, sucrose, glycerol and lactate as substrate but lactate is used preferentially to other substrate (Jefferson, 2008). The G+C content of their genome ranges from 53 to 67 mol% (Cummins and Johonson, 1986 and Kusano *et al.*, 1997) which places them in the high G+C group of bacteria (Olsen *et al.*, 1994). Distinguishing between the species of *Propionibacterium* sp. is not always an easy task, but there are some species-specific characteristics as follows (Lind, 2010):

Propionibacterium freudenreichii: None haemolytic, colonies usually white or beige, sub species *freudenreichi* and *shermanii* can be separated by lactose fermentation (freudenreichi - , shermanii +).

Propionibacterium jensenii: *Both* aerobic and anaerobic haemolytic, colonies creamy to yellow, some strains produce extracellular slime.

Propionibacterium theoni: *Â*-haemolytic, slowest growing species, colonies yellow, orange, red or brown.

Propionibacterium acidipropionici: *Weak* or negative catalase reaction, grows aerobically, colonies beige to orange, some species produce extracellular slime.

Propionibacterium cyclohexanicum: *Negative* catalase reaction, produce lactic acid, colonies white or creamy.

Johns (1951a) reported that *V. gazogenes* colonies grow on agar plate were of same kind, clear, covex, smoth, circular and entire, greish white by reflected light and of characteristic honey colour by transmitted light. The cells were spherical, 0.3-0.6 μ in diameter, occurring singly, in pairs and in large irregular groups, Gram-positive in young cultures but Gram-negative after 12-18 hr. growth. The organism did not liquefy gelatine, was non-haemolytic, H_2S positive, indole negative and reduced nitrate to nitrite. Most strains appeared to be weakly catalase positive but this varied from strain to strain.
Veillonella gazogenes can ferment DL-lactate, pyruvate, L-malate, fumarate, succinate and D-tartrate. It ferments lactate to propionate, acetate, nitrogen and CO_2 .The following compounds can not be fermented by *V. gazogenes*: glucose, fructose, arabinose, L - xylose, galactose, mannose, maltose, L-sorbose, sucrose, lactose, raffinose, rhamnose, inulin, starch, glycerol, dextrin, mannitol, salicin, citrate, erythritol, DL-serine, alanine, glutamicacid, glycine, methionine, L-cystine, tryptophan ,and sodium formate (Johns, 1951a). It cannot ferment glucose because it lacks the enzyme hexokinase.

Table 3: Characteristics of propionic acid producing bacteria

Character-istics	*Propioni-ba-cterium* spp	*Veillonella* spp	*Megas-phaera* spp	*Selenomonas* spp	*Succiniclast-cum* spp
Respiration	Facultative anaerobe	Obligate anaerobe	Obligate anaerobe	Obligate anaerobe	Obligate anaerobe
Colony colour	Creame, yel-low, orange, red, brown	Greyish white, honey colour	Yellowish white	Creamy white	Colourless
Colony shape	Circular	Convex, smooth, circu-lar and entire	Circular	circular	Irregularly round
Gram's staining	Gram pos-itive	Gram positive in young culture, Gram negative after 2-18 hours of growth.	Gram neg-ative	Gram neg-ative	Gram neg-ative
Shape of cells	Pleomorphic rod	Speri-cal,0.3-0.6µ	Cocci, 2.0µm	Curved rods	Rods
Arrange-ment of cells	V or Y con-figuration	Single,pairs, large irregular groups	Single, pairs, chains	Single	Groups
Tempera-ture	32°C	37°C	39°C	39°C	39°C
pH	6-7	6-8	5.5-6.5	6.7	6.8
References	Seshadri, 1995	Johns, 1951a	Rogossa, 1971	Prins, 1971	Gylswyk, 1995

Megasphaera elsdenii is an obligately anaerobic Gram-negative bacterium known as a normal inhabitant of the rumen of cattle and sheep. It can use both carbohydrate and organic acids and is considered to be one of the main organisms involved in lactic acid catabolism and amino acid deamination. It is a ruminal soluble-substrate-fermenting species. Its nutritional strategy is based on utilization of lactate and soluble products of polymer hydrolysis, supplied by other bacteria (Hungate, 1966). Rogossa (1971) describes *M. elsdenii* as Gram-negative cocci, 2.0µm or more in diameter that occurs as single cells, pairs and occasionally in chains.

Prins (1971) reported that the deep colonies of *Selenomonas* spp. in agar were creamy white, lenticular, and sometimes umbonate at both sides. They were visible after 1 to 2 days of incubation as glassy, round, translucent colonies with the appearance of tiny droplets of water. After 2 to 3 days, colonies changed from translucent to opaque and grew in 5 or 6 days to a maximum diameter of 6 mm. Growth in agar or gelatin stabs was filiform. Colonies usually not appearing before the third day of incubation.The cells measured 2 to 3, µm in width and 5 to10 µm in length. In liquid culture, cells were slightly thicker than in agar, were more uniform in size, and helicoidal filaments were never seen. Cells were curved, crescentic, gram-negative, and rapidly motile with bluntly tapered or rounded ends, occurring mostly as singles. The organisms moved by means of flagella, attached to the concave side of the cells in one or what seemed sometimes two tufts (dividing organisms). In some electron micrographs they were distributed over the concave side of the cell.

Succiniclasticum ruminis is gram-negative, anaerobic, none motile, non-spore-forming, rod-shaped bacterium that ferment succinate quantitatively to propionate. It cannot ferment carbohydrates, amino acids and mono-, di-, and tricarboxylic acids other than succinate, which is converted to propionate. They are non proteolytic and do not produce catalase or urease or reduce nitrate. Cells are not extensively branched. They are short rods which are 0.3 to o.5µ m wide and up to 1.8µm long. Large clumps of cells are often observed in liquid media .Colonies are irregularly round with smooth edges, covex, practically colourless and slightly opaque with a glistening appearance. Submerged colonies that develop after inoculation into molten agar are lens-shaped discs (Gylswyk, 1995).

Growth conditions of propionic acid producing bacteria

Propionibacteria can utilise many different carbon sources, for example glucose, fructose and glycerol, but the preferred substrate is lactate (Jan *et al.*, 2007). The pH value also influences the growth of propionibacteria. The optimal pH range for growth of propionibacteria is 5.1 to 8.5 (Jan *et al.,* 2007). The temperature optimum for dairy propionibacteria is normally 30°C, but growth is possible between 15°C and 40°C depending on strain and other growth condition (Jan *et al.,* 2007). Davidson *et al.* (1995) also used 32°C to grow rumen propionibacteria under anaerobic conditions (GasPak). Seshadri (1993) reported that optimal growth conditions are pH range of 6-7 and a temperature range of 30-37°C. If pH value is below 4.5, there is practically no growth and low organic acid formation (playne, 1985).

Veillonella gazogenes grow under strictly anaerobic conditions, with temperature and pH at 37°C and 6.0-8.0 respectively (Johns, 1951a).

Megasphaera elsdenii grow under strictly anaerobic conditions, at pH 5.5-6.5 and temperature 39°C (Horn *et al*, 2009).

Prins (1971) reported that *Selenomonas* are strict anaerobes; no growth occurred in media in which the resazurin was oxidized. Optimal growth was obtained at 39° C. No growth occurred at or below 33°C, at 42°C or at 45°C in 40 hr. Growth in a liquid medium fortified with fermentable sugars was heavily turbid with slight sediment formation. Gylswyk (1995) reported that *Succiniclasticum* spp. grow better at pH 6.8, temperature 39°C under anaerobic condition. There is no growth at 22°C and at 45°C growth is reduced to half.

Table 4: Fermentation characteristics of propionic acid producing bacteria

Characteristics	***Propionibacterium* spp**	***Veillonella* spp**	**Megasphaera spp**	***Selenomonas* spp**	***Succiniclas-tic-um* spp**
Catalase production	+	+	–	–	–
Nitrate reduction	+	+	–	+	–
Esculin hydrolysis	+				
Gelatin liquidation	+	–			–
H_2S production		+			
Glucose fermentation	+	–	+	+	–
Sucrose fermentation	+	–			–
Maltose fermentation			+	+	–
Lactose fermentation	+	–			–
Xylose fermentation	+	–			–
Lactate fermentation	+	+	+	+	–
Succinate fermentation	+	+	–	+	+
Fermentation end products	Propionic acid, Acetic acid and CO_2	Propionic acid, Acetic acid and CO_2	Propionic acid, Acetic acid Butyric acid and Valeric acid	Propionic acid, Acetic acid and CO_2	Propionic acid, CO_2

Propionic acid

The usual name, propionic acid, is from the greek words *protos* = 'first' and *pion* = 'fat', because it was the smallest carboxylic acid that exhibited the properties of the other fatty acids, such as producing an oily layer when salted out of water and having a soapy potassium salt.

Table 5: Chemical and physical properties of propionic acid (Jefferson, 2008)

Molecular formula	$C_3H_6O_2$
Molar mass	74.08 g/mol
Appearance	Colourless liquid
Melting point	-21°C
Boiling point	141°C
Density	0.992g/cm^3
Solubility in water	miscible

Propionic acid is a naturally-occurring carboxylic acid with chemical formula CH_3CH_2COOH ($C_3H_6O_2$), which is systematically named propionic acid. In the pure state, it is a colourless, corrosive liquid with a pungent odor (Playne, 1985). It is a water soluble liquid at any proportion, because its molecules can be attracted by hydrogen bonds. Furthermore, propionic acid reacts with alcohols rendering esters and with basis rendering organic salts (Playne, 1985).

Routes of propionic acid production

Two mechanisms have been proposed for the formation of propionic acid from lactate by propionibacteria (Werkman & Wood, 1942)

a) The dehydration of lactate to acrylate which is then reduced to propionate.

b) The formation of succinic acid and its subsequent decarboxylation, i.e. lactate → pyruvate → oxalate ↔ malate ↔ fumarate → succinate → propionate + CO_2.

Two groups of bacteria have been reported to decarboxylate succinic acid: *Veillonella gazogenes* (Johns, 1951a, b) and species of genus propionibacterium (Delwiche, 1948; Johns, 1951c) to produce propionic acid. They cannot ferment acrylic acid, so the first mechanism is untenable for this organisms and second mechanism is the probable pathway for propionate formation in these bacteria (Johns, 1952). In the rumen succinic acid is produced as intermediate product from pyruvate during the fermentation of carbohydrate by *Ruminococcus flavefaciens,* isolated by Sijpesteijn (1951) and *Bacteroides succinogenes* isolated by Hungate (1950). Propionic acid is produced mainly by decarboxylic acid pathway (Succinate pathway) by decarboxylation of succinic acid by propionibacteria and *V. gazogenes* and optimum pH for the reaction was found to be about 6.0 (Sijpesteijn and Elsden, 1951).*S. ruminantium* also forms propionate via succinate pathway (Paynter& Elsden,1970). Krebs and Eggleston (1941) showed that the set of reactions from oxalate to succinate occurred in the propionibacteria (oxalate ↔ malate ↔ fumarate → succinate).

About 10% Propionic acid may also be formed from lactate through acrylate pathway (direct reduction pathway). But sometimes upto 30% of the total propionate may be produced through this pathway (Swenson and Reece, 1996). In this pathway pyruvate is converted to lactate by lactic acid producing bacteria. Lactate and pyruvate are readily converted into propionate by Propionibacteria, where lactate is also produce as intermediate during the conversion of pyruvate to propionate, i.e. pyruvate '! lactate '! propionate (Kluyver, 1931*). M. elsdenii* produce propionate via acrylate pathway. The pathway through lactate and acrylate predominates when the ruminant's diet includes a high proportion of concentrates and pathways through succinateare employed when the diet consists mainly of fibrous roughages (McDonald *et al.*, 2002). With concentrate diets, lactate produced by the acrylate pathway may accumulate in the rumen and threaten the animal with acidosis. In this situation supplementation of lactate utilising bacteria as DFM may be beneficial.

Propionic acid as an energy source

Although ruminants have a major need for glucose, little of the need is met by glucose absorbed from the gastrointestinal tract. Less than 10% of the total glucose need may be met by absorbed glucose, whether ruminants are fed a predominantly hay diet (Armstrong, 1965) or a diet consisting largely of grain (Otchere *et al*, 1974). The ruminant's large glucose need, coupled with the failure to absorb major amounts of glucose from the gastrointestinal tract, emphasizes the importance of gluconeogenesis in ruminant metabolism and the large continuous demand for carbon sources to maintain gluconeogenesis. Much of the carbon used to support ruminant gluconeogenesis is derived from either propionate or glucogenic amino acids (William and Young, 1977).

Propionate is quantitatively the most important single precursor of glucose synthesis among VFA, and therefore has a major impact on hormonal release and tissue distribution of nutrients (Nagaraja *et al*, 1997). In ruminant animals, considerable amounts of propionate are produced from carbohydrate breakdown in the rumen. The acid then passes across the rumen wall, where a little is converted to lactate to glucose through pyruvate. The remainder is carried to the liver, where 90% is changed into glucose. The first stage in this process is conversion to succinyl-coenzyme A. This enters the tricarboxylic acid cycle and is converted to malate and equivalent of ATP are produced. The malate is transported into the cytosol where it is converted to oxaloacetate and then phosphoenolpyruvate. The phosphoenolpyruvate may then be converted to fructose diphosphate by reversal steps in the glycolytic sequence. This is then converted to fructose -6-phosphate by hexosediphosphatase and then to glucose -6-phosphate and finally to glucose by the reverse steps of glycolytic pathway. The glucose may be used eventually to provide energy (Table 6). The propionate which are not utilized in the liver, enter the peripheral circulation where it is converted to Acetyl coA instead of glucose and provide energy to the animals (Table 7).

Table 6: Energy balance sheet of propionic acid utilization through glucose (McDonald *et al.*, 2002)

Reactions	Moles ATP (+)	Moles ATP (-)
2 moles propionate to 2 moles succinyl-coenzyme A	-	6
2 moles succinyl-coenzyme A to 2 moles malate	6	-
2 moles malate to 2 moles phosphoenolpyruvate	6	2
2 moles phosphoenolpyruvate to 1 mole glucose	-	8
1 mole glucose to carbon dioxide and water	38	-
Total	50	16

Table 7: Energy balance sheet of propionic acid utilization through Acetyl coA (Mc Donald *et al.*, 2002)

Reactions	Moles ATP (+)	Moles ATP(-)
2 moles propionate to 2 moles succinyl-coenzyme A	-	6
2 moles succinyl-coenzyme A to 2 moles malate	6	-
2 moles malate to 2 moles phosphoenolpyruvate	6	2
2 moles phosphoenolpyruvate to 2 moles acetyl coA	8	-
2 moles acetyl coA to carbon dioxide and water	24	-
Total	44	8

Thus, there is a net gain of 18 moles of ATP per moles of propionic acid through acetyl coA and 17 moles of ATP per moles of propionic acid through glucose. For growing ruminants and lactating cows, propionate has been estimated to account for 61 (Reynolds *et al*, 1994) to 67 % (Huntington, 2000) of glucose release. Therefore, propionic acid producing bacteria can be used as direct fed microbial in growing cattle and in lactating cows when they are in negative energy balance especially during early lactation.

References

Beecham, T. J., Chambers, J. V. and Cunningham, M. D. 1977. Influence of Lactobacillus acidophilus on performance of young dairy calves. *J. Dairy Sci.* 60 (Suppl.1):74 (abstract).

Bruce , B. B., Gilliland, S. E., Bush. L. J and Staley, T. E. 1979. Influence of feeding cells of *Lactobacillus acidophilus* on the fecal flora of young dairy calves. *Oklahoma Animal Science Research report*. Stillwater, OK, 207.

Cheng, K. J., Stewart, C. S., Dinsdale, D. and Costerton, J. W. 1984. Electron microscopy of bacteria involved in the digestion of plant celwalls. *Anim. Feed Sci. Technol.*10: 93-120

Cummins C. S. and Johnson J. L. 1986. Genus I. *Propionibacterium* Orla-Jensen 1909. pp. 1346-1353. In: Krieg N. R., Holt J. G. (eds.). *Bargey's manual of systematic bacteriology* Williams and Wilkins. Baltimore. USA.

Cummins, C. S. and Johnson J. L. 1992. The genus *Propionibacterium*. In. Balows A.,Truper H. G., Dworkin M., Harder W., Schleifer K. H. (eds). The prokaryotes. Springer. New York, USA. pp. 834-849.

Davidson, C. A., and Rehberger, T. G. 1995. Characterization of *Propionibacterium* isolated from the rumen of lactating dairy cows. Presented at American Society for Microbiology.

Davidson, C. A. 1998. The isolation, characterization and utilization of *Propionibacterium* as a direct-fed microbial for beef cattle. M. S. Thesis, Oklahoma State Univ., Stillwater.

Delwiche, E. A. 1948. Mechanism of propionic acid formation by *Propionibacterium pentosaceum*. *J. Bacteriol.* 56: 811-820.

Douglas, H. C. and Gunter, S. E. 1946. The taxonomic position of Corynebacterium acnes. *J. Bact.* 52: 15.

Elsden, S. R.1945. The fermentation of carbohydrate in the rumen of sheep. *J. Exp. Biol.* 22: 51-62.

Elsden, S. R. and Lewis, D.1953. The production of fatty acids by a Gram-negative coccus. *Biochem. J.* 55:183-189.

Elsden, S. R., Gilchrist, F. M., Lewis, D and Volcani, B. E. 1956. Properties of a fatty acid forming organism isolated from the rumen of sheep. *J. Bacteriol.* 72: 681-689.

Fuller. 1989. Probiotics in man and animals: A review: *J. Appl. Bacteriol.* 66: 365-378.

Fuller, 1999. Probiotics for farm animals. In: G. W. Tannock (ed.) Probiotics-A critical Review. p 15. Horizon Scientific Press, Wymondham, England.

Ghorbani, G. R., Morgavi, D. P., Beauchemin, K. A. and Leedle, J. A. Z. 2002. Effects of bacterial direct-fed microbial on ruminal fermentation, blood variables, and the microbial populations of feedlot cattle.

Grinstead, D. A., and Barefoot, S. F. 1992. Jenseniin G, a heat-stable bacteriocins produced by *Propionibacterium jensenii* P126. *Appl. Environ. Microbiol.* 58: 215-220.

Gutierrez, J., Davis, R. E., Lindahl, I. L. and Warwick, E. J. 1959. Bacterial changes in the rumen during the onset of feed-lot bloat and characteristics of *Peptostreptococcus elsdenii*. 1959. *Appl. Microbiol.* 7:16-22.

Gylswyk, N. O. V. 1995. Succiniclasticum ruminis gen. nov., sp. nov., a ruminal bacterium converting succinate to propionate as the sole energy–yielding mechanism. International *Journal of Systemic Bacteriology*, Apr. p. 297-300.

Horn, C. H., Kistner, A., Greyling, B. J. and Smith, A. H. 2009. *Megasphaera elsdenii* and its effect. Patent application number: 2009246177.

Hugenholtz, J., Hunik, J. H. Santos, and Smid, E. 2002. Nutraceutical production by propionibacteria. Lait. 82: 103-112. Hungate, R. E. (1950). *Bact. Rev.*14: 1.

Hungate, R. E. (1966) .The rumen and its microbes. New York: Academic press.

Huntington, G. B. 2000. High-starch rations for ruminant production discussed. *Feedstuffs* 12-13:23.

Jan, G., A. Lan,. and Leverrier, P. 2007. Dairy propionibacteria as probiotics. In: Saarela, M. (Ed.). Functional dairy products. Wood Head Publishing 2. pp 165-194.

Jarvis, G. N., Strompl, C., Moore, E. R. B. and Thiele, J. H. 1998. Isolation and characterization of obligately anaerobic, lipolytic bacteria from the rumen of red deer. *Systematic and Applied Microbiology.* 21:135-143.

Jefferson Coral. 2008. Propionic acid production by *Propionibacterium* species using low–cost carbone sources in submerged fermentation. A master degree thesis, Biotechnology and Bioprocesses Engineering Division, Federal University of Parana, Curitiba, Brazil.

John, A. T. 1951a. Isolation of bacterium, producing propionic acid, from the rumen of sheep. *J. Gen. Microbiol.* 5: 317-325.

Johns, A. T. 1951b. The mechanism of propionic acid formation by Veillonella gazogens. *J. Gen. Microbiol.* 5: 326.

Johns, A. T. 1951c. The mechanism of propionic acid formation by Propionibacteria. *J. Gen. Microbiol.* 5: 337-345.

Johns, A. T. 1952. The mechanism of propionic acid formation by *Clostridium propionicum. J. Gen. Microbiol.* 6: 123-127.

Kluyver, A. J. 1931. The chemical activities of micro-organisms. London: University of London Press.

Koike, S., Yoshitani, S., Kobayashi, Y. and Tanaka, K. 2003b. Phylogenetic analysis of fiber-associated rumen bacterial community and PCR detection of uncultured bacteria. FEMS *Microbiol. Lett.* 229: 23-30.

Koike, S., H. Yabuki and Y. Kobayashi. 2007. Validation and application of real-time polymerase chain reaction assays for representative rumen bacteria. *Anim. Sci. J.* 78: 135-141.

Kung, L., Hession, A., Tung, R.S. and Maciorirowski, K. 1991. Effect of *Propionibacterium shermanii* on ruminal fermentations. Proc. 21st Biennial Conf. on Rumen Function. Chicago, IL, pp. 31.

Kung, L., Jr., and Hession, A. O. 1995. Preventing in vitro lactate accumulation in ruminal fermentations by inoculation with *Megasphaera elsdenii. J. Anim. Sci.* 73: 250-256.

Kusano K., Yamada H., Niwa M., Yamasato K. 1997. *Propionibacterium cyclohexanicum* sp. Nov., a new acid–tolerant ù-cyclohexyl fatty acid-containing *Propionibacterium* isolated from spoiled orange juice. *Int. J. Syst. Bacteriol.* 47: 825-831.

Langa, R. L. S. 2010. Optimisation of cell growth and shelf life stability of *Megasphaera elsdenii* NCIMB 41125. Master degree thesis, Deptt. of Microbiology and Plant pathology, Faculty of Natural and Agricultural Sciences, University of Pretoria, South Africa.

Mackie, R. I., and Gilchrist, F. M. C. 1979. Changes in lactate-producing and lactate-utilizing bacteria in relation to pH in the rumen of sheep during stepwise adaptation to a high concentrate diet. *Appl. Environ. Microbiol.* 38: 422-430.

Mc Donald, P; Edwards, R. A.; Greenhalgh, J. F. D. and Morgan, C. A. 2002. Digestion of carbohydrates. In Animal Nutrition, sixth edition.pp-183.

Melo, M. R. 2007. Producao de acidopropionicoemsoro de leitepor bacteria do rumen bovino. 54f. Dissertacao (Mastradoemzootech)-Universidade Federal de Vicosa ,Vicosa, MG.

Mirobn, J. and Ben-Ghedallia, D. 1992. The degradation and utilization of monosaccharide components by defined ruminal cellulolytic bacteria. *Appl. Microbiol. Biotechnol.* 38: 432–437.

Olsen, G. J., Woese, C. R., Overbeek, R. 1994. The winds of (evolutionary) change: Breathing new life into microbiology. *J. Bacteriol.*176: 1-6.

Orla-Jensen, S. 1909. Die Hauptlinien des naturlichen Bacterrien systems. Zbl. Bakt. Abt. II 22: 305-346.

Playne, M. J. 1985. Propionic and Butyric acids. in Moo-Young, M. Comprehensive Biotechnology. Great Britain: Pergamon Press. PP. 731-759.

Prins, R. A. 1971. Isolation, culture, and fermentation characteristics of *Selenomonas ruminantium* var. Bryanti var. n. from the rumen of sheep. *Journal of Bacteriology*.105 (3): 820-825.

Rehberger T. G. 1993. Genome analysis of Propionibacterium freudenreichi by pulsed field gel electrophoresis.*Curr.Micrbiol.* 27: 21-25.

Rehberger, T., Hibberd, C. and Swartzlander, J. 1993. Identification and application of a Propionibacteria strain for nitrate and nitrite reduction in the rumen. Available online at http://www.beefextension.com/proccedings/nitrates93/nitrates93_e.pdf.

Reynolds, C. K.; Harmon, D. L. and Cecava, M. J. 1994. Absorption and delivery of nutrients for milk protein synthesis by portal–drained viscera. *J. Dairy Sci.* 77: 2787-2808.

Rinta-Koshi, M., Beasley, S., Montonen, L. And Mantere-Alhonen, S. 2001. Propionibacteria isolated from rumen used as possible probiotics together with bifidobacteria. *Milchwissenschaft* 56 (1): 11-13.

Rogossa, M. 1971. Family: *Veillonaceae*. In Bergeys Man. Syst. Bacteriol. 685.

Rogosa, M. (1971). Transfer of *Peptostreptococcus elsdenii* to a new genus, *Megasphaera*. *Int. J. Syst. Bacteriol.* 21:187-189.

Rossi, F., Sammartino, M., Torriani, S. 1997. 16S-23S ribosomal spacer polymorphism in dairy Propionibacteria. *Biotechnol. Techniques.* 11: 159-161.

Rossi, F., Torriani, S., Zapparoli, G, Dellaglio, F. 1998. Use of different genetic methods for typing and identification of dairy Propionibacteria. Poster presentation. 2nd international symposium on Propionibacteria. Cork. Ireland.

Rossi, F., Torriani, S. And Dellaglio, F. 1999. Genus and species–specific PCR-Based Detection of Dairy propionibacteria in environmental samples by using Primers targeted to the Genes encoding 16S rRNA. *Applied and Environmental Microbiology* 65(9): 4241-4244.

Russell, J. B. 1985. Fermentation of cellodextrins by cellulolytiand noncellulolytic rumen bacteria. *Appl. Environ. Microbiol.* 49: 572-576.

Russell, J. B. and Baldwin, R. L. 1978. Substrate preference in rumen bacteria: Evidence of catabolite regulatory mechanisms. *Appl. Environ. Microbiol.* 36: 319-329.

Sawanon, S., Shinkai, T., Koike, S., Kobayashi, Y. and Tanaka, K. 2003. Indication of a novel group of *Selenomonas ruminantium* with high cellulase and fiber-attaching activities from the rumen, pp. 363–368 In: K. Ohmiya, K. Sakka, S. Karita, T. Kimura, M. Sakka,Y. Ohnishi (Eds): Biotechnology of Lignocellulose Degradation and Biomass Utilization. Uni Publishers, Tokyo

Sawanon, S. and Kobayashi, Y. 2006. Synergistic fibrolysis in the rumen by cellulolytic *Ruminococcus flavefaciens* and non-cellulolytic *Selenomonas ruminantium*: evidence in defined cultures. *Anim. Sci. J.* 77, 208–214.

Scheifinger, C. C. and Wolin, M. J. 1973. Propionate formation from cellulose and soluble sugars by combined cultures of *Bacteroides succinogenes* and *Selenomonas ruminantium*. *Appl. Microbiol.* 26: 789-795.

Seshadri, N.; Mukhopadhyay, S. N. 1993. Influence of environmental parameters on propionic acid upstream bio processing by Propionibacterium acidi-propionici. *J. Biotechnology.* 29: 321-328.

Sijpesteijn, A. K. and Elsden, S. R. 1951.The metabolism of succinic acid in the rumen of the sheep. *J. Gen. Microbiol.* 5 (2): 326-336.

Sijpesteijn, A.K. 1951. On *Ruminococcus flavefacience*, a cellulose-decomposing bacterium from the rumen of sheep and cattle. *J. Gen. Microbiology*. 5: 869.

Sun Guo Quan; Liu GuoWen; Xing Xin; Pang XiaoYang; Long Miao; Yuan Xue; Yang Wen Yang; Wang Zhe. 2009. Isolation and identification of *Megasphaera elsdenii* in cow rumen and effect on lactic fermentation. *Chinese Journal of Veterinary Science.* 29(9): 1115-1119.

Swenson, M. J. and Reece, W. O. 1996. Digestion, absorption and metabolism. Dukes' physiology of domestic animals. First Indian reprint, Panima Publishing Corporation, New Delhi: pp. 387-427.

Vorobjeva, L. I. 1999. Propionibacteria. Kluwer Academic Publishers.

Wolin, M. J. 1960. A theoretical rumen fermentation balance. *J. Dairy Sci.* 43:1452-1459.

Werkman, C. H. and Wood, H. G. 1942. Heterotrophic assimilation of carbon dioxide. *Advances in Enzymology*. 2: 135.

Wiederhold, A. H. 1994. Isolation, selection and cultivation of lactic acid utilising rumen bacteria for the treatment of acute and chronic acidosis. MSc thesis, University of Natal, Pietermaritzburg.

Wolin, M. J., Miller, T. L. and Stewart, C. S. 1997. Microbe-microbe interactions. In: The Rumen Microbial Ecosystem (Ed.P.N. Hobson and C. S. Stewart). pp. 467-491. Blackie Academic and Professional Publishers, London.

Wu. L, Zhao, M., Xia, C., Ni, H. and Zhang, H. 2009. Isolation, Identification and rumen fermentation characteristics of *Propionibacterium acnes*. *Wei Sheng Wu Xue Bao*. 49 (2): 168-173.

17

Impact of Defaunating Agents on Ruminant Production

Kaushalendra Kumar, Pankaj Kumar Singh and Kamdev Sethy

Introduction

The process of making the rumen of animals free of rumen protozoa is called defaunation and the animal is called defaunated animal. The role of rumen ciliate protozoa on the performance of host animals became debatable issue when Becker and Everett (1930) demonstrated that rumen protozoa were non-essential for growth in lambs. Nevertheless, the reports of recent years reflect that though protozoa may be non essential for ruminant, still they have significant role to play in the rumen metabolism specially to stabilize the rumen pH (Santra and Karim, 2002a). Rumen protozoa are the largest in size among rumen microbes and contribute 40-50% of the total microbial biomass and enzyme activities in the rumen (Agarwal *et al.*, 1991).

Methods of defaunation

There are several ways to defaunate the animals and to obtain a ruminant animal free from rumen ciliate protozoa. The different methods of defaunation are as follows:

Isolation of New Born Animals

One of the method of producing defaunated animals is the separation of newborn animals from their dams after birth and preventing them from any contact with the adult ruminant animals. The newborn animals should be separated 2 to 3 days after birth (Jouany, 1978). During this time the newborn animals gets contaminated with the native bacterial population but do not get rumen ciliate protozoa (Fonty *et al.*, 1984). Although this is one of the most reliable, safe and sure methods of getting defaunated animals. However, once the animal is separated, proper care should be taken so that the isolated animals do not come in contact with any adult animals as well as any contamination from the handlers who look after faunated and defaunated animals.

The second physical method used for defaunation has been standardised. The rumen is emptied and the rumen contents are heated at 50^0 C for 15 minutes. This heat treatment helps in eliminating the protozoa but the bacteria and fungi are either not affected or only slightly affected. Their number may be resumed again after sometimes. Alternatively, the rumen contents may be frozen and thawed again before filling it back into the rumen. This also is efficient in killing of protozoa. In the meantime the mucosa of the rumen is completely washed and treated with 1 % formaldehyde solution which is rinsed off soon and again washed with water. The physically treated rumen contents are filled back into the rumen and the animal is given normal diet. This treatment is also very effective, but this can be done only in the fistulated animals. This treatment is not possible in intact animals.

Chemical Treatment

Another method of defaunation is by use of chemicals and majority of researchers has used this method for obtaining animals free from rumen ciliate protozoa. The chemicals which have been widely used to defaunate the animals are copper sulphate (Ramprasad and Raghavan, 1981), manoxol (Chaudhary *et al.*, 1995) and sodium lauryl sulphate (Santra and Karim, 1999). Becker *et al.* (1929) recommended the use of 50 ml of 2 % copper sulphate solution for 2 consecutive days following a 72 hr fast. Sheep are very sensitive to copper toxicity and the animals died even after administration of zinc and molybdenum in dose of 100 mg/day. Hydrochloric acid and acetic acid may be better defaunating agents than copper sulphate because these acids reduce the pH of rumen liquor and kill the protozoa, but do not have harmful effect on the physiology of animals. The use of surfactant detergents which lyse the protozoa and help in defaunating the animals. The surfactant detergents can be of two types i.e. anionic surfactants (Dioctyl sodium sulfosuccinate- Manoxal OT), alkanates (calcium alkyl benzene sulfonate, alkyl phenoxy poly-oxyethylene suphate, sodium lauryl diehoxy sulphate) and non-ionic surfactants (polypropylene glycol, poloxalene etc). Calcium peroxide can also be used as a defaunating agent (Demeyer, 1981). This also needs starvation of the animals for 2-3 days. Calcium peroxide releases hydrogen peroxide and leads to its accumulation in the rumen. As the enzyme catalase, required for the breakdown of hydrogen peroxide, is absent the rumen. Hydrogen peroxide accumulates to such an extent that it is toxic for the protozoa and leads to defaunation of the animals.

For using these chemicals as defaunating agents the animals are kept off feed for 48-72 hrs and are introduced in the rumen of animals either orally by a stomach tube or through rumen fistula. However, these chemicals are not only toxic to the rumen protozoa but also kill the other rumen microorganism like bacteria. These chemicals are also toxic to the animals resulting in depressed feed intake, dehydration, pulmonary congestion and some time mortality also reported (Jouany *et al.*, 1988).

Dietary Manipulation

The ciliate protozoa are very much sensitive to change in rumen pH. The activity of ciliate protozoa is adversely affected when the pH of the rumen falls below 5.8 and if the rumen pH falls below 5.0, the ciliate protozoa are being completely eliminated. Therefore, offering high energy feed (especially cereal grains like barley, maize etc) to the starved (for 24 hours) animals creates acidic condition in the rumen and rumen pH fall below 5.0. This fall in rumen pH eliminates the ciliate protozoa completely and the animal becomes defaunated. However one serious disadvantage of this method is that chances of developing acidosis in treated animal is more. Once rumen acidosis develops the animals will suffer from various secondary complications. The drenching of vegetable oils eliminates ciliate protozoa and hence can be used as a defaunating agent (Nhan *et al.*, 2001).

Ionophores Compounds

Ionophores are used commercially throughout the world in the beef and poultry industries. Production efficiency of cattle is increased through alteration of rumen fermentation and control of protozoa that cause coccidiosis. Ionophores act by interrupting transmembrane movement and intracellular equilibrium of ions in certain classes of bacteria and protozoa that inhabit the gastrointestinal tract. The actions of ionophores provide a competitive advantage for certain microbes at the expense of others. In general, the metabolism of the selected microorganisms favors the host animal. Energy metabolism is enhanced through increased production of propionate among ruminal fatty acids with a concomitant reduction in methane. Ruminal degradation of peptides and amino acids is reduced, thereby increasing the flow of protein of dietary origin to the small intestine. Total flow of protein to the lower tract is often increased with ionophore feeding. Risk of digestive disorders such as bloat and acidosis that result from abnormal rumen fermentation is reduced, as are certain conditions caused by toxic products of fermentation, e.g., 3-methyl indole. Dry matter and nitrogen digestibilities are increased with ionophores, thereby providing environmental benefits.

Monensin and lasalocid have been the most studied in research demonstrating benefits to the dairy cow. Ionophores enhance the glucose status of dairy cows through increased production of propionate. Many of the demonstrated benefits of ionophores are associated with enhancement of the energy status of the cow in the transition period and during early lactation. The benefits include less mobilization of body fat as evidenced by reduced blood non-esterified fatty acids and ketones and increased glucose. Animal manifestations include lower incidence of ketosis and displaced abomasum, reduced loss of body condition, increased milk production, and improved milk production efficiency.

Ionophores attach to bacteria and protozoa and probably rumen fungi (Habib and Leng, 1987). Attachment facilitates the movement of cations across cell membranes, exchanging a monovalent cation for a proton (monensin, laidlomycin propionate, salinomycin, lasalocid) or a divalent cation for two protons (lasalocid) (Pressman, 1976). The ionophore-cation complex involves hydrogen bonding between the ether oxygens, the hydroxyl oxygens, and the carboxylic acid oxygen of the polyether, and the cation (monovalent or divalent). This hydrogen bonding is most effective when the ionic diameter of the cation permits it to be positioned closely to the oxygens in the cavity being formed in the ionophore or, in the case of lasalocid, in the cavity between two ionophore molecules.

Rumen protozoa have been reported to be sensitive to monensin *in vivo* (Habib and Leng, 1987), but the degree of sensitivity may vary with species. Wakita *et al.* (1987) observed that salinomycin reduced the number of protozoa in cattle consuming a forage diet, although this effect seemingly disappeared between 180 and 245 d of the feeding period. As with the effect on fungi, ionophore effects on protozoa appear to be diet dependent. Protozoa modify their fermentation patterns in the presence of ionophore (Hino, 1981). Generally, any reduction in protozoa numbers may actually be beneficial if the protozoa are replaced in the biomass by bacteria (Klopfenstein *et al.*, 1966).

The beneficial role of protozoa is largely confined to the period when diets are abruptly changed from high fiber to high starch. In this circumstance, the engulfment by rumen protozoa of starch granules and adherent lactic acid-producing bacteria may reduce the amount of substrate available for the rapid production of lactate (Nagaraja, *et al.*, 1986).

Effect of Defaunation on the Rumen Ecosystem

Rumen microbes

Defaunation causes both qualitative and quantitative change in rumen bacterial population. After defaunation the bacterial population increased (Chaudhary *et al.*, 1995) since rumen protozoa feed on the rumen bacteria to meet out their nitrogen requirement. A total of 4 to 45 g bacterial dry matter is engulfed by rumen protozoa per day per sheep (Coleman, 1975). Defaunation increase the number of amylolytic bacteria due to elimination of nutritional competition between bacteria and protozoa for using starch whereas the cellulolytic bacterial population become decreased. Fungal population in the rumen also increase due to defaunaton. Orpin and Letcher (1983/84) reported a predation of fungal zoospore by rumen protozoa but till today it is unclear whether the increase in fungal population following defaunation is a consequence of reduction in the predation by the protozoa or increased availability of nutrients for fungal growth in the absence of protozoa.

Rumen pH

The buffering capacity of the rumen seems to be better in presence of protozoa on a wide variety of diets. The rumen pH starts falling immediately after ingestion of feed, both in faunated and defaunated animals whereas, the drop in pH was much higher in defaunated than in faunated animals (Santra and Karim, 2002a). Rumen protozoa engulf the readily fermentable carbohydrate (starch) which is stored in their body as amylopectin and thus decrease the rate of carbohydrate (starch degradation) fermentation resulting in a lower pH in the rumen of defaunated compared to faunated animals (Williams and Coleman, 1992).

Volatile fatty acid (VFA) production

The effect of defaunation on the production and composition of VFA is variable. The VFA production rate and its composition are greatly influenced by experimental diet. Higher VFA concentration in the rumen of faunated animals may be due to higher hydrolytic enzyme activity in the rumen protozoa because about 40- 60% of hydrolytic enzyme activity is found in the rumen protozoa (Agarwal *et al.*, 1991) and also due to stimulatory effect of protozoa over bacteria. The ciliate protozoa engulfed the feed particle and degrade it to acetic acid and butyric acid during carbohydrate metabolism. Defaunation should, therefore, increase the molar proportion of propionic. Moreover higher propionate production in the rumen of defaunated animals may be due to increase and shift in ruminal bacterial population. It has been reported that in defaunated animals number of acetate producing species such as *Ruminococcus* spp. are not predominant while succinate producing bacteria such as *Bacteroides* spp. are predominant. These changes in ruminal bacterial population may stimulate more propionate production in the rumen of ciliate free animals.

Ammonia nitrogen concentration

The significant reduction in ammonia-N concentration in the rumen of defaunated animals was reported by several workers (Chaudhary *et al.*, 1995; Santra and Karim, 2002a). Ammonia is utilized by bacteria to meet their nitrogen requirement for body protein synthesis while ciliate protozoa does not use it. In defaunated animals, the number as well as activity of rumen bacteria increase (Eadie and Gill, 1971) resulting in more uptake/utilization of ammonia by bacteria and as a result, ruminal ammonia concentration is reduced. Further, low production of free amino acid from the degradation of protein or peptide in absence of ciliates and/or lower rate of recycling of microbial nitrogen in the rumen of defaunated animals (Demeyer and Van Nevel, 1979), could have contributed to lower ruminal ammonia nitrogen concentration. The recycling of bacterial nitrogen in the rumen is higher in presence of ciliate protozoa (Hristova *et al.*, 2001) and the number of ruminal bacteria capable to utilize ammonia decrease with increased ruminal break down of dietary protein.

Microbial protein synthesis

Microbial protein synthesis in the rumen of defaunated animals is higher than faunated animals. It is now generally accepted that in absence of rumen ciliate protozoa, the efficiency of rumen bacterial growth is enhanced and more microbial protein flows from reticulo-rumen to duodenum (Bird and Leng, 1985). Although bacteria and protozoa are active in synthesis of microbial protein, outflow of microbial protein in to duodenum is primarily of rumen bacterial origin. About half of the microbial protein in the rumen can be of protozoal origin while as a proportion of the microbial protein leaving the rumen, protozoal protein is usually under 10% because of higher rate of bacterial was out from reticulo-rumen (Owens and Zinn, 1988). Additionally, the absence of rumen protozoa is known to increase the efficiency of net bacterial growth due to elimination of protozoal predation and increasing rumen bacterial turn over. This could have resulted in more microbial protein flow to the duodenum in the defaunated animals.

Enzyme profile

Rumen ciliate protozoa secrete various hydrolytic enzyme which are responsible for breakdown of the plant cell wall polysaccharides. The ciliate protozoa and fungi are most important microbial groups of the rumen organisms required for the ruminal digestion of plant fibre. Carboxymethyl cellulase enzyme activity was lower in the rumen of defaunated than faunated animals (Santra and Karim, 2002a). About 62% of the total rumen cellulase enzyme activity is associated with rumen protozoal population (Coleman, 1986). Hence, elimination of ciliate protozoa decreases ruminal cellulase enzyme activity. The activity of other carbohydrate degrading enzymes like amylase, xylanase and â-glucosidase are not affected by the presence or absence of ciliate protozoa in the rumen. Protease enzyme activity was lower in the rumen of faunated than defaunated animals (Santra *et al.*, 1996). The specific activity of protease enzyme from bacterial fraction is 6-10 times higher than that from protozoal fraction (Brock *et al.*, 1982). Defaunation increases the number of ruminal bacterial population, resulting in higher ruminal protease enzyme activity. Bacteria are the only source of ruminal urease enzyme (Cook, 1976) while ciliate protozoa have no urease enzyme activity (Onodera *et al.*, 1977) and hence ciliates can not utilize urea for their body protein synthesis.

Methane production

Defaunation is reported to considerably decrease the methane production compared with the normal faunated animals (Williams and Coleman, 1992). The reduction in methane production in absence of rumen protozoa has been attributed to various reasons. Rumen protozoa contribute hydrogen moiety for the production of methane by the methanogenic bacteria (Van Hoven and Prins, 1977). Further, ectosymbiotic attachment methanogens have with ciliate protozoa and elimination of their symbiotic partner on defaunation results in reduced methane production. Reviewing the published literatures on the topic, Kreuzer (1986) calculated that defaunation decreased energy losses through methanogenesis by 5.5 to 7.9% of gross energy intake.

Rumen volume, flow rate of digesta and physical characteristics

The effect of defaunation on rumen physiological characteristics (e.g. motility, absorption) are lacking in literature. Contradictory reports have appeared on the effect of defaunation on rumen fluid volume and digesta flow rate. Veira *et al.* (1983) reported no significant difference in the rumen volume of faunated and defaunated sheep whereas, Orpin and Letcher (1983/84) observed higher rumen fluid volume and lower fractional outflow of liquid digesta from reticulo-rumen in defaunated sheep. Chaudhary *et al.* (1995) reported higher rumen volume in defaunated buffaloes whereas liquid outflow rate remained unchanged irrespective of presence or absence of ciliate protozoa. Kayouli *et al.* (1983/84) reported no difference in rumen volume and liquid fractional outflow rate in defaunated and faunated animals while, the particle outflow rate was significantly higher in the absence of ciliate protozoa.

Animal performance

Feed intake and nutrient digestibility

The first and the most easily detectable influence of defaunation (by chemical method) on the animal is the loss of appetite and therefore decreased feed intake for a few days after defaunation of the animals. In case the animals does not regain its appetite even after 5 to 7 days of defaunation it can be induced by offering highly palatable feed. Once stabilized, daily dry matter intake in defaunated animals attend the level similar to that of faunated animals. Daily feed intake of the animals is not influence by the presence or absence of ciliate protozoa is a consistent finding in several studies. Protozoa play an important role in the digestion of plant cellwalls, especially cellulose and hemicellulose. The digestibility of fibre fractions like, NDF, ADF, hemicellulose and cellulose are decreased by defaunation. The reduced digestibility of fibre fraction in defaunated animals may be due to elimination of large entodiniomorphid ciliates which have higher cellulolytic activity (Ushida and Jouany, 1990). Moreover, better digestibility of cell wall constituents in faunated animals might be due to increased retention time of feed particles in the rumen, stabilization of rumen environment favoring development of cellulolytic microbes and stimulatory effect of rumen ciliate protozoa on rumen bacteria for cellulolysis (Onodera *et al.*, 1988). When rumen fibre digestion is impaired in absence of ciliate protozoa, it is often increased in hindgut resulting in similar total GI tract digestibility of fibre in presence or absence of ciliate protozoa. However, the effect of defaunation on the digestibility of starch and sugars is negligible (Jouany, 1978). Bacteria and free amino acids may provide nitrogen to satisfy the requirement of protozoa for their body protein synthesis. Bacteria play a dominant role in degradation of most soluble proteins while rumen protozoa help in ruminal degradation of relatively insoluble protein (Ushida and Jouany, 1985). Protozoa utilize bacterial and feed protein available in the rumen. Hence, on defaunation, the degradability of protein is decreased in the rumen. However, defaunation had no effect on apparent protein digestibility in the rumen (Santra *et al.*, 1994a).

Degradation of toxic substances

Defaunated animals are more susceptible to the bloat than normal animals. Ochratoxin A produce more toxicity in the defaunated animals than in faunated animals (Jouany *et al.*, 1988). Further, defaunated animals are more sensitive to copper toxicity. The rumen protozoa induce the complexion of the Cu^{2+} in sulphide form resulting the toxic copper become unavailable for absorption from the intestine.

Growth and feed conversion efficiency

The results of effect of defaunation on the animal performance are mainly related to animal feed composition and feeding schedule (nature of feed stuff, its presentation and distribution). The heart girth of ciliate free lambs was found to be larger than that of faunated lambs (Kamra *et al.*, 1987). Bird and Leng (1984) observed increased wool growth in the defaunated lambs. The availability of metabolizable energy was higher in defaunated animals due to reduction of energy loss in methane production. Total heat production of animal was also significantly lower in absence of rumen ciliate protozoa. Therefore, defaunation has a positive effect on the growth rate and feed conversion efficiency of the animals. On the contrary, Ramprasad and Raghavan (1981) observed reduced feed conversion efficiency in the absence of ciliate protozoa while (Chaudhary *et al.*, 1995) reported that defaunation had no effect on feed conversion efficiency. The effect of defaunation on feed conversion efficiency seems to be diet dependent. On high roughage fed animals the protozoa do not seem to have any specific function to perform and the general function of feed degradation is taken over by the increased population of bacteria and fungi in the absence of ciliate protozoa. The reduction in methanogenesis results in better feed conversion efficiency on such feed. But when the animals are fed on high grain diet, the ciliate protozoa have a specific function to perform i.e., the pH stabilization by controlling the degradation of easily fermentable sugars by protecting them in their bodies as amylopectin. Thus in absence of ciliate protozoa, the pH drops below the optimum level required for various enzymes activity in the rumen. This results in poor feed utilization and decreased feed conversion efficiency on such feeds. Therefore, defaunation protocol can be used as an important tool to improve the productivity of animals in tropical countries, where majority of livestock are maintained on sole diet on low grade roughage. Sen *et al.* (2000) reported that carcass composition of defaunated, refaunated and normal finisher lambs was similar.

Summary

Rumen is a natural fermentative anaerobic system which should be manipulated essentially by altering the composition of rumen microflora. There is ample scope to manipulate the rumen by feeding local plants or tree leaves or agro industrial by products to defaunate the animals for improving its productivity. Introduction of naturally occurring microorganism from digestive system of one species to another species for efficient degradation of plant toxins as well as for efficient utilization of nutrients will be one of the major thrust area in near future for rumen manipulation. Genetically manipulation of rumen microorganism for efficient ruminal fermentative digestion has an enormous biotechnological potential. However in tropical countries, more emphasis should be given for manipulating the rumen to increase cellulolytic activity for efficient utilization of low grade roughage. This results in poor feed utilization and decreased feed conversion efficiency on such feeds. Therefore, defaunation protocol can be used as an important tool to improve the productivity of animals in tropical countries, where majority of livestock are maintained on sole diet on low grade roughage.

References

Agarwal, N., N. Kewalramani, D. N. Kamra, D. K. Agrawal and K. Nath. 1991. Hydrolytic enzymes of buffalo rumen: Comparison of cell free rumen fluid, bacterial and protozoal fractions. *Buff. J.* 2:203-207.

Becker, E. R., Schultz, J.A and Emerson, M.A. 1929. Experiments on the physiological relationships between the stomach infusoria of ruminants and their hosts, with a bibliography Iowa State College. *Journal of Scien*ce. 4:215-251.

Becker, E. R. and R. C. Everett. 1930. Comparative growth of normal and infusoria free lambs. *Am. J. Hyj.* 11:362-370.

Bird, S. H. and R. A. Leng. 1985. Productivity response to eliminating protozoa from the rumen of sheep. *Rev. Rural Sci.* 6:109-117.

Brock, F. M., C. W. Forsberg and S. J. G. Buchman. 1982. Proteolytic activity of rumen micro-organisms and effect of proteinase inhibitors. *Appl. Environ. Microbiol.* 44:561-569.

Chaudhary, L. C., A. Srivastava and K. K. Singh. 1995. Rumen fermentation pattern and digestion of structural carbohydrate in buffalo (*Bubalus bubalis*) calves as affected by ciliate protozoa. *Anim. Feed. Sci. Technol.* 56:111-117.

Coleman, G. S. 1975. The inter relationship between rumen ciliate protozoa and bacteria. In: Digestion and Metabolism in the Ruminant (Ed. W. Mc Donald and A. C. I. Warner). The University of New England Publishing Unit, pp. 149-164.

Coleman, G. S. 1986. The distribution of carboxymethyl cellulose between fractions taken from the rumen of sheep containing no protozoa or one of five different protozoal population. *J. Agric. Sci.* 106:121-127.

Cook, R. A. 1976. Urease activity in the rumen of sheep and distribution of ureolytic bacteria. *J. Gen. Microbiol.* 92:32-49.

Demeyer, D. I. 1981. Rumen microbes and digestion of plant cell walls. *Agric. Environ.* 6:295-337.

Demeyer, D. I. and C. J. Van Nevel. 1979. Effect of defaunation on the metabolism of rumen microorganisms. *Br. J. Nutr.* 42:515-524.

Eadie, J. M. and J. C. Gill. 1971. The effect of the absence of rumen ciliate protozoa on growing lambs fed on roughage and concentrate diet. *Br. J. Nutr.* 26:155-167.

Fonty, G., J. P. Jouany, J. Senaud, Ph. Gouet and J. Grain. 1984. The evolution of microflora, microfauna and digestion in the rumen of lambs from birth to four months. *Can. J. Anim. Sci. Supp.*:165-169.

Habib, G., and R. A. Leng. 1987. The effects of monensin on rumen fungi population in sheep given diets based on oaten chaff or wheat straw. *In* Recent Advances in Animal Nutrition in Australia, University of New England, Armidale, NSW, Australia.

Hino, T. 1981. Action of monensin on rumen protozoa. *Jpn. J. Zootech. Sci.* 52:171-179.

Hristova, A. N., M. Ivan, L. M. Rode and T. A. McAllister. 2001. Fermentation characteristics and rumen ciliate protozoal populations in cattle fed medium or high barley based diet. *J. Anim. Sci.* 79:515-524.

Jouany, J. P. 1978. Contribution a l'etude des protozoaries cilies de rumen: leur dynamique, leur role dans la digestion et leurinterct pour le ruminant. These de Doctorat, Universite de Clermont II, no d'Ordre 256, Vol. 2, pp. 195.

Jouany, J. P., D. I. Demeyer and J. Grain. 1988. Effect of defaunating the rumen. *Anim. Feed. Sci. Technol.* 21:229-265.

Kamra, D. N., N. N. Pathak and R. Singh. 1987. Role of ciliate protozoa on digestion in ruminants. *Livestock Adviser.* 12:9-12.

Kayouli, C., D. I. Demeyer and R. Dendooven. 1983/84. Effect of defaunation on straw digestion *in sacco* and on particle retention in the rumen. *Anim. Feed. Sci. Technol.* 10:165-172.

Klopfenstein, T. J., D. B. Purser, and W. J. Tyznik. 1966. Effects of defaunation on feed digestibility, rumen metabolism, and blood metabolites. *J. Anim. Sci.* 25:765-773.

Kreuzer, H. 1986. Methods and application of defaunation in the growing ruminant. *J. Vet. Med. Sci.* 33:723-745.

Nagaraja, T. G, S. M. Dennis, S. J. Galitzer, and D. L. Harmon. 1986. Effect of lasalocid, monensin, and thiopeptin on lactate production from in vitro rumen fermentation of starch. *Can. J. Anim. Sci.* 66:129-139.

Nhan, N. T. H., N. V. Hon, M. T. Ngu, N. T. Von, T. R. Preston and R. A. Leng. 2001. Practical application of defaunation of cattle on farms in Vietnam: Response of young cattle fed rice straw and grass to a single drench of ground nut oil. *Asian-Aus. J. Anim. Sci.* 14:485-490.

Onodera, R., N. Yamasaki and K. Murakami. 1988. Effect of inhabitation by ciliate protozoa on the digestion of fibrous materials *in vivo* in the rumen of goats and *in vitro* rumen microbial ecosystem. *Agric. Biol. Chem.* 52:2635-2637.

Onodera, R., Y. Nakagawa and M. Kandatsu. 1977. Ureolytic activity of the washed suspension of rumen ciliate protozoa. *Agric. Biol. Chem.* 41:2177-2182.

Orpin, C. G. and Letcher, A. J. 1983/84. Effect of absence of ciliate protozoa on rumen fluid volume, flow rate and microbial population in sheep. *Anim. Feed. Sci. Technol.* 10:145-153.

Owens, N. F. and R. Zinn. 1988. Protein metabolism of ruminant animals. In: The Ruminant Animal Digestive Physiology and Nutrition (Ed. D. C. Chruch), Prentice-Hall, Englewood Cliffs, NJ, pp. 227-249.

Pressman, B. C. 1976. Biological applications of ionophores. *Annu. Rev. Biochem.* 45:501-503.

Ramprasad, J. and G. V. Raghavan. 1981. Note on the growth rate and body composition of faunated and defaunated lambs. *Indian J. Anim. Sci.* 51:570-572.

Santra, A. and S. A. Karim. 1999. Efficacy of sodium laurel sulphate as defaunating agent in sheep and goats. *International J. Anim. Sci.* 14:167-171.

Santra, A., D. N. Kamra and N. N. Pathak. 1996. Influence of ciliate protozoa on biochemical changes and hydrolytic enzymes in the rumen of Murrah buffalo (*Bubalus bubalis*). *Buffalo J.* 1:95-100.

Santra, A., Kamra, D. N. and Pathak, N. N. 1994a. Effect of defaunation on nutrient digestibilty and growth of Murrah buffalo (*Bubalus bubalis*) calves. International *J. Anim. Sci.* 9:185-187.

Santra., A and S. A. Karim. 2002a. Influence of ciliate protozoa on biochemical changes and hydrolytic enzyme profile in the rumen ecosystem. *J. Appl. Microbiol.* 92:801-811.

Sen. A. R., S. A. Karim and A. Santra. 2000. Effect of defaunation on carcass and meat characteristics of finisher lambs. Indian *J. Anim. Sci.* 70:659-661.

Ushida, K. and J. P. Jouany. 1985. Effect of protozoa on rumen protein degradation in sheep. *Reprod. Nutr. Dev.* 25:1075-1081.

Ushida, K. and J. P. Jouany. 1990. Effect of defaunation on fibre digestion in sheep given two isonitrogenous diets. *Anim. Feed. Sci. Technol.* 29:153-158.

Van Hoven, W. and R. A. Prins. 1977. Carbohydrate fermentation by the rumen ciliate *Dasytricha ruminantium*. *Protistologiaca.* 13:599-606.

Veira, D. M., M. Ivan and P. Y. Jui. 1983. Rumen ciliate protozoa: effects on digestion in the stomach of sheep. *J. Dairy Sci.* 66:1015-1022.

Wakita, M., Y. Kobayashi, S. Hoshino, Y. Kitabayaski, M. Hashimura, and H. Kudo. 1987. Effects of salinomycin and monensin on feed conversion of concentrate and ruminal fluid characteristics in fattening Holstein steers. *Jpn. J. Zootech. Sci.* 58:396-402.

Williams, A. G. and G. S. Coleman. 1992. The rumen protozoa. Springer-Verlag, New York.

18

Utilization of Exogenous Enzymes in Ruminant Feeding

Ravindra Kumar and Shalini Vaswani

Introduction

Enzymes are globular proteins that act as biological catalysts. They are produced by living cells to bring about specific biochemical reactions. Digestive enzymes catalyze the degradative reactions by which substrates (i.e., feedstuffs) are digested into their chemical components (e.g., simple sugars, amino acids, fatty acids) which in turn, are used for different body functions. Different enzymes are utilized for the complete digestion of complex feed. The enzymes are being used in the feeding industry from last several decades. Various digestive enzymes have been studied for use as feed additives to enhance animal performance They are commonly used to improve the nutritive value of feeds for non-ruminants (especially, poultry and swine). They are not routinely used in adult ruminant diets because the fibrolytic activity within the rumen environment is normally very high, and it is assumed that exogenous enzymes cannot survive proteolysis in the rumen (Kopency *et al.*, 1987). Forages are the main feed component that serve as the major source of energy available to the animal in ruminant production system. However, due to slow or incomplete digestion of fibrous substrates only 10 to 35% of energy intake is available as net energy that significantly limits livestock performance and profits in production systems. To improve the ruminal fibre degradability many strategies have been developed to stimulate the digestion of the fibrous components in ruminant feeds. That includes, use of various feed additives which stimulates fiber digestion, physical and chemical processing of feeds in order to enhance the rate and extent of fiber digestion. However, the use of "natural" product to enhance production efficiency is highly recommended that can potentially play some role in replacing performance benefits vacated by antibiotics and hormones. Enzymes are proteins that are ultimately digested or excreted by the animal, leaving no residues in products, which was a major drawback in the use of antibiotics and hormones.

The use of supplemental exogenous enzymes to improve fibre digestibility and feed utilization was first examined for ruminants in 1960s (Beauchemin, 2004). Feeding of enzymes in ruminants was a questionable practice in past decades due to poor characterization of enzyme products, variable animal response and high cost. But now a day, with recent advancement in fermentation technology and biotechnology, economic production of large quantities of biologically active enzymes preparations as animal feed additives has become possible. It is acknowledged that enzyme preparations with specific activities can be used to drive specific metabolic and digestive processes in the gastrointestinal tract and may increase natural digestive processes to improve the availability of nutrients and feed intake thereafter (Colombatto *et al.*, 2003). The exogenous enzymes should aid in the digestive process in the ruminant animals. It should assist in the breakdown of fibre or fibre components. The result of extensive fiber degradation is greater release of energy and other nutrients. Thus, the positive results in terms of increase animal performance have been attributed to increase in feed digestion due to the use of exogenous feed enzymes.

Benefits of feed enzyme

Improving efficiency and reducing cost of production – by breakdown of anti-nutrients, enhancing the digestion that leads to increase feed conversion efficiency.

For a better environment – by improving digestion and absorption of nutrients, reducing the volume of manure produced and lowering phosphorus and nitrogen excretion.

Improving consistency – by reducing the nutritional variation in feed ingredients that results in more consistent feed for more uniform growth and production.

Maintains gut health – by improving nutrient digestibility, thus fewer nutrients will be available in the animal's gut for the potential growth of disease-causing bacteria.

Increases the availability of nutrients – due to improved digestion and extensive fibre degradation.

Classification of enzymes and enzyme preparations

Various enzyme products have been marketed for livestock. These are all derived mostly from four bacterial species (*Bacillus subtilis, Lactobacillus acidophilus, L. plantarum, and Streptococcus faecium)* and three fungal species *(Aspergillus oryzae, Trichoderma reesei and Saccharomyces cerevisiae)* (McAllister *et al.*, 2001). Commercial enzymes are produced as a result of microbial fermentation. They are produced by a batch fermentation process, beginning with seed culture and growth media (Cowman, 1994). After the fermentation enzyme protein is separated from residue and source organism. They do not contain microbial cells, because they are removed from fermentation and finally concentrated and purified. The type and the activity of the enzyme depends on the strain selected, growth media used and culture conditions. Complete digestion of complex feeds requires hundreds of enzymes. These enzymes break certain specific bonds in feedstuffs which are not usually degraded by endogenous enzymes, thus releasing more nutrients (Sheppy, 2001). The prime focus of enzymes research for ruminants is given on cell wall degrading enzymes. Enzyme preparations for ruminants are mainly marketed based on their capacity to degrade plant cell walls (cellulose and xylan) and are referred to as *cellulases* and *xylanases* (Morgavi *et al.*, 2000a). In addition to these fibre degrading enzymes certain products also contain many secondary enzymes activities such as, *amylases, proteases and pectinases*. Differing proportions and activities of these enzymes has an impact on the efficacy of cell wall degradation by the product. A blend of cellulase and xylanase is more effective than cellulase alone (Schingoethe *et al.*, 1999). The type and activity of enzymes produced can vary a lot, even within a single microbial species. Enzyme products for ruminants are usually standardized by blending crude enzyme extracts to obtain specified levels of one or two enzyme activities. Depending on the type of enzyme and preparations, the stability, rate of passage and the efficacy of enzymes differ. Because of the difference in commercial preparations, suppliers should provide the information such as activity rates and type of enzyme (Zo Bell *et al.*, 2000).

Feed enzymes for ruminants

Cellulase

The degradation of cellulose is complex and requires cellulases, and numerous specific enzymes also contribute to its activity. The major enzymes involved in cellulase system contains mainly endoglucanase, exoglucanase and β- glucosidase or cellobiase. Most endoglucanases hydrolyze cellulose chains at random to produce cellulose oligomers of varying degrees of polymerization, exoglucanases hydrolyze the cellulose chain from the non reducing end, producing cellobiose. β-1, 4 glucosidase hydrolyze short chain cellulose oligomers and cellobiose to glucose. (Beauchemin *et al.*, 2003)

Xylanase

The main enzymes involved in degrading the xylan core polymer to soluble sugars are xylanase and ß-1, 4 xylosidase (Bhat and Hazlewood, 2001). The xylanase include endoxylanase which yield xylo oligomers, and β-1, 4 xylosidase, which in turn yield xylose.

Hemicellulase

The two main enzymes involved in hemi cellulose breakdown are endo xylanase and endo mannanases, which attack the backbone structure of xylans and mannans (Viikari *et al.*, 1993). Other hemicellulase enzymes involved primarily in the digestion of side chains include β-mannosidase α L - arabinofuranosidase, α-D-glucuronidase, α-D-galactosidase, acetyl xylan esterases and ferulic acid esterases (Bhat and Hazlewood, 2001).

Enzyme Activity

Enzyme activities can be assayed by measuring the rate of release of end products (e.g. reducing sugars) from pure substrate. Enzyme unit is expressed as the quantity of reducing sugars released per unit time per unit of enzyme. Various methods are available to assay the enzymatic activities like DNSA (Dinitrosalicylic acid) method (Miller, 1959), Nelson Somogyi copper method (Somogyi, 1952) and also some viscosimetric method. The most commonly used substrate for measuring cellulase activity is carboxymethyl cellulose, which measures endo-β-1, 4-glucanase activity (Wood and Bhat, 1988). Exoglucanase activity can be measured using crystalline cellulose preparations, such as Avicel. β-glucosidase activity is determined by measuring the release of glucose from cellobiose, or the release of *p*-nitrophenol from *p*-nitrophenyl-β-D-glucoside (Bhat and Hazlewood 2001). Xylanase activity is most commonly measured by determining the release of reducing sugars from prepared xylan, such as oat spelt or birchwood xylan. Xylanases are specific for the internal β-1,4 linkages within the xylan backbone, and are generally considered endoxylanases (Bhat and Hazlewood, 2001). Endoxylanases can be considered to be debranching or non debranching based on their ability to release arabinose in addition to hydrolyzing the main chain of xylan. β-xylosidase activity can be determined by using *p*-nitrophenyl derivatives.

Most enzyme preparation having predominant self activity may also contain substantial array of minor activities. For example, an enzyme preparation with predominant xylanase activity may also have sufficient protease, cellulase, pectinase, beta glucanase or other enzymatic activities (Beauchemin, 1998).

Enzyme activity measurements must be conducted under certain specific conditions of controlled temperature, pH, ionic strength, substrate concentration, and substrate type, since all of these factors will affect the activity of an enzyme. Enzyme activities of commercial enzyme products are typically measured at the manufacturers' recommended optima. A temperature of approximately 60°C and a pH between 4 and 5 are the optimal conditions for most commercial enzymes (Coughlan, 1985).

Mode of Action

In order to ensure effective and consistent results in animal performances an understanding of mode of action of exogenous enzymes in ruminant system is essential. The mode of action of enzymes is quite complex and several potential modes of action have been proposed, that are interdependent.

Pre- ingestive effects

The effectiveness of exogenous enzymes increases when applied to feed before ingestion. It is through pre-ingestive attack of the enzymes upon the plant fibre. The enzyme adsorb onto the substrate which is an important prerequisite for hydrolysis. Applying exogenous enzymes directly to feed causes a release of reducing sugars (Hristov *et al.* 1996), and in some cases, partial solubilization of neutral detergent fibre and acid detergent fibre (Gwayumba and Christensen 1997; Krause *et al.* 1998). Forsberg *et al.* (2000) reported that this release of soluble sugars would provide sufficient additional available carbohydrates that will encourage rapid microbial growth, and will shorten the lag time required for microbial colonization. The major constraint to digestion is the slow and limited colonization, and penetration of cellulolytic microbes and their hydrolytic enzymes onto the exposed surfaces of feed particles. There is evidence that applying fibrolytic enzymes to feed prior to feeding alters the structure of the feed, thereby making it more amenable to degradation (Beauchemin *et al.*, 2004). Enzymes applied to feed prior to ingestion are particularly stable because on applying the enzyme on feed the binding between enzyme and feed increases, thereby increasing the resistance of the enzymes to proteolysis in the rumen (Fontes *et al.*, 1995).

Ruminal effects

Stability of enzymes

Before it was thought that the proteolytic activity in the rumen ecosystem would rapidly inactivate unprotected exogenous enzyme feed additives (Chesson, 1994; Graham and Balnave, 1995). However, more recent studies have shown that exogenous enzymes in the rumen are generally more stable than previously assumed (Hristov *et al.*, 1998a; Morgavi *et al.*, 2000b), particularly when added to feed pre-ingestion (Fontes *et al.*, 1995). Various *in vitro* and *in vivo* studies have been conducted to assess the resistance of enzymes to rumen inactivation. Hristov *et al.* (1998a) and Morgavi *et al.* (2001) observed the cellulose and xylanase activities of some enzyme products to be remarkably stable with little or no decline in enzymatic activity when incubated in rumen fluid for up to 6 hours. Fontes *et al.* (1995) postulated that the stability of xylanases and cellulases in the rumen may be due to their glycosylation, which may protect them from inactivation from temperature and proteases. It was found that exogenous enzymes are generally stable in the rumen, although there can be differences depending on the source organism, the enzyme activity, and whether the enzyme product is applied to feed. Exogenous enzymes are likely to be more susceptible to gastrointestinal proteases in the abomasum and intestines than to ruminal proteases (Morgavi *et al.*, 2001).

Direct hydrolysis

The enzyme additives degrade the substrate that would be naturally degraded by the ruminal microbes. It is predicted that in the rumen total enzyme activity will increases when exogenous enzymes are added in the diet and hence digestion is improved .It may be due to increase in hydrolytic capacity. Beauchemin and Rode (1996) calculated that adding exogenous enzymes to feed could potentially increase cellulose activity in the rumen by up to 15%. Hristov *et al.* (1998b) administered two enzyme products through the rumen cannula of a dairy cow and reported that cellulase activity increased by 169 and 198%, xylanase activity increased by 49 and 375%, and β-glucanase activity increased by 61 and 107%, for the two products, respectively. It can be concluded that increase in the hydrolytic capacity is a function of amount of enzyme applied to the feed.

Synergism with ruminal microbes

The concept of synergism between exogenous and endogenous enzymes implies, the increased net additive effect of each of the individual components. There has been considerable increase in the total enzymic activity in the rumen.

Synergy among cellulases and xylanases has been extensively documented by enzymologists (Bhat and Hazlewood 2001). Morgavi *et al.* (2000a) demonstrated substantial synergism between exogenous enzymes and ruminal enzymes such that the net combined hydrolytic effect in the rumen was much greater than estimated from the individual activities (Fig. 1). Thus, it can be hypothesized that increase enzymic activity due to synergy of exogenous enzymes and ruminal endogenous enzymes enhances fibre digestion in the rumen.

Sub-optimal ruminal conditions for fibre digestion

Under the intensive rearing conditions of ruminants the ruminal conditions are usually not optimum for growth of fibrolytic bacteria. However, certain exogenous enzymes are even active at lower pH also. When the pH conditions are not optimum for the growth of fibrolytic ruminal bacteria than these exogenous enzymes can play important role and their addition might be of greater benefit. Colombatto *et al.* (2003b) used the same continuous culture system maintained at high (pH 6.0 to 6.6) or low (pH 5.4 to 6.0) pH. Enzyme addition increased neutral detergent fibre degradability by 43% and 25% at high and low pH, respectively, largely due to an increase in hemicelluloses degradation. Thus, it can be predicted that exogenous enzymes improve fibre digestion when the pH conditions are sub-optimal.

Bacterial attachment and colonization

The exogenous enzymes are known to stimulate the attachment and colonization of rumen microbes on plant fibre. The application of enzymes to feed causes the release of soluble sugars from feed, that increases the chemotactic attraction and which in turn increases the attachment of ruminal bacteria to the surface of substrate. It also increases the "roughness" of the feed surface making it more suitable for microbial colonization (Morgavi *et al.* 2000c). It should also be taken care of that if the enzymes are used at higher levels than it will compete with the rumen population for cellulose's binding sites available on feed surface. This probably explains the reason for lack of response reported when enzymes are used at high doses. Yang *et al.* (1999) reported that improvements in digestibility caused by enzymatic treatment were related to increased microbial colonization as there is a positive relationship exists between microbial colonization and dry matter disappearance. Attachment of ruminal microorganisms to the substrate is a perquisite for digestion of feed particles in the rumen and increased microbial colonization would increase attachment and thereafter increase digestion. These enzymes also expose certain other additional cell wall sites for bacterial attachment that causes better digestion of the feed.

Stimulation of rumen microbes

There is some evidence that adding feed enzymes to the diet increases the total viable bacteria in the rumen. Wang *et al.* (2001) used the Rumen Simulation Technique (RUSITEC) and observed that applying enzyme onto feed increased cellulolytic bacteria 10-fold. Stimulation of total microbial numbers due to the use of enzymes can result in greater microbial biomass and increased flow of microbial protein to the duodenum. Increased microbial flow would have a significant impact on the supply of metabolizable protein to the animal, which would benefit animals fed for high levels of production (Beauchemin *et al.*, 2004).

An Integrated Mode of Action

It is clear that numerous modes of actions for exogenous enzymes are suggested that are inter dependent. Therefore, an integrated mode of action has been proposed that can better explain the observed increases in feed digestion. When enzymes are applied prior to ingestion their efficacy increases by removal of structural barriers, release of soluble sugars and increasing the stability of enzymes. Application of enzymes to feed enhances the binding of the enzyme with the substrate, which may increase the resistance of the enzymes to proteolysis. The ruminal effects following enzyme addition includes increase in the hydrolytic capacity of the rumen which may be due to increased bacterial attachment, stimulation of rumen microbial populations and increased enzyme activity within the rumen, which enhances the digestibility of the feed.

Enzyme Level

The supplementation of the exogenous enzymes in the ruminant's diet should be optimum depending on the diet. Response to level of enzyme supplementation is not linear, thus the optimum concentration needs to be investigated (Beauchemin *et al.*, 2000). Both high and low levels of enzymes will limit the production. Lack of response to low levels of enzyme addition may indicate an insufficient supply of enzyme activity and when excess enzymes are applied than it will compete with the rumen population for cellulose's binding sites available on feed surface and restrict microbial attachment and limit the digestion of feed. Kung *et al.* (2000) offered forage treated with increasing levels (0, 1, 2.5 mL/kg of total mixed ration) of an enzyme product to cows. Cows fed the low level of enzyme tended ($P < 0.10$) to produce more milk (39.5 kg/d) than those fed the control diet (37.0 kg/d) or those fed the high level of enzyme (36.2 kg/d). The study demonstrated that high levels of enzyme addition can be less effective than low levels, and the optimal level of enzyme supplementation depend on the diet. Certain factors such as, substrate specificity, moisture level of feed, pH, and temperature of the feed etc should also be taken care to assess the optimum enzyme level as they also affect the binding of enzyme with the substrate. (Beauchemin *et al.*, 1995).

Enzyme-Feed specificity

The understanding of enzyme-feed specificity is necessary for formulating the ruminant feed enzyme products because the diet of ruminant contains several types of forages and concentrates. Therefore, best approach is to use different enzyme sources in a basic diet. Use the enzyme that is relatively suitable for most of the feeds.

Method of providing enzyme to animals

A positive response is shown when the fibrolytic exogenous enzymes are applied on feed in a liquid form before the ingestion (Yang *et al.*, 1999). On the other hand, infusion of enzymes into the rumen has not shown any positive effect because the enzyme could pass from the rumen before sufficient contact with the forage has taken place or it will be degraded by bacterial proteases (Lewis *et al.*, 1996). Application of enzyme on feed prior to ingestion enhances the binding between enzyme and feed, thereby making the enzyme more stable by increasing the resistance of the enzymes to proteolysis in the rumen.

Enzymes have been applied to TMR (Yang *et al.*, 2000), hay, ensiled forages (Beauchemin *et al.*, 1995), concentrate (Rode *et al.*, 1999) and supplement or premix (Bowman, 2001). Exogenous enzymes may be expected to be more effective when applied to high moisture feeds (such as silages) compared to dry feeds because of the higher moisture content as water is required for hydrolysis of soluble sugars from polymers . Furthermore, silage pH values are usually at, or around, the optimal pH for most fungal enzymes. Yang *et al.* (2000) reported that the digestibility and milk production was increased, when the enzymes were added to the concentrate portion, but not when they were added directly to TMR. The reduced efficacy of exogenous enzymes applied to ensiled feeds may be due to inhibitory compounds in fermented feeds.

Proportion of the diet to which the enzymes are applied

To ensure maximal beneficial response the proportion of the diet to which the enzyme is applied must be considered (Bowman *et al.*, 2002a). Beauchemin *et al.* (1999b) suggested that enzyme should be applied to a larger proportion of the diet in order to ensure better digestion which could be due to better dispersion of the enzyme in the rumen. Adding enzyme to a small portion of the diet may allow rapid passage of enzyme from the rumen, lessening the enzyme effect in the rumen.

Digestion is increased when the enzyme is added to the concentrate portion of the diet but not if only to the supplement or premix. The reason for the ineffectiveness of the enzyme applied to the supplement could be that the high temperatures of the pelleting process could reduce the activity of enzymes.

Practical considerations

The use of exogenous enzyme in ruminants increases the forage utilization and improves the production economically. In contrast to use of antibiotics and hormones there is no risk of with drawls /residues in tissues or animal products by using the enzymes in ruminant diet. (Beauchemin *et al.*, 1999a). The concept of enzymes and feed specificity must be understood for formulating newer and effective enzyme products. In order to produce enzyme preparations to improve feed digestion, the factors that limit the rate and extent of digestion should be well known e. g, in maize the protein matrix surrounding the starch granules dictates the rate and extent of starch digestion in the grain. Thus exogenous enzymes designed to improve the utilization of maize should contain proteases, rather than amylases, so that the protein matrix can be digested, exposing the starch granules to digestion by endogenous ruminal or host enzymes (McAllister *et al.*, 2001). Detailed knowledge regarding the mode of action, method of application, optimal concentrations of enzyme need to be established for formulated the specific enzyme supplements that can be well commercialized and popularized. With better understanding of the basic concepts regarding the use of enzymes, we will be able to reduce the cost of production of enzyme preparations.

Production responses to use of enzymes

A number of studies have examined the effects of fibrolytic exogenous enzyme on digestibility and production responses. The beneficial effects appears to be a result of, in part, improvements in feed digestibility (Beauchemin *et al.*, 1995; Feng *et al.*, 1996). Lewis *et al.* (1999) treated forage with a cellulase/xylanase mixture @ (1 mL/kg of total mixed ration [TMR], DM basis) and observed that cows in early lactation produced 6.3 kg/d (16%) more milk. However, higher and lower levels of the same enzyme product were less effective. Supplementing dairy cow diets with a fibrolytic enzyme mixture has the potential to enhance milk yield and nutrient digestibility of cows in early lactation without changing feed intake (Rode *et al.*, 1999). The positive response was attributed to increase in digestibility. Fibrolytic enzymes can be used to improve milk production in lactating cows, the cows fed forages treated with cellulase D and xylanase B tended to produce more 3.5% FCM than did cows fed the untreated forages (Kung *et al.*, 2002). (Gaafar, 2010) concluded that use of high fiber total mixed ration (high roughage) was economically effective in feeding buffaloes and fibrolytic enzyme supplementation increase the actual milk and 7% FCM yield significantly.

Several studies on supplementation of exogenous enzyme had shown positive response while others have shown no effect (Yang *et al.*, 2000). These inconsistencies could be due to factors such as type of enzyme preparations, enzyme activities, diet composition, method of application and physiological status of animal etc. Several studies have examined the use of exogenous enzyme products on beef cattle. Supplementing fibrolytic enzyme mixture with alfalfa hay diets improved live weight gain of cattle by as much as 35% and feed conversion by up to 10% as a result of enhanced total tract digestion (Beauchemin *et al.*, 1995). A similar enzyme mixture was added to a high grain feedlot finishing diet resulted in an improvement of 11% in feed conversion ratio, 6% in weight gain, and a 5% decrease in feed intake (Beauchemin *et al.*, 1997). In another study, an enzyme mixture added to a high concentrate diet of growing feedlot heifers increased average daily gain by 9% and numerically improved feed to gain ratio by 10% but had no effect on dry matter intake (Beauchemin *et al.*, 1999a). Inspite of greater benefits of exogenous enzyme on beef industry the adoption of this technology is not wide, that may be due to high cost of enzyme products in comparison to antibiotics, hormones, ionophores and implants etc. Thakur *et al.* (2010) conducted the study to show the effect of supplementing exogenous fibrolytic enzymes (cellulase 4050 μM glucose/g/h and xylanase 8600 μM xylose/g/min mixed equally w/w) at 2 levels @ 1.5 and 3.0 g mixture/kg DM of TMR on the growth performance and nutrient utilization in Murrah buffalo calves. It was concluded that average daily gain (ADG), dry matter intake (DMI) and the digestibility coefficient of neutral detergent fibre (NDF), cellulose and hemicellulose were higher in the group that were fed fibrolytic enzymes @ 1.5 g mixture/kg DM.

Regulations for use of enzyme as additives

On January 1, 1998, the AAFCO Enzyme and Microbial Task Force that includes members of the AAFCO, FDA, and Agriculture and Food Canada have provided following guidelines for the use of enzymes in animal feeds:

1. Manufacturers must mention the source of the enzyme (organism), type of enzyme and its activity on their products.

2. The source of enzymes should be non-pathogenic organisms.
3. Enzymes from genetically altered organisms are acceptable if the amino acid sequence of the enzyme has not been significantly altered and if no altered organisms are in the formulation and no transformable antibiotic resistant DNA is present.
4. Products must be safe for animal, human and environmental concerns.
5. Functionality must be proven via *in vitro* tests (Kung, 2002).

Summary

Supplementation of exogenous fibrolytic enzymes to ruminants has the potential to increase the efficiency of feed utilization. Various results in different studies have dictated positive responses, although some studies showed no effect. Better understanding of enzyme levels, enzyme fed specificity and mode of action is demanded in order to assure consistent results. With growing concern of consumers over the use of antibiotics and hormones as feed additives and high cost involved in livestock production, the use of exogenous enzymes supplementation in ruminant ration will gain popularity in near future.

References

Beauchemin, K. A., and L. M. Rode. 1996. Use of feed enzymes in ruminant nutrition. *In* Animal Science Research and Development–Meeting Future Challenges. L. M. Rode, ed. Ministry of Supply and Services, Ottawa, Canada. pp. 103.

Beauchemin, K. A., Colombatto, D. Morgavi, D. P. and Yang, W. Z. 2003a. Use of exogenous fibrolytic enzymes to improve feed utilization by ruminants. *J. Anim. Sci.* 81: (E. Suppl.): E37–E47.

Beauchemin, K. A., Colombatto, D., Morgavi, D. P., Yang, W. Z. and Rode, L. M. 2004. Mode of action of exogenous cell wall degrading enzymes for ruminants. *Can. J. Anim. Sci.* 84: 13–22.

Beauchemin, K. A., L. M. Rode, and D. Karren. 1999b. Use of feed enzymes in feedlot finishing diets. *Can. J. Anim. Sci.* 79:243–246.

Beauchemin, K. A., L. M. Rode, and V. J. H. Sewalt. 1995. Fibrolytic enzymes increase fiber digestibility and growth rate of steers fed dry forages. *Can. J. Anim. Sci.* 75:641–644.

Beauchemin, K. A., L. M. Rode, M. Maekawa, D. P. Morgavi, R. Kampen. 2000. Evaluation of a non-starch polysaccharidase feed enzyme in dairy cow diets. *J. Dairy Sci.* 83:543–553.

Beauchemin, K. A., S. D. M. Jones, L. M. Rode, and V. J. H. Sewalt. 1997. Effects of fibrolytic enzyme in corn or barley diets on performance and carcass characteristics of feedlot cattle. *Can. J. Anim. Sci.* 77:645–653.

Beauchemin, K. A., W. Z. Yang, and L. M. Rode. 1999a. Effects of grain source and enzyme additive on site and extent of nutrient digestion in dairy cows. *J. Dairy Sci.* 82:378–390

Beauchemin, K.A. 1998. Use of feed enzymes in ruminant nutrition. Proceedings of the Pacific Northwest Nutrition Conference, Vancouver, BC. pp. 121-135.

Bhat, M.K. and G.P. Hazlewood. 2001. Enzymology and other characteristics of cellulases and xylanases In: M.R. Bedford and G.G. Partridge (eds.). Enzymes in farm animal nutrition. CABI Publishing, Oxon, U.K.

Bowman, G. R. 2001. Digestion, ruminal pH, salivation, and feeding behaviour of lactating dairy cows fed a diet supplemented with fibrolytic enzymes. M.S. Thesis, Univ. of British Colombia, Vancouver, Canada.

Bowman, G. R., Beauchemin, K. A. and Shelford, J. 2002a. The proportion of the diet to which fibrolytic enzymes are added affects nutrient digestion by lactating dairy cows. *J. Dairy Sci.* 85:3420–3429.

Chesson, A. 1994. Manipulation of fibre degradation: an old theme revisited. *In* T. P. Lyons and K.A. Jacques, eds. Biotechnology in the feed industry. *Proceedings of Alltech's Tenth Annual Symposium*, Nottingham University Press, Loughborough. pp. 83–98.

Colombatto, D., F. L. Mould, M. K. Bhat and Owen, E. 2003. Use of fibrolytic enzymes to improve the nutritive value of ruminant diets: A biochemical and *in vitro* rumen degradation assessment. *Anim. Feed Sci. Technol.* 107: 201-209.

Coughlan, M. P. 1985. The properties of fungal and bacterial cellulases with comment on their production and application. *Biotechnol. Genet. Eng. Rev.* 3:39–109.

Cowan, W. D. 1994. Factors affecting the manufacture, distribution, application and overall quality of enzymes in poultry feeds. *In:* Joint Proc. 2nd Int. Roundtable on Anim. Feed Biotechnology-Probiotics, and Workshop on Animal Feed Enzymes. Ottawa. pp. 175–184.

Feng, P., C. W. Hunt, G. T. Pritchard, and W. E. Julien. 1996. Effect of enzyme preparations on in situ and in vitro degradation and *in vivo* digestive characteristics of mature cool-season grass forage in beef steers. *J. Anim. Sci.* 74:1349–1357.

Fontes, C. M. G. A., Hall, J., Hirst, B. H., Hazlewood, G. P. and Gilbert. H. J. 1995. The resistance of cellulases and xylanases to proteolytic inactivation. *Appl. Microbiol. Biotechnol.* 43: 52–57.

Forsberg, C. W., Forano, E. and Chesson, A. 2000. Microbial adherence to the plant cell wall and enzymatic hydrolysis. *In* P. B. Cronje (ed.) Ruminant physiology: Digestion, metabolism, growth and reproduction. CABI Publishing, Wallingford, UK. pp. 79–97.

Gaafar, H.M.A., Raouf. E.M.and El-reidy, K.F.A. 2010. Effect of fibrolytic enzyme supplementation and fibre content of total mixed rartion on productive performance of lactating buffaloes. *Slovak. J. Anim. Sci.*, 43(3): 147-153.

Graham, H. and Balnave, D. 1995. Dietary enzymes for increasing energy availability. *In* R. J. Wallace and A. Chesson, eds. Biotechnology in animal feeds and animal feeding. VCH, Weinheim, Germany. pp. 295–309.

Gwayumba, W. and D. A. Christensen. 1997. The effect of fibrolytic enzymes on protein and carbohydrate degradation fractions in forages. *Can. J. Anim. Sci.* 77: 541–542.

Hristov, A. N., L. M. Rode, K. A. Beauchemin, and R. L. Wuerfel. 1996. Effect of a commercial enzyme preparation on barleysilage *in vitro* and in sacco dry matter degradability. *In Proc. Western Section, American Society of Animal Sciences*.. pp. 282–284.

Hristov, A. N., McAllister, T. A. and Cheng, K.J. 1998a. Stability of exogenous polysaccharide-degrading enzyme in therumen. *Anim. Feed Sci. Technol.* 76: 161–168.

Hristov, A. N., McAllister, T. A., Treacher, R. J. and Cheng, K. J. 1998b. Effect of dietary or abomasal supplementation of exogenous polysaccharide-degrading enzymes on rumen fermentation and nutrient digestibility. *J. Anim. Sci.* 76: 3146–3156.

Hristov, A. N., T. A. McAllister and K. J. Cheng. 1998. Effect of dietary or abomasal supplementation of exogenouspolysaccharide- degrading enzymes on rumen fermentation and nutrient digestibility. *J. Anim. Sci.* 76:3146-3156.

Hristov, A. N., T. A. McAllister, and K.J. Cheng. 1998. Stability of exogenous polysaccharide-degrading enzyme in the rumen. *Anim. Feed Sci. Technol.* 76:161–168.

Kopecny, J., Marounek, M. and Holub, K. 1987. Testing the suitability of the addition of *Trichoderma viride* cellulases to feedrations for ruminants. *Zivocisna Vyroba.* 32: 587–592.

Krause, M., Beauchemin, K. A., Rode, L. M., Farr, B. I. and Nørgaard, P. 1998. Fibrolytic enzyme treatment of barley grain and source of forage in high-grain diets fed to growing cattle. *J. Anim. Sci.* 76: 1010–1015.

Kung, L. 2002. Direct fed microbials and enzymes for ruminants. Department of Animal & Food Sciences University of Delaware.

Kung, L., Jr., R. J. Treacher, G. A. Nauman, A. M. Smagala, K. M. Endres, and M. A Cohen. 2000. The effect of treating forages with fibrolytic enzymes on its nutritive value and lactation performance of dairy cows. *J. Dairy Sci.* 83:115–122.

Lewis, G. E., C. W. Hunt, W. K. Sanchez, R. Treacher, G. T. Pritchard, and P. Feng. 1996. Effect of direct-fed fibrolytic enzymes on the digestive characteristics of a forage-based diet fed to beef steers. *J. Anim. Sci.* 74:3020–3028.

Lewis, G. E., W. K. Sanchez, C. W. Hunt,M. A. Guy, G. T. Pritchard, B. I. Swanson, and R. J. Treacher. 1999. Effect of direct-fed fibrolytic enzymes on the lactational performance of dairy cows. *J. Dairy Sci.* 82:611–617.

McAllister, T. A., A. N. Hristov, K. A. Beauchemin, L. M. Rode and K-J. Cheng. 2001. Enzymes in ruminant diets. Recent Advances in Enzymes in Farm Animal Nutrition (Ed. M. R. Bedford and G. G. Partridge). CABI Publishing,Wallingford, Oxon, UK. pp. 273-298.

Miller, G. L. 1959. Use of dinitrosalicylic acid reagent for determination of reducing sugar. *Anal. Chem.* 31:426–428.

Morgavi D.P., Beauchemin K.A., Nsereko V.L., Rode L.M., Iwaasa A.D., Yang W.Z., McAllister T.A. and Wang Y. 2000. Synergy between ruminal fibrolytic enzymes and en-zymes from *Trichoderma Longibrachiatum*. *J. Dairy Sci.* 83, 1310-1321.

Morgavi, D. P., C. J. Newbold, D. E. Beever, and R. J. Wallace. 2000b. Stability and stabilization of potential feed additive enzymes in rumen fluid. *Enzyme Microb. Technol.* 26:171–177.

Morgavi, D. P., K. A. Beauchemin, V. L. Nsereko, L. M. Rode, T. A. McAllister, A. D. Iwaasa, Y. Wang, and W. Z. Yang. 2001. Resistance of feed enzymes to proteolytic inactivation by rumen microorganisms and gastrointestinal proteases. *J. Anim. Sci.* 79:1621–1630.

Rode, L. M., W. Z. Yang, and K. A Beauchemin. 1999. Fibrolytic enzyme supplements for dairy cows in early lactation. *J. Dairy Sci.* 82:2121–2126.

Schingoethe, D. J., G. A. Stegeman and R. J. Treacher. 1999. Response of lactating dairy cows to a cellulase and xylanase enzyme mixture applied to forages at the time of feeding. *J. Dairy Sci.* 82:996-1003.

Sheppy. C. 2001. The current feed enzyme market and likely trends. *In:* M.R. Enzymes in farm animal nutrition, CABI Publishing, Oxon, U.K., pp.1-10.

Somogyi, M. 1952. Notes on sugar determination. *J. Biol. Chem.* 195: 19–23.

Thakur, S.S., Verma, M.P., Ali. B. Shelke.S.K., and Tomar. S.K. 2010. Effect of exogenous fibrolytic enzyme supplementation on growth and nutrient utilization in murrah buffalo calves. *Ind. J. Anim. Sci.* 80 (12): 1217-19.

Viikari, L., M. et.al. .1993. Hemicellulases for industrial applications. In: Saddler, J.N. (ed.) Bioconversion of forest and agriculture plant residues. CAB International, Wallingford, UK, pp.131-182.

Wang, Y., T. A. McAllister, L. M. Rode, K. A. Beauchemin, D. P. Morgavi, V. L. Nsereko, A. D. Iwaasa, and W. Yang. 2001. Effects of an exogenous enzyme preparation on microbial protein synthesis, enzyme activity and attachment to feed in the Rumen Simulation Technique (Rusitec). *Br. J. Nutr.* 85:325–332.

Wood, T.M., and K. M. Bhat. 1988.Methods for measuring cellulose activities. Page 87 in Methods in Enzymology. Vol. 160. W. A. Wood and S. T. Kellogg, ed. Academic Press Inc.

Yang, W. Z., K. A. Beauchemin, and L. M Rode. 1999. Effects of enzyme feed additives on extent of digestion and milk production of lactating dairy cows. *J. Dairy Sci.* 82: 391–403.

Yang, W. Z., K. A. Beauchemin, and L. M. Rode. 2000. A comparison of methods of adding fibrolytic enzymes to lactating cow diets. *J. Dairy Sci.* 83:2512–2520.

ZoBell, D. R., R. D. Weidmeier, K. C. Olson, and R. J. Treacher. 2000. The effect of an exogenous enzyme treatment on production andcarcass characteristics of growing and ûnishing steers. *Anim. Feed Sci. Technol.* 87:279–285.

19

Metabolic Modifiers for Quality Meat Production

Avijit Dey

Introduction

While world population is expected to increase from 6 billion to about 8.3 billion, India is likely to be the most populous nation with 1.6 billion in the year 2030. Therefore, it is essential to be prepared to produce sufficient food for the increased population with the application of newer technologies. The consumption of animal food was 10 kg/yr in the 1960s increasing to 26 kg/yr in 2000 and is expected to be 37 kg/yr by 2030 (FAO, 2008; 2009). Large increases in per capita and total demand for meat, milk and eggs are forecast for most developing countries for the next few decades (Delgado *et al.*, 1999). In developed countries, per capita intakes are forecast to change slightly, but the increases in developing countries, with their larger populations and more rapid population growth rates, will generate a very large increase in global demand.

Moreover, an increasing number of consumers demanding healthy and quality foods have pushed researchers into development of technologies that can control the proportion of lean and fat muscle deposition in meat producing animals. A number of technologies have been developed that led to remarkable improvements in the efficiency of animal production and in carcass composition and meat quality. There are some dietary additives and vitamins that may increase protein and muscle deposition, while often simultaneously reducing fat deposition by modifying nutrient metabolism in individuals.

Metabolic Modifiers

Metabolic modifiers are a group of compounds that can improve the efficiency of milk, meat and egg production and, in certain cases, enhance the quality of animal products by altering physiology and metabolism of animals. They can be supplemented with feed, injected or implanted in animals and, in general, these metabolic modifiers increase protein synthesis and decrease fat synthesis and deposition (Dunshea *et al.*, 2005).

These technological advancements benefited the producers, processors and consumers by enhancing efficiency of production and processing. At the same time, consumers are getting desired quality of animal products. The metabolic modifiers can be categorized as follows:

1. Anabolic steroids, 2. Somatotropin, 3. Beta agonists, 4. Vitamins and minerals, 5. Other modifiers

1. Anabolic steroids

Anabolic steroids are estrogenic, androgenic or both and used widely on cattle to enhance growth rate and feed efficiency of animals and profitability of the livestock industry. There are many compounds that are used either alone or in combination including naturally occurring steroids such as progesterone and testosterone, 17 α-oestradiol and their synthetic counterparts like trenbolone acetate, melengestrol acetate and zeranol (Moon *et al.*, 2009). Breeds or type of animals that have high potential of muscle growth show the greatest response to these hormones. Again the responses are more during first few weeks of implantation, suggesting a peak and then a decline in circulating concentration of the hormones (Hayden *et al.*, 1992). Study of various researchers revealed 10-30% increase in growth rate of cattle with an improvement of feed efficiency of 15% by application of anabolic steroid implants.

The carcass leanness is increased, which reduces the marbling and tenderness of meat. Therefore, meat quality and palatability is decreased with anabolic steroid implantation. The Warner-Bratzler shear force values are more for meat from anabolic steroid treated animals due to reduced tenderness. However, the types of implant, time and length of application, category of animals have significant roles on these parameters. A meta-analysis revealed that shear force was increased by 2.55 N and sensory tenderness score was decreased by 5 units on a 100 point scale. However, there are differences among hormone growth promotants in their effect on tenderness; some have a distinct negative effect, whereas others have a rather modest effect (Morgan, 1997). Implanting anabolic steroids to grass-fed cattle has more negative effects on meat quality than feedlot cattle. Barham *et al.* (2003) reported that implanting with Synovex-

S twice or Synovex-S followed by Revalor-S to *Bos indicus* crossbred cattle increased shear force values compared to non-implanted controls after only 14 days of aging, but not after 21 days of aging. These authors determined that consumers failed to detect any differences in steak samples related to implant treatment after 7 or 14 days of aging. Although there is limited study on the effects of anabolic implants in poultry, sheep and pigs, Lee *et al.* (2002) deliberated the effects of implanting castrate pigs with Revalor-H (trenbolone acetate plus estradiol-17b) on performance, carcass composition and meat quality. Back fat thickness was reduced with implantation without affecting overall pork quality. Therefore, implantation of anabolic steroids should be used judiciously to improve meat production with minimum effects on its quality.

2. Somatotropin

Somatotropin, also known as growth hormone is a peptide hormone secreted by somatotropic cells within the lateral wings of the anterior pituitary gland and stimulates growth, cell reproduction and regeneration in humans and other animals. It plays important role in protein synthesis, lipolysis, mineral metabolism, gluconeogenesis and several other functions. There are also recombinantly produced somatotropins which have been approved for use in swine (porcine somatotropin, rpST) and ruminants (bovine, rbST and ovine somatotropin, roST) in different countries (Dikeman, 2007). These exogenous somatotropins are normally injected to animals once to thrice per week and responses are dose-dependent. Injection of rpST (150μg/kg body weight/day) to pigs was reported to increase average daily gain (+20%) and improvement of feed conversion efficiency. Carcass protein accretion rates are increased (+74%) concurrent with a decrease (-82%) in lipid accretion rate when rpST is administered from 30 to 90 kg body weight (Beermann, 1994). Increased protein synthesis results in increase in protein deposition; however, dressing percentage remains same due to higher organ weight. A meta-analysis of various experiments conducted on rpST showed a decrease in intramuscular fat (-12%), drip loss (-6%) and tenderness (-9%) with increased shear force value (+9%). However, the effects are highly dependent on dose, genotype and managemental variables (Dunshea *et al.*, 2005). Responses of ruminants to rbST are lesser than pigs. Somatotropin decreases intramuscular lipid content in both pork and beef in a dose-dependent manner by a magnitude of 20–50% (McKeith *et al.*, 1994). Administration of rbST and roST to growing sheep also reduces carcass fatness and the effect is greater in ewe lambs than wethers (Beermann *et al.*, 1990). Somatotropin is efficient in improving growth performance and meat production. However, it reduces marbling and tenderness by decreasing intramuscular fat deposition.

3. Beta agonists

The beta agonists are naturally occurring and synthetic organic compounds that enhance lean muscle gain, increase growth rate, and feed efficiency. There are differences between specific beta-agonists, but those approved by most of the countries include ractopamine and zilpaterol for use in swine and beef productions. The beta-agonist binds to receptors in a muscle cell where it initiates an increase in protein synthesis, resulting in an increase in muscle fiber size. They also decrease lipogenesis, while increasing lipolysis. Beta-agonists are different than hormone implants in that the effects occur at a cellular level and do not affect the hormone status of the animal; whereas an implant, made of natural and synthetic hormones, does affect the hormone status of the animal to promote growth.

Ractopamine hydrochloride (PayleanÒ) for pigs has positive effects on growth rate, feed efficiency, dressing percent, and carcass composition when fed at dosages of 5.0 and 9.9 mg of ractopamine per kg of feed for 30–50 days before slaughter (Dikeman, 2007). The visual meat quality and sensory traits remain neutral to positive. Feeding of ractopamine (OptaflexxÒ) to feed lot cattle at 20 and 30 ppm, 46 days before slaughter increases growth rate, feed efficiency, and final live and carcass weights over the controls (Schluter *et al*., 1991). However, feeding of ractopamine to steer at 300 mg per day for the last 33 days on feed increased shear force besides decrease in feed intake and improvement in feed efficiency and dressing percent (Avendano-Reyes *et al*., 2006).

Cattle, supplemented with zilpaterol hydrochloride (ZilmaxÒ) improved growth performance, dressing percent, and carcass weight (Dikeman, 2003; Avendano-Reyes *et al*., 2006). Feeding diets with 0.15 mg zilpaterol for the final 15, 30 or 45 days of the feedlot period until 48 h before slaughter reported to reduce sensory evaluated tenderness and juiciness, and increase shear force of the *longissimus* muscle in animal fed zilpaterol for 45 days but not for shorter durations (Strydom and Nel, 1999). Zilpaterol is more effective than ractopamine in improving meat production without much affecting meat quality attributes. Therefore, it suggests that 30 days supplementation of zilpaterol or ractopamine in finishing diets of cattle would be optimum and economical for improving meat production performance without negatively affecting its quality.

4. Vitamins and minerals

A number of micronutrients (vitamins and minerals) regulate the biochemical pathways to modify metabolism of nutrients towards improvement in animal production. Oxidation of myoglobin pigment in fresh meat to brown colour lowers consumers' acceptability. Lipid oxidation in meat and meat products contribute to the deterioration in quality and reduction of shelf life. Vitamin E supplementation improves meat quality by reducing lipid and myoglobin oxidation of meat and meat products. Positive effects of feeding vitamin E to improve shelf life and meat quality in pigs, poultry and cattle have been reported (Tsai *et al.*, 1978; Bartov *et al.*, 1983; Faustman *et al.*, 1989). Pig diets incorporated with vitamin E (100 - 700 mg/kg feed) reduce lipid oxidation (-20 to 80%), drip loss (-40%), and improvement of colour of pork and its products. The colour improvement is more evident in beef and lamb due to higher myoglobin content than pork and poultry products (Chan *et al.*, 1996; Guidera *et al.*, 1997; Guo *et al.*, 2006).

Vitamin D_3 has the potential to improve meat tenderness. It increases blood and muscle calcium levels. Supplementation of vitamin D_3 increases the tenderness in beef and improves firmness and colour and decrease drip loss in pork (Dikeman, 2003). However, feeding vitamin D_3 as metabolic modifier to improve meat quality needs more research as it reduces feed intake and potential toxicity in humans.

Vitamin A is involved in adipocyte differentiation and marbling development. However, the effects seem to be dependent on age and type of animals and level of vitamin A supplementation. Marbling scores have been negatively correlated with concentrations of vitamin A in blood and liver. Steers fed on diets containing low levels of vitamin A have produced carcasses with increased marbling scores (Kruk *et al.*, 2004). However, high levels of vitamin A supplementation decreases marbling in young cattle (Oka *et al.*, 1998). This suggests that high vitamin A might negatively affect differentiation of adipocytes and, consequently, marbling. Lambs fed diets with 1,365 IU vitamin A per kg feed had higher percentages of intramuscular fat in the *longissimus* muscle than non-supplemented control (Arnett *et al.*, 1996). A decreased concentration of saturated fat (C18:0, stearate) and increased unsaturated fat (C18:1, oleate) was reported in lambs fed a diet containing low beta-carotene (Seibert *et al.*, 2000). Colour and shelf life of meat are improved by supplementing (7500 mg/day/head) beta-carotene (Muramoto *et al.*, 2003).

Niacin plays a significant role in biochemical pathways, thereby modifying body metabolism. Supplementation of niacin (110-550 ppm) to finishing pigs fed corn-soybean meal based diet improves feed efficiency and meat quality (Real *et al.*, 2002).

Chromium has several biological functions, including roles in nuclear protein and RNA synthesis, but its predominant physiological role seems to be as an integral component of glucose tolerating factors (GTF) to potentiate the action of insulin. Trivalent chromium alters the direction of energy accumulation towards muscles rather than fats. Boleman *et al.* (1995) found that feeding elevated levels of chromium picolinate to pigs increased percentage of muscle, decreased backfat, and had no effect on tenderness or sensory traits. Feeding chromium (1.5 mg/day) to goats down regulates the expression of the acetyl CoA carboxylase 1 (ACC1) gene, which is responsible for secretion of the enzyme acetyl CoA carboxylase1 that catalyzes the carboxylation of acetyl CoA to form malonyl CoA, the key intermediate metabolite in fatty acid synthesis (Najafpanah *et al.*, 2014). Dietary supplementation (1200 ppb) of chromium-yeast to broiler birds improves meat quality by reducing lipid content and oxidation of meat (Toghyani *et al.*, 2010).

Supplementation of magnesium to finishing pigs decreases fat thickness and increases muscle percentage (Apple *et al.*, 2000). It also reduces the drip loss and increases muscle pH and colour scores of pork. Feeding 1% magnesium oxide to sheep 4 days before slaughter are reported to increase muscle glycogen concentrations (Gardner *et al.*, 2001).

Selenium supplementation has beneficial effect on growth and carcass quality traits in pigs, poultry and cattle (Close, 1999). Feeding selenium either in inorganic form (sodium selenite) or organic form (yeast-bound selenium) to pigs at 0.1 and 0.3 ppm results increased productivity from both the sources, however, organic selenium supplementation results higher selenium levels in the muscle (Mahan and Kim, 1996).Selenium, an essential constituent of the antioxidant enzyme glutathione peroxidase, has also been shown to significantly improve meat quality by decreasing cell membrane oxidation leading to reduced muscle drip loss in poultry (Edens, 1996).

5. Other modifiers

Betaine, an active methyl donor with a lipotropic effect has been reported to increase protein deposition and carcass leanness and to decrease back fat (Suster *et al.*, 2004). Besides improvement in growth performance by reducing the maintenance energy requirement of animals, betaine supplementation improves carcass quality, meat colour and water holding capacity and decrease drip loss, cooking loss (Matthews *et al.*, 2001).

L-carnatine is a vitamin-like compound, which can modify metabolism of long chain fatty acids by abetting their transport to mitochondrial matrix. Supplementation of L-carnatine to pigs decreases backfat thickness and increases lean deposition without affecting growth performance (Owen *et al.*, 1994).

Conjugated linoleic acid (CLA) is a mixture of positional and geometric isomers of linoleic acid with conjugated double bonds on the carbon chain with several biological effects. While *cis-9 trans-11* CLA has anticancer and other health properties, *trans-10 cis-12* isomer increases body weight gain, feed efficiency and muscle deposition. Despite of reduction in backfat thickness in pigs, intramuscular fat and marbling increases with CLA supplementation (Dunshea *et al.*, 2002; Ostrowska *et al.*, 2003b). Cook *et al.* (1998) demonstrated a reduction (-20%) in backfat for pigs fed CLA and an increase (+7%) in lean muscle mass. Dietary CLA is incorporated into adipose tissue and to a lesser extent into intramuscular fat in a dose dependant manner in pigs (Eggert *et al.*, 2001; Ostrowska *et al.*, 2003a), which offers the opportunity to improve the amount of CLA in fat. However, *cis-9 trans-11* CLA is incorporated more efficiently than *trans-10 cis-12* isomer (Whigham *et al.*, 2000). CLA enriched meat can be produced from ruminants since CLA occurs naturally in ruminant fat. Trans-vaccenic acid and CLA *cis-9 trans-11* are products of the normal biohydrogenation of linoleic acid to stearic acid in the rumen. McGuire *et al.* (1998) observed that increasing both forage level and high oil corn in feedlot diets of cattle increased CLA content by 24% in intramuscular fat. Feeding linseed oil (6% of feed dry matter) to the diet of finishing steers increases CLA content of *longissimus* muscle (Enser *et al.,* 1999). A meta-analysis of collated data showed an average reduction in backfat (-6%) and an increase in intramuscular fat (+7%) and marbling score (+11%) by supplementation of CLA to swine. However, reduction in consumer perception of flavour (-3%), juiciness (-12%) and tenderness (-2.5%) was noticed (Dunshea *et al.,* 2005).

Potential risk of using metabolic modifiers

Use of metabolic modifiers to improve growth rate and meat quality has a threat by their residual effect in animal products. Accordingly, approval has been issued by different international agencies for different types of metabolic modifiers. There is a clear worldwide scientific consensus to support the use of these approved and licensed metabolic modifiers when used according to good veterinary practices. The world's scientific community has agreed that these products are safe when used according to label directions in food producing animals. Beta-agonists are approved feed additives and are deemed safe by the U.S. Food and Drug Administration (FDA). There are differences between specific beta-agonists, currently Ractopamine is approved in 26 countries including Canada, Mexico, Australia and Brazil. In July 2012, the Codex Alimentarius Commission, developed in part by the World Health Organization, confirmed Ractopamine's safety by establishing a maximum residue limit. The Codex decision provides a worldwide standard.

Recombinantly produced porcine somatotropin (rpST) has been approved for use in swine production in several countries, but not in the US. Currently, rbST and roST are not approved for beef cattle or lamb production due to potential human health hazards.

Summary

Technological development improves animal productivity and product quality. The use of metabolic modifiers in farm animals has a significant advantage of improving quality and quantity of meat. In general, they increase protein synthesis and decrease fat synthesis and deposition. Production of lean meat sometimes lessens the quality of meat by reducing tenderness, juiciness, marbling and enhancement of shear force. However, as these metabolic modifiers increase the growth rate and feed efficiency considerably; their judicious application would enhance the profitability of animal production. Safety measures are required to be considered while selecting metabolic modifiers and their doses to assure quality animal products.

References

Apple, J. K., Maxwell, C. V., deRodas, B., Watson, H. B. and Johnson, Z. B. 2000. Effect of magnesium mica on performance and carcass quality of growing-finishing swine. *J. Anim. Sci.* 78: 2135–2143.

Arnett, A., Dikeman, M. E., Spaeth, C., Johnson, B. J. and Hildabrand, B. 2006. Effects of vitamin A supplementation in young lambs in performance, serum lipid, and carcass adipose tissue attributes. *J. Anim. Sci.,* 84(Suppl. Abst.): 243.

Avendano-Reyes, L., Torres-Rodriguez, V., Meraz-Murill, F. J., Perez- Linares, C., Figueroa-Saavedra, F. and Robinson, P. H. 2006. Effects of two adrenergic agonists on finishing performance, carcass characteristics, and meat quality of feedlot steers. *J. Anim. Sci.* 80: 3259.

Barham, B. L., Brooks, J. C., Blanton, J. R., Jr., Herring, A. D., Carr, M.A. and Kerth, C. R. 2003. Effects of growth implants on consumer perceptions of meat tenderness in beef steers. *J. Anim. Sci.* 81(12): 3052–3056.

Bartov, I., Basker, D. and Angel, S. 1983. Effect of dietary vitamin E on the stability and sensory quality of turkey meat. *Poult. Sci.* 62: 1224.

Beermann, D. H., Hogue, D. E., Fishell, V. K., Aronica, S., Dickson, H. W. and Schricker, B. R. 1990. Exogenous human growth hormone releasing factor and ovine somatotropin improve growth performance and composition of gain in lambs. *J. Anim. Sci.* 68: 4122.

Beermann, D. H. 1994. Carcass composition of animals given partitioning agents. In: Low fat meats, H. D. Hafs and R. G. Zimbleman (*Eds.*), Academy Press, San Diego, pp. 203–222.

Boleman, S. L., Boleman, S. J., Bidner, T. D., Southern, L. L., Ward, T. L. and Pontif, J. E. 1995. Effect of chromium picolinate on growth, body composition, and tissue accretion in pigs. *J. Anim. Sci.* 73: 2033–2042.

Chan, W. K. M., Hakkarainen, K., Faustman, C., Schaefer, D. M., Scheller, K. K. and Liu, Q. 1996. Dietary vitamin E effect on color stability and sensory assessment of spoilage in three beef muscles. *Meat Sci.* 42: 387–399.

Close, W. H. 1999. Organic minerals for pigs: An update. In: Biotechnology in the feed industry, T. P. Lyons and K. A. Jacques (*Eds.*), Proc. 15th annual symposium, Nottingham, UK. Pp. 51–60.

Cook, M. E., Jerome, D. L., Crenshaw, T. D., Buege, D. R., Pariza, M. W. and Albright, K. J. 1998. Feeding conjugated linoleic acid improves feed efficiency and reduces carcass fat in pigs. Federation of American Society for Experimental Biology.

Delgado, C. L., Rosegrant, M., Steinfeld, H., Ehui, S. and Courbois, C. 1999. Livestock to 2020: the next food revolution. Food agriculture and environment discussion paper 28. Washington DC: International Food Policy Research Institute.

Dikeman, M. E. 2003. Metabolic modifiers and genetics: Effects on carcass traits and meat quality. *Br. J. Food Techol.* (Special issue), 6: 1–38.

Dikeman, M. 2007. Effects of metabolic modifiers on carcass traits and meat quality. *Meat Sci.*, 77: 121–135.

Dunshea, F. R., Ostrowska, E., Luxford, B., Smits, R. J., Campbell, R. G. And D'Souza, D. N. 2002. Dietary conjugated linoleic acid can decrease backfat in pigs housed under commercial conditions. *Asian- Aust. J. Anim. Sci.* 15: 1011–1017.

Dunshea, R. R., D'Souza, D. N., Pethic, D. W., Harper, G. S. and Warner,R. D. 2005. Effects of dietary factors and other metabolic modifiers on quality and nutritional value of meat. *Meat Sci.* 71: 8–38.

Edens, F. W. 1996. Organic selenium: From feathers to muscle integrity to drip loss. Five years onwards: No more selenite! In: Biotechnology in the feed industry, T. P. Lyons and K. A. Jacques (*Eds.*), Proc. 12th annual symposium, Nottingham, UK: Nottingham University Press. Pp. 165–185.

Eggert, J. M., Belury, M. A., Kempa-Steczko, A., Mills, S. E. and Schinckel, A. P. 2001. Effects of conjugated linoleic acid on the belly firmness and fatty acid composition of genetically lean pigs. *J. Anim. Sci.* 79: 2866–2872.

Enser, M., Scollan, N. D., Choi, N. J., Kurt, E., Hallet, K. and Wood, J. D. 1999. Effect of dietary lipid on the content of conjugated linoleic acid in beef muscle. *Br. J. Anim. Sci.* 69: 143–146.

FAO. 2008. STAT database 2008. Food and Agriculture Organization. Rome Italy. Available online: www.fao.org.

FAO. 2009. Food security and agricultural mitigation in developing countries: options for capturing synergies. Food and Agriculture Organization, Rome, Italy.

Faustman, C., Cassens, R. G., Schaefer, D. M., Buege, D. R. and Scheller, K. K. 1989. Vitamin E supplementation of Holstein steer diets improves sirloin steak colour. *J. Food Sci.* 54: 485–486.

Gardner, G. E., Jacob, R. and Pethick, D. W. 2001. The effect of magnesium oxide supplementation on muscle glycogen metabolism before and after exercise and at slaughter in sheep. *Aust. J. Agric. Res.* 52: 723–729.

Guidera, J., Kerry, J. P., Buckley, D. J., Lynch, P. B. and Morrissey, P. A. 1997. The effect of dietary vitamin E supplementation on the quality of fresh and frozen lamb meat. *Meat Sci.* 45: 33–43.

Guo, Q., Richert, B. T., Burgess, J. R., Webel, D. M., Orr, D. E. and Blair, M. 2006. Effect of dietary vitamin E supplementation and feeding period on pork quality. *J. Anim.Sci.* 84(11): 3071–3078.

Hayden, J. M., Bergen, W. G. and Merkel, R. A. 1992. Skeletal muscle protein metabolism and serum growth hormone, insulin, and cortisol concentrations in growing steers implanted with estradiol-17, trenbolone acetate, or estradiol-17 plus trenbolone acetate. J. *Anim. Sci.* 70: 2109.

Kruk, Z. A., Siebert, B. D., Pitchford, W. S., Davis, J., Harper, G. S. and Bottema, C. D. K. 2004. The effect of vitamin A on fatness in lot-fed cattle: (I) Fat Quantity. In: Proc. 50th international congress of meat science and technology, Helsinki, Finland.

Lee, C. Y., Lee, H. P., Jeong, J. H., Baik, K. H., Jin, S. K. and Lee, J. H. 2002. Effects of restricted feeding, low-energy diet, and implantation of trenbolone acetate plus estradiol on growth, carcass traits, and circulating concentrations of insulin-like growth factor (IGF)-I and IGF-binding protein-3 in finishing barrows. *J. Anim. Sci.* 80: 84–93.

Mahan, D. C. and Kim, Y. Y. 1996. Effect of inorganic or organic selenium at two dietary levels on reproductive performance and tissue selenium concentrations in first-parity gilts and their progeny. *J. Anim. Sci.* 74: 2711–2718.

Matthews, J. O., Southern, L. L., Higbie, A. D., Persica, M. A. and Bidner, T. D. 2001. Effects of betaine on growth, carcass characteristics, pork quality, and plasma metabolites of finishing pigs. *J. Anim. Sci.* 79: 722–728.

McGuire, M. A., Duckett, S. K., Andrae, J. G., Giesy, J. G. and Hunt, C. W. 1998. Effect of high-oil corn on conjugated linoleic acid (CLA) in beef. *J. Anim. Sci.* 76 (Suppl. 1, Abstr.): 301.

McKeith, F. K., Lan, Y. H. and Beermann, D. H.1994. Sensory characteristics of meat from animals given partitioning agents. In: Low fat meats, H. D. Hafs and R. G. Zimbleman (*Eds.*), San Diego, CA: Academic Press, Inc. pp. 233–252.

Moon, H., Kang, T., Kim, T., Kang, I., Ki, H., Kim, S. and Han, S. 2009. OECD validation of phase 3 Hershberger assay in Korea using surgically castrated male rats with coded chemicals. *J. Appl. Toxicol.*, 29: 350-355.

Morgan, J. B. 1997. Implant program effects on USDA beef carcass quality grade traits and meat tenderness. In: Proc. Oklahoma State University Implant Symposium, Stillwater.

Muramoto, T., Nakanishi, N., Shibata, M. and Aikawa, K. 2003. Effect of dietary b-carotene supplementation on beef colour stability during display of two muscles from Japanese Black steers. *Meat Sci.* 63: 39–42.

Najafpanah, M.J., Sadeghi, M., Zali, A., Moradi-shahrebabak, H. and Mousapour, H. 2014 Chromium downregulates the expression of acetyl CoA carboxylase 1 gene in lipogenic tissues of domestic goats: a potential strategy for meat quality improvement. *Gene.* 543: 253–258.

Oka, A., Maruo, Y., Miki, T., Yamasaki, T. and Saito, T. 1998. Influence of vitamin A on the quality of beef from the Tajima strain of Japanese Black cattle. *Meat Sci.* 48: 159–167.

Owen, K. Q., Smith, J. W., Nelssen, J. L., Goodband, R. D., Tokach, M. D. and Friesen, K. G. 1994. The effect of L-carnitine on growth performance and carcass characteristics of growing-finishing swine. *J. Anim. Sci.* 72(Suppl. 1, Abstr.): 274.

Ostrowska, E., Cross, R. F., Muralitharan, M., Bauman, D. E. and Dunshea, F. R. 2003a. Dietary conjugated linoleic acid differentially alters fatty acid composition and increases conjugated linoleic acid content in porcine adipose tissue. *Br. J. Nutr.* 90: 915–928.

Ostrowska, E., Suster, D., Muralitharan, M., Cross, R. F., Leury, B. J. and Bauman, D. E. 2003b. Conjugated linoleic acid decreases fat accretion in pigs: Evaluation by dual-energy X-ray absorptiometry. *Br. J. Nutr.* 89: 219–229.

Real, D. E., Nelssen, J. L., Unruh, J. A., Tokach, M. D., Goodband, R. D. and Dritz, S. S. 2002. Effects of increasing dietary niacin on growth performance and meat quality in finishing pigs reared on two different environments. *J. Anim. Sci.* 80: 3203–3210.

Schluter, A. R., Preston, R. L., Davis, G. W., Ramsey, C. B. and Miller, M. F. 1991. Ractopamine effects on feedlot performance, carcass traits and chemical composition of feedlot steers and heifers. In: Proc. Reciprocal meat conference, 44(Abstract): 206.

Seibert, B. D., Pitchford, W. S., Kuchel, H., Kruk, Z. A. and Bottema, C. D. K. 2000. The effect of beta-carotene on desaturation of ruminant fat. *Asian-Aust. J. Anim. Sci.* 13: 185–188.

Suster, D., Leury, B. J., King, R. H., Mottram, M. and Dunshea, F. R. 2004. Interrelationships between porcine somatotropin (pST), betaine, and energy level on body composition and tissue distribution of finisher boars. *Aust. J. Agric. Res.* 55: 983–990.

Strydom, P. E. and Nel, E. 1999. The effect of supplementation period of a beta-agonist (zilpaterol), electrical stimulation, and ageing period on meat quality characteristics. *In: Proc. 45th International Congress of Meat Science and Technology*, Yokohama, Japan, 2:58–59.

Toghyani, M., Gheisari, A. A., Khodami, A.,Toghyani, M., Mohammadrezaei, M. and Bahadoran, R. 2010. Effect of dietary chromium yeast on thigh meat quality of broiler chicks in heat stress condition. *World Acad. Sci. Eng.Technol.* 4: 2010-12-22.

Tsai, T. C., Wellington, G. H. and Pond, W. G. 1978. Improvement in the oxidative stability of pork by dietary supplementation of swine rations. *J. Food Sci.* 43: 193–196.

Whigham, L. D., Cook, M. E. and Atkinson, R. L. 2000. Conjugated linoleic acid: implications for human health. *Pharmacol. Res.* 42: 503-510.

20

Hormones as Feed Additive for Livestock

Papori Talukdar and Goutam Mondal

Introduction

For raising animal and its products especially protein, food and nutrition plays an important role. Feed cost comprises about 60% of total recurring expenditure. Feed additives play an increasingly important role in animal nutrition by augmenting the performance by maintaining the needs of essential nutrients, improving feed conversion therefore optimize feed utilization (Wenk, 2000). Feed additives are a group of feed ingredients that can cause a desired animal response in a non-nutrient role, such as rumen modifier, growth or metabolic modifier (Hutjens, 1991).These are used in micro quantities and require careful handling and mixing. The development and introduction of various new feed additives is a major and practical solution to improve the efficiency of animal production, to alleviate environmental pollution, to maintain animal health, welfare and to ensure the safe and of good quality food production.

Hormones are chemicals synthesized and secreted by specific organs or glands by animals to co-ordinate their physiological activities. They act as messengers, produced and released from one kind of tissue to gradually stimulate or inhibit some process in a different tissue over a long period (Norman and Litwack, 1997). In essence, it is a chemical messenger that transports a signal from one cell to another. All multi-cellular organisms produce hormones. Hormones in animals are often transported through blood. Cells respond to a hormone when they express a specific receptor for that hormone. The hormone binds to the receptor protein, resulting in the activation of a signal transduction mechanism that ultimately leads to cell type-specific responses. Endocrine hormones are secreted (or released) directly into the blood stream, or simply diffuse through the interstitial spaces to nearby target tissues (paracrine signaling). A variety of exogenous chemical compounds, both natural and synthetic, have hormone-like effects on both humans and animals. Their interference with the synthesis, secretion, transport, binding, action, or elimination of natural hormones in the body can change the homeostasis, reproduction, development, and/or behavior, just as endogenously produced hormones do. Steroid hormones fulfill an important role at different stages of mammalian development comprising prenatal development, post natal growth, reproduction and sexual and social behavior. The importance of individual hormones varies between sexes and age and a disruption of the endocrine equilibrium may result in multiple biological adverse effects. One hormone can have multiple actions, e.g. the male hormone testosterone controls many processes from the development of the foetus, to libido in the adult. Alternatively, one function may be controlled by multiple hormones, e.g. the estrus cycle involves oestradiol, progesterone, follicle-stimulating hormone and luteinizing hormone.

History of hormone used as feed additive

The use of growth hormones in animals began in the 1950s with the introduction of diethylstilbestrol (DES) as a feed additive (Preston 1997). The idea was that less feed consumed by animals, more money farmers can save. Later, DES was produced as a pellet that was implanted between the cow's jaw and the ear, where it would slowly release the hormone over a maximum of 120 days. Once the 120 days had elapsed, a new pellet would be implanted. These pellets were seen as a great advancement in beef industry. Unfortunately, by the 1970s, it was discovered that continued use of DES could cause cancer in fetuses. The use of DES and growth hormone implants was quickly banned. Eventually, a new type of implant was developed but these were not so popular due to some earlier reported effects.

In recent years, there have major increases in additives and hormones in food manufacturing, specifically with poultry and dairy production. Hormones as additives are being added in poultry ration to help the animals gain more weight faster and fulfill consumer needs and desires. They help to reduce the waiting time and the amount of feed eaten by an animal before slaughter in meat industries. In dairy cows, hormones can be used to improve milk let down and production. Thus, hormones can increase the profitability of the meat and dairy industries. Recently, some developed countries banned the use of hormone implants due to their residual effect on humans.

Mechanism of action of hormones

Hormones are secreted in variable amounts according to need, and there is a constant turnover by inactivation and excretion of the hormones. They cause a trigger effect to modulate the activity of the target tissue. The effects of hormones are seen long after levels of hormone return to basal levels. Nervous signals can regulate hormone production but are short lasting and more immediate. Hormones are present in trace amount in plasma, usually ranging from 10^{-9} to 10^{-6} g/ ml (Squires, 2003). Anabolic steroids increase the retention of dietary nitrogen as body protein and increase muscle mass through hypertrophy rather than hyperplasia. Increased body protein can be the result of increased protein synthesis or decreased protein degradation. So, it can act by direct mechanism through receptors or indirectly by modulating the production of other hormones (Sheffield- Moore, 2000). As regards the anabolic androgens, evidence exist indicating competition with glucocorticoids for receptor sites on the muscle cell membrane since glucocorticoids have a catabolic effect on and when combined with oestradiol-17β, causes a marked decrease in the concentration of total thyroxin in plasma of steers indicates that the external use of hormones for long time may affect homeostasis in body.

Role of hormones in animal production

A) Effect of hormones on carcass quality

Anabolic agents are used as carcass modifiers which enhance nitrogen retention in the body and particularly in the muscle, by way of significantly decreasing blood and urinary urea and urinary nitrogen (i.e., controlling partition of nutrients absorbed) and results in the production of leaner carcasses (Sejrsen *et al.*, 1995). Male animals are castrated to reduce the undesirable aggressive behavior, to produce a more consistent carcass with improved tenderness and finish, to prevent objectionable odour and flavour in meat whereas in females tend to produce carcasses with less muscle mass and increased fat content. For these reasons, animals have been treated with a variety of steroid hormones to improve lean growth or to produce more consistent, high quality carcasses. Androgens are mainly used in females to improve muscle growth and decrease carcass fat while in castrated males oestrogenic compounds are used in males to increase carcass fat and meat tenderness and to decrease aggressive behavior (Squires, 2003). Combination of oestrogens and androgens may result in higher average daily gain than when a single agent used. The mode of action of androgens and oestrogens in increasing nitrogen retention and average daily gain is different (Dikeman, 2007). Anabolic implants may improve feed efficiency, growth and meat quality (Gebre *et al.*, 2012). Steroid hormones currently used as anabolic agents include testosterone, oestradiol and progesterone as well as steroid analogues such as zeranol and synthetic steroids such as trenbolone acetate (TBA), melengestrol acetate (MGA), combination of Trenbolone acetate (TBA) with Oestradiol, Zeralenone + TBA etc.

Anabolic steroids

Natural estrogens are secreted by the ovaries of all vertebrates, and the principal active compound is oestradiol-17β. In actual practice, estradiol is generally not applied for growth promotion on its own, but it gives good results in combination with other compounds, such as androgens and gestagens. Natural estrogens exhibit a strong anabolic effect, depending on the dose. At lower doses they are anabolic, whereas at higher doses they become lipogenic and catabolic. The effect is also influenced by the species e.g., in cattle and sheep, they are generally anabolic. In veal male calves (90 to 100 kg), implantation of estradiol increased nitrogen retention, weight gain, and food conversion efficiency.

Zeralenone

It is non-steroido estrogenic compound derived from *Fusarium* mould toxin produced by *Gibberella zeaezeralenone* (fermentation estrogenic substance, FES). Zeralenone implants (steers and lambs) increase apparent absorption and retention of Ca, P, Mg and Zn. It is marketed as zeranol. This has the role in natural testosterone and collagen synthesis and function in body.

Trenbolone acetate (TBA)

It is a synthetic androgen, which has 10-50 times the anabolic properties of testosterone with reduced secondary androgenic effects. TBA enhances the rate of protein synthesis by interfering the catabolic action of glucocorticoids on muscle protein.

Melengestrol acetate (MGA)

It is an orally active progestagen used for suppressing oestrus and improving the rate of gain and feed efficiency in female animals.

Oxandrolone

It is a synthetic analogue of testosterone which has been used in human studies.

Somatotropin

Growth hormone (GH) or Somatotropin (ST) is a single polypeptide chain consisting of 191 amino acids, varying considerably between species. GH increases weight gain and stimulates protein accretion concurrent with a reduction in fat deposition (Courtheyn *et al.*, 2002). Recombinant porcine somatotropin (rpST) has been approved for use in swine production in several countries. GH cannot be administered orally, but it is delivered as an injection one to three times per week. Average daily gain is increased by 20% with 150 µg rpST/ kg body weight/ day and feed conversion efficiency was improved (Beermann, 1994). Carcass protein accretion rates are increased up to 74% concurrent with an 82% decrease in lipid accretion rate when rpST is administered from 30 to 90 kg body weight. On the other hand, dressing percentage is not increased with rpST administration, primarily because of increased organ weights. Dikeman (2003) concluded that the large majority of studies show that shear force is increased and sensory panel tenderness is reduced for chops from pigs treated with rpST. Beermann *et al.* (1990) found that ewe lambs exhibited greater reductions in fat accretion and greater responses in growth rate than withers when rpST was administered over an 8-week period prior to slaughter. There are number of anabolic steroids used in different trade names listed below (Table 1).

Table 1.Steroid anabolic agents used as implants

Trade Name	Chemical Name
Compudose	Oestradiol 17β
Ralgro	Zeralenone
Synovex-S	Oestradiol 17β benzoate + progesterone
Synovex-H	Oestradiol 17β benzoate + testosterone propionate
Finaplix	Trenbolone acetate
Revalor	Oestradiol 17β + trenbolone acetate
Forplix	Zeranol + trenbolone acetate

(Unruh, 1986)

B. Growth and Feed Efficiency

In 1956, FDA first approved estrogenic and androgenic growth promoting agents in beef production in the United States. In order to compete in the domestic and export markets, beef producers must be able to keep the costs of running their operation as low as possible. Implanting hormonal growth promoters is currently widespread in the beef cattle industry of many non-EU countries for the better performance in growth and improvement of feed efficiency. These hormonal implants may enhance growth during suckling, growing and finishing stages of production (Hafs *et al.*, 1971). Growth hormones are implanted under the skin (usually behind the ear) of the animal in the form of depot capsules, where they release a specific dose of hormones over a fixed period of time. These hormones can act in two ways:

1. Growth hormones act as replacements for substances naturally produced by the animal that are deficient or no longer present. For example, male cattle that have been castrated (steers) do not produce enough testosterone, an essential component for animal's growth and development. Cattle are castrated to help control dangerous behaviour and to prevent random breeding. By supplementing the animal with testosterone, it is allowed to grow at a more natural rate using a substance that would otherwise be deficient.
2. Growth hormones act as supplements for substances that are naturally produced by the animal. For example, cattle need hormones like estrogen for growth and development. Supplementing these hormones, or adding to what is already present, allows the animal to grow and develop more quickly. The hormones approved for use in beef production are estradiol (estrogen), progesterone, testosterone, and their synthetic alternatives zeranol, melengestrol acetate, and trenbolone acetate.

Diethylstilbesterol (DES)

In 1956 DES implants approved for cattle and sheep by FDA. It is a synthetic estrogen with anabolic activity which increase rate of gain and feed efficiency. Increased GH secretion has been proposed as the mode of action through which DES improves the rate of gain and feed conversion of ruminants (Preston, 1999).It has been reported that traces of oestrogenic activity remained in meat of cockrels (caponettes) animals implanted with DES. These traces are harmful. Hence DES has been banned since 1979. This was based on its known carcinogenicity as well as lack of an acceptable analytical method for the compound in meat. Hexesterol (a synthetic oestrogen) and melengesterol (a synthetic progesterone) have also been tried in animal production.

Zeranol (Ralgro)

Zeranol was implanted (12 mg pellet) subcutaneously on the backside of the ear. It stimulates pituitary gland to secrete increased amounts of somatotropin growth hormone. It is approved for growth promotion in cattle in an implantable form.

Trenbolone acetate (TBA)

It is a synthetic androgen having strong anabolic and very effective growth promoter especially in ruminants. Implanting steers with TBA and oestradiol-17 β made their growth rate comparable to that of bulls but, rather surprisingly, their carcass composition was still essentially that of steer.

β-adrenergic agonists *(*β-agonists*)*

These are structural analogues of the naturally occurring catecholamines, i.e., adrenaline (epinephrine) and noradrenaline. These are orally active. These are used as metabolic modifiers to enhance the lean content and reduce the fat content of animals by partition of energy deposition for growth in form of protein and fat (Wenk, 2000). Thus, they increase feed efficiency and growth promotion. β agonists react with special cell receptors and increase concentration of cyclic AMP which in turn results in reduced lipogenesis and increased lipolysis as reported by Dunshea *et al.*, 2005, Mersmann, 1998. Ractopamine hydrochloride for pigs (Paylean®), cattle (Optalexx®) and zilpaterol hydrochloride (Zilmax®) for cattle are the only approved β-agonists. Three analogs of norepinephrine, namely, clenbuterol, cimaterol, and ractopamine, have been found to increase muscle mass and decrease adipose tissue mass when fed to livestock (Anderson *et al.,* 1987).

Clenbuterol

It encourages lipolysis and these fatty acids in farm animals are reported to be utilized for protein synthesis. In lambs clenbuterol (2 mg/kg) increased live weight gain and protein content of the hind quarter with a corresponding reduction in fat content. In cattle clenbuterol (10 mg/day) had no effect on growth rate but affected carcass composition in a similar way, as in sheep, It has an effect on immune function in ewe lambs. It may inhibit humoral antibody response to infection.

Cimaterol

It markedly stimulates skeletal muscle hypertrophy in mammals. It did not affect dry matter intake (DMI), but improved feed efficiency (FE) and average daily gain (ADG).

Ractopamine

It was recognized as the first β- adrenergic agonist to be approved for use in meat animals as a production enhancer by FDA in 2000, under Elanco's trade name of 'Paylean'. In 2003 it was approved for use in finishing cattle, under the trade name of 'Optaflexx'. It has the action of redirecting nutrients, favoring protein synthesis at the expense of fat deposition in the carcass (Avendano-Reyes *et al.*, 2006). This is approved by USDA/FDA for use to increase weight gain, feed efficiency and carcass leanness in finishing pigs from 68 kg to 108 kg. It requires no withdrawal period. Formerly a repartitioning agent, now it is termed as 'leanness enhancer'. Improved feed conversion was reported by Garbossa *et al.*, 2013 at doses up to 20 ppm of ractopamine since it stimulates the so-called beta-receptors in the body responsible for repartitioning nutrients into lean and fat tissue. It repartitions nutrients away from the normal accumulation of fat before marketing and towards more muscle.

C. Milk Production

Hormones basically regulate the partitioning of nutrients between muscle and fat. Steroid hormones are generally involved and are orally active not species specific. Peptide or protein hormones *viz.* somatotropin is a polypeptide and therefore not orally active and it is species specific. Hence the little, if any, somatotropin present in animal products get degraded in the gastrointestinal tract and thus has no influence on human health.

Bovine Growth Hormone (BGH)

The hormone is secreted by the anterior pituitary and is called the growth hormone (GH).GH basically takes part in intermediary metabolism for growth promotion process either directly at tissue level or through the release of insulin like growth factor-I (IGF-I) (Breir and Gluckman, 1991). In 1954, at the International Symposium on the Hypophyseal Growth Hormone, it was proposed that the hormone be designated as somatotropin. Somatotropin and GH continue are used interchangeably. The secretion of this hormone is enhanced following feeding of protein. Growth hormone is a more powerful lactogenic stimulant than anabolic agent. There are some of the attributes of somatotropin

- Somatotropin stimulates various biosynthetic processes. It influences the biosynthesis of protein in muscle cells directly and indirectly by supplying amino acids to the sites of biosynthesis as it facilitates the transportation of amino acids and their absorption into the muscle cells.
- The activity of somatotropin is manifested in increased nitrogen retention in the body.
- It increases the growth of skeletal tissues as well.

- Growth hormone is homothetic because it manifests its actions by chronic influence on metabolism and involves partitioning of nutrients for selected processes such as growth, milk production.

Biotechnologically derived somatotropin

Includes recombinant somatotropin (rST) and its effective use in both dairy cattle as bST and in growing pigs as pST has provoked several studies in which the mechanism of action of the hormone has been investigated. Effects include improved maintenance of secretory cells. The net result is that bST supplemented cows have higher daily milk yield and produce higher quantity of milk throughout the lactation cycle. In general, bST improves the availability of glucose to tissues and mammary glands and increases the use of stored fat for energy if nutrient supply is inadequate, thereby improving the productive efficiency of animals. rbST administration to dairy animals increases the daily milk yield by 1-2 kg and significant positive effect on persistency of lactation.

Exogenous bST exerts positive energy balance in cows during the first few weeks after beginning the administration with gradual increase feed intake to obtain the required nutrients for body weight gain and increased milk production. Long term studies with high yielding cows treated with daily subcutaneous administration of bST (12-50 mg) prepared from the pituitary gland of cows or by rDNA technology gave increases in the milk yield in the range of 22 to 41% (Bauman, 1992). However, at the higher levels of bST administration certain difficulties may be encountered in subsequent reproductive performance. The increases in the milk yield may be associated with increase in feed intake. The availability of a slow-release device of bST has eliminated the necessity for daily injection.

D. Role in Reproduction

Exogenous hormonal combinations can also be used for various reproductive performances in large animals. Although genetic, nutritional and environmental variables plays important role puberty and could be induced in heifers by the application of hormonal regimens such as progesterone and oestradiol-17β (Gonzalez-Padilla *et al.*, 1975), oestrogens and progesterones (Peter, 1979), gonadotropins (Glencross, 1980) and GnRH (McLeod *et al.*, 1984). Oestrus synchronization could be achieved in cyclic animals with prostaglandin ($PGF_2\alpha$ or its analogues) or progesterone implants such as norgestomet (Seguin *et al.*, 1989). Oestrus synchronization was shown to be a good technique for advancing and/or shortening the breeding season in beef cattle. Oestrus synchronization could be achieved in cyclic animals with prostaglandin ($PGF_2\alpha$ or its analogues) or progesterone implants such as norgestomet (Seguin *et al.* 1989). Norgestomet and estradiol valerate were effectively utilized to synchronize oestrus and to control ovulation in cattle (Kazmer *et al.* 1981). Norgestamet and pregnant mare serum gonadotropin (PMSG) treatment was effective in induction of estrus and advancing the age of first conception in prepubertal heifers for augmenting production under field conditions (Sarmah *et al*., 2014). β-agonists like clenbuterol causes relaxation of uterine smooth muscle, and this has led to itsapplication in controlling parturition and managing dystocia and to attempts to stimulated contractions of the uterus during nonsurgical embryo transfer (Walton *et al*., 1986). In gilts treated with lower doses of porcine somatotropin (pST) during the finishing phase, during the peri- puberal period, and during early pregnancy reproduced normally. In ruminants, enhanced growth due to bovine somatotropin (bST) or growth hormone releasing factor (GRF) leads to a larger skeletal size and a larger pelvic opening, a potential benefit at parturition (Day and Britt, 1991).

Metabolism, routes and rates of elimination

The general patterns of metabolism and elimination of endogenous hormones in farm animals have been outlined. The rate of absorption of the agent is determined by formulation and site of administration and is best achieved by the use of slow release implants (Heitzman, 1983).The lowest concentrations of residues are found in muscle and fat, higher concentrations are present in liver and kidney and the highest concentrations are in the bile, urine and feces. The metabolism of some of the anabolic agent occurs in the blood, liver and kidney. In ruminants, testosterone and oestradiol-17β are rapidly converted to their epimers, biologically much less active, epitestosterone and oestradiol-17β. Progesterone is partially converted to androgens before excretion (Rico, 1983). In the pig, epimerization of testosterone and oestradiol-17β does not appear to take place to a significant degree. The faecal route of elimination dominates in ruminants, while in the pig urinary excretion is more prevalent.

Economic implications of the use of hormones in animal production

In the production of meat for human consumption, a hormonally-induced increase in growth rate of the order of 10% evidently has major economic implications. The improvement in feed conversion efficiency (FCE) which usually accompanies the increase in gain adds to the economic benefits, and at the same time makes possible greater production of edible protein per unit energy used, which is of paramount importance today to the world community. Some of the hormones that have become available to improve the performance of livestock, should be approved for use in animal production but the risk/ benefit analysis must take this fact into account.

Impact on Human and Animal Health

The health effects were extensively studied before rbST was approved by the FDA. Extensive trials extending over several lactations have not revealed any adverse effects attributable to somatotropin administration. Indeed the cows previously subjected to long-term treatment with somatotropin are less likely to develop metabolic disorders associated with lipid mobilization after calving. However, recent reports showed fertility problems and impairment of reproduction during post calving period in multiparous cows. The level of bST in milk is only a small fraction of the blood concentration i.e., after injecting a very high dose of bST, very small but significant increase in milk concentration has been achieved. Pasteurization of milk destroys 85-90% of immune reactive bST. Also levels of somatotropin in milk produced by treated cows took exceedingly low and constitute no threat to the human consumer (Schams, 1989). Intact proteins which survive protease activity in the stomach and small intestine are not normally absorbed. Even if small amount of somatotropinis absorbed, species specificity would ensure that the absorbed hormone would not influence growth. The availability of slow-release preparation of somatotropin has eliminated the need for daily injections. In addition to bST, there are number of somatotropins available to improve productivity in livestock without affecting the health of the animals and consumers.

Use of Hormones in Animal Production and Food Safety

Several hormone implants *viz,*oestradiol, zeranol, trenboloneacetate and testosterone are used for economic and improved animal production. Use of hormones in animal production has drawn much attention over the years, particularly in the European Union (EU) where anabolic steroids and growth hormone use has long been banned and it has also banned meat imported from countries, such as the United States (US), where growth hormone is used to achieve improved efficiency and leaner carcasses. Many consumers are concerned over the long- range effects of the hormones fed to animals that might be residual in meat, eggs, milk, or other products. However, there is no concrete report of any human health problem from the use of any natural or synthetic hormones fed to livestock, yet to be tested under Indian context.

References

Anderson, D. B., Vecnhuizm, E.L.,Waitt,W.P., Ston, R.E. and Mowrey, D.H. 1987. Effect of ractopamine on nitrogen retention, growth performance and carcass Composition of pigs. *J. Anim. Sci.* 65(1): 130.

Avendano-Reyes L, Torres-Rodriguez V, Meraz-Murill F, Perez-Linares Billy, N. D. and Jack, H. B. 2006. Impact of metabolism modifiers on reproductive function. *J. Anim. Sci.*, 69:76-87.

Bauman, D. E. 1992. Bovine Somatotropin: Review of An Emerging Animal Technology. *J. Dairy Sci.* 75(12):3432-3451.

Beermann, D. H. 1994. Carcass composition of animals given partitioning agents. In: H. D. Hafs & R. G. Zimbleman (Eds.), Low Fat Meats. San Diego: Academic Press, Inc. pp. 203–222.

Beermann, D. H., Hogue, D. E., Fishell, V. K., Aronica, S., Dickson, H. W., and Schricker, B. R. 1990. Exogenous human growth hormone releasing factor and ovine somatotropin improve growth performance and composition of gain in lambs. *J. Anim. Sci.*68: 4122.

Breir, B.H. and Gluckman, P.D. 1991. The regulation of post-natal growth: Nutritional influences on endocrine pathways and function of the somatotrophic axis. *Livestock Prod. Sci.* 27: 77-94.

Courtheyn, D., Le, B. B., Brambilla, G., Brabander, H.F., Cobbaert, E., Wiele, M.V.,Vercammen, J., and Wasch, K. 2002. Recent developments in the use and abuse of growth promoters.*Analytica Chimica Acta*. 473: 71–82.

Day, B.N. and Britt, J.H. 1991. Impact of metabolism modifiers on reproductive function. *J. Anim. Sci.* 69:76-87.

Dikeman, M. E. 2003. Metabolic modifiers and genetics: Effects on carcass traits and meat quality. *Brazilian J. Food Tech.* 6: 1–38.

Dikeman, M. E. 2007. Effects of metabolic modifiers on carcass traits and meat quality. *Meat Science*. 77: 121–135.

Dunshea, R., D'Souza, D., Pethic, D., Harper, G., and Warner, R. 2005. Effects of dietary factors and other metabolic modifiers on quality and nutritional value of meat. *Meat Sci.*, 71: 8–38.

Garbossa, C.A.P., D'Sousa, R.V., Cantarelli,V.S., Pimenta, M.E., Zangeronimo, M.G.,Silveira, H., Kuribayashi, T.H.,Cerqueira, L.G. 2013. Ractopamine levels on performance, carcass characteristics and quality of pig meat. *R. Bras. Zootec.,* 42(5): 325-333.

Gebre, B.A., Gebretsadik, Z.T. and Anal, A.K. 2012. Effect of metabolic modifiers on meat quantity and quality. *African J. Food Sci.* 6(11): 294-301.

Glencross, R. G. 1980. Attempts to induce normal ovarian function in prepubertal heifers. *Anim. Breeding Abstract* 48: 442.

Gonzalez-Padilla E, Niswender, G. D. and Wiltbank, J. N. 1975. Puberty in beef heifers. II. Effect of injections of progesterone and estradiol-17 ß on serum LH, FSH and ovarian activity. *J. Anim. Sc.* 40: 1105–09.

Hafs, H.D., Purchas, R.W. and Pearson, A.M. 1971. A Review: relationships of some hormones to growth and carcass quality of ruminants. *J. of Anim. Sci.* 33(1): 34-42

Heitzman, R. J. 1983. The Absorption, Distribution and Excretion of Anabolic Agents.*J. Anim. Sci.,* 57:233-238.

Hutjens, M. F. 1991. Feed additives. Vet Clinics North America. *Food Anim.Prac.* 7(2): 525.

Kazmer, G. W., Barnes, M. A. and Halman, R. D. 1981. Endogenous hormone response and fertility in dairy heifers treated with norgestomet and estradiol valerate. *J. Anim. Sci.* 53: 1333–40.

McLeod, B. J., Haresing, W., Peter, A. R. and Lamming, G. E. 1984. Plasma LH and FSH concentration in prepubertal beef heifers before and in response to repeated injections of low doses of GnRH. *J. Repro. Fertility* 70: 137–44.

Mersmann, H.J. 1998. Overview of the effect of beta-adrenergic receptor agonists on animal growth including mechanisms of action. *J. Anim. Sci.*, 76: 160-172.

Norman, A.W. and Litwack, G. 1997. Hormones, 2ndedn. Academic Press, Toronto.

Peter, J B. 1979. Induction of puberty and subsequent reproductive performance. *Theriogenology.* 12: 215–21.

Preston, R.L. 1999. Hormone containing growth promoting implants in farmed livestock. *Adv. Drug Delivery Rev.* 38,123–138.

Raun, A. P. and Preston, R. L. 1997. History of hormonal modifier use. In: symposium 'Impact of implants on performance and carcass value of beef cattle. *Oklohama Expt. Station.* P-957: 1-9.

Rico, A.G. 1983. Metabolism of Endogenous and Exogenous Anabolic Agents in Cattle. *J. Anim. Sci.* 57:226-232.

Sarmah, B.K., Biswas, R.K., and Dutta, D.J. 2014. Effect of norgestomet on induction of oestrus in cattle: an overview, *Indian J. Anim. Sci.* 84 (3): 223–230.

Schams, D. 1989. Use of Somatotropin in livestock production., Elsevier Applied Sci. N.Y.

Seguin, B. E., Momomt, H. W., Fahmi, H., Fortin, M. and Tibary, A. 1989. Single appointment insemination for heifers after prostaglandin or progestin synchronization of estrus. *Therio.*31: 1233–38.

Sejrsen, K., Oksbjerg, N., Vetsergaard, M., Sorensen, M.T. 1995. Growth hormone and related peptides as growth promoters. *In: Proceedings of the Scientific Conference on Growth Promotion in Meat Production*, Brussels, Belgium, 29 Nov.–1 Dec., pp. 87–119.

Sheffield- Moore, M. 2000. Androgens and the control of skeletal muscle protein synthesis. *Annals of Medicine*, 32: 181-186.

Squires, E.J. 2003. Applied animal endocrinology, CABI Publishing, Cambridge, MA, USA.

Stoller, G.M., Zerby, H.N., Moeller, S.J., Baas, T.J., Johnson, C. and Watkins, L.E. 2003. The effect of feeding ractopamine (Paylean) on muscle quality and sensory characteristics in three diverse genetic lines of swine. *J. Anim. Sci.* 81: 1508-1516.

Unruh, J.A. 1986. Effects of endogenous and exogenous growth promoting compounds on carcass composition, meat quality and meat nutritional value. *J. Anim. Sci.* 62: 1441-1448.

Wenk, C. 2000. Recent advances in animal feed additives such as metabolic modifiers, antimicrobial agents, probiotics, enzymes and highly available minerals. *Asian-Aust. J. Anim. Sci.* 13 (1): 86-95.

21

Implications on the use of Beta Agonists in Animal Feeding

Nirbhay Kumar and Rashmi Ranjan

Introduction

Growth enhancement technologies have been widely embraced in the livestock industry to improve growth, efficiency and carcass traits. For over 50 years, estrogenic and androgenic hormone implants were widely used in the cattle feeding industry, and it is estimated that approximately 97 percent of feedlot cattle in the U.S. receive one or more implants during the finishing phase (Barham *et al.*, 2003; Tatum, 2006). There is a large body of literature that indicates the positive growth enhancement effects which implants impart on growing and finishing cattle and producers have extensively used implants to increase live body weight, improve average daily gain and feed efficiency, and reduce the number of days cattle are on feed (Apple *et al.*, 1991; Duckett *et al.*, 1997; Milton and Horton, 1996; Perry *et al.*, 1991). Likewise, there is a growing body of literature that indicates the positive response on growth and carcass characteristics that finishing cattle have to beta-adrenergic agonists. The hog industry, and more recently, the cattle feeding industry have increased utilization of beta-adrenergic agonists to improve the efficiency of production. While the use of Ractopamine hydrochloride (Paylean®, Elanco Animal Health, Greenfield, Indiana) has been used extensively in hog production since its approval by the Food and Drug Administration (FDA) in 1999, similar pharmaceutical feed additives have only recently been implemented in commercial cattle feeding operations. Beta-adrenergic agonist use in fed cattle has substantially increased following the FDA approval of Ractopamine hydrochloride (Optaflexx®, Elanco Animal Health, Greenfield, Indiana) in 2003, and the approval of Zilpaterol hydrochloride (Zilmax®, Merck Animal Health, Summit, New Jersey) in 2006. The use of both Ractopamine hydrochloride (RH) and Zilpaterol hydrochloride (ZH) have shown to elicit similar improvements in live weight gain, average daily gain, and feed efficiency as those reported in hormone implants (Allen *et al.*, 2009; Avendano-Reyes *et al.*, 2006; Beckett *et al.*, 2009; Elam *et al.*, 2009; Gruber *et al.*, 2007; Kellermeier *et al.*, 2009; Scramlin *et al.*, 2010). In addition, these compounds have been reported to have profound effects on hot carcass weight, Longissimus muscle area and carcass cutability (Gruber *et al.*, 2007; Hilton *et al.*, 2010; Rathmann *et al.*, 2009; Scramlin *et al.*, 2010; Shook *et al.*, 2009; Vogel *et al.*, 2009). It is because of this that cattle feeders have drastically increased use of beta-agonists since their FDA approval, and a growing portion of the fed cattle population receive a dietary beta-agonist supplement that is provided within the final 20 to 30 days of finishing.

While the efficacy of these compounds is well documented, so too are the negative effects on meat quality. The use of both RH and ZH has corresponded to increased toughness in meat from cattle given a dietary supplement of either compound (Avendano-Reyes *et al.*, 2006; Garmyn *et al.*, 2010; Gruber *et al.*, 2008; Leheska *et al.*, 2009; Woerner *et al.*, 2011). Likewise, beta-agonists have been reported to decrease marbling scores, while also negatively affecting the percentage of cattle qualifying for quality-based premiums (Kellermeier *et al.*, 2009; Vasconcelos *et al.*, 2008). It is because of these factors that the use of beta-agonists has come under recent scrutiny. The positive effects of dietary beta-agonist supplementation has allowed cattle feeders to improve efficiency and packers to improve yields; however, packers also face the issue of lower quality carcasses and retailers are required to market products that could provide consumers with a less desirable eating experience. Researchers continue to analyze the effects of beta-agonist supplementation at different doses, potencies, and feeding periods, as well as differential effects on cattle of varying biological type. Determining the optimal dose, as well as the preferential cattle type for beta-agonist supplementation, will aid in optimizing the positive effects on growth while mitigating the negative effects on quality across the population at-large.

Mechanism of Action

Beta agonists act directly through β-adrenergic receptors on skeletal muscle and adipose cell membranes and generate signals that control metabolic activities in the cells. When Ractopamine or other β-agonists bind to the β-adrenergic receptors on fat cells, biochemical signals are initiated, activating several enzymes in the pathways that lead to decreased rates of lipogenesis (lipid synthesis and storage) and to increased lipolysis (lipid mobilization in the cell) (Dunshea, 1993; Mersmann, 1989; Mills and Liu, 1990). The rate of fat accumulation or growth in the animal slows, resulting in a leaner animal. The magnitude of these changes is influenced by the dose (amount) and the length of time the β-agonist is consumed, the type of β-agonist, and the target species (Beermann 1993; Mersmann, 1998; Moody *et al.*, 2002).

Skeletal muscle cells also contain β-adrenergic receptors. Interaction of a β-agonist with the receptor stimulates similar signaling pathways as in fat cells, altering muscle metabolism in a dose-dependent manner (Byrem *et al.*, 1996). Direct infusion of the β-agonist cimaterol, a β-agonist that has not been approved as a metabolic modifier, into the hind limb of growing steers increases the rate of amino acid extraction from the blood and results in increased rates of muscle protein synthesis and muscle growth (Byrem *et al.*, 1998), independent of any systemic endocrine changes. The muscle growth enhancement results from hypertrophy (an increase in cell size) without any increase in cell number. The total number of muscle fibers in a muscle generally is set at birth in most domestic animal species. The changes that occur in skeletal muscle and adipose tissue are progressive over short periods of time, but they are not sustained over long periods because desensitization of receptors on target tissues occurs. For example, there is a marked down-regulation in adipose tissue of swine within 4 days after the commencement of feeding of Paylean® (Dunshea and King, 1995). Therefore, the recommended time of feeding is near the end of the finishing period. Longer feeding time has little or no effect on muscle or adipose tissue growth and would result in markedly decreased economic benefit. Less energy per weight is required to grow muscle than to grow adipose tissue. Use of feed for growth in animals fed β-agonists is more efficient overall. The β-agonists stimulate muscle growth and decrease the rate of nutrient use for adipose tissue growth, resulting in less feed required to produce an animal of the same weight. Less animal waste is produced, decreasing environmental impact when β-agonists are fed to meat-producing animals.

Beta-adrenergic agonists

Beta-adrenergic agonists (βAAs) are a multifaceted pharmacological agent utilized in both human medicine and livestock production. In human medicine, βAA are used as a bronchodilator to treat asthma, and to stimulate cardiac contraction strength and rate (Hossner, 2005). Alternatively, βAA are used in livestock to stimulate skeletal muscle growth, increase live and carcass weight gains and improve efficiency in cattle and swine. Evaluation of the use of βAA started to build momentum in the 1970's and 80's with the use of clenbuterol, cimaterol, L664, 969, salbutamol and fenterol for use in livestock (Anderson *et al.*, 2005; Mersmann, 1998). However, these compounds were never given FDA approval for commercial use in food producing animals. Clenbuterol is associated with adverse human health effects when tissues from those animals were consumed or inhalation of the compound during feed mixing. Several public health issues were associated with clenbuterol residues through the 1990's; hence, the use of the substance and associated βAA's were made illegal in the livestock industry. RH and ZH are the only βAA approved for use in livestock in the United States. Ractopamine (Paylean®; Elanco Animal Health, Greenfield, Indiana) was approved for use in swine in 1999, and later for cattle (Optaflexx®, Elanco Animal Health, Greenfield, Indiana) in 2003. Zilpaterol (Zilmax®; Merk Corp., Summit, New Jersey) was approved for use in cattle for many years in South Africa and Mexico, and later received FDA approval for use in cattle in the United States in 2006.

The catecholamines epinephrine and norepinephrine serve as biological βAA's in mammalian species, which bind to β-adrenergic receptors (βAR) to elicit a response on the sympathetic nervous system. These are of primary interest to human medicine, as they relate to bronchodilation and effect cardiovascular function (Mersmann, 1998). These hormones are inactivated and reabsorbed by specific uptake mechanisms to prevent the βAR from remaining active (Mersmann, 1998). Beta-adrenergic agonists have the ability to bind to any of three βAR (β_1-AR, β_2-AR, or β_3-AR) which are present in most mammalian cells, but are present in different concentration depending upon the tissue and species (Johnson, 2004; Mersmann, 1998). In skeletal muscle and lipid cells, β_1-AR and β_2-AR are the most common receptors, and have the greatest affinity for pharmacological βAA (Mersmann, 1998). These receptors are members of the Gs protein coupled receptors. The βAR contain a seven membrane-spanning domain that forms loops on the cell membrane and are exposed on both the intra- and extracellular surface (Johnson, 2004). The physiological response to βAA is initiated when the βAA binds on the extracellular surface to the βAR, causing a conformational change to the receptor. This activates adenylate cyclase to synthesize cyclic-adenosine monophosphate (cAMP) from adenosine triphosphate (ATP). Cyclic AMP regulates the activity of protein kinase A (PKA), which is responsible for the phosphorylation of necessary enzymes involved in lipid and protein synthesis, as well as regulation of DNA transcription factors (Anderson *et al.*, 2005; Johnson, 2004). Furthermore, PKA targets intracellular domains on the βAR to render it inactive.

In more general terms, βAA and the intracellular signaling cascade increases protein synthesis and decreases fat synthesis in livestock. It is because of this that βAA's are referred to as *"repartitioning agents"* due to repartitioning normal metabolic processes from synthesis of fat to synthesis of muscle. Beta-adrenergic receptors present on fat cells decrease lipid synthesis and increase lipid degradation; likewise, βAR present on muscle cells elicit a response which increases protein synthesis and a decrease in protein degradation (Anderson *et al.*, 2005). When fed to ruminant livestock, it is suggested that the compound leaves the rumen intact, and is absorbed in the lower parts of the gastrointestinal tract; after which it enters the blood stream, is degraded in the liver, and is then delivered to the target tissues to elicit a response (Johnson, 2004).

Effects of Beta-Adrenergic Agonist

It is widely understood that βAA supplementation improves live animal performance and efficiency, carcass weight and yield, and has negative effects on muscle tenderness and eating quality. Efficacy varies depending upon βAA-type, animal age, cattle biological type, and if it is used in combination with other growth promoting strategies. The following section will highlight research focusing on varying βAA's used in cattle production and its effects on production traits and carcass characteristics.

Live Animal Performance

There are a multitude of recent studies investigating RH and ZH effects on cattle growth. Gruber *et al.* (2007) reported RH supplementation at 200 mg/head/day increased final body weight (BW), average daily gain (ADG), and gain to feed ratio (G:F ratio) of Brahman, Continental, or British influenced cattle. Gruber *et al.* (2007), however, did not report a RH × breed effect, indicating that RH has no differential effect on cattle of differing biological types. Likewise, Abney *et al.* (2007) reported that RH supplementation improved final BW, ADG, and G:F ratio. Both Gruber *et al.* (2007) and Abney *et al.* (2007) indicated RH supplementation improved final BW by 7.3 and 5.9 kg, respectively; ADG was improved by approximately 2 kg/day. Abney *et al.* (2007) also reported a RH feeding duration effect when fed for 28, 35, and 42 days. Gains from RH supplementation were optimized at 35 day of feeding for final BW, ADG, and G:F ratio compared to shorter or longer feeding periods (Abney *et al.*, 2007). Vogel *et al.* (2009) reported similar results feeding RH to calf-fed Holstein steers.

Comparatively, Quinn *et al.* (2008) reported minimal effects of treatment on live animal performance in feedlot heifers supplemented RH at 200 mg/head/day. Allen *et al.* (2009) reported results similar to Quinn *et al.* (2008) that live performance was not affected when feeding market dairy cows RH at 312 mg/head/day. These results were attributed to the inherent variability associated with feeding market cows versus feedlot cattle that are more consistent in their biological type and age (Allen *et al.*, 2009). The majority of experiments assessing the effects of Zilpaterol on live animal performance have been conducted in the last five years due to the recent FDA approval for use in meat producing livestock. Early research conducted in Mexico indicated advantages for growth and performance feeding ZH in comparison to cattle receiving none (Avendano-Reyes *et al.*, 2006; Plascencia *et al.*, 1999). A growing body of literature from the United States indicates ZH supplemented feedlot cattle possess similar growth and performance characteristics to those fed RH. A study published by Vasconcelos *et al.* (2008) reported minimal differences in final BW of steers supplemented ZH for the final 20, 30 and 40 days of finishing compared to steers fed no β-agonist. The same study also reported that steers supplemented with ZH had significantly higher ADG and G:F ratio than steers fed no beta-agonist (Vasconcelos *et al.*, 2008). In a similar study that evaluated ZH supplementation and duration of feeding, Elam *et al.* (2009) indicated that ZH supplementation increased final BW and improved ADG and G:F ratio versus non-supplemented steers. Elam *et al.* (2009) reported an 8, 9.3, and 10.8 kg increase in final BW when ZH is supplemented for the final 20, 30 and 40 d of the finishing period, respectively. In addition, other studies reported improvements in ADG and G:F ratio due to dietary ZH supplementation (Beckett *et al.*, 2009; Montgomery *et al.*, 2009; Scramlin *et al.*, 2010).

Studies comparing efficacy of RH and ZH have resulted in varied results with respect to live animal growth and performance. Avendano-Reyes *et al.* (2006) reported that ZH supplementation improved final BW, ADG and G:F ratio of beef steers versus those supplemented RH. Conversely, Scramlin *et al.* (2010) reported an advantage of 4.35 kg of final BW for RH supplemented beef steers over ZH supplemented steers, as well as increases in ADG and G:F ratio when supplemented with RH. Strydom *et al.* (2009) indicated that RH supplementation imparted significant increases in ADG and G:F in beef steers versus those supplemented ZH, while also reporting an numerical advantage in final BW, albeit not statistically significant.

Carcass Characteristics

Research investigating effects of βAA focused primarily on the compounds clenbuterol and cimaterol. Ricks *et al.* (1984) reported that cattle supplemented with clenbuterol had lower 12th-rib fat thickness (FT), larger longissimus muscle area (LMA), and lower numerical yield grade (YG). The study also reported changes in overall carcass composition due to clenbuterol supplementation, indicating an increase in percent protein and a decrease in percent fat due to treatment (Ricks *et al.*, 1984). More contemporary studies that evaluated effects of RH supplementation on carcass characteristics indicate improvements in carcass cutability and yield. Gruber *et al.* (2007) reported an increase in hot carcass weight (HCW) of 5.5 kg and larger LMA when steers were supplemented RH at 200 mg/head/day. The results showed no difference in dressing percent, 12th-rib FT, or USDA yield grades (YG) (Gruber *et al.*, 2007). However, RH did increase the percentage of YG2 and decrease the percentage of YG3 carcasses (Gruber *et al.*, 2007). Likewise, RH did not affect the distribution of carcasses in respective USDA YG categories (Gruber *et al.*, 2007). In a study analyzing effects of RH fed at 200 and 300 mg/head/day to calf-fed Holstein steers, a study by Vogel *et al.* (2009) generated results similar to those reported by Gruber *et al.* (2007). Holstein steers that were supplemented with RH at 200 and 300 mg/head/day had heavier HCW and larger LMA compared to control steers, while steers supplemented at 300 mg/head/day had leaner 12th-rib FT (Vogel *et al.*, 2009). In addition, steers supplemented 300 mg/head/day of RH had a lower calculated YG, while steers supplemented RH at 200 mg/head/day had lower marbling scores and a lower percentage of carcasses grading USDA Prime and Choice (Vogel *et al.*, 2009). Similar results were reported in studies comparing treatment with either RH or ZH (Avendano-Reyes *et al.*, 2006; Scramlin *et al.*, 2010). A more exhaustive body of literature indicates improvements in carcass traits and yield in swine supplemented with RH at varying doses and potencies (Armstrong *et al.*, 2004; Uttaro *et al.*, 1993; Watkins *et al.*, 1990).

Zilpaterol supplementation in finishing beef cattle has proven to be highly effective at improving carcass composition and yield. Vasconcelos *et al.* (2008) reported that ZH supplementation for the final 20, 30, or 40 days of finishing increased HCW, LMA, and decreased 12th-rib FT and USDA YG in comparison to cattle receiving no βAA supplementation. These results were consistent with other studies investigating the effects of ZH on carcass characteristics and were indicative of the normal response in ZH supplemented beef steers (Elam *et al.*, 2009; Kellermeier *et al.*, 2009; Rathmann *et al.*, 2009). A similar trend in improved HCW, LMA, and YG also were reported in calf-fed Holstein steers supplemented with ZH for differing lengths of time (Beckett *et al.*, 2009; Garmyn *et al.*, 2010).

An effect of ZH treatment of fed cattle that is not reported in RH literature was an improvement in dressing percent. Multiple studies reported a higher mean dressing percent in ZH fed cattle compared to controls, while also indicating a greater increase in HCW than in live weight (Beckett *et al.*, 2009; Elam *et al.*, 2009; Vasconcelos *et al.*, 2008). There is no literature which supports a likely hypothesis for the reason behind the differential partitioning of nutrients between fat and lean tissues. Vasconcelos *et al.* (2008) reported a linear increase in lean tissue accretion with increased days of ZH supplementation with little variation in carcass fat between feeding periods. Further research is necessary to determine reasons for the disparity in live and carcass weights, as well as to determine what causes increases in dressing percent recognized via ZH supplementation, but not by feeding RH. Carcass cutability and subprimal yields have been widely investigated in cattle supplemented with ZH, but there is currently no literature addressing differences in beef carcass yield due to RH supplementation. Kellermeier et al. (2009) indicated an increase in subprimal yield from nearly all fabricated carcass subprimals compared to controls. Carcass cutability studies analyzing both beef-type feedlot cattle and calf-fed Holsteins have reported increases in subprimal yield as well (Boler *et al.*, 2009; Garmyn *et al.*, 2010; Hilton *et al.*, 2009; Rathmann *et al.*, 2009). These results indicate the additional value added for packers due to larger subprimals-specifically higher value cuts from the rib and loin – due to Zilpaterol supplementation. The greatest effect of dietary ZH supplementation is recognized in cuts from the round (Shook *et al.*, 2009). This is largely attributed to an increased concentration of Type IIa, glycolytic fibers, which are more responsive to βAA supplementation (Miller *et al.*, 1988). Along with this, a study analyzing carcass composition of steers and heifers receiving dietary ZH supplementation reported steers and heifers had an increase in soft tissue protein percentage and weight, as well as an increased protein to bone ratio (Leheska *et al.*, 2009).

Supplementation of ZH also has been shown to not only negatively affect marbling score, but also to cause a shift in the distribution of USDA quality grades in feedlot cattle. Kellermeier *et al.* (2009) reported a decrease in marbling score between ZH supplemented steers and controls of approximately 40 degrees. More importantly though, when compared to controls, ZH supplementation decreased the frequency of carcasses grading Premium Choice by nearly 17 percent (20% vs. 3.3%); decreased the percentage of carcass grading Choice by approximately 13 percent (36.67% vs. 23.33%); and increased the frequency of carcasses grading Select by nearly 20 percent (Kellermeier *et al.*, 2009). A study conducted by Vasconcelos *et al.* (2008) reported results consistent with those of Kellermeier *et al.* (2009), to the extent that ZH supplementation for 20, 30, or 40 d decreased the frequency of carcasses grading Choice and Prime by 16.1, 18.4, and 22 percent, respectively; the frequency of carcasses grading Select increased by 8.2, 9.8, and 19.8 percent, respectively. Vasconcelos *et al.* (2008) also reported that mean marbling score decreased by 31.1, 46.0, and 54.4 degrees for each feeding period. These results indicate that while ZH supplementation decreases marbling scores by less than half of a total marbling score, the deleterious effects on quality can drastically effect the distribution of carcasses receiving a quality based premium.

Tenderness and Palatability

The largest detriment of βAA supplementation is likely to be the consequential reduction in beef tenderness. A growing body of literature suggests that supplementation of βAA causes a decrease in objective tenderness measurements and are consistently rated less desirable by trained and consumer sensory panelists.

Studies to assess the effects of RH supplementation on postmortem tenderness have shown mixed results. In a study in which RH was supplemented at 200 mg/head/day for the final 28 days of the finishing period, longissimus muscle (LM) steaks from supplemented steers were on average 0.38 and 1.4 kg tougher for Warner Bratzler Shear Force (WBSF) and Slice Shear Force (SSF), respectively (Gruber *et al.*, 2008). Furthermore, the study reported that 3, 7, 14, and 21 days of post mortem aging did not diminish the effect of RH supplementation (Gruber *et al.*, 2008). The difference in tenderness was further recognized by trained sensory panelists, who rated steaks from RH supplemented cattle lower for tenderness and juiciness attributes (Gruber *et al.*, 2008). Scramlin *et al.* (2010) reported that RH supplementation increased WBSF values at 3 and 7 d aging versus steaks from control carcasses; however, the disparity in tenderness was negated at the 14 and 21 d aging interval. These results agreed with Quinn *et al.* (2008) who reported no differences in WBSF values for LM steaks aged 14 days from heifers supplemented 200 mg/head/day of RH for the final 28 days of the finishing period. Boler *et al.* (2012) reported that steers supplemented with RH at 200 mg/head/day for the final 28 days of the finishing period produced LM steaks with WBSF values that differed from control steers only at 4 day postmortem; however, steers receiving RH at 300 mg/head/day for the final 28 days of the finishing period produced LM steaks that were tougher at 7, 14, and 21 days postmortem than both control and 200 mg/head/day supplemented steers. At 28 days postmortem aging, steaks from cattle in the control, 200, and 300 mg/head/day treatments did not differ (Boler *et al.*, 2012). While these results vary, it appeared that increased post mortem aging for steaks from RH supplemented cattle provides the opportunity to mitigate negative effects on tenderness.

Zilpaterol supplementation in feedlot cattle has been shown to have more pronounced effects on beef tenderness. Leheska *et al.* (2009) reported that carcasses of steers and heifers supplemented with ZH for the final 20 and 40 days of finishing generated LM steaks with greater WBSF values than controls after 28 days post mortem aging. Zilpaterol supplementation for the final 20 days of finishing accounted for a 0.72 and 0.84 kg increase in WBSF in steers and heifers, respectively (Leheska *et al.*, 2009). Likewise, trained sensory panelists rated steaks from carcasses of ZH supplemented steers significantly tougher than controls; while LM steaks from carcasses of ZH supplemented heifers tended to be rated tougher by trained panelists (Leheska *et al.*, 2009). Rathmann *et al.* (2009) reported similar increases in WBSF values for steers fed dietary ZH for the final 20, 30, and 40 days of finishing. Increased time of ZH supplementation increased WBSF at 7, 14, and 21 days post mortem aging, and WBSF values for all feeding and post mortem aging periods exceed those from control carcasses (Rathmann *et al.*, 2009). In the study conducted by Rathmann *et al.* (2009), a study by Miller *et al.* (2001) was cited for determination of tenderness thresholds. Miller *et al.* (2001) determined that at a WBSF of < 3.0 kg, 100 percent of consumers found New York Strip steaks to be acceptable for tenderness (Miller *et al.*, 2001). The study also determined at a WBSF value of 4.3 kg, 86 percent of consumers found steaks acceptable, and that 4.9 kg was a major point of distinction from consumers between tough and tender steaks (Miller *et al.*, 2001). In the study published by Rathmann *et al.* (2009), frequency distributions for ZH vs. control treatments at each of the WBSF values referenced by Miller *et al.* (2001) showed that LM steaks from control carcasses had a significantly greater percentage of steaks at 3.0 and 4.3 kg of WBSF. Similarly, steaks from each of the three ZH treatments (fed for 20, 30, and 40 days) had a significantly greater percentage of steaks at or above the 4.9 kg threshold for toughness (Rathmann *et al.*, 2009). Results from these studies are consistent with others that have evaluated tenderness from ZH supplemented cattle, and illustrate the profound effect dietary ZH has on postmortem tenderness and consumer acceptability (Avendano-Reyes *et al.*, 2006; Hilton *et al.*, 2009; Kellermeier *et al.*, 2009).

Physiological effects of Beef Tenderness

While it is widely recognized that βAA have deleterious effects on beef tenderness, there are various theories on the physiological modes of action which effect post mortem tenderness.

A generally accepted theory for the effects of βAA on beef tenderness is an effect of changes in muscle fiber diameter. Calkins *et al.* (1981) suggested that fiber type, and subsequently fiber diameter, is associated with beef tenderness. Type I, β-red, or slow twitch oxidative fibers are understood to have a smaller fiber diameter and higher concentrations were more associated with improved muscle tenderness (Klont *et al.*, 1998). Likewise, type IIa, α-white, or fast twitch glycolytic fibers are understood to have a larger fiber diameter and an increased concentration of these fibers are more associated with negative tenderness attributes; Type IIx, α-red, or intermediate fast twitch oxidative glycolytic fibers are intermediate to Type I and Type IIa fibers in fiber diameter

(Klont *et al.*, 1998). Research associating βAA with fiber type report a shift in fiber type, with decreased concentrations of Type I fibers and increased concentrations of Type II fiber types (Gonzalez *et al.*, 2010; Gonzalez *et al.*, 2009; Kellermeier *et al.*, 2009). Gonzalez *et al.* (2009) reported that the *Adductor, gracilis, Longissimus laborum, and Vastus lateralis* muscles decreased in Type I fiber concentration while increasing in Type II fiber concentration due to dietary RH supplementation. However, the Type I and Type II fiber diameter was unchanged due to RH supplementation (Gonzalez *et al.*, 2009). Baxa *et al.* (2010) reported no changes in concentration of Type I and Type IIa fibers due to dietary ZH supplementation; however, an increase in concentration of Type IIx fibers was reported in ZH supplemented steers.

Differential effects on proteolytic enzyme activity due to βAA supplementation also have been studied to determine their effects on post mortem tenderness. The calpain system (m-calpain and ì-calpain) is generally recognized to improve post mortem tenderness due to the degradation of structural proteins (Goll, 1991). Likewise, calpastatin is an endogenous inhibitor of calpains, and increased concentration of calpastatin has been proven reduced beef tenderness (Goll, 1991). Hilton et al. (2009) reported no effect on μ- or m-calpain activity or calpastatin activity in steers supplemented with dietary ZH. Likewise, Rathmann *et al.* (2009) reported no effect on calpastatin mRNA abundance in steers supplemented dietary ZH. Walker *et al.* (2010) also reported that calpastatin mRNA expression was not affected by dietary RH supplementation at 200 mg/head/day at day 14 or 28 of the feeding period. Studies which reported differences in calpain activity due to βAA supplementation involved supplementation of cattle with cimaterol, a βAA now banned by the FDA for use in livestock (Bardsley *et al.*, 1992; Parr *et al.*, 1992).

Considering the effect of βAA's on the proteolytic enzyme system are negligible, extended postmortem aging periods should have minimal effect on disparities in tenderness associated with βAA treatments. Gruber *et al.* (2008) reported that steaks from steers supplemented 200 mg/head/day of dietary RH for the final 28 days of the feeding period had similar aging curves to steaks from control steers, and the difference in WBSF value between treatments was only slightly diminished from 3 to 21 days post mortem aging. These results were similar to those of Woerner *et al.* (2011), who determined that the effect of RH on LM WBSF was unaffected by postmortem aging out to 28 days. However, other studies evaluating the effect of postmortem aging on steaks from RH supplemented steers have reported that, while these steaks had tougher WBSF values early in the postmortem aging period, by 21 days, the disparities in tenderness were largely diminished (Boler *et al.*, 2012; Scramlin *et al.*, 2010). Scramlin *et al.* (2010) reported that while tenderness was improved after a 21 day aging period, WBSF values for steaks from ZH supplemented steers was still significantly tougher after postmortem aging. These results were supported in other ZH studies (Brooks *et al.*, 2009; Kellermeier *et al.*, 2009; Rathmann *et al.*, 2009), and indicated that, while the differences in tenderness between steaks from ZH supplemented steers and controls is diminished, ZH steaks are still considerably tougher after 21 days post mortem aging. Garmyn *et al.* (2010) determined that at a tenderness threshold of 4.6 kg WBSF, 100 percent of steaks from steers receiving no βAA supplementation were under the threshold value at 7, 14, 21, 28, and 35 days post mortem; by 35 days post mortem, 100 percent of steaks from ZH supplemented steers were also under the 4.6 kg threshold. More research is needed to evaluate the influence of extended postmortem aging periods on tenderness improvement for beef cattle supplemented with βAA past 21 days of aging.

Additional areas of research to determine βAA effects on postmortem tenderness have evaluated protein accretion and degradation. Proteins are continually degraded in skeletal muscle in the living animal, and rate of protein degradation versus protein accretion is related to an animal's ability to undergo muscle hypertrophy (Wheeler and Koohmaraie, 1992). Early research using the βAA L644,969 reported that muscle protein degradation was only reduced after 3 weeks of supplementation, however, muscle protein accretion was increased after 1, 3, 5, and 6 weeks of supplementation (Wheeler and Koohmaraie, 1992). More contemporary studies agree with these findings, and have indicated that βAA supplementation has a greater effect on protein accretion compared to protein degradation (Kellermeier *et al.*, 2009; Leheska *et al.*, 2009).

Safety of feeding β-agonists to Meat Animals

Extensive testing using stringent criteria to address the safety of feed additives and animal health products is mandated by the FDA before a product is allowed to be marketed for animals intended for food. Tests include chronic toxicity, mutagenicity, lifetime carcinogenicity, and effects on reproduction over two generations in animal model systems. The food safety for humans consuming foods of animal origin when feed additives are used also is tested. The safety evaluation of ractopamine hydrochloride – the active ingredient in Paylean® and Optaflexx®- included extensive tests in both laboratory animals and meat animals to establish safety of the approved dosages and conditions under which Paylean® and Optaflexx® are produced and used.

In 1989 the EU banned all β-agonists for use in meat animals, but no full scientific evaluation of β-agonists was conducted to support the ban, particularly for those approved for use elsewhere. Illegal and unsafe use of β-agonists was, and continues to be, a problem in many parts of the world, and this practice may have contributed to the EU ban. In this context, an important distinction must be made between ractopamine and other β-agonists. Ractopamine is what might be described as a recent-generation β-agonist specifically designed to meet the desired properties of an in-feed ingredient. These properties include rapid clearance from the body, targeting of specific tissues such as muscle and fat, no residues at slaughter, and no conversion of the compound to other active compounds. By contrast, early-generation β-agonists such as cimaterol, salbutamol, and clenbuterol were designed to act like pharmaceutical medicines used in the treatment of respiratory and other diseases, and the desired properties of such medicines are different from those required for an in-feed ingredient for finisher animals. For example, these compounds have a longer retention time in the body; act on a number of tissues rather than targeting specific ones; and result in residues in the body, particularly in the liver and kidney. Persons eating organs containing significant residues of an illegal β-agonist risk developing symptoms of β-agonist toxicity (heart fibrillations and bronchial spasms). In addition, certain β-agonists used illegally (e.g., clenbuterol) also are converted to more active or toxic compounds, which are retained in tissues.

Ractopamine is not converted to more active or toxic compounds, and hence it has been approved for use in meat animals. Ractopamine also has been evaluated for possible effects on animal welfare. Animal health, behavior, and well-being were evaluated extensively at dosages at least ten times higher than dosages approved for use in meat animals. There were subtle changes in certain behaviors in some pigs fed 10 ppm ractopamine hydrochloride, making them more difficult to handle and potentially more susceptible to handling and transport stress (Marchant-Forde *et al.*, 2003). No effects were noted, however, in most studies.

Global Scenario

Although 27 countries, including the United States, Canada, Mexico and Japan, allow ractopamine use, 160 countries, including the EU and China, do not. Countries that use ractopamine generally cannot sell meat to countries that do not allow it. The EU and China combined represent 70 percent of global pork consumption, a huge potential market for U.S. pork producers. The United States began pressuring the international forum on food safety standards, the United Nations' Codex Alimentarius Commission (Codex), to set a residue limit for ractopamine in meat shortly after the FDA first approved the drug. A Codex-approved ractopamine residue limit would effectively force countries to allow the import of meat treated with ractopamine because Codex is the accepted standard under international trade rules. Codex remained deadlocked for four years over the issue and only narrowly set an amount of ractopamine residue that can remain in meat and be considered safe to eat in 2012. For countries that ban ractopamine, the Codex decision represents a weakening of international rules. The new Codex standard is more stringent than the FDA allows, meaning that the United States falls outside the Codex standard. Nonetheless, the EU, Russia and China all insisted that they would maintain their ractopamine bans despite the Codex standard. Russia has demanded that the United States certify that its meat ractopamine-free. Since the United States tests so little meat for residues, and the meat tested for ractopamine frequently contains small residues, Russia has banned U.S. beef and pork imports.

To access the Russian market, Brazil has banned ractopamine use in its livestock until it develops a dual-track supply chain to ensure that it can provide ractopamine-free meat to Russia, the largest customer for Brazil's beef. Since the Codex decision, Taiwan has set its own maximum allowable ractopamine residue levels for beef, but not pork. In 2011, Taiwan found ractopamine residues in U.S. meat imports, resulting in recalls and a steep drop in U.S. meat imports. In the spring of 2012, the United States threatened to discontinue any new trade agreements with Taiwan over its refusal to establish maximum allowable ractopamine residue levels. Taiwanese citizens, including livestock producers, have held repeated public demonstrations, opposing any imports of meat from animals treated with ractopamine.

Summary

Growth enhancement technologies have been widely used in livestock production for many years. While hormonal implants have been utilized for decades in cattle production, recent advancements in dietary βAA supplements have offered swine and cattle producers the opportunity to further improve their production efficiency. Dietary supplementation of βAA's has been shown to improve live animal growth and have profound effects on carcass traits and carcass cutability. However, the improvements in growth and carcass traits are accompanied by reduction in marbling score, and have been reported to reduce the frequency of carcasses qualifying for premiums in quality based grid marketing. Furthermore, the most profound negative effect of both RH and ZH supplementation has been to reduce postmortem tenderness. Growth enhancement technologies offer a tremendous amount of potential to cattle feeders, yet judicious use is required to reduce deleterious effects on postmortem beef quality and eating quality. Further research is necessary to fully understand the physiological effects on beef tenderness. However, selection of the biological type of cattle that are inherently more tender with βAA supplementation is vital to upholding consumer acceptance of beef from aggressive growth promoting strategies.

Cattle which inherently produce higher quality carcasses and more tender beef (e.g. greater British breed influence) are ideal subjects for more aggressive growth enhancement regimens. Similarly, faster growing breeds that produce inherently tougher beef (e.g. greater Continental or Brahman influence) should be limited to less aggressive implanting and dietary βAA supplementation as this can impart additive effects on increasing beef toughness. Thus, βAAs are, and will continue to be widely used in the cattle feeding industry. While their effectiveness is undeniable, optimizing animal growth, carcass traits, and beef quality is necessary for maintaining consumer preference for beef.

References

Abney, C. S., Vasconcelos, J. T., McMeniman, J. P., Keyser, S. A., Wilson, K. R., Vogel, G. J. and Galyean, M. L. 2007. Effects of ractopamine hydrochloride on performance, rate and variation in feed intake, and acid-base balance in feedlot cattle. *J. Anim. Sci.* 85: 3090-3098.

Allen, J. D., Ahola, J. K., Chahine, M., Szasz, J. I., Hunt, C. W., Schneider, C. S., Murdoch, G. K. and Hill, R. A. 2009. Effect of preslaughter feeding and ractopamine hydrochloride supplementation on growth performance, carcass characteristics, and end product quality in market dairy cows. *J. Anim. Sci.* 87: 2400-2408.

Anderson, D. B., Moody, D. E. and Hancock, D. L. 2005. Beta Adrenergic Agonists. In: W. G. Pond and A. W. Bell (eds.) Encyclopedia of Animal Science. p 104-107. Marcel Dekker, New York, NY.

Apple, J. K., Dikeman, M. E., Simms, D. D. and Kuhl, G. 1991. Effects of synthetic hormone implants, singularly or in combinations, on performance, carcass traits, and longissimus muscle palatability of Holstein steers. *J. Anim. Sci.* 69: 4437-4448.

Armstrong, T. A., Ivers, D. J., Wagner, J. R., Anderson, D. B., Weldon, W. C. and Berg, E. P. 2004. The effect of dietary ractopamine concentration and duration of feeding on growth performance, carcass characteristics, and meat quality of finishing pigs. *J. Anim. Sci.* 82: 3245-3253.

Avendano-Reyes, L., Torres-Rodriguez, V., Meraz-Murillo, F. J., Perez-Linares, C., Figueroa-Saavedra, F. and Robinson, P. H. 2006. Effects of two β-adrenergic agonists on finishing performance, carcass characteristics, and meat quality of feedlot steers. *J. Anim. Sci.* 84: 3259-3265.

Bardsley, R. G., Allock, S. M. J., Allcock, S. M. J., Dumelow, N. W., Higgins, J. A., Lasslett, Y. V., Lockley, A. K., Parr, T. and Buttery, P. J. 1992. Effect of β-agonists on expression of calpain and calpastatin activity in skeletal muscle. *Biochimie.* 74: 267-273.

Barham, B. L., Brooks, J. C., Blanton, J. R., Herring, A. D., Carr, M. A., Kerth, C. R. and Miller, M. F. 2003. Effects of growth implants on consumer perceptions of meat tenderness in beef steers. *J. Anim. Sci.* 81: 3052-3056.

Baxa, T. J., Hutcheson, J. P., Miller, M. F., Brooks, J. C., Nichols, W. T., Streeter, M. N., Yates, D. A. and Johnson, B. J. 2010. Additive effects of a steroidal implant and zilpaterol hydrochloride on feedlot performance, carcass characteristics, and skeletal muscle messenger ribonucleic acid abundance in finishing steers. *J. Anim. Sci.* 88: 330-337.

Beckett, J. L., Delmore, R. J., Duff, G. C., Yates, D. A., Allen, D. M., Lawrence, T. E. and Elam, N. 2009. Effects of zilpaterol hydrochloride on growth rates, feed conversion, and carcass traits in calf-fed Holstein steers. *J. Anim. Sci.* 87: 4092-4100.

Beermann, D. H. 1993. ß-adrenergic agonists and growth. pp. 345–366. In M. P. Schriebman, C. G. Scanes, and P. K. T Pang (eds.). The Endocrinology of Growth, Development and Metabolism in Vertebrates. Academic Press, San Diego, California.

Boler, D. D., Shreck, A. L., Faulkner, D. B., Killefer, J., McKeith, F. K., Homm, J. W. and Scanga, J. A. 2012. Effect of ractopamine hydrochloride (Optaflexx) dose on live animal performance, carcass characteristics and tenderness in early weaned beef steers. *Meat Sci.* 92: 458-463.

Boler, D. D., Holmer, S. F., McKeith, F. K., Killefer, J., VanOverbeke, D. L., Hilton, G. G., Delmore, R. J., Beckett, J. L., Brooks, J. C., Miller, R. K., Griffin, D. B., Savell, J. W., Lawrence, T. E., Elam, N. A., Streeter, M. N., Nichols, W. T., Hutcheson, J. P., Yates, D. A. and Allen, D. M. 2009. Effects of feeding zilpaterol hydrochloride for twenty to forty days on carcass cutability and subprimal yield of calf-fed Holstein steers. *J. Anim. Sci.* 87: 3722-3729.

Brooks, J. C., Claus, H. C., Dikeman, M. E., Shook, J., Hilton, G. G., Lawrence, T. E., Mehaffey, J. M., Johnson, B. J., Allen, D. M., Streeter, M. N., Nichols, W. T., Hutcheson, J. P., Yates, D. A. and Miller, M. F. 2009. Effects of zilpaterol hydrochloride feeding duration and postmortem aging on Warner-Bratzler shear force of three muscles from beef steers and heifers. *J. Anim. Sci.* 87: 3764-3769.

Byrem, T. M., Beermann, D. H. and Robinson, T. F. 1996. Characterization of dose-dependent metabolic responses to close arterial infusion of cimaterol in the hindlimb of steers. *J. Anim. Sci.* 74:2907–2916.

Byrem, T. M., Beermann, D. H. and Robinson, T. F. 1998. The β-agonist cimaterol directly enhances chronic protein accretion in skeletal muscle. *J. Anim. Sci.* 76: 988–998.

Calkins, C. R., Dutson, T. R., Smith, G. C., Carpenter, Z. L. and Davis, G. W. 1981. Relationship of Fiber Type Composition to Marbling and Tenderness of Bovine Muscle. *J. Food Sci.* 46: 708-710.

Duckett, S. K., Owens, F. N. and Andrae, J. G. 1997. Effects of Implants on Performance and Carcass Traits of Feedlot Steers and Heifers. *In:* Symp.: Impact of Implants on Performance and Carcass Value of Beef Cattle, Oklahoma State University, Stillwater. p 63-82.

Dunshea, F. R. 1993. Effects of metabolism modifiers on lipid metabolism in the pig. *J. Anim. Sci.* 71:1966–1977.

Dunshea, F. R. and King, R. H. 1995. Responses to homeostatic signals in ractopamine-treated pigs. *Brit. J. Nutr.* 73: 809– 818.

Elam, N. A., Vasconcelos, J. T., Hilton, G., VanOverbeke, D. L., Lawrence, T. E., Montgomery, T. H., Nichols, W. T., Streeter, M. N., Hutcheson, J. P., Yates, D. A. and Galyean, M. L. 2009. Effect of zilpaterol hydrochloride duration of feeding on performance and carcass characteristics of feedlot cattle. *J. Anim. Sci.* 87: 2133-2141.

Garmyn, A. J., Shook, J. N., VanOverbeke, D. L., Beckett, J. L., Delmore, R. J., Yates, D. A., Allen, D. M. and Hilton, G. G. 2010. The effects of zilpaterol hydrochloride on carcass cutability and tenderness of calf-fed Holstein steers. *J. Anim. Sci.* 88: 2476-2485.

Goll, D. E. 1991. Role of proteinases and protein turnover in muscle growth and meat quality. *In:* Proc. 44th Annu. Recip. Meat Conf., Natl. Livest. and Meat Board, Chicago, IL. p 25-36.

Gonzalez, J. M., Johnson, S. E., Stelzleni, A. M., Thrift, T. A., Savell, J. D., Warnock, T. M. and Johnson, D. D. 2010. Effect of ractopamine–HCl supplementation for 28 days on carcass characteristics, muscle fiber morphometrics, and whole muscle yields of six distinct muscles of the loin and round. *Meat Sci.* 85: 379-384.

Gonzalez, J. M., Johnson, S. E., Thrift, T. A., Savell, J. D., Ouellette, S. E. and Johnson, D. D. 2009. Effect of ractopamine-hydrochloride on the fiber type distribution and shelf-life of six muscles of steers. *J. Anim. Sci.* 87: 1764-1771.

Gruber, S. L., Tatum, J. D., Engle, T. E., Prusa, K. J., Laudert, S. B., Schroeder, A. L. and Platter, W. J. 2008. Effects of ractopamine supplementation and postmortem aging on longissimus muscle palatability of beef steers differing in biological type. *J. Anim. Sci.* 86: 205-210.

Gruber, S. L., Tatum, J. D., Engle, T. E., Mitchell, M. A., Laudert, S. B., Schroeder, A. L. and Platter, W. J. 2007. Effects of ractopamine supplementation on growth performance and carcass characteristics of feedlot steers differing in biological type. *J. Anim. Sci.* 85: 1809-1815.

Hilton, G. G., Garmyn, A. J., Lawrence, T. E., Miller, M. F., Brooks, J. C., Montgomery, T. H., Griffin, D. B., VanOverbeke, D. L., Elam, N. A., Nichols, W. T., Streeter, M. N., Hutcheson, J. P., Allen, D. M. and Yates, D. A. 2010. Effect of zilpaterol hydrochloride supplementation on cutability and subprimal yield of beef steer carcasses. *J. Anim. Sci.* 88: 1817-1822.

Hilton, G. G., Montgomery, J. L., Krehbiel, C. R., Yates, D. A., Hutcheson, J. P., Nichols, W. T., Streeter, M. N., Blanton, J. R. and Miller, M. F. 2009. Effects of feeding zilpaterol hydrochloride with and without monensin and tylosin on carcass cutability and meat palatability of beef steers. *J. Anim. Sci.* 87: 1394-1406.

Hossner, K. L. 2005. Hormonal Regulation of Farm Animal Growth. CABI Publishing, Cambridge, MA. USA.

Johnson, B. J. 2004. Beta-adrenergic agonists: Efficacy and potential mode of action in cattle. *In:* Plains Nutrition Council Spring Conference, Texas A&M Research and Extension Center, Amarillo, TX. p 51-61.

Kellermeier, J. D., Tittor, A. W., Brooks, J. C., Galyean, M. L., Yates, D. A., Hutcheson, J. P., Nichols, W. T., Streeter, M. N., Johnson, B. J. and Miller, M. F. 2009. Effects of zilpaterol hydrochloride with or without an estrogen-trenbolone acetate terminal implant on carcass traits, retail cutout, tenderness, and muscle fiber diameter in finishing steers. *J. Anim. Sci.* 87: 3702-3711.

Klont, R. E., Brocks, L. and Eikelenboom, G. 1998. Muscle fibre type and meat quality. *Meat Sci.* 49, (Supplement 1): S219-S229.

Leheska, J. M., Montgomery, J. L., Krehbiel, C. R., Yates, D. A., Hutcheson, J. P., Nichols, W. T., Streeter, M., Blanton, J. R. and Miller, M. F. 2009. Dietary zilpaterol hydrochloride. II. Carcass composition and meat palatability of beef cattle. *J. Anim. Sci.* 87: 1384-1393.

Marchant-Forde, J. N., Lay, D. C., Pajor, E. A., Richert, B. T. and Schinckel, A. P. 2003. The effects of ractopamine on the behaviour and physiology of finishing pigs. *J Anim Sci* 81:416–422.

Mersmann, H. J. 1989. Inhibition of porcine adipose tissue lipogenesis by β-adrenergic agonists. *Comp. Biochem. Physiol.* 94 (C): 619–623.

Mersmann, H. J. 1998. Overview of effects of β-adrenergic receptor agonists on animal growth including mechanisms of action. *J. Anim. Sci.* 76:160–172.

Miller, M. F., Garcia, D. K., Coleman, M. E., Ekeren, P. A., Lunt, D. K., Wagner, K. A., Procknor, M., Welsh, T. H. and Smith, S. B. 1988. Adipose Tissue, Longissimus Muscle and Anterior Pituitary Growth and Function in Clenbuterol-Fed Heifers. *J. Anim. Sci.* 66: 12-20.

Miller, M. F., Carr, M. A., Ramsey, C. B., Crockett, K. L. and Hoover, L. C. 2001. Consumer thresholds for establishing the value of beef tenderness. *J. Anim. Sci.* 79: 3062-3068.

Mills, S. E. and Liu, C. Y. 1990. Sensitivity of lipolysis and lipogenesis to dibutyryl-cAMP and β-adrenergic agonists in swine adipocytes in vitro. *J. Anim. Sci.* 68:1017–1023.

Milton, C. T., and Horton, D. 1996. Implant strategies for finishing calves. In: Kansas State University Cattlemen's Field Day, Kansas State Univ., Manhattan, KS. p 108.

Montgomery, J. L., Krehbiel, C. R., Cranston, J. J., Yates, D. A., Hutcheson, J. P., Nichols, W. T., Streeter, M. N., Bechtol, D. T., Johnson, E., TerHune, T. and Montgomery, T. H. 2009.

Dietary zilpaterol hydrochloride. I. Feedlot performance and carcass traits of steers and heifers. *J. Anim. Sci.* 87: 1374-1383.

Moody, D. E., Hancock, D. L. and Anderson, D. B. 2002. Phenethanoloamine repartitioning agents. Pp. 65–96. In J. P. F. D'Mello (ed.). Farm Animal Metabolism and Nutrition. CAB International, New York.

Parr, T., Bardsley, R. G., Gilmour, R. S. and Buttery, P. J. 1992. Changes in calpain and calpastatin mRNA induced by β-adrenergic stimulation of bovine skeletal muscle. *Eur. J. Biochem.* 208: 333-339.

Perry, T. C., Fox, D. G. and Beermann, D. H. 1991. Effect of an implant of trenbolone acetate and estradiol on growth, feed efficiency, and carcass composition of Holstein and beef steers. *J. Anim. Sci.* 69: 4696-4702.

Plascencia, A., Torrentera, N. and Zinn, R. A. 1999. Influence of the beta-agonist, Zilpaterol, on the growth performance and carcass charactersistics of feedlot steers. *In:* Western Section Meetings, American Society of Animal Science. p 331-334.

Quinn, M. J., Reinhardt, C. D., Loe, E. R., Depenbusch, B. E., Corrigan, M. E., May, M. L. and Drouillard, J. S. 2008. The effects of ractopamine-hydrogen chloride (Optaflex) on performance, carcass characteristics, and meat quality of finishing feedlot heifers. *J. Anim. Sci.* 86: 902-908.

Rathmann, R. J., Mehaffey, J. M., Baxa, T. J., Nichols, W. T., Yates, D. A., Hutcheson, J. P.,. Brooks, J. C, Johnson, B. J. and Miller, M. F. 2009. Effects of duration of zilpaterol hydrochloride and days on the finishing diet on carcass cutability, composition, tenderness, and skeletal muscle gene expression in feedlot steers. *J. Anim. Sci.* 87: 3686-3701.

Ricks, C. A., Dalrymple, R. H., Baker, P. K. and Ingle, D. L. 1984. Use of a β-Agonist to Alter Fat and Muscle Deposition in Steers. *J. Anim. Sci.* 59: 1247-1255.

Scramlin, S. M., Platter, W. J., Gomez, R. A., Choat, W. T., McKeith, F. K. and Killefer, J. 2010. Comparative effects of ractopamine hydrochloride and zilpaterol hydrochloride on growth performance, carcass traits, and longissimus tenderness of finishing steers. *J. Anim. Sci.* 88: 1823-1829.

Shook, J. N., VanOverbeke, D. L., Kinman, L. A., Krehbiel, C. R., Holland, B. P., Streeter, M. N., Yates, D. A. and Hilton, G. G. 2009. Effects of zilpaterol hydrochloride and zilpaterol hydrochloride withdrawal time on beef carcass cutability, composition, and tenderness. *J. Anim. Sci.* 87: 3677-3685.

Strydom, P. E., Frylinck, L., Montgomery, J. L. and Smith, M. F. 2009. The comparison of three β-agonists for growth performance, carcass characteristics and meat quality of feedlot cattle. *Meat Sci.* 81: 557-564.

Tatum, J. D. 2006. Pre-Harvest Cattle Management Practices for Enhancing Beef Tenderness. http://www.beefresearch.org/executivesummaries.aspx.

Uttaro, B. E., Ball, R. O., Dick, P., Rae, W., Vessie, G. and Jeremiah, L. E. 1993. Effect of ractopamine and sex on growth, carcass characteristics, processing yield, and meat quality characteristics of crossbred swine. *J. Anim. Sci.* 71: 2439-2449.

Vasconcelos, J. T., Rathmann, R. J., Reuter, R. R., Leibovich, J., McMeniman, J. P., Hales, K. E., Covey, T. L., Miller, M. F., Nichols, W. T. and Galyean, M. L. 2008. Effects of duration of zilpaterol hydrochloride feeding and days on the finishing diet on feedlot cattle performance and carcass traits. *J. Anim. Sci.* 86: 2005-2015.

Vogel, G. J., Duff, G. C., Lehmkuhler, J., Beckett, J. L., Drouillard, J. S., Schroeder, A. L., Platter, W. J., Van Koevering, M. T. and Laudert, S. B. 2009. Effect of Ractopamine Hydrochloride on Growth Performance and Carcass Traits in Calf-Fed and Yearling Holstein Steers Fed to Slaughter. *Prof. Anim. Sci.* 25: 26-32.

Walker, D. K., Titgemeyer, E. C., Baxa, T. J., Chung, K. Y., Johnson, D. E., Laudert, S. B. and Johnson, B. J. 2010. Effects of ractopamine and sex on serum metabolites and skeletal muscle gene expression in finishing steers and heifers. *J. Anim. Sci.* 88: 1349-1357.

Watkins, L. E., Jones, D. J., Mowrey, D. H., Anderson, D. B. and Veenhuizen, E. L. 1990. The effect of various levels of ractopamine hydrochloride on the performance and carcass characteristics of finishing swine. *J. Anim. Sci.* 68: 3588-3595.

Wheeler, T. L., and Koohmaraie, M. 1992. Effects of the beta-adrenergic agonist L644,969 on muscle protein turnover, endogenous proteinase activities, and meat tenderness in steers. *J. Anim. Sci.* 70: 3035-3043.

Woerner, D. R., Tatum, J. D., Engle, T. E., Belk, K. E. and Couch, D. W. 2011. Effects of sequential implanting and ractopamine hydrochloride supplementation on carcass characteristics and longissimus muscle tenderness of calf-fed steers and heifers. *J. Anim. Sci.* 89: 201-209.

22

Recent Advances in Feed Additives for Sustainable Fish Production

Ashutosh Mishra

Introduction

Supplementary feeding is an essential practice in fish farming operation which accounts for over 60% in total input cost. The dependence of candidate species on balanced supplementary feed increases with increase in stocking population in ponds targeting per hectare higher production as the standing crop of cultured species exceeds the natural feeding capacity of ponds. Therefore, feeding of artificial feed balanced in all nutrients such as protein, lipid, carbohydrate, vitamins and minerals containing optimum protein and energy ratio has assumed foremost importance in aquaculture industry. The use of balanced feed however, is directed towards optimum realization of genetic potentials of farmed species for survival, immunity, growth and reproduction (Mohanty, 2009). Consequent to researches done in India and abroad, a strong data base has been developed about the nutritional behavior of great majority of these cultivable fish and shell fish species and several feed formulations have been successfully undertaken. With optimum pond management, feeding of formulated balanced feed, it has been possible to increase aquaculture production as high as 15- 17 t/ ha/ yr for carps and 8-12 t/ ha/ yr in semi intensive prawn and shrimp farming in India (Mohanty, 2009).

Nutritional requirements

There are about 40 essential dietary nutrients required by fish, prawn and terrestrial animals. The requirement of nutrients varies throughout the life cycle of an individual. At early stage the requirements of nutrients is comparatively high which declines with age. In addition, the requirements depend upon the feeding habits that change accordingly to the morphology of digesting system. The nutritional requirements of important cultivable fishes are as follows-

Proteins are the major organic material in fish tissue constituting about 65-75% of the total on dry weight basis and are needed for replacement of worn out and synthesis of new tissues and growth as several proteineous products like intestinal epithelial cells, enzymes and hormones which are required for proper body function (De Silva and Anderson, 1995). Since protein acts both as structural component and as an energy source, its requirement for fish is 2-3 times higher than those mammals. Dietary protein requirement has a linear relationship with the specific growth rate and requirement data are obtained from dose response curves in fish. The optimal protein requirement in fish is affected by the nutritional qualities of dietary protein and the level of energy from non protein sources, fish size, water temperature, dissolved oxygen and pH, feeding rate *etc*. (De Silva and Anderson, 1995). It is seen that there exists a large variation in the gross protein requirement, which decrease with increase in age and size of fish. However, generally 25-30% protein is optimum for inclusion in practical feeds for herbivorous and omnivorous fishes for pond feedings. A protein deficit is known to occur when absolute protein requirement of the fish population stocked per unit area is higher than the available protein from natural fish food organisms needing protein supplementation from exogenous source. The information on protein requirement is of limited value unless the requirement of amino acids of fish is under stood, because the protein quality is largely dependent on its amino acid composition (Ali *et al.*, 2011). The fish do not have true protein requirement but need a well balanced mixtures of essential and non essential amino acids

Lipids are important source of energy, essential fatty acids and phospholipids provide a vehicle for absorption of fat soluble sterols and vitamins. These also play a vital role in the structure of cell and cellular membrane and serve as a precursor of several hormones in addition to their function for prostaglandin synthesis (Mohanty, 2009). These are highly digestible in fish and are reported to spare protein. Also the feeding of excess lipid is known to produce fatty fish that has deleterious effect on the flavor, consistency and storage life of finished products. Although several authors reported wide range of 4-15% as gross lipid requirements for several species, 7-9% dietary lipids are generally considered optimum for practical feeds of carps (Mohanty, 2009). Supplementation of both n-3 and n-6 PUFA is required for brood stock feed which is done by providing vegetable oil and fish oil of marine source and this greatly influence gonadal maturation, breeding efficiency and sperm recovery of carps.

Like protein and lipids, carbohydrates are also another source of energy. Fish do not have specific dietary requirement but carbohydrates are always included in fish feeds as they are inexpensive energy source and acts as pellet binder (De Silva and Anderson, 1995). Carbohydrates also serve as precursors for formation of various metabolic intermediates needed for growth. Carbohydrates have a sparing effect on the utilization of dietary protein in many aquaculture species. Proper dietary balance of carbohydrates would enable fiber to move other nutrients in gastro- intestinal tracts for proper digestion. Carps efficiently utilize carbohydrate as energy. However, the ability of fish to utilize dietary carbohydrate varies considerably and most carnivorous species have the limited ability to metabolize carbohydrates. The optimum dietary requirements of carbohydrates are 22-26% for Indian major carps and 30-40% for common carps (Mohanty, 2009). However, carbohydrate levels generally don't exceed 30% in carp feeds.

Vitamins are required in trace amounts. These micro- nutrients are essential for fish growth and to fight against diseases. The vitamins are required for the metabolism of other nutrients in tissue components. Many of the water soluble vitamins also act as coenzymes. Fishes derive the required vitamins from natural food which become limited in the intensive fish culture due to high stocking densities (Gopal, 2011). Fish require 11 water soluble (thiamine, riboflavin, pyridoxine, niacin, pantothenic acid, inositol, folic acid, choline, biotin, ascorbic acid and vitamin B_{12}) and 4 fat soluble (Vitamins A, D, E and K) vitamins. Excess water and fat soluble vitamin results in abnormal growth and liver disease.

Minerals are required for osmotic balance of various metabolic processes and for structural functions in fish. Some minerals such as calcium are directly obtained by the fish through the gills and skins or both, while others are made available from food and detritus ingested. The vitamins and minerals are therefore provided as premix in balanced artificial feeds. There are about 20 recognized inorganic elements which perform essential functions in the body. The minerals required by fish are calcium, chlorine, magnesium, phosphorous, sodium and potassium along with a number of trace elements such as cobalt, copper, iodine iron, manganese, selenium, zinc, aluminium, chromium and vanadium. Phosphorus and calcium are required for bone formation. Calcium plays a major role in blood clotting, muscle function, proper nerve impulse transmission, etc. Phosphorus is involved in energy transformation, permeability of cellular membrane, genetic coding and general control of reproduction.

Practical feeds

The supplementary feed used in fish farming consists of rice bran and oil cakes (mustard oil cake or groundnut oil cake) which are compounded in 1:1 ratio. The feeding of the cake- bran mixture in conjunction with natural fish feed organisms is still a practice for semi intensive carp farming in the country. It is not a nutritionally balanced and is normally used to supplement protein and energy deficiencies in ponds. It is observed that stocked fish species receive about 50% protein, 8% lipid and 27% carbohydrate and 4 Kcal/ g gross energy from natural food organisms (Giri, 2011). The bottom dweller fishes appear to take maximum advantage of energy supplements added in practical diets that spare protein for their growth because benthic fish food organisms are inadequately available for the fish (Nandeesha, 1993). The phytoplankton provides high quantities of ω-3 and ω-6 PUFA's. Due to non- availability of commercial feeds and economic reasons more than 90% farmers use farm made feed of cake- bran mixture or improved version of feed mixture. Artificial feeding in the presence of natural fish food organisms has been found beneficial as later influences dietary efficiency and economic utilization of the former. There has been rapid shift from traditional feeding cake bran mixture to pellet feeding of "nutritionally complete feed". It may be noted that a variety of feed additives such as antioxidants, binders, stimulants, attractants, immunizer, pigments and hormones are also used in preparation of pellet feeds to improve feed and feeding efficiency.

Advances in feed additives

Feed additives or supplements are substances which are added in trace amounts to a diet or feed ingredient either to preserve its nutritional characteristics prior to feeding (i.e. antioxidants and mould inhibitors), to facilitate ingredient dispersion or feed pelleting (i.e. emulsifiers, stabilizers and binders), to facilitate growth (i.e. growth promotants, including antibiotics and hormones), to facilitate feed ingestion and consumer acceptance of the product (i.e. feeding stimulants and food colourants), or to supply essential nutrients in purified form (i.e. vitamins, minerals, amino acids, cholesterol and phospholipids) (De Silva and Anderson, 1995). The recent advances taken place in feed additives used for aqua farming are as follows-

Feed preservatives

A major problem faced by the animal feed compounder is the susceptibility of individual feed ingredients and formulated feeds to oxidative damage (oxidative rancidity) and microbial attack on storage. For example, in the absence of natural antioxidant protection (i.e. absence of vitamin E, selenium, soy lecithin, active β-carotene) feedstuffs rations rich in polyunsaturated fatty acids (i.e. fish oils, fish meals, rice bran, and some expeller oil seed cakes) are highly prone to oxidative decomposition which in turn may cause a reduction in the nutritive value of the constituent lipids, protein and vitamins. Similarly, feedstuffs and rations possessing an elevated moisture content (> 15%) are prone to microbial attack and decomposition with a consequent loss in nutritional value for non-ruminant animals and deleterious mycotoxin production. There are many antioxidants and chemical preservatives used by the animal feed manufacturing industry to combat the development of oxidative rancidity and microbial infestation within stored agricultural feedstuffs and rations (De Silva and Anderson, 1995).

Pellet Binders

Binders are substances which are used within aquaculture feeds to improve the efficiency of the feed manufacturing process, to reduce feed wastage, and/or to produce a water-stable diet. For example, binders such as bentonites, lignosulphonates, hemicellulose and carboxy methylcellulose are used primarily within feed rations to improve the efficiency of the feed manufacturing process (i.e. during pelleting by reducing the frictional forces of the feed mixture through the pellet dies and thereby increasing the output and horse power efficiency of the feed mill) and for the production of a durable pellet (i.e. by increasing pellet hardness and reducing wastage in the form of 'fines' during the pelleting process and during handling/transportation) (Jain, 1998). The dietary inclusion level of these binding agents generally varies between 1 and 2% of the dry diet. By contrast, for those aquaculture species which have a slow feeding habit and require to masticate their food externally prior to ingestion (i.e. marine shrimp and freshwater prawns), it is essential that specific binders be used to delay the physical disintegration of the pellet or feed mash within the water until ingestion is complete. Under these circumstances additional dietary binding agents will be required such as starchy plant products (i.e. sago palm starch, cassava starch, potato starch, bread or wheat flour, rice and maize; binding being achieved through heat treatment and consequent starch gelatinization), alginates (i.e. salts of alginic acid extracted from seaweeds), carrageenin, plant gums (i.e. guar gum, locust bean gum, gum arabic), agar, high-gluten wheat flour, chitosan, propylene glycol alginate and gelatin. Alternatively, binding agents such as polymethylolcarbamide (i.e. Basfin) and urea-formaldehyde/calcium sulphate mixtures (i.e. Maxi-Bond) with dual binding properties may be used in conjunction with steam pelleting (i.e. these binders being activated by the injection of steam during pelleting; the condensation reaction can be controlled during pelleting to give the required durability and water stability).

The dietary inclusion level of these two binding agents is generally below 0.5% of the dry diet. For moist or semi-moist diets the use of alginates or salt should be mentioned. Alginates are valuable binding agents through their ability to produce water-stable diets. Consequently, for diets containing ingredients such as fish meal and fish protein hydrolysates which have a high soluble calcium content, gelling occurs rapidly during mixing (i.e. trash fish/fish silage moist diet preparations). However, for dry diet preparations the premature gelling of the feed mixture prior to pelleting has to be controlled by the use of a sequestering agent such as sodium polyphosphates. The dietary level of alginates used in feed rations generally varies between 0.5 and 5%, with the level of sequestering agents (typically sodium hexametaphosphate) varying from 0.5 to 1.5% depending on diet composition. For those diets which have a low soluble calcium content for adequate gel formation by alginates, the addition of soluble calcium salt such as calcium sulphate, calcium carbonate or calcium phosphate will be required. Alternatively, salt can be used as an effective and low-cost binding agent for raw trash fish/moist diet combinations (1% dietary inclusion level) (De Silva and Anderson, 1995).

Food colourants

Food colourants are substances which are added in trace amounts to a diet or feed mixture to facilitate its ingestion (through improved visibility of feed particles) or to impart a desired colouration within the carcass of the cultured fish or shrimp (Sahu and Pal, 2011).

An important factor governing the consumer acceptance and market value of many cultivated fish and shrimp species is the pink or red colouration of their flesh or boiled exoskeleton. In the wild, this colouration is derived through the ingestion of carotenoid pigments contained within invertebrate food organisms; the characteristic pink colouration of salmonids and red sea bream, and the red colour of the boiled crustacean exoskeleton, being mainly due to the carotenoid pigment astaxanthin. The carotenoid pigments represent a widespread group of plant-synthesized polyene pigments, which vary in colour from yellow and orange to red. Although animals, including fish and shrimp, are believed to be unable to synthesize carotenoids, certain aquaculture species (i.e. crustaceans, omnivorous/herbivorous fish) are capable of transforming ingested carotenoids such as β-carotene and depositing the astaxanthin in their tissues. By contrast, carnivorous fish species such as salmonids and red sea bream are believed to be incapable of carotenoid transformation; the ingested carotenoid being deposited in its unaltered state within the body tissues. Since this natural carotenoid pigmentation is usually lacking within intensively farmed fish and shrimp with no access to live (carotenoid containing) food organisms, it is necessary to fortify practical rations with carotenoid - rich ingredient sources (i.e. marine animal by-products such as shrimp waste, fish and crustacean oils, marine yeast and algae) or with purified carotenoid preparations (i.e. canthaxanthin powders) if the farmer is to achieve a marketable product with the desired carcass pigmentation. For example, in Scandinavian countries (i.e. Norway), shrimp offal meals are commonly used as natural carotenoid sources within commercial salmonid rations. For those countries where natural carotenoid-containing feedstuffs are in short supply, adventitious artificially produced carotenoids must be added to the ration; commercially available canthaxanthin being the predominant pigment source for salmonid fish in the United Kingdom. Although these industrially produced microbial pigments are costly, they usually impart the same visual colouration of muscle as natural astaxanthin sources, and as a dry powder can be included within rations. Within the United Kingdom, canthaxanthin (in the form of carophyll red) is routinely added to salmonid rations at a dietary level of 50–100 mg carotenoid/kg for a two or three month feeding period prior to harvesting or restocking. However, the success of the pigmentation programme will depend upon a variety of factors, including: dietary pigment source and concentration, length of feeding and continuity of feeding the pigmented diet, size of the fish or shrimp, water temperature, developmental status of the fish or shrimp (i.e. sexual maturity), hormonal status of the animal, chemical form of the carotenoid fed, photoperiod regime, dietary lipid/emulsifier concentration, and the availability of natural food organisms. In addition to the role of carotenoids in carcass pigmentation and as dietary sources of provitamin A, these compounds may play other important biological functions within the animal body (Kannappan, 2011).

Feeding stimulants

Maximum benefit from feeding can only be achieved if the food provided is ingested. An understanding of the feeding behaviour of the fish or shrimp is, therefore, essential. The diet presented must have the correct appearance (i.e. size, shape and colour), texture (i.e. hard, soft, moist, dry, rough or smooth), density (buoyancy) and attractiveness (i.e. smell or taste) to elicit an optimal feeding response. However, the relative importance of these individual factors will depend on whether the fish or shrimp species in question is mainly a visual feeder or a chemosensory feeder. For example, although marine fish held in captivity generally rely on sight to locate their food, they also rely on chemoreceptors located in the mouth or externally on appendages such as lips, barbels and fins; the feed being carefully 'sensed' before ingestion. A similar situation also exists with marine shrimp and freshwater prawns. The use of dietary feeding stimulants for these cultivated species is therefore essential to elicit an acceptable and rapid feeding response. The practical importance of feed attractants and diet palatability is particularly critical during the weaning of marine fish larvae from a live to a non-living diet. Similarly, as attempts are made to replace the fishmeal component of practical fish feeds with unconventional protein sources of an alien nature, the problem of diet texture and palatability will become even greater. In addition, by using feeding stimulants and improving feed palatability, the period of time the feed remains in the water can be reduced, thus minimizing nutrient leaching (Ali *et al,* 2011).

Two types of feeding stimulants may be considered for use within aquaculture feeds; natural ingredient sources which exhibit attractant or feeding stimulant properties or the use of the purified or synthetic chemical derivatives which are responsible for the attractant property of natural ingredient sources. For example, feed ingredients which have been found to impart specific attractant properties for shrimp and marine fish include squid meal, mussel flesh, shrimp meal and waste, short-necked clam flesh, marine polychaete worms, blood worms, certain terrestrial oligochaete worms, marine fish oils, fish meal, fish solubles, fish protein hydrolysates and soybean protein hydrolysates. Purified or synthetic substances which have been found to act as dietary feeding stimulants include mixtures of L-amino acids (particularly amino acid mixtures including glycine, alanine, proline and histidine and possibly the freshwater prawn, mixtures of L-amino acids and the quaternary amine glycine betaine, the nucleosides inosine and inosine-5-monophosphate. Good dietary sources of betaine and soluble nucleotide bases include mussels, polychaetes, squid, shrimp waste and shrimp blanch water, fish products, polychaetes, respectively.

Microbial feed supplements in aquaculture

Due to increased intensification in aquaculture practices, the disease outbreaks may occur which may leads to economic loss of fish farmer. To combat these diseases, wide spread use of broad-spectrum chemotherapeutics has led to drug resistance problems in aquaculture and created human health hazards. In order to rectify this situation, greater emphasis has been placed on improving water quality and lowering stocking density, using vaccines and biological control agents which depend on the fact of microbial antagonism. The bacteria present in aquatic environment influence the composition of gut biota and vice- versa as the host and micro- organisms share the ecosystem. The use of natural prophylactic supplements in place of chemotherapeutics in aquaculture has received a great deal of attention in the past decade; such preventive products include probiotics. The probiotics can be applied through external bathing or dietary supplementation and have been demonstrated to improve growth performance, feed utilization, digestibility of dietary ingredients, disease resistance and stimulate the immune response of aquatic animals. Aquatic Probiotics are live microbial supplements that can modulate microbial communities and improve microbial balance thus providing benefits to the host. The beneficial effects of these probiotics include higher growth and feed efficiency, prevention of intestinal disorders and pre-digestion of anti nutritional factors present in the ingredients (De and Ghosal, 2011).

The probiotics that currently used in aquaculture industry include *Lactobacillus,Bifidobacterium, Pediococcus, Streptococcus, Carnobacterium, Bacillus, Flavobacterium, Cytophaga, Pseudomonas, Altomonas, Enterococcus, Nitrosomonas, Nitrobactor, Vibrio, Saccharomyces, Debaryomyces etc.* These probiotics are available in dry as well as in liquid form. Dry probiotics can be given with feed or applied to water. Liquid probiotics are used in hatcheries which gives positive results in lesser time. There is no reports of any harmful effect for probiotics but it is found that the biological oxygen demand may temporarily be increased on its application. Therefore, subsurface aeration should be provided to expedite the establishment of probiotic organism.

Immunomodulators

Nutrients are important modulator of immune function and can often tip the balance between health and disease. The immune system benefits greatly from proper nutrition of animal. The immune system uses many of the same types of glucose and amino acid transporters that nervous and other high priority tissue use. When leukocytes become activated, they express high levels of nutrient transporters, which allows them to easily obtain necessary nutrients from muscle and other tissues. Trace elements such as iron, copper and zinc are problematic because of their low concentration in muscle and their relatively high need within the immune system. Evidence is accumulating that the dietary requirement for some trace nutrients may be higher for optical immune function than it is for maximal growth or reproduction performance (Dayal and Ambashankar, 2011).

Nutritional immunity is the process whereby the body with hold essential nutrients from pathogens to reduce their rate of replication. For example, it is well documented that orally feeding baby pigs with iron is contradictory in case of enteritis. In birds, a similar, situation exists within the egg to deprive nutrients to bacteria so that they are unable to colonize the albumen and infect the developing embryo. Immune cells sequester trace minerals, such as manganese and iron, when they engulf pathogens and this action serves to starve pathogens and prevent their replication.

Dietary factors such as type of fat can change the proportion of prostaglandins and other eicosanoids that their responses to disease challenge. Fish oil is high in eicosapentanoic acid, which causes macrophages to be predisposed to release interleukins that drive T helper cells towards a th2 type of response and predisposed to a th1 type of response, especially the inflammatory response.

The role of essential amino acid, arginine to providing immunity to providing immunity to chicks as a modulator of cellular metabolic pathway producing nitric oxide in macrophages. The role of Vitamin E as immune system "booster" along with selenium and zinc is well established. Vitamin C as an "antistress" agent plays a central role in the synthesis of corticosterone hormone. The role of vitamin A is maintain proper epithelial tissue differentiation and maintaining cellularity of lymphoid organs is well established. Vitamin A has reduced mortality in chicks infected with *Coccidial oocytes from E. tenella* and *E. acervulina.*

The role of probiotics in livestock in feeds is widely studied. Their role as immune modulators is also being established. These probiotics appear to affect the establishment of CD 8 + cells in the gut and they protect the host against bacterial translocation (Swain, 2009). Probiotics could be considered as a reliable tool to prevent intestinal bacterial infections (Alavandi *et al*, 2011). The feeds have specific effects on immune-competency when they contain non nutrient substances that influence the function of leukocytes, the integrity of the intestinal epithelia or the population of commensal microflora found in the intestines.

Enzymes as feed additives

Enzymes are an important supplements to be added in aquafeed to enhance its digestibility. In Europe and Australia feed enzymes have been used for nearly a decade but their usage in Indian diets has been only in recent years. Some of the enzymes that have been used and are still in use in the feed industry include cellulase (β- glucanases), xylanases and associated enzymes viz. phytases, proteases, lipases and glactosidases (Table 1) (Valli, 2011).

Indian farmers need to explore usage of alternate feed sources like sunflower, rapeseed, safflower etc. for protein sources and sorghum, millets or rice bran for energy sources in order to reduce feed cost. However, the cell wall composition of these ingredients show that they contain large amounts of arabino- xylans, pectic polysaccharides and some cellulose. Use of specific enzymes like xylanase, pectinase and cellulase could breakdown the fibre releasing energy as well as increasing the protein digestibility due to better accessibility of the protein when the fibre gets break down. This can reduce the feed cost as well as the protein levels in feed also be reduced.

Table 1: Common enzymes used in aquaculture

Enzymes	Substrate	Function	Benefits
β- glucanases	Barley	Viscosity reduction	Enhanced digestion and utilization of nutrients
Xylanases	Wheat, rye, triticale, sorghum, rice bran	Viscosity reduction	Enhanced digestion and utilization of nutrients
β- galactosidases	Grain legumes	Viscocity reduction	Enhanced digestion and utilization of nutrients
Phytases	Plant origin feed stuffs	Release of phosphorus from phytate- P	Enhanced phosphate absorption
Proteases	Proteins	Hydrolysis of protein	Increased digestion of protein
Lipases	Lipids	Hydrolysis of fats	Used in young animals
Amylases	Starch	Hydrolysis of starch	Supplemental amylase for young animals

Another enzyme, which is quite common in Indian Feed Industry, is phytase. Phytase has been known for several decades however, feed industry could not use it economically due to its high production costs. Phytase breaks down the indigestible phytate in cereals and oil seeds and releases digestible phosphorus. This reduces the use of supplemental inorganic phosphorus like dicalcium phosphate. Phytase also releases minerals (Ca, Mg, Zn, and K) amino acids and proteins which are complexed with the phytate molecule (Valli, 2011). Phytase can also controls the algal bloom due to reduction in ground water phosphorus levels. Enzymes like protease and xylanases are also being introduced now in aquafeed when the trend of replacing fish meal with soya is on the rise. Aquafeed manufactures are also looking into the possibility of using phytase for releasing the non available phosphorus from deoiled rice bran.

Nutraceuticals

Nutraceuticals are a food product that provides health and medical benefits, including the prevention and treatment of disease. The dietary ingredients in these may include: vitamins, minerals, herbs, or other botanicals, amino acids, fatty acids and substances such as enzymes, organ tissues, glandular and metabolites (Sahu and Pal, 2011).

Functional foods are designed to allow consumers to eat enriched foods close to their natural state, rather than by taking dietary supplements manufactured in liquid or capsule form.

Antibiotics and several other chemical drugs have been tested in aquaculture operations for various remedies. Even though they give positive effects, they can not be recommended due to their residual and other side effects. The alternative neutraceuticals/ herbal medicinal products used in the aquaculture operations have the characteristics of growth promoting ability and improve the immune system. They increase feed consumption, induce maturation, and have antimicrobial capability and also anti-stress characteristics that will be of immense use in the culture of shrimps and other fin fishes without any environmental and hazardous problems. In aquaculture, application of nutraceuticals includes addition of feed additives in feed such as antioxidants, vitamins, minerals and carotenoids etc. Most widely used nutraceuticals in aquaculture are as follows:

(a) Herbal extract

Plants have been used as traditional medicine since time immemorial to control bacterial, viral and fungal diseases. Recently, research has been initiated to evaluate the feasibility of using herbal medicines in fish disease management. Many herbal extracts are found to have novel compounds benefiting the health of human and fish as well. Various immunostimulants, especially of herbal origin for maintaining the health of fish and shell fishes directly activate the innate defence mechanisms acting on receptors and trigger intracellular gene activation that may result in the production of antimicrobial molecules. A number of plant material/ products such as saponin, glycyrrhizin, aloe, *Ocimum sanctum,* azardirachtin etc. have been reported to enhance the immunity of fish against disease. It has been shown that herbal based immunostimulants are capable of enhancing immune responses and/ or reducing losses from viruses, bacteria and/ or parasitic infections in carp. The nonspecific immune response is often reported as a function of macrophage activity such as phagocytosis and chemotaxis. Immunostimulants can be applied injection, bathing or oral administration, the latter seems to be the most practicable (Sahu and Pal, 2011).

(b) Polysaccharides

Polysaccharides are considered natural immunostimulants that have been shown to promote the secretion of cytokines and antibodies, as well as enhance the function of natural killer cells, T and B lymphocytes. Though both oral and parental administration of polysaccharides are reported to be successful but oral administration is the preferred way because of its convenience, but it is argued that oral administration may decrease the immunostimulating effects of polysaccharides because of the possibility of being destroyed by intestinal enzymes. Among the polysaccharides most promising results on health benefits of fish have been reported for glucans, pentanes and some oligosaccharides.

β-glucan, a group of glucose polymers, is the main structural component of cell wall in fungi, plants and some bacteria. It may induce activation of leucocytes, phagocytic activity, production of inflammatory cytokines and chemokines, elimination nd killing of micro- organisms and initiate the development of adaptive immunity, all of which contribute to anti- infective and anti- tumorigenic properties of glucan. Chitin is a polysaccharide and is the main component of insect and crustacean exoskeletons, and the cell walls of certain fungiSome of the researches have reported on its immunostimulant properties in fish (Sahu and Pal, 2011).

(c) N- 3 Fatty Acids

PUFAs of n-3 family acts as nutraceutical in fish diet. It serves as precursors of biologically active eicosanoides, which are in turn important mediators in inflammatory reactions and partly also in the regulation of immune response. Increasing the level of n-3 PUFAs in the diet with fish oils influenced the immune response and hence disease resistance in cultured fish. It has been shown for fish that HUFAs enriched diet increased the resistance of larvae to different stresses such as handling and variations in temperature and salinity. Non- gelatinized carbohydrate supplemented with 1% n-3 PUFA is found to be optimum to enhance the immunity in *Labeo rohita* juveniles (De Silva and Anderson, 1995).

(d) Nucleotides, peptide and amino acids

It has been reported that addition of RNA or Uracil to the diet restores most of the immune functions that's why acts as nutraceuticals in aquaculture. Feeding nucleotide also enhances the adaptive immunity. One of the possible mechanisms by which dietary nucleotides beneficially influence the fish immune system is by partially offsetting the inhibitory effects of cortisol release associated with stress. Antimicrobial peptides are key components of the innate immune system of most multicellular organisms. They have antimicrobial activities but with no resistance potential and they show potential to act as alternatives to antibiotics. Among them, apidaecins refer to a series of small proline rich 18 to 20 residue peptides produced by the insects. Apidaecins are active against a wide range of gram- negative bacteria (Fish pathogens are usually gram negative bacteria), through a bacteriostatic rather than a lytic process, but it shows no activity against a wide range of gram positive bacteria. Some of the amino acids augment the immunity by blocking the cortisol level as prolonged exposure to cortisol causes immune- suppression in fish. Among them tryptophan and pyridoxine are most common. Though tryptophan or pyridoxine are nutrients but they exhibited augmenting the immunity of fish during stress condition. Hence, inclusion of these amino acids beyond the normal range specified for a species having some health benefit may be considered as nutraceuticals, but this needs further debate (Sahu and Pal, 2011).

(e) Carotenoids

β- carotene extracted from *Dunaliella salina* are among nature's best antioxidants, containing a variety of carotenoids including Zeaxanthin, Cryptoxanthin and Lutein. β-carotene is pro vitamin A, a potent antioxidant, able to protect the body from oxidative damage. It can be converted to vitamin A which is essential for bone development and retinol which is important for eye health.

Astaxanthin is a carotenoid found in microalgae consumed by fish and crustaceans. It is a powerful antioxidant. It may maintain healthy cholesterol levels by reducing the effects of oxidative stress. Astaxanthin is shown to protect cellular membranes and ocular tissue against photo- oxidation.

Fucoxanthin is a marine carotenoid responsible for the brown colour in seaweeds. Fucoxanthin may promote the formation of DHA, an ω-3 fatty acid found in fish oil attenuates the weight gain of white adipose tissue and decreases blood glucose (Kannappan, 2011).

(f) Anti- stress nutraceuticals

The herbal antioxidant effect has been shown to be similar to that of superoxide dismutase, metal ion chelators and xanthine oxidase inhibitors. The best example of the herb *Picrorhiza kurroa* is used as the antistress compound for shrimps. Rutin is a bioflavonoid extracted from *Toona sinensis* with strong antioxidant and antistress activity in crustaceans. The herbal extracts, *Astragalus membranaceus, Portulaca oleracea, Flavescent sophora* and *Andrographis paniculata*, act as an antistress and induce the immunological parameters such as serum lysozyme activity, SOD, NOS and levels of total serum protein, globulin and albumin (Sahu and Pal, 2011).

Summary

Feed additives are used in aquaculture for improving growth and immune functions of fish. However, the area of feed supplements in aquaculture is still in infancy stage, but future scope is bright. The immune system of fish is not too strong to protect them from harmful pathogens, leading to complete washout of the crops. In aquaculture some ingredients are used to maintain the water quality of the pond, which indirectly protects the health of fish. On other hand some enzymes like phytase are included in fish feed due to several advantages. There are several advantages of feed additives, which ultimately enhance the sustainable fish production.

References

Alavandi, S.V., Poornima, M and Kalaimani, N. 2011. use of microbes as probiotics and feed additives in aquaculture. *In:* Advances in aquaculture nutrition and feed processing technology, CIBA, Chennai.pp. 57- 62.

Ali, S. A., Ambasankar ,K and Syama Dayal, J. 2011. Protein, lipid and energy requirements of shrimp and fish. *In*: Advances in aquaculture nutrition and feed processing technology, CIBA, Chennai. pp.1- 6.

Dayal, J. Syama and Ambasankar, K. 2011. Immunity enhancement through feed additives. *In:* Advances in aquaculture nutrition and feed processing technology, CIBA, Chennai. pp.76-77.

De Silva, S.S. and Anderson, T.A. 1995. Fish Nutrition in aquaculture. Chapman & Hall. p 319.

De, Debasis and Ghosal. T.K. 2011. Microbial feed supplements in aquaculture. *In:* Advances in aquaculture nutrition and feed processing technology, CIBA, Chennai.pp. 48-51.

Giri, S. S. 2011. Nutritional requirements of candidate species for freshwater aquaculture. *In:* Advances in aquaculture nutrition and feed processing technology, CIBA, Chennai. pp.151-156.

Gopal, C. 2011. Vitamin and mineral requirements of brackish water finfish and shrimp. *In:* Advances in aquaculture nutrition and feed processing technology, CIBA, Chennai. pp. 220-225.

Jain, K. K. 1998. Selection of ideal binder for making stable aquafeed. *In:* Modern approach to aquafeed formulation and on-farm feed management, CIFE, Mumbai. pp.109-111.

Kannappan, S. 2011. Role of carotenoids in crustaceans. *In:* Advances in aquaculture nutrition and feed processing technology, CIBA, Chennai.pp.109-111.

Mohanty, S.N. 2009. Nutritional requirements and feed formulation of carps. *In:* Recent trends in seed production of freshwater finfishes and shellfishes, CIFA, Bhubaneshwar. pp. 97-104.

Nandeesha, M. C. 1993. Aquafeeds and feeding stragies in India. Proc. of the Regional Expert Consultation on Farm- made Aquafeeds (Ed. New M.B., Tocon, A.G.J. and Csavas, I.), 14- 18 Dec., 1992, Bangkok, Thailand. pp. 213-254.

Sahu, N.P. and Pal, A.K. 2011. Nutraceuticals use in aquaculture. *In:* Advances in aquaculture nutrition and feed processing technology, CIBA, Chennai. pp. 156-160.

Swain, P. 2009. Immunity of brood and young fishes. *In:* Recent trends in seed production of freshwater finfishes and shellfishes, CIFA, Bhubaneshwar. pp. 123-126.

Valli, C. 2011. Current concepts and future trends in use of enzymes as feed additives. *In:* Advances in Aquaculture Nutrition and Feed Processing Technology. CIBA, Chennai. pp. 105-108.

23

Essential Oils as Feed Additives for Livestock and Poultry

Sanjay Kumar, Rajni Kumari, Kaushalendra Kumar and A.K Srivastava

Introduction

Public concern over use of antibiotics in livestock production has increased in recent years because of their possible contribution to emergence of antibiotic resistant bacteria, and their transmission from livestock to humans. Accordingly, ruminant microbiologists and nutritionists have been exploring alternative methods of favorably altering ruminal metabolism to improve feed efficiency and animal productivity. Plant extracts contain secondary metabolites, such as essential oils (EO) that have antimicrobial properties that make them potential alternatives to antibiotics to manipulate microbial activity in the rumen. Essential oils are naturally occurring volatile components responsible for giving plants and spices their characteristic essence and color. Over the last few years, a number of studies have examined effects of EO, and their active components, on rumen microbial fermentation. However, many of these studies are laboratory based (*i.e.*, *in vitro*) and of a short-term nature. Nevertheless, results from *in vitro* batch culture studies provide evidence that EO and their components have the potential to improve N and/or energy utilization in ruminants. Effects of EO on ruminal N metabolism is more likely mediated by their impact on hyper-ammonia producing (HAP) bacteria resulting in reduced deamination of amino acids (AA) and production of ammonia N.

Essential oils hold promise as feed additives in ruminant nutrition to improve feed efficiency and control the spread of pathogens in livestock. However identification of EO, or their active components, that favorably alter fermentation without resulting in broad overall inhibition of rumen fermentation, continues to be a major challenge for researchers. Plant extracts offer a unique opportunity in this regard (Wallace, 2004), as many plants produce secondary metabolites, such as saponins and tannins, which have antimicrobial properties. These compounds have been shown to modulate ruminal fermentation to improve nutrient utilization in ruminants (Wang *et al.*, 1996; Hristov *et al.*, 1999). Similarly, the well documented antimicrobial activity of essential oils (EO), and their active components, has prompted a number of scientists to examine the potential of these secondary metabolites to manipulate rumen microbial fermentation to improve production efficiency in ruminants. Contrary to their name, EO are not true oils (*i.e.*, lipids) and are commonly derived from the components responsible for fragrance, or *Quinta essentia*, of plants. Essential oils are considered safe for human and animal consumption, and are categorized as generally recognized as safe (GRAS; FDA, 2004) in the USA. The antimicrobial properties of EO have been demonstrated against a wide range of microorganisms, including bacteria, protozoa and, fungi (Dean and Ritchie, 1987; Sivropoulou *et al.*, 1996; Chao *et al.*, 2000). Essential oils have also been exploited for their activity against a wide variety of food-borne pathogens. For example, *Escherichia coli* O157:H7 was inhibited by oregano oil and its two main components carvacrol and thymol (Helander *et al.*, 1998; Elgayyar *et al.*, 2001).

Definition and chemistry

Essential oils, also known as volatile or ethereal oils, occur in edible, medicinal, and herbal plants. As these aromatic compounds are largely volatile, they are commonly extracted by steam distillation or solvent extraction (Simon, 1990; Greathead, 2003). Essential oils can be extracted from many parts of a plant, including the leaves, flowers, stem, seeds, roots and bark. However, the composition of the EO can vary among different parts of the same plant (Dorman and Deans, 2000). For instance, EO obtained from the seeds of coriander (*Coriandrum sativum* L.) have a different composition from the EO of cilantro, which is obtained from the immature leaves of the same plant (Delaquis *et al.*, 2002). Chemical differences among EO extracted from individual plants, or different varieties of plants, also exist and are attributed to genetically determined properties, age of the plant, and the environment in which the plant grows (Cosentino *et al.*, 1999). For instance, Mart´ýnez *et al.* (2006) observed that the concentration of carvacrol, thymol, *p*-cymene and terpinene in thyme EO varied widely depending on the species of the thyme plant (Table 1).

Table. 1: Concentrations of carvacrol, thymol, *p*-cymene, and y-terpinene in thyme essential oils (adapted from Martýnez *et al.*, 2006)

Component (%)	Source of thyme essential oils	
	Thymus hyemalis Lange	*Thymus zygis sub sp. gracilis*
Carvacrol	24.3	3.13
Thymol	4.79	62.1
p-Cymene	20.93	17.03
y-Terpinene	18.0	3.13

Chemically, EO are variable mixtures of principally terpenoids, mainly monoterpenes (C10) and sesquiterpenes (C15), although diterpenes (C20) may also be present, and a variety of low molecular weight aliphatic hydrocarbons, acids, alcohols, aldehydes, acyclic esters or lactones and exceptionally N- and S-containing compounds, coumarins and homologues of phenylpropanoids (Dorman and Deans, 2000). The major chemical components of some common EO are in Table 2 and examples of the chemical structures of some EO are in Fig. 1.

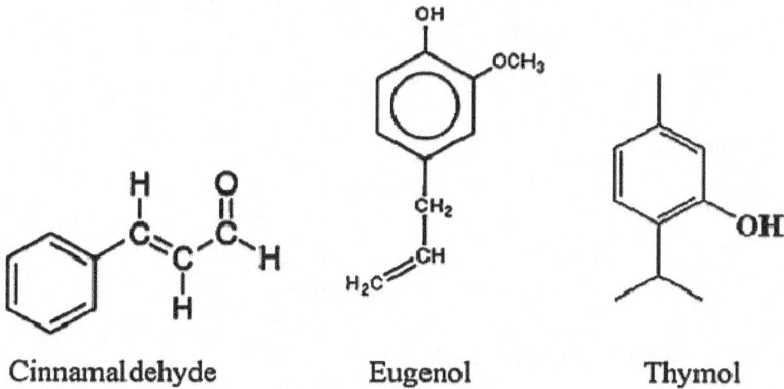

Fig. 1: Examples of chemical structures of some essential oil compounds

Table 2: Examples of some essential oils and their main components (Chao *et al.*, 2000)

Essential oil	Plant part	Botanical source	Main components	% of total
Angelica	Roots	*Angelica archangelica L.*	α- Pinene Delta-3-Carene α-Phellandrene + myrcene	24.7 10.5 10.8 12.9 10.4 7.7
Bergamot	Fruits	*Citrus bergamia* Risso et Poit	β - Pinene Limonene + Limonene + β-Phellandrene y-Terpinene Linalool Linalyl acetate	7.7 3 9.4 8.6 11.1 28.0
Cinnamon	Inner bark	*Cinnamomum zeylanicum Blu.*	(E)-Cinnamaldehyde Eugenol	77.1 7.2
Coriander	Seeds	*Coriandrum sativum* L.	p-Cymene Linalool	6.1 72.0
Dill (Indian)	Seeds	*Anethum sowa* Roxb	Limonene trans-Dihydrocarvone Carvone Dillapiole	50.9 10.4 20.3 36.6
Eucalyptus	Leaves	*Eucalyptus citriodora*	Citronellal Citronellol	72.8 14.5
Ginger	Roots	*Zingiber officinale* Rosc.	Camphene Neral Geranial + bornyl acetate β-Bisabolene ar-Curcumene β-Eudesmol	14.1 4.9 8.1 22.1 14.5 5.4
Juniper	Berries	*Juniperus communis* L.	α- Pinene Sabinene Myrcene	33.7 27.6 5.5
Pepper	Fruits	*Piper nigrum* L.	α- Pinene β- Pinene Sabinene Delta-3-Carene Limonene β-Caryophyllene	9.0 10.4 19.4 5.4 17.5 14.7
Rosemary	Whole plant	*Rosemarinus officinalis* L.	α- Pinene β- Pinene 1,8-Cineole Camphor	7.4 5.0 43.6 12.3
Tea tree	Branches	*Melaleuca alternifolia* L.	α- Terpinene 1,8-Cineole Terpinene-4-ol y-Terpinene	10.4 5.1 40.1 23.0

Antimicrobial properties

Plant secondary metabolites have traditionally held an important role in human health and wellness (Bode and Muller, 2003). Exploited for their essence, flavor, antiseptic and/or preservative properties, plants and their extracts have been used by mankind since early history (Burt, 2004). Early accounts of plant-based, traditional medicine date back to Mesopotamia, approximately 2600BC (Greathead, 2003). Plant-based medicine was widely practised until the early 20th, century at which point it was rapidly phased out due to the introduction of novel and effective synthetic medicines that could be produced more economically and had easily identifiable benefits to human health (Greathead, 2003). In recent years however, the emergence of multi-drug resistant bacteria, and the risk it represents to human health, has renewed interest in plant extracts.

Antimicrobial activities of EO have been demonstrated against a wide variety of microorganisms, including Gram-positive and Gram-negative bacteria. The antimicrobial activity of EO has been attributed to a number of terpenoid and phenolic compounds (Chao *et al.*, 2000), as well as the chemical constituents and functional groups contained in the EO, the proportions in which they are present and the interactions between them (Dorman and Deans, 2000). Additive, antagonistic, and synergistic effects have been observed between components of EO (Burt, 2004). By assessing the minimum inhibitory concentration of oregano EO and its two main constituents, thymol and carvacrol, against *Staphylococcus aureus* and *Pseudomonas aeruginosa*, Lambert *et al.* (2001) observed that the combination of thymol and carvacrol exhibited higher antibacterial activity than either compound alone and that the inhibitory effect of oregano EO is mainly due to the additive antibacterial action of these two compounds. Delaquis *et al.* (2002) examined the antibacterial activity of crude oils and the distilled fractions of dill (*Anethum graveolens* L.), coriander (seeds of *Coriandrum sativum* L.), cilantro (leaves of immature *C. sativum* L.), and eucalyptus (*Eucalyptus dives*) against some common Grampositive and Gram-negative food spoilage bacteria (*i.e.*, *Salmonella typhimurium*, *Listeria monocytogenes*, *S.aureus*, *P.fragi*, *Serratia grimesii*, *Enterobacter agglomerans*, *Yersinia enterocolitica*, *Bacillus cereus*). Results showed that the magnitude and the spectrum of antibacterial activity of these individual fractions frequently exceeded those of the crude oils. For instance, crude dill EO had weak antimicrobial activity, while distilled fractions of dill EO contained higher concentrations of the main chemical constituents, d-limonene and carvone, and exhibited higher antimicrobial activity. Mixing distilled fractions of cilantro and eucalyptus resulted in additive, synergistic, and antagonistic effects against the species of bacteria examined. Mourey and Canillac (2002) evaluated the antibacterial activity of six main components of conifer EO (*i.e.*, α- and β-pinene, *R*- and *S*-limonene, 1,8 cineole, and borneol) against the bacterium *L. monocytogenes*. With the exception 1,8 cineole, all individual components had higher bacteriostatic activity than fir or pine oil (Canillac and Mourey, 1996), suggesting that other components within whole oil reduce, or dilute, the antimicrobial activity of the individual components. In contrast, a comparison of the minimum inhibitory concentration of individual components with that of whole spruce oil (Canillac and Mourey, 2001) showed that individual components were less active, or similar in activity, than whole spruce oil suggesting that the antimicrobial activity of spruce EO is the result of the synergistic effects between individual components contained in the EO.

Mode of action

A number of theories have been proposed to explain the mechanism by which EO exert antibacterial activity. Given that EO comprise a large number of components, it is most likely that their antibacterial activity is not due to one specific mode of action but involves several targets in the bacterial cell (Skandamis *et al.*, 2001; Carson *et al.*, 2002; Burt, 2004). Acamovic and Brooker (2005) suggested that because plant secondary metabolites, including EO, interact with a wide variety of cellular components and can modulate a response at their targets, these compounds have the ability to modulate a large number of cellular targets. It is believed that most EO exert their antimicrobial activities by interacting with processes associated with the bacterial cell membrane, including electron transport, ion gradients, protein translocation, phosphorylation, and other enzyme-dependent reactions (Ultee *et al.*, 1999; Dorman and Deans, 2000). Helander *et al.* (1998) showed that thymol from thyme oil (*Thymus vulgaris*) and carvacrol from oregano oil (*Origanum vulgaris*) both disrupt the cell membrane thereby decreasing the intracellular ATP pool and increasing the extracellular ATP pool in *E. coli*.

Essential oils have a high affinity for lipids of bacterial cell membranes due to their hydrophobic nature, and their antibacterial properties are evidently associated with their lipophilic character. Dorman and Deans (2000) noted that this mechanism is a function of the lipophilic properties of the constituent component of EO and the potency of their functional group. Burt (2004) suggested that Gram-positive bacteria appear to be more susceptible to the antibacterial properties of plant EO compounds than Gram-negative bacteria. This may be expected as Gram-negative bacteria have an outer layer surrounding their cell wall that acts as a permeability barrier, limiting the access of hydrophobic compounds. However, Helander *et al.* (1998) reported that the phenolics thymol and caravacrol also inhibited growth of Gram-negative bacteria by disrupting the outer cell membrane. It appears that the small molecular weight of EO allows them to penetrate the inner membrane of Gram-negative bacteria (Nikaido, 1994; Dorman and Deans, 2000). Trombetta *et al.* (2005) reported that the monoterpenes linalyl acetate, menthol, and thymol were active against Gram-positive *Staphlococcus aureus* and Gram-negative *E. coli.*, and suggested that antimicrobial effect of these monoterpenes is due to the disruption of the plasma membrane of bacteria, thereby interfering with membrane permeability causing intracellular leakage.

Effects on rumen microbial fermentation

The well-documented antimicrobial activity of EO has recently prompted a number of researchers to examine their potential to manipulate ruminal fermentation as a means of improving feed efficiency and nutrient utilization by ruminants. The limited number of EO and EO compounds evaluated to date show some promise in this regard. However, the range of EO and their component compounds available is extensive and many of them have yet to be examined for this purpose. Furthermore, most studies have been conducted *in vitro* and research is needed to determine effects *in vivo*, the mode of action of the various EO and their compounds and the concentrations that favorably modify ruminal fermentation. Various studies have been conducted to determine effects of EO and their components on rumen microbial fermentation. These studies used a wide range of EO and EO compounds, dose rates and diets and, not surprisingly, results have been inconsistent. The varied response among EO products evidently reflects differences in chemical structure, which influences their effects on microbial activity. Initially, ruminant nutritionists were interested in EO mainly because of their role in reducing the palatability of some plant species. Oh *et al.* (1967, 1968) and Nagy and Tengerdy (1968) were the first to investigate effects of EO on ruminal microbial fermentation, as gas production, *in vitro.* Nagy and Tengerdy (1968) observed that EO extracted from Sagebrush (*Artemesia tridentata*) markedly inhibited activity of ruminal bacteria *in vitro*. Oh *et al.* (1967) showed that EO extracted from Douglas fir needles (*Pseudotsuga menziesii*) exerted a general inhibitory effect on ruminal bacteria activity *in vitro*. The degree of inhibition depended, however, on the chemical structure of the EO compound added. Of the compounds evaluated, oxygenated monoterpenes, particularly monoterpene alcohols and aldehydes, strongly inhibited growth and metabolism of rumen microbes, whereas monoterpene hydrocarbons slightly inhibited and, sometimes, stimulated activity of rumen microbes. These findings were some of the first to demonstrate that the chemical composition of EO greatly influences their effects on activity of ruminal microorganisms. Much of the current research with EO in ruminant nutrition has centered upon their potential to improve ruminal N and energy utilization.

Effects on protein metabolism

Symbiosis between ruminants and their microflora instils ruminants with the unique advantage of being able to utilize non-protein sources of N as nutrients. The microbial protein that flows from the rumen to the small intestine provides the host with an excellent source of amino acids (AA) for synthesis of milk and meat proteins. However, the microbial proteins synthesized in the rumen are not sufficient to support the AA requirements of high producing ruminants. Consequently, diets are usually supplemented with sources of feed protein, but such practices can increase feed costs. Furthermore, inefficient N utilization by ruminants results in excretion of N-rich wastes to the environment. Lapierre *et al.* (2005) estimated that about 0.3 of the N consumed by the dairy cow is excreted in urine. Therefore, improving N utilization has a positive impact on efficiency of animal production and on the environment. Several *in vitro* (batch or continuous cultures) studies have been conducted to determine effects of EO and their components on ruminal N metabolism. Early work by Borchers (1965) showed that the addition of thymol to ruminal fluid (1 g/l) containing casein resulted in an accumulation of AA and a decrease in ammonia N (NH_3–N) concentration, suggesting inhibition of AA deamination by ruminal bacteria. Broderick and Balthrop (1979) also observed that thymol inhibited deamination of AA to NH_3–N. More recently, McIntosh *et al.* (2003) observed a 9% reduction in the rate of AA deamination when casein acid hydrolysate was incubated *in vitro* for 48 h in batch cultures of ruminal fluid collected from cows fed a silage-based diet supplemented with 1 g/day of a commercial mixture of EO compounds (MEO; Crina® ruminants; Akzo Surface Chemistry Ltd., Herfordshire, UK). The Crina® supplement contains 100–300 g/kg of phenolic compounds including cresol, resorcinol, thymol, guaiacol and eugenol (Rossi, 1994). Newbold *et al.* (2004) also reported a reduction, "24%, in the *in vitro* rate of AA deamination when casein acid hydrolysate was incubated for 24 h with ruminal fluid collected from sheep fed diets containing 110 mg of Mixture of Essential Oil (MEO). In both studies, peptidolytic and proteolytic activities in ruminal fluid were unaffected by MEO. In the same study, McIntosh *et al.* (2003) observed no additional reduction in the rate of deamination when the ionophore monensin was added to ruminal fluid, suggesting that the bacterial species affected by MEO were the same as those inhibited by monensin. In further work, McIntosh *et al.* (2003) demonstrated that MEO inhibited growth of some (*i.e.*, *Clostridium sticklandii* and *Peptostreptococcus anaerobius*) hyper-ammonia producing (HAP) bacteria, but other HAP bacteria (e.g., *Clostridium aminophilum*) were less sensitive. Hyper-ammonia producing bacteria are present in low numbers in the rumen (<0.01 of the rumen bacterial population), but they possess a very high deamination activity (Russell *et al.*,

1988). Wallace (2004) reported that the number of HAP bacteria was reduced by 77% in sheep receiving a low protein diet supplemented with MEO at 100 mg/day, but that MEO had no effect on HAP bacteria when sheep were fed a high-protein diet. Collectively, results of the studies by McIntosh *et al.* (2003), Newbold *et al.* (2004), and Wallace (2004) suggest that effects of EO on ruminal protein metabolism are on AA degradation (*i.e.*, deamination) and these effects are likely due to inhibition of HAP bacteria. Others have used continuous culture systems to explore effects of EO and their constituents on ruminal N metabolism. Using dual-flow continuous culture fermenters for 8 days of incubation, and maintained at constant pH, Castillejos *et al.* (2005) observed that addition of MEO at 1.5 mg/l had no effect on NH_3–N concentration, bacterial and dietary N flows, degradation of crude protein, or efficiency of microbial protein synthesis. The lack of effect of MEO on N metabolism was attributed to the dose of 1.5 mg/l, which may have been too low to alter activity of ruminal bacteria. However, when Castillejos *et al.* (2007) used the same MEO at higher concentrations (*i.e.*, 5, 50, and 500 mg/l) there was still no effect of MEO on ruminal N metabolism (*i.e.*, ruminal concentrations of NH_3–N, small peptides + AA, and large peptides) in continuous fermenters for 9 days of incubation maintained at constant pH. McIntosh *et al.* (2003) and Newbold *et al.* (2006) suggested that concentrations of MEO above 35 mg/l would be required to substantively alter N metabolism in the rumen, a level that may be difficult to achieve *in vivo*. Indeed, Benchaar *et al.* (2006b, 2007) observed no change in ruminal NH_3–N concentration, N retention, and N digestibility when lactating dairy cows were supplemented with MEO at doses of 0.75 or 2 g/day. Assuming a rumen volume of 100 litres and an outflow rate of 0.1/h for an adult dairy cow, ruminal concentration of MEO would have been 3.1 and 8.3 mg/l for each of the doses, respectively. These ruminal concentrations are indeed much lower than the range of concentrations (*i.e.*, 35–360 mg/l) required for MEO to alter N metabolism of ruminal bacteria (McIntosh *et al.*, 2003). However, McIntosh *et al.* (2003) speculated that local concentrations of EO, some of which are often sparingly soluble, may be higher on the surface of ingested plant materials, which may increase the bactericidal effects of EO *in vivo*. Other studies used the rumen *in situ* bag technique to investigate effects of MEO on ruminal N metabolism. For example, Molero *et al.* (2004) used growing heifers to evaluate the influence of MEO (700 mg/day) on *in situ* ruminal degradability of proteins in soybean meal, corn gluten feed, fish meal, green peas, sunflower meal, and lupin seeds. Of the five protein supplements examined, MEO only tended to reduce the effective ruminal protein degradabilities of lupin seeds, green peas, and soybean meal, and the reductions were too small to have any likely nutritional impact on ruminal protein metabolism in the animal. Recent studies by Newbold *et al.* (2004) and Benchaar *et al.* (2006b) reported no change in the kinetics of protein degradation from soybean meal incubated in the rumen of sheep or dairy cows supplemented daily with 110 mg or 2 g of MEO, respectively. The lack of effects of MEO on N metabolism in long-term (*i.e.*, continuous culture) *in vitro*, ruminal *in situ*, and *in vivo* studies compared to short-term *in vitro* batch culture studies may be related to the duration that ruminal bacteria are exposed to EO. Longer duration of exposure may result in shifts in microbial populations, and it is possible that some of the EO compounds are subject to degradation by ruminal bacteria. Cardozo *et al.* (2004) and Busquet *et al.* (2005c) observed that some of the effects of EO and their main components on rumen microbial fermentation dissipated after 6–7 days of fermentation in a dual flow continuous-culture system, suggesting that rumen microbial populations may adapt to EO. Therefore, results from *in vitro* batch culture studies must be interpreted with caution as they report effects over a set incubation time (*e.g.*, 24 or 48 h) and do not account for possible shifts in microbial populates that may occur as a result of exposure of rumen microbes to EO. More recently, a number of studies have shown that factors such as the chemical composition and dosage rate of EO could influence effects of EO on ruminal N metabolism. Busquet *et al.* (2005c) observed that the addition of clove bud EO (*Syzygium aromaticum*) to a continuous culture fermenter at 2.2 mg/l strongly reduced (*i.e.*, "80%) the concentration of large peptides, but had no effect on NH_3–N concentration, suggesting that clove bud EO reduced the peptidolytic activity of ruminal bacteria. However, addition of the principal component of clove bud EO, eugenol, at the same concentration had no effect on N metabolism, suggesting that the anti-peptidolytic activity of clove bud EO is not due to its main component, but results from unidentified compounds within the oil fraction. In contrast, Busquet *et al.* (2006) reported that, when supplied at the same concentration (*i.e.*, 3000 mg/l), both oregano EO and its major constituent carvacrol reduced the concentration of NH_3–N in *in vitro* batch cultures, indicating that carvacrol accounts for the majority of the antimicrobial activity in oregano EO.

The chemical composition of EO may also affect the manner in which they alter ruminal N metabolism. Castillejos *et al.* (2006) observed variation in the effects of increasing dosage levels (5, 50, 500, and 5,000 mg/l) of different EO compounds on fermentation products in 24 h *in vitro* batch cultures of rumen fluid. The aldehyde vanillin was ineffective at altering NH_3–N concentration at doses of 5, 50 and 500 mg/l while the monoterpene limonene reduced NH_3–N concentration at the dose of 500 mg/l. The phenolic eugenol decreased NH_3–N concentration at concentrations of 5, 50, and 500 mg/l, whereas the phenolic guaiacol decreased NH_3–N concentration at all concentrations. These results illustrate that effects of EO components on N metabolism vary with chemical structure. In general, phenolic compounds have been shown to have high antimicrobial activity due to the presence of a hydroxyl groups within the phenolic structure (Dorman and Deans, 2000; Ultee *et al.*, 2002; Burt, 2004). In a study investigating the antibacterial properties in berries, Puupponen-Pimi¨a *et al.* (2001) determined that the phenolic compounds cinnamic acid, 3-coumaric acid, caffeic acid, and ferulic acid extracted from various berries (*i.e.*, blueberry, cranberry, raspberry, strawberry, and black currants) inhibited Gram-negative bacteria, but were ineffective in inhibiting Gram-positive bacteria. Compounds with phenolic structures have a broad spectrum of activity against a variety of both Gram-positive and Gram-negative bacteria (Helander *et al.*, 1998; Dorman and Deans, 2000; Lambert *et al.*, 2001).

Essential oils and their components have been shown to affect ruminal N metabolism in a dose-dependent manner. For instance, Busquet *et al.* (2006) demonstrated that some EO (*i.e.*, anise oil, cade oil, capsicum oil, cinnamon oil, clove, bud oil, dill oil, garlic oil, ginger oil, oregano, oil, and tea tree oil) and their main components (*i.e.*, anethol, benzyl salicylate, carvacrol, carvone, cinnamaldehyde, and eugenol) markedly inhibited NH_3–N concentration at high concentrations (*i.e.*, 3000 mg/l), but effects were marginal at moderate doses (*i.e.*, 300 mg/l) and nonexistent at low doses (*i.e.*, 3 mg/l). The decreased ruminal NH_3–N concentration was, however, associated with a reduction in total VFA concentration, suggesting a reduction in overall fermentation of the diet. Because VFA are the principal source of energy for ruminants, decreasing ruminal VFA production could have adverse nutritional consequences if this effect was expressed *in vivo*. Information on effects of EO and their constituents on ruminal bacterial N escape are scarce. Flow of bacterial N was not changed by addition of garlic EO or cinnamaldehyde (Busquet *et al.*, 2005a), but it was reduced by cinnamon leaf EO (Fraser *et al.*, 2007) in continuous culture fermenters. Newbold *et al.* (2004) and Benchaar *et al.* (2006b) observed no change in duodenal bacterial N flow of sheep and dairy cows fed daily dosages of 110 mg and 2 g of MEO, respectively. The discrepancy between studies is probably due to the dose, chemical composition of EO, and experimental conditions. Ruminal protozoa have a negative role on utilization of N by ruminants. Protozoa engulf and digest large numbers of ruminal bacteria thereby decreasing net microbial protein flow from the rumen to the duodenum (Ivan *et al.*, 2000). Protozoa also possess proteolytic and deaminating activities (Williams and Coleman, 1992). Thus, removal of protozoa from the rumen (*i.e.*, defaunation) prevents recycling of N between bacteria and protozoa, which results in increased flow of microbial N from the rumen. For instance, bacterial protein flow to the intestine in defaunated sheep was 35% higher than the microbial protein flow in faunated sheep (Ivan *et al.*, 1992).

Increased bacterial protein synthesis in the rumen due to defaunation could benefit the host by supplying additional AA for absorption. Moreover, improved efficiency of N metabolism in the rumen could reduce N losses in feces and urine. Due to the lack of a suitable defaunating agent, and spontaneous refaunation, defaunation has not been practical in commercial ruminant production systems. Plant extracts, such as condensed tannins and steroidal saponins, have been extensively investigated for their inhibitory effects on ciliate ruminal protozoa (Wallace *et al.*, 1994; Wang *et al.*, 1996). However, few studies to date have evaluated the effects of EO and their compounds on ruminal protozoa. Ando *et al.* (2003) reported that feeding 200 g/day (*i.e.*, 30 g/kg of total dietary DM) of peppermint (*Menthaxpiperita* L.) to Holstein steers decreased the total number of protozoa, and the numbers of *Entodinum*, *Isotrica*, and *Diplodium*. Mohammed *et al.* (2004) observed no change in the number of protozoa in ruminal fluid of steers supplemented with cyclodextrin encapsulated horseradish oil at 20 g/kg of dietary DM. McIntosh *et al.* (2003) observed that the bacteriolytic activity of rumen ciliate protozoa was unaffected in dairy cows supplemented with 1 g/day of MEO. Newbold *et al.* (2004) and Benchaar *et al.* (2007) reported that ruminal protozoa counts were not affected when sheep and dairy cows were fed 110 and 750 mg/day of MEO, respectively. Supplementation of dairy cows diets with 1 g/day of cinnamaldehyde had no effect on the number, or generic composition, of ciliate protozoa (Benchaar *et al.*, 2005). Cardozo *et al.* (2006) observed that addition of a mixture of cinnamaldehyde (180 mg/day) and eugnol (90 mg/day) to the diets of beef heifers increased numbers of holotrichs and had no effect on entodiniomorphs, but that there was no effect on numbers of these protozoal species when the mixture contained higher concentrations of cinnamaldehyde (600 mg/day) and eugenol (300 mg/day). In contrast, feeding 2 g/day of anise extract containing 100 g/kg of anethol to beef heifers decreased counts of holotrichs and entodiniomorphs (Cardozo *et al.*, 2006). Overall, EO and their components have no marked effect on numbers and/or activity of ruminal ciliate protozoa.

Effects on volatile fatty acid production

Supplementation with EO or EO compounds has increased ruminal total VFA concentration, which may indicate improved feed digestion. In one such study, addition of 1.5 mg/l of MEO increased total VFA concentration in continuous cultures maintained at constant pH, although there was no concomitant increase in organic matter digestibility (Castillejos *et al.*, 2005). Two *in vivo* studies with MEO reported no effects when fed to sheep (110 mg/day) or cattle (1 g/day) on total VFA concentration or proportions (Newbold *et al.*, 2004; Beauchemin and McGinn, 2006). It is possible that effects of MEO on total

VFA concentration may depend on the composition of the diet. Benchaar *et al.* (2007) reported that MEO (750 mg/day) tended to increase total VFA concentration in the rumen of lactating cows when the diet contained alfalfa silage, but tended to decrease total VFA concentration when the diet contained corn silage. Mohammed *et al.* (2004) reported that increasing levels (*i.e.*, from 0.17 to 1.7 g/l) of cyclodextrin encapsulated horseradish linearly increased total VFA concentration in batch cultures. When the same product was fed to cattle, there was a very small increase in total VFA, but no change in feed digestibility. Overall, supplementation with EO or their components has caused either a decrease or no change in total VFA concentration in most studies. Whether VFA concentration decreases as a result of the antimicrobial effects of EO may be dose dependent. For example, Busquet *et al.* (2006) studied effects of various plant extracts (*i.e.*, anise oil, cade oil, capsicum oil, cinnamon oil, clove, bud oil, dill oil, fenugreek, garlic oil, ginger oil, oregano, oil, tea tree oil, and yucca), and secondary plant metabolites (*i.e.*, anethol, benzyl salicylate, carvacrol, carvone, cinnamaldehyde, and eugenol) on ruminal fermentation in a 24 h batch culture. Each treatment was supplied at varying doses up to 3 g/l of culture fluid. None of the EO or EO compounds increased total VFA concentration but, at the highest concentration, most treatments decreased total VFA concentration, a possible reflection of decreased feed digestion. Similar effects were reported by Castillejos *et al.* (2006) for eugenol, guaiacol, limonene, thymol, and vanillin using doses up to 5 g/l. These EO compounds generally had no effect on total VFA concentration, with the exception of the highest dose, which decreased total VFA concentration in cultures for all compounds. Lack of a change in total VFA concentration could be viewed as desirable if it was accompanied by changes such as decreased NH_3–N concentration, decreased methane production, or a change in molar proportions of VFA. However, a reduction in total VFA production as a result of EO supplementation would generally be viewed as nutritionally unfavorable. The challenge is to identify the dose rates for various EO or EO active components that favorably alter aspects of rumen metabolism without reducing total VFA concentrations. A number of studies have shown that certain EO and their components shift molar proportions of VFA in a manner similar to that of monensin (*i.e.*, decreased acetate and increased propionate proportions; McGuffey *et al.*, 2001), which is viewed as a favorable outcome of EO supplementation. Mohammed *et al.* (2004) reported decreased acetate proportion, and increased propionate proportion, with cyclodextrin encapsulated horseradish both *in vitro* and *in vivo*. Busquet *et al.* (2005a) used cinnamaldehyde and garlic oil added at two doses (*i.e.*, 31.2 and 312 mg/l of culture fluid) in a continuous culture study. At the low dose of cinnamaldehyde, and the high dose of garlic oil, acetate proportion decreased and that of propionate increased. Molar proportion of butyrate also increased at the high dosage rate. Thus the EO, or the active components and dose rates used in that study, resulted in effects similar to monensin, with the exception of the increase in butyrate concentration. In a subsequent study by this group (Busquet *et al.*, 2006), garlic oil (300 and 3000 mg/l) and benzyl salicylate (300 and 3000 mg/l) also reduced acetate and increased propionate and butyrate proportions. High butyrate concentration as a result of supplementation with some EO compounds suggests that the mode of action of these compounds differs from that of monensin. While many studies have shown beneficial changes in VFA profiles, some EO produce undesirable changes in proportions of individual VFA. For example, Castillejos *et al.* (2006) reported that eugenol (500 mg/l) reduced the proportion of propionate, without affecting total VFA concentration. Cardozo *et al.* (2005) showed *in vitro* that effects of EO and their components on VFA profile are pH-dependent. For instance, at pH 7.0, cinnamon EO and its main component cinnamaldehyde resulted in higher acetate to propionate ratio whereas, at pH 5.5, the acetate to propionate ratio was lower with cinnamon EO and cinnamaldehyde. There also appears to be an adaptive response to EO supplementation in the rumen at the bacterial and/or population level. Such a response is a major challenge to developing EO feed additives with long-lasting effects. The adaptive response is particularly evident when low levels of EO are supplemented. Cardozo *et al.* (2004) used cinnamon, garlic and anise oils (7.5 mg/kg DM or 0.22 mg/l) in continuous culture and observed changes in VFA profiles during the first 6 days of microbial adaptation but no effects thereafter. Similarly, Busquet *et al.* (2005b) studied effects of garlic oil on *in vitro* rumen microbial fermentation in a 24 h batch culture. Total VFA concentration decreased at a dose of 300 mg/l but, in a subsequent continuous culture study, there were no effects of garlic oil, at doses up to 312 mg/l, on total VFA concentration (Busquet *et al.*, 2005a). These studies are compelling evidence that microbial populations are able to adapt to EO over time, which presents a challenge for commercial application of this feed additive technology.

Methane production

There is growing worldwide interest in reducing methane emissions from domestic ruminants. Methane is a potent greenhouse gas and its release into the atmosphere is directly linked with animal agriculture, particularly ruminant production. The antimicrobial activity of EO has prompted interest in whether these compounds could be used to inhibit methanogenesis in the rumen. The challenge is to identify EO that reduce methane production without a concomitant reduction in feed digestion. Evans and Martin (2000) observed that thymol (400 mg/l), a main component of EO derived from *Thymus* and *Origanum* plants, was a strong inhibitor of methane *in vitro*, but acetate and propionate concentrations also decreased. Kamra *et al.* (2005) investigated methanol and ethanol extracts of various spices, including fennel, clove, garlic, onion, and ginger for effects on methane production *in vitro*. Among the extracts tested, methanol extract of garlic was the most effective suppressant of methane, with a 64% reduction *in vitro* and no adverse effects on feed digestibility. Similarly, Busquet *et al.* (2005a) reported that garlic oil (312 mg/l) reduced acetate and increased propionate proportions in a manner consistent with decreased methane production *in vitro*, although methane was not directly measured. Evaluating the effects of garlic oil and four of its main components (diallyl sulfide, diallyl disulfide, allylmercaptan, and allicin), Busquet *et al.* (2005b) observed, in batch culture, that garlic oil and diallyl disulfide (300 mg/l of ruminal fluid) reduced methane production by 74 and 69% respectively, without altering digestibility. Interestingly, monensin did not reduce methane production to the same extent as garlic oil or diallyl disulfide. Busquet *et al.* (2005b) suggested that garlic oil and diallyl disulfide did not exert their effects through the same mode of action as monensin, but rather that inhibition of methane production by these compounds was due to the direct inhibition of rumen methanogenic archaea. Patra *et al.* (2005) reported that ethanol and methanol extracts of cloves and the methanol extract of fennel also inhibited methane production *in vitro*, but digestibility of the feed was also reduced. The ground root from Rhubarb (*Rheum officinale*, 1.6 g/l) was reported to reduce methane production *in vitro* by 20% without affecting digestibility, whereas the bark from buckthorn (*Rhamnus frangula*, 1.6 g/l) had no effect on methane (Garcýa-Gonzalez *et al.*, 2005). There may be potential to select EO compounds that reduce methane by selectively inhibiting protozoal numbers, which would be expected to decrease methane production because ruminal protozoa provide a habitat for methanogens that live on and within them. However, the anti-protozoal effects of EO have been inconsistent and variable among EO and EO active components. Few studies have evaluated effects of EO and their main components *in vivo* for effects on methane emissions, and no studies have assessed long-term effects of EO and their constituents on methane production. In one study, MEO was fed (1 g/day) to beef cattle consuming a high forage diet. Methane emissions were not affected, although feed digestibility decreased (Beauchemin and McGinn, 2006). Using the same commercial product in an *in vitro* study using pure cultures, McIntosh *et al.* (2003) reported that growth of the methanogen *Methanobrevibacter smithii* was inhibited, but only when the concentration of the product was 33 fold higher than that fed *in vivo*, as reported by Beauchemin and McGinn (2006), a feeding rate that is impractical due to potentially deleterious effects on diet digestibility. Using another high supplementation rate (*i.e.*, 20 g/kg of DM intake) of encapsulated horseradish, Mohammed *et al.* (2004) observed a 19%

decrease in methane production in steers that was not accompanied by a reduction in protozoal numbers or feed digestibility. It seems clear that there is potential to select EO compounds that selectively reduce methane when used at levels that do not depress feed utilization, but further research is necessary to evaluate these compounds *in vivo*.

Effects on ruminant performance

Few studies have been published on effects of EO or their constituents on milk production and composition of dairy cows. Benchaar *et al.* (2006b, 2007) observed no changes in DM intake, milk production, and milk components when dairy cows were fed 750mg or 2g of MEO daily. Similarly, supplementation of dairy cows with peppermint at 20 g/kg DM had no effect on milk yield and milk composition (Hosoda *et al.*, 2005). More recently, Yang *et al.* (2006) observed that addition of garlic (*Allium sativa*, 5 g/day) and juniper berry (*Juniperus communis*, 2 g/day) oils to dairy cow diets had no effect on DM intake, milk production or milk composition. In these studies, the lack of effect of EO and their active components on milk performance was consistent with the absence of effects of these plant extracts on feed intake and ruminal fermentation. Essential oils have an antibacterial activity against Gram-negative and Gram-positive bacteria (Helander *et al.*, 1998). Several Gram-positive bacteria are involved in ruminal biohydrogenation of unsaturated dietary fatty acids (Harfoot and Hazlewood, 1988). Therefore, feeding EO could lower biohydrogenation of fatty acids by reducing the number, and the activity, of bacteria involved in the biohydrogenation of unsaturated fatty acids. Benchaar *et al.* (2007) reported no change in milk fatty acid profile when cows were supplemented daily with 750 mg of MEO. However, supplementing the same mixture at a higher concentration (*i.e.*, 2 g/day) increased the concentration of conjugated linoleic acid (CLA), a health-promoting fatty acid, in milk fat. Data on effects of EO and their compounds on beef cattle performance are almost nonexistent. In one study, Benchaar *et al.* (2006a) evaluated growth performance of beef cattle fed a silage base diet supplemented with 2 or 4 g/day of a commercial mixture of EO compounds (Vertan®, IDENA, Sautron, France) consisting of thymol, eugenol, vanillin and limonene. Results showed that DM intake and average daily gain were not affected by the addition of this EO compounds mixture. However, the gain to DM intake ratio was affected quadratically with a dose of 2 g/day maximizing feed efficiency.

Control of pathogens

Although it has been shown that EO and its main constituents can inhibit several food borne pathogens including *E. coli* O157:H7, *S. aureus*, *L. monocytogenes*, and *Salmonella* spp. (Burt and Reinders, 2003; Friedman *et al.*, 2004; Penalver *et al.*, 2005; Oussalah *et al.*, 2007) the broad spectrum antimicrobial activity of these compounds may makes it difficult to use them to specifically target pathogens within the ruminant digestive tract. Those EO and EO compounds that exhibit the highest antimicrobial activity against ruminal bacteria (*e.g.*, carvacrol, oregano and thyme oils) are also generally the most potent against pathogens (Oussalah *et al.*, 2007). In some instances, EO may also increase the sensitivity of bacteria to other antimicrobials (Rafii and Shahverdi, 2007; Santiesteban-Lopez *et al.*, 2007) and it is possible that they may also enhance the sensitivity of pathogens within the digestive tract to other microbialcides. Recently, EO from pepper mint was shown to exhibit activity against *Giardia* (Vidal *et al.*, 2007) a protozoal parasite that is highly prevalent in cattle (Olson *et al.*, 2004). It is possible that EO, or their active components, may also have activity against other parasites that reside in the intestine such as *Cryptosporidium*, coccidia or nematodes. Effects of EO on parasites in the lower digestive tract would be dependent on the ability of the antimicrobial components that they contain to remain active after passage through the rumen. To date, the extent to which EO escape from the rumen and flow to the lower digestive tract has not been examined.

Summary

Plant-derived EO may be a useful means to improve efficiency of nutrient utilization in ruminants and reduce the impact of their production on the environment. Most studies to date have been laboratory based (*i.e.*, *in vitro*) and of a short-term nature, but indicate that EO and their active components may favorably alter ruminal fermentation. At high doses, EO and their constituents may inhibit deamination of AA and reduce methane production in the rumen. However, long-term *in vitro* (*i.e.*, continuous cultures) and *in vivo* studies suggest that benefits associated with EO diminish over time due to shifts in microbial populations or adaptation of individual microbial species to EO. Consequently, it may be difficult to continue to realize benefits from EO throughout the feeding or lactation period. Several studies have shown that EO possess strong bactericidal activity against several food borne pathogens. However, the extent to which microbial pathogens exhibit a similar adaptive response to EO is presently unknown. The range of EO and their components is complex in terms of nature and activity and variability in their composition may make it difficult to obtain consistent positive responses in ruminant production from these complex mixtures.

References

Acamovic, T., Brooker, J.D. 2005. Biochemistry of plant secondary metabolites and their effects in animals. *Proc.Nutr. Soc.* 64: 403–412.

Ando, S., Nishida, T., Ishida, M., Hosoda, K., Bayaru, E., 2003. Effect of peppermint feeding on the digestibility, ruminal fermentation and protozoa. *Livest. Prod. Sci.* 82: 245–248.

Beauchemin, K.A., McGinn, S.M. 2006. Methane emissions from beef cattle: Effects of fumaric acid, essential oil, and canola oil. *J. Anim. Sci.* 84: 1489–1496.

Benchaar, C., McAllister, T.A., Chouinard, P.Y. 2005. Effects of cinnamaldehyde, yucca saponins extract and condensed tannins on fermentation characteristics, and ciliate protozoal populations in the rumen of lactating dairy cows. *J. Dairy Sci.* 83: (Suppl.1):304.

Benchaar, C., Duynisveld, J.L., Charmley, E., 2006a. Effects of monensin and increasing dose levels of a mixture of essential oil compounds on intake, digestion and growth performance of beef cattle. *Can. J. Anim. Sci.* 86: 91–96.

Benchaar, C., Petit, H.V., Berthiaume, R., Whyte, T.D., Chouinard, P.Y., 2006b. Effects of addition of essential oils and monensin premix on digestion, ruminal fermentation, milk production and milk composition in dairy cows. *J. Dairy Sci.* 89: 4352–4364.

Benchaar, C., Petit, H.V., Berthiaume, R., Ouellet, D.R., Chiquette, J., Chouinard, P.Y., 2007. Effects of essential oils on digestion, ruminal fermentation, rumen microbial populations, milk production, and milk composition in dairy cows fed alfalfa silage or corn silage. *J. Dairy Sci.* 90: 886–897.

Bode, H., M¨uller, R., 2003. Possibility of bacterial recruitment of plant genes associated with the biosynthesis of secondary metabolites. *Plant Physiol.* 132: 1153–1161.

Borchers, R., 1965. Proteolytic activity of rumen fluid *in vitro*. J. Anim. Sci. 24: 1033–1038.

Broderick, G.A., Balthrop, J.E., 1979. Chemical inhibition of amino acid deamination by minal microbes *invitro*. *J. Anim. Sci.* 49: 1101–1111.

Burt, S.A., Reinders, R.D., 2003. Antibcaterial activity of selected plant essential oils gainst *Escherichia coli*O157:H7. *Lett. Appl. Microbiol.* 36: 162–167.

Burt, S., 2004. Essential oils: Their antibacterial properties and potential applications in foods- a review. *Int. J.Food Microb.* 94: 223–253.

Busquet, M., Calsamiglia, S., Ferret, A., Cardozo, P.W., Kamel, C., 2005a. Effects of cinnamaldehyde and garlic oil on rumen microbial fermentation in a dual flow continuous culture. *J. Dairy Sci.* 88: 2508–2516.

Busquet, M., Calsamiglia, S., Ferret, A., Carro, M.D., Kamel, C., 2005b. Effect of garlic oil and four of its compounds on rumen microbial fermentation. *J. Dairy Sci.* 88: 4393–4404.

Busquet, M., Calsamiglia, S., Ferret, A., Kamel, C., 2005c. Screening for the effects of natural plant extracts and secondary plant metabolites on rumen microbial fermentation in continuous culture. *Anim. Feed Sci. Technol.* 123: 597–613.

Busquet, M., Calsamiglia, S., Ferret, A., Kamel, C., 2006. Plant extracts affect *in vitro* rumen microbial fermentation. *J. Dairy Sci.* 89: 761–771.

Canillac, N., Mourey, A., 1996. Comportement de *Listeria monocytogenes* en pr´esenced'huile essentielle de sapin et de pin. *Sci. Alimen.* 16: 403–411.

Canillac, N., Mourey, A., 2001. Antibacterial activity of the essential oil of *Picea excelsa* on *Listeria*, *Staphlococcus aureus* and coliform bacteria. *Food Microbiol.* 18: 261–268.

Cardozo, P.W., Calsamiglia, S., Ferret, A., Kamel, C., 2004. Effects of natural plant extracts on ruminal protein degradation and fermentation profiles in continuous culture. *J. Anim. Sci.* 82: 3230–3236.

Cardozo, P.W., Calsamiglia, S., Ferret, A., Kamel, C., 2005. Screening for the effects of natural plant extracts at different pH on in vitro rumen microbial fermentation of a high-concentrate diet for beef cattle. *J. Anim.* Sci. 83: 2572–2579.

Cardozo, P.W., Calsamiglia, S., Ferret, A., Kamel, C., 2006. Effects of alfalfa extract, anise, capsicum, and a mixture of cinnamaldehyde and eugenol on ruminal fermentation a n d protein degradation in beef heifers fed a high-concentrate diet. *J. Anim. Sci.* 84: 2801–2808.

Castillejos, L., Calsamiglia, S., Ferret, A., Losa, R., 2005. Effects of a specific blend of essential oil compounds and the type of diet on rumen microbial fermentation and nutrient flow from a continuous culture system. *Anim. Feed Sci. Technol.* 119: 29–41.

Castillejos, L., Calsamiglia, S., Ferret, A., 2006. Effect of essential oils active compounds on rumen microbial fermentation and nutrient flow in *in vitro* systems. *J. Dairy. Sci.* 89: 2649–2658.

Castillejos, L., Calsamiglia, S., Ferret, A., Losa, R., 2007. Effects of dose and adaptation time of a specific blend of essential oils compounds on rumen fermentation. *Anim. Feed Sci. Technol.* 132: 186–201.

Carson, C.F., Mee, B.J., Riley, T.V., 2002. Mechanism of action of Melaleuca alternifolia (t e a tree) oil on Staphylococcus aureus determined by time-kill, lysis, leakage and salt tolerance assays and electron microscopy. *Antimicrob. Agents Chemoth.* 46: 1914–1920.

Chao, S.C., Young, D.G, Oberg, C.J., 2000. Screening for inhibitory activity of essential oils on selected bacteria, fungi and viruses. *J. Essent. Oil Res.* 12: 639–649.

Cosentino, S.,Tuberoso, C.I.G., Pisano, B., Satta, M., Mascia,V., Arzedi, E., Palmas, F., 1999. In vitro antimicrobial activity and chemical composition of Sardinian thymus essential oil. *Lett. Appl. Microbiol.* 29: 130–135.

Dean, S.G, Ritchie, G, 1987. Antibacterial properties of plant essential oils. *Int. J. Food Microbiol.* 5: 165– 180.

Delaquis, R.J., Stanich, K., Girard, B., Massa, G., 2002. Antimicrobial activity of individual and mixed fractions of dill, cilantro, coriander and eucalyptus essential oils. *Int. J. Food Microbiol.* 74: 101–109.

Dorman, H.J.D., Deans, S.G., 2000. Antimicrobial agents from plants: antibacterial activity of plant volatile oils. *J. Appl. Microbiol.* 88: 308–316.

Elgayyar, M., Draughon, F.A., Golden, D.A., Mount, J.R., 2001. Antimicrobial activity of essential oils from plants against selected pathogenic and saprophytic microorganisms. *J. Food Prot.* 64: 1019–1024.

Evans, J.D., Martin, S.A., 2000. Effects of thymol on ruminal microorganisms. *Curr. Microbiol.* 41: 336–340.

FDA, 2004. Food and Drug Administration of the US, 21 CFR 184. Online. Available at: http://www.cfsan.fda.gov/eafus.html.

Fraser, G.R., Chaves, A.V.,Wang, Y.,McAllister, T.A., Beauchemin, K.A., Benchaar, C., 2007. Assessment of the effects of cinnamon leaf oil on rumen microbial fermentation using two continuous culture systems. *J. Dairy Sci.* 90: 2315–2328.

Friedman, M., Henika, P.R., Levin, C.E., Mandrell, R.E., 2004. Antibacterial activities of plant essential oils and their components against *Escherichia coli* O157:H7 and *Salmonella enterica* in apple juice. *J. Agric. Food Chem.* 52: 6042–6048.

Garcia-Gonzalez, R., Lopez, S., Fernandez, M., Gonzalez, J.S. 2005. Effects of the addition of some medicinal plants on methane production in a rumen simulating fermentor (RUSITEC). In: Soliva, C.R., Takahashi, J., Kreuzer, M. (Eds.), Proceedings of the 2nd International Conference of Greenhouse Gases and Animal Agriculture. ETH Zurich, Zurich, Switzerland, pp. 444–447.

Greenhouse Gases and Animal Agriculture. ETH Zurich, Zurich, Switzerland, pp. 115–118.

Harfoot, C.G., Hazlewood, G.P., 1988. Lipid metabolism in the rumen. In: Hobson, P.N. (Ed.), The Rumen Microbial Ecosystem. Elsevier Applied Science Publishers, London, UK, pp. 285–322.

Greathead, H., 2003. Plant and plant extract for improving animal productivity. *Proc. Nutr. Soc.* 62: 279–290.

Helander, I.M., Alakomi, H.-L., Latva-Kala, K., Mattila-Sandholm, T., Pol, L., Smid, E.J., Gorris, L. G. M., von Wright, A., 1998. Characterization of the action of selected essential oil components on Gram negative bacteria. *J. Agric. Food Chem.* 46: 3590–3595.

Hosoda, K., Nishida, T., Park, W.Y., Eruden, B., 2005. Influence of Menthaxpiperita L. (peppermint) supplementation on nutrient digestibility and energy metabolism in lactating dairy cows. *Asian-Aust. J. Anim. Sci.* 18: 1721–1726.

Hristov, A.N., McAllister, T.A., Van Herk, F.H., Cheng, K.-J., Newbold, C.J., Cheeke, P.R., 1999. Effect of Yucca schidigera on ruminal fermentation and nutrient digestion in heifers. *J. Anim. Sci.* 77: 2554–2563.

Ivan, M., Dayrell, M.D., Mahadevan, S., Hidiroglou, M., 1992. Effect of bentonite on wool growth and nitrogen metabolism in fauna-free and faunated sheep. *J. Anim. Sci.* 70: 3194–3202.

Ivan, M., Neill, L., Forster, R., Alimon, R., Rode, L.M., Entz, T., 2000. Effects of Isotricha, Dasytricha, Entodinium, and total fauna on ruminal fermentation and duodenal flow in wethers fed different diets. *J. Dairy Sci.* 83: 776–787.

Kamra, D.N., Agarwal, N., Chaudhary, L.C., 2005. Inhibition of ruminal methanogenesis by tropical plants containing secondary plant compounds. In: Soliva, C.R., Takahashi, J., Kreuzer, M. (Eds.), Proceedings of the 2nd International Conference of Greenhouse Gases and Animal Agriculture. ETH Zurich, Zurich, Switzerland, pp. 102–111.

Lambert, R.J.W., Skandamis, P.N., Coote, P., Nychas, G.-J.E., 2001. A study of the minimum inhibitory concentration and mode of action of oregano essential oil, thymol and carvacrol. *J. Appl. Microbiol.* 91: 453–462.

Lapierre, H., Berthiaume, R., Ragio, G., Thivierge, M.C., Doepel, L., Pacheco, D., Dubreuil, P., Lobley, G.E., 2005. The route of absorbed nitrogen into milk protein. *Anim. Sci.* 80:11–22.

Mart´ýnez, S., Madrid, J., Hern´andez, F., Meg´ýas, M.D., Sotomayor, J.A., Jord´an, M.J., 2006. Effect of thyme essential oils (Thymus hyemalis and Thymus zygis) and monensin on in vitro ruminal degradation and volatile fatty acid production. *J. Agric. Food Chem.* 54: 6598–6602.

McIntosh, F.M., Williams, P., Losa, R., Wallace, R.J., Beever, D.A., Newbold, C.J., 2003. Effects of essential oils on ruminal microorganisms and their protein metabolism. *Appl. Environ. Microbiol.* 69:5011–5014.

McGuffey, R.K., Richardson, L.F., Wilkinson, J.I.D., 2001. Ionophores for dairy cattle: Current status and futureoutlook. *J. Dairy Sci.* 84 (E. Suppl.): E194–E203.

Mohammed, N., Ajisaka, N., Lila, Z.A., Mikuni, K., Hara, K., Kanda, S., Itabashi, H., 2004. Effect of Japanese horseradish oil on methane production and ruminal fermentation *in vitro* and in steers. *J. Anim. Sci.* 82:1839–1846.

Molero, R., Ibara, M., Calsamiglia, S., Ferret, A., Losa, R., 2004. Effects of a specific blend of essential oil compounds on dry matter and crude protein degradability in heifers fed diets with different forage to concentrate ratios. *Anim. Feed Sci. Technol.* 114:91–104.

Mourey, A., Canillac, N., 2002. Anti-listeria monocytogenes activity of essential oils components of conifers. *Food Control* 13:289–292.

Newbold, C.J., McIntosh, F.M., Williams, P., Losa, R., Wallace, R.J., 2004. Effects of a specific blend of essential oil compounds on rumen fermentation. *Anim. Feed Sci. Technol.* 114: 105–112.

Newbold, C.J., Duval, S.M., McEwan, N.R., Y´a˜nez-Ruiz, D.R., Hart, K.J., 2006. New feed additives for ruminants—a European perspective. In: Proceedings of the Pacific Northwest Animal Nutrition Conference and Virtus Nutrition Pre-conference, Vancouver, BC, Canada, pp. 81–90.

Nikaido, H., 1994. Prevention of drug access to bacterial targets: permeability barriers and active efflux. *Science*. 264: 382–388.

Nagy, J.G., Tengerdy, R.P., 1968. Antibacterial action of essential oils of Artemisia as an ecological factor. II. Antibacterial action of the volatile oils of Artemisia tridentata (big sagebrush) on bacteria from the rumen of mule deer. *Appl. Microbiol.* 16: 441–444.

Oh, H.K., Sakai, T., Jones, M.B., Longhurst, W.M., 1967. Effect of various essential oils isolated from Douglas Fir Needles upon sheep and deer rumen microbial activity. *Appl. Microbiol.* 15: 777–784.

Oh, H.K., Jones, M.B., Longhurst,W.M., 1968. Comparison of rumen microbial inhibi resulting from various essential oils isolated from relative unpalatable plant species. *Appl. Microbiol.* 16: 39–44.

Olson, M.E., O'Handley, R.M., Ralston, B.J., McAllister, T.A., Thompson, R.C.A., 2004. Update on Cryptosporidium and Giardia infections in cattle. *Trends Parsitol.* 20: 185–191.

Oussalah, M., Caillet, S., Saucier, L., Lacroix, M., 2007. Inhibitory effects of selected essential oils on the growth of four pathogenic bacteria: E. coli O157:H7, Salmonella typhimurium, Staphylococcus aureus and Listeria monocytogenes. *Food Control* 18: 414–420.

Patra, A.K., Kamra, D.N., Agarwal, N., 2005. Effect of spices on rumen fermentation, methanogenesis and protozoa counts in in vitro gas production test. *In:* Soliva, C.R., Takahashi, J., Kreuzer,M. (Eds.), *Proceedings of the 2nd International Conference of Greenhouse Gases and Animal Agriculture*, ETH Zurich, Switzerland, pp. 115-118.

Penalver, P., Huerta, B., Borge, C., Astorga, R., Romero, R., Perea, A., 2005. Antimicrobial activity of five essential oils against strains of the Enterobacteriaceae family. *APMIS* 113: 1–6.

Puupponen-Pimi¨a, R., Nohynek, L., Meier, C., Kahkonen, M., Heinonen, M., Hopia, A., 2001. Antimicrobial properties of phenolic compounds from berries. *J. Appl. Microbiol.* 90: 494–550.

Rafii, F., Shahverdi, A.R., 2007. Comparison of essential oils from three plants for enhancement of antimicrobial activity of nitrofurantion against enterobacteria. *Chemotherapy*. 53: 21–25.

Rossi, J., 1994. Composition pour am´eliorer la digestibilit´e des aliments destin´es aux animaux ruminants. Demandede brevet europ´een. EP 0,630,577, A1.

Russell, J.B., Strobel, H.J., Chen, G., 1988. Enrichment and isolation of a ruminal bacterium with a very high specific activity of ammonia production. *App. Environ. Microbiol.* 54: 872–877.

Santiesteban-Lopez, A., Palou, E., Lopez-Malo, A., 2007. Suceptibility of food-borne bacteria to binary combinations of antimicrobials selected *a*w and pH. *J. Appl. Microbiol.* 102: 486–497.

Simon, J.E., 1990. Essential oils and culinary herbs. In: Janick, J., Simon, J.E. (Eds.), Advances in New Crops. Timber Press, Portland, OR, pp. 472–483.

Sivropoulou, A., Papanikolaou, E., Nikolaou, C., Kokkini, S., Lanaras, T., Arsenakis, M., 1996. Antimicrobial and cytotoxic activities of Origanum essential oils. *J. Agric. Food Chem.* 44: 1202–1205.

Skandamis, P.,Koutsoumanis, K., Fasseas, K., Nychas, G.-J.E., 2001. Inhibition of oregano essential oil and EDTA on Escherichia coli O157:H7. *Italian J. Food Sci.* 13 (1): 65–75.

Trombetta, D., Castelli, F., Sarpietro, M.G,Venuti,V., Cristani, M., Daniele, D., 2005. Mechanisms of antibacterial action of three monoterpenes. *Antimicrob. Agents Chemother.* 49: 2474-2478.

Ultee, A., Kets, E.P.W., Smid, E.J., 1999. Mechanisms of action of carvacrol on the food-borne pathogen *Bacillus cereus*. *Appl. Environ. Microbiol.* 65: 4606–4610.

Ultee, A., Bennik, M.H.J., Moezelaar, R. 2002. The phenolic hydroxyl group of carvacrol is essential for action against the food-borne pathogen *Bacillus cereus*. *Appl. Environ. Microbiol.* 68: 1561–1568.

Vidal, F., Vidal, J.C., Gadelha, A.P.R., Lopes, C.S., Coelho, M.G.P., Monteiro-Leal, L.H., 2007. *Giardia lamblia*: the effects of extracts and fractions from *Mentha×piperita* Lin. (*Lamiaceae*) on trophozoites. *Exp. Parastiol.* 115: 25–31.

Wang, Y., Douglas, G.B., Waghorn, G.C., Barry, T.N., Foote, A.G., Purchas, R.W., 1996. Effects of condensed tannins upon the performance of lambs grazing *Lotus corniculatus* and lucerne (*Medicago sativa*). *J. Agric. Sci. (Camb.)* 126: 87–98.

Wallace, R.J., Arthaud, L.C., Newbold, J., 1994. Influence of *Yucca schidigera* extract on ruminal ammonia concentrations and ruminal microorganisms. *Appl. Environ. Microbiol.* 60: 762–1767.

Wallace, R.J., 2004. Antimicrobial properties of plant secondary metabolites. *Proc. Nutr. Soc.* 63: 621–629.

Williams, A.G. and Coleman, G.S., 1992. The Rumen Protozoa. Springer-Verlag, New York, NY, USA.

Yang, W.Z., Chaves, A.V., He, M.L., Benchaar, C., McAllister, T.A., 2006. Effect of monensin and essential oil on feed intake, milk yield and composition of lactating dairycows. *Can. J. Anim. Sci.* 86: 598 r.

24

Phytogenic Feed Additives

Pankaj Kumar Singh, A.K. Srivastava, Sanjay Kumar and Rajni Kumari

Introduction

Plant products have been used for centuries by humans as food and to treat ailments. Natural medicinal products originating from herbs and spices have also been used as feed additives for farm animals in ancient cultures for the same length of time. Keeping farm animals healthy is necessary to obtain healthy animal products. The main scope in animal husbandry is to ensure good performance of farm animals and get quality animal products, which can be achieved only with the effort to keep the animals healthy. Only quality feed together with proper hygiene, potable water and management can ensure the production of nutritious animal products with desired organoleptic properties (Saxena, 2008). A ban of antibiotics as feed additives in animal nutrition is realized since 1986 in Sweden and later on to various countries. A general ban is foreseen in some years from now, because of the increased occurrence of pathogens resistant against therapeutical antibiotics used in animals and humans. With the restricted use or ban of dietary antimicrobial agents we must explore new ways to improve and protect the health status of farm animals, to guarantee animal performance and to increase nutrient availability (Wenk, 2003). In this aspect, herbs and spices are not just appetite and digestion stimulants, but can, with impact on other physiological functions, help to ensure good health and welfare of the animals, what can positively affect their performance.

Herbs, spices and their extracts were already used thousands of years ago in Mesopotamia, Egypt, India, China and old Greece, where they were appreciated for their specific aroma and various medicinal properties (Greathead, 2003). For the last decade the use of additives of natural origin in animal and human nutrition has been encouraged. Numerous researches focused on the clarification of the biochemical structures and physiological functions of various feed additives like probiotics, prebiotics, organic acids and plant extracts. When discussing the use of herbs and spices as feed additives, we can hardly rely only on old believes about health impact of certain herbs and spices or their active components. We need a scientific proof of their beneficial effect on health and performance of the animals to justify their use. The technological progress enables us to more easily determine the structure and function of yet unidentified active molecules of plant origin.

What are Herbs, Spices or Botanicals?

Beside the feed enzymes, probiotics (for monogastric animals mainly lactobacilli), prebiotics (oligosaccharides), organic acids, the herbs and botanicals can be used as feed additives. In the last years the modern western world has been learning what many Asians and native Americans (Bye and Linares, 1999) have known for centuries, namely that plant extracts and spices can play a significant role in health and nutrition.

A definition can be derived from Webster's Encyclopedic Unabridged Dictionary of the English Language (1989) as follows:

Herb: A flowering plant whose stem above ground does not become woody and persistent. A plant when valued for its medical properties, flavor, scent, or the like.

Spices: Any of a class of pungent or aromatic substances of vegetable origin, as pepper, cinnamon, cloves, and the like, used as seasoning, preservatives, etc.

Botanical: A drug made from part of a plant, as from roots, leaves, bark etc. Essential oils are any of a class of volatile oils obtained from plants, possessing the odour and other characteristic properties of the plant, used chiefly in the manufacture of perfumes, flavours and pharmaceuticals (extracts after hydro - distillation).

To gain advantageous effects of herbs and spices, they can be added to feed as dried plants or parts of plants and as extracts. The composition of extracts from the same plant depends on the method of extraction and the properties of the extraction solvent used. Depending on the chemical characteristics of extraction solvents we can extract only certain molecules. There is also a difference between purified and unpurified extracts. Unpurified extracts contain a number of different molecules extracted with certain solvent, which can affect the action of each other, while purified extracts contain only one active component. The purified active molecules extracted from plants can be sometimes substituted by synthetic naturally identical molecules. Plants mainly contain one or some predominant active molecules (secondary metabolites), which are responsible for certain biological effects. The amount of these molecules varies depending on the variety of plant, growing conditions, harvest time etc. When we need the effect of a specific active component, it is more efficient to use a purified molecule alone than a dried plant or unpurified extract. But we have to be aware, that the potency of an unpurified extract often exceeds the potency of a purified one because of synergistic effect among the molecules in it. The effect of active components from herbs and spices depends largely on the dosage used. No effect whatever can be observed at small doses; on the other hand, large amounts can be even toxic.

Plants have evolved a wide range of low molecular weight secondary metabolite. Generally these compounds enable the plants to interact with the environment and may act in a defence system against physiological and environmental stress as well as predators or pathogens. Beside compounds with toxic properties, several of these secondary plant metabolites have been reported to show beneficial effects in food products and also in mammalian metabolism. They are of main relevance in herbs and are specifically enriched and eventually standardized in botanicals. In the last years the number of studies with herbs or botanicals has significantly increased.

Mode of action of phytobiotics

The herbs and plant extracts used as feed additives include many different bio-active ingredients such as essential oils, alkaloids, flavonoids, glycosides, mucilage, saponins and tannins etc. (Wang *et al.*, 1998; Wenk, 2003). Therefore, the effects of herbs and plant extracts also vary. Beneficial effects of herbs or botanicals in farm animals may arise from activation of feed intake and secretion of digestive secretions, immune stimulation, anti-bacterial, coccidiostatic, anthelmintic, antiviral or anti-inflammatory activity and antioxidant properties. The mode of action of most phytogenic feed additives is still not fully understood. However, the following are some of the potential mechanisms by which they may improve performance.

Antimicrobial action

There are a large number of herbs which have antibacterial activities (Cowman, 1999). In recent years, the use of phytogenic compounds has gained momentum for their potential role as natural alternatives to antibiotic growth promoters in animal nutrition (Lee *et al.,* 2014). Numerous secondary metabolites formed by plants serve as defence agents against physiological and environmental stressors, predators and pathogenic microorganisms. Several *in vitro* studies showed strong antimicrobial activity of certain plant extracts against Gram" and Gram+ bacteria. Pasqa *et al.* (2006) found a change in long chain fatty acid profile in the membranes of *E. coli* grown in the presence of limonene or cinnamaldehyde. Similar observations were made with *Salomonella enterice* grown in the presence of carvacrol or eugenol and with *Bronchotrix thermosphacta* grown in the presence of either limonene, cinnamaldehyde, carvacrol or eugenol. In the case of *Pseudomonas fluorescens* in *Staphylococcus* aureus none of the tested phytochemicals changed the fatty acid profile. The changes in fatty acid composition can affect surviving ability of microorganisms.

Studies measuring hydrophobicity of *E. coli* (test for measuring the ability of microbial attachment) showed a large increase of hydrophobicity of *E. coli* grown in the presence of St. John's wort or Chinese cinnamon and a moderate increase when medium was supplemented with thyme or Ceylon cinnamon. The differences in hydrophobicity were in good correlation with MIC50 values (minimal inhibitory concentration). This confirms the fact that herbs and spices act as antimicrobial agents by changing the characteristics of cell membranes, and causing ion leakage, thus making microbes less virulent (Windisch *et al.*, 2008). The exact antimicrobial action of herbs and spices in *in vivo* situations is hard to evaluate, because of the very complex and balanced microbial populations in gastrointestinal tract and the interaction of active components from herbs and spices with other nutrients. Feed supplements with growth promoting activity increase stability of feed and beneficially influence the gastrointestinal ecosystem mostly through growth inhibition of pathogenic microorganism's growth. Due to improved health status of digestive system, animals are less exposed to the toxins of microbiological origin. Consequently herbs and spices help to increase the resistance of the animals exposed to different stress situations and increase the absorption of essential nutrients, thus improving the growth of the animals (Windisch *et al.*, 2008). Castillo *et al.* (2006) reported that the mixture of cinnamaldehyde, capsicum oleoresin and carvacrol enhances the growth of lactobacilli, and so increases the ratio of lactobacilli to enterobacteria. So herbs and spices do not posses only the antimicrobial activity, but also modulate the composition of microbial population.

Anti - Inflammatory action

Extracts of curcuma, red pepper, black pepper, cumin, cloves, nutmeg, cinnamon, mint and ginger showed anti-inflammatory effect in the studies on rats (Srinivasan, 2005; Manjunatha and Srinivasan, 2006). The major active molecules with anti-inflammatory action are terpenoids and flavonoids. These molecules suppress the metabolism of inflammatory prostaglandins.

Antioxidative action

Many active components of herbs and spices can prevent lipid peroxidation through quenching free radicals or through activation of antioxidant enzymes like superoxide dismutase, catalase, glutathione peroxidise and glutathione reductase. Main molecules responsible for the antioxidative properties of herbs and spices are phenolic substances (flavonoids, hydrolysable tannins, proanthocianidins, phenolic acids, phenolic terpenes) and some vitamins (E, C and A). Often used herbs rich in phenolics are: rosemary, thyme, garlic, oregano, sage, green tea, chamomile, ginko, dandelion and marigold. The health promoting effect of antioxidants from plant is thought to arise from their protective effects by counteracting reactive oxygen species. Antioxidants are compounds that help delay and inhibit lipid oxidation and when added to foods tend to minimize rancidity, retard the formation of toxic oxidation products, and help maintain the nutritional quality (Aghsaghali, 2012). Herbs and spices can protect the feed against oxidative deterioration during storage. This is a widely used practice in pet food and human food industry. The herb commonly used for feed/food preservation is rosemary (*Rosmarinus officinalis*). It can be used alone or in combination with tocopherols or synthetic antioxidants (Jacobsen *et al.*, 2008).

Immunostimulant function

The immune system generally benefits from the herbs and spices rich in flavonoids, vitamin C and carotenoids. The plants containing molecules which possess immunostimulatory properties are echinacea, liquorice, garlic and cat's claw. These plants can improve the activity of lymphocytes, macrophages and NK cells, they increase phagocytosis or stimulate the interpheron synthesis (Craig, 1999).

Increased feed intake

The stimulatory effect of phytogenics on feed intake is due to the improvement in palatability of the diet resulting from the enhanced flavour and odor, especially with the use of essential oils (Papatisiros *et al.*, 2009). A product from rhizomes of Sanguinaria canadensis is frequently used in Europe for pigs and poultry. With other herb mixtures or botanicals effects on feed intake as well as growth performance in piglets and growing finishing pigs as well as poultry could be achieved. In has been concluded that in both species (piglet and broiler) a slight increase in feed intake could be observed at least partly at low levels and then a dramatic decrease at higher levels of a particular herb. Similar observations were made in laying hens with turmeric, a powder of the rhizome of Curcuma longa, a spice that is frequently used in the south Asian kitchen (Wenk, 2003) 0.25% Turmeric improved feed intake. At higher levels up to 1% feed intake returned to the control treatment without supplementation.

Appetite and digestion stimulants

Herbs develop their initial activity in the feed of farm animals as flavour and can therefore influence the eating pattern, secretion of digestive fluids and total feed intake. Herbs or the phytochemicals can influence selectively the micro-organisms by an anti-microbial activity or by a favourable stimulation of the eubiosis of the microflora. The consequence can be a better nutrient utilization and absorption or the stimulation of the immune system. Finally herbs can contribute to the nutrient requirements of the animals and stimulate the endocrine system and intermediate nutrient metabolism.

When considering supplementing the feed with herbs and spices or their extracts to stimulate the appetite, we have to know the taste preferences of different animal species. Janz *et al.* (2007) found that pigs preferred the feed supplemented with garlic or rosemary over the feed supplemented with oregano or ginger. Furthermore, Jugl-Chizzola *et al.* (2006) noticed that weaned pigs consumed significantly less feed if it was supplemented with thyme or oregano. If pigs in this experiment had the possibility to choose among feed with or without above mentioned spices, they had chosen the unsupplemented feed. The spices known for their appetite stimulant effect are cinnamon, cloves, cardamom, laurel and mint.

Due to the wide variety of active components, different herbs and spices affect digestion processes differently. Most of them stimulate the secretion of saliva. Curcuma, cayenne pepper, ginger, anis, mint, onions, fenugreek, and cumin enhance the synthesis of bile acids in the liver and their excretion in bile, what beneficially effects the digestion and absorption of lipids. Most of the prelisted spices stimulate the function of pancreatic enzymes (lipases, amylases and proteases), some also increase the activity of digestive enzymes of gastric mucosa (Srinivasan, 2005). Besides the effect on bile synthesis and enzyme activity, extracts from herbs and spices accelerate the digestion and shorten the time of feed/food passage through the digestive tract (Platel and Srinivasan, 2001; Suresh and Srinivasan, 2007).

Commonly used Herbs

Turmeric (*Curcuma longa*)

Turmeric (*Curcuma longa*) is a rhizomatous herbaceous perennial plant of the ginger family, Zingiberaceae with medicinal properties known to possess antimicrobial and anti-oxidant properties. It is native to tropical South Asia and needs temperatures between 20°C and 30°C and a considerable amount of annual rainfall to thrive. Rhizome is the portion of medicinal importance and is usually boiled, cleaned, dried and powdered before usage. In medieval Europe, turmeric became known as Indian saffron, since it was widely used as an alternative to the far more expensive saffron spice. Turmeric is commonly called "Pasupu" in Telugu, Kaha in Sinhala, Manjal in Tamil, "Arisina" in Kannada, "Haridra" in Sanskrit and haldar or Haldi in Hindi (Eevuri and Putturu, 2012).

Chemical Composition: Dried turmeric contains 6.3% protein, 5.1% fat, 3.5% minerals, 69.4% carbohydrates and 13.1% moisture. Turmeric contains up to 5% essential oils and up to 5% curcumin, a polyphenol. Curcumin is the active substance of turmeric which is known as C.I. 75300, or Natural Yellow 3. The systematic chemical name is (1*E*, 6*E*)-1,7-bis(4-hydroxy-3-methoxy- phenyl)-1,6-heptadiene-3,5-dione. The active ingredients of turmeric are tetrahydro curcuminoids, curcumin, dimethoxy curcumin and bismethoxy curcumin (HMPC, 2009).

Uses: The continuing research indicates that turmeric and its active compound "Curcumin" are unique anti-oxidants, antimutagenic, antitumorigenic, anticarcinogenic, anti-inflammatory, antiarthritis, antimicrobial and hypocholesterolemic properties (Miquel *et al.*, 2002; Gowda *et al.*, 2009). Therapeutic properties of turmeric includes anti-oxidant, anti-diabetic, antibacterial, antifungal, antiprotozoal, antiviral and hypocholesteremic activities (Abbas *et al.*, 2010; Ahmadi, 2010).

Turmeric and curcumin have been shown to protect liver against a variety of toxicants including carbon tetrachloride, aflatoxin B_1 and cyclophospha-mide in mouse, rat and duckling (Soudamini and Kuttan, 1992). The curcuminoids (yellowish pigments) present in turmeric powder have shown protective effect against aflatoxin B_1 (Gowda *et al.*, 2008). The traditional usage of turmeric in various conditions like biliary disorders, anorexia, cough, diabetes, wounds, hepatic disorders, rheumatism and sinusitis. Kirtikar and Basu (1996) reported that Rhizome is also used as carminative, diuretic, hepatoprotective and in treatment of urinary tract and skin diseases like scabies, leech bites and bruises.

Tulsi (*Ocimum tenuiflorum*)

Ocimum tenuiflorum, Holy Basil (also *Tulsi*, *tulasî*), is an aromatic plant in the family Lamiaceae, which is native throughout the Old World tropics and widespread as a cultivated plant and an escaped weed. The two main morphotypes cultivated in India and Nepal are green-leaved (Sri or Lakshmi *Tulsi*) and purple-leaved (Krishna *Tulsi*).

Chemical composition: Tulsi contains eugenol (1-hydroxy 2-methoxy 4 allyl benzene) a phenolic compound and ursolic acid having pharmacological effects. Some of the main chemical constituents of *Tulsi* are: oleanolic acid, ursolic acid, rosmarinic acid, eugenol, carvacrol, linalool, β-caryo-phyllene, β-elemene (c.11.0%), β-caryophyllene (about 8%), and germacrene D (about 2%) (Prakash and Gupta, 2005).

Uses: Tulsi is a well known therapeutic agent for several pathological conditions having antistress and antioxidant properties and also possess remarkable biological activities like antimicrobial, immunomodulatory, anti-cancerous, anti-oxidant, anti-inflammatory, hepatoprotective and cardioprotective etc. Prakash and Gupta (2005) reported that other therapeutic potential action of tulsi includes antifungal, antispasmodic, antiemetic, analgesic, hypolipidemic and antiviral also. Furthermore, tulsi could inhibit the lipid peroxidation. It is an effective treatment for reducing blood glucose levels (Rai *et al.*, 1997) and total cholesterol levels (Suanarunsawat *et al.*, 2011). Tulsi also shows some promise for protection from radiation poisoning and cataracts. The fixed oil has demonstrated antihyperlipidemic and cardio-protective effects in rats fed a high fat diet. Experimental studies have shown an alcoholic extract of *Tulsi* modulates immunity, thus promoting immune system function (Mondal et.al, 2011). *O. sanctum* extracts are antibacterial against *E. coli*, *S. aureus* and *P. aeruginosa*).

Tulsi is considered to be an adaptogen, balancing different processes in the body, and helpful for adapting to stress. Marked by its strong aroma and astringent taste, it is regarded in Ayurveda as a kind of "elixir of life" and believed to promote longevity. *Tulsi* extracts are used in ayurvedic remedies for common colds, headaches, stomach disorders, inflammation, heart disease, various forms of poisoning, and malaria. Traditionally, *tulsi* is taken in many forms: as herbal tea, dried powder, fresh leaf, or mixed with *ghee*. Essential oil extracted from Karpoora *Tulsi* is mostly used in skin preparations due to its antibacterial activity. The dried leaves have been mixed with stored grains to repel insects (Biswas and Biswas, 2005).

Tulsi is a sacred plant for Hindus and is worshipped by Hindus as the avatar of goddess Lakshmi and plants are grown in front of or near their home. Water mixed with the petals is given to the dying to raise their departing souls to heaven. The ritual lighting of lamps each evening during *Kartika* includes the worship of the *tulsi* plant, which is held to be auspicious for the home. Vaishnavas traditionally use *japa malas* made from *tulsi* stems or roots, which are an important symbol of initiation. *Tulsi malas* are considered to be auspicious for the wearer, and believed to put them under the protection of Hanuman.

Amla (*Phyllanthus emblica*)

Phyllanthus emblica (syn. *Emblica officinalis*), the Indian gooseberry, or *aamla*, is a deciduous tree of the Phyllanthaceae family and is known for its edible fruit. Common name of this tree include Usiri (in Telugu), Nellikai (in Tamil and Kannada).

Chemical Composition: Amla powder contains 5.05 to 6.78 per cent moisture, 0.23 to 0.59 per cent fat and minerals like Calcium 79.6mg, Phosphorous 12.38 mg and Iron 88.03mg/100g (Mishra et.al, 2009). Amla is one of the richest sources of Vit-C and contains 700mg. In addition to this several active tannoid principles (Emblicanin A, Emblicanin B, Punigluconin and pedunculagin) have been identified for their health benefits (Kim *et.al*, 2005).

Uses: Medical studies conducted on Amla fruit suggest that it has anti-viral properties and also functions as an anti-bacterial and anti-fungal agent. Amla has been particularly indicated for anemia, asthma, bleeding gums, diabetes, chronic lung disease, hyperli-pidaemia, yeast infections, scurvy and cancer. Amla has been known in Ayurvedic medicine for its tonifying, anti-ageing and immune enhancing properties (Lalla *et al.*, 2001). Animals fed on amla powder showed better ability for uptake and killing of bacteria, which might be due to the presence of tannins which stimulates phagocytic cells.

Indian gooseberry has undergone preliminary research, demonstrating *in vitro* antiviral and antimicrobial properties. There is preliminary evidence *in vitro* that its extracts induce apoptosis and modify gene expression in osteoclasts involved in rheumatoid arthritis and osteoporosis (Pennolazzi *et al.*, 2008). It also promoted the spontaneous repair and regeneration process of the pancreas occurring after an acute attack (Sidhu *et al.*, 2011). Experimental preparations of leaves, bark or fruit have shown potential efficacy against laboratory models of disease, such as for inflammation, cancer, age-related renal disease, and diabetes. It has antioxidant property even though it has high density of tannins. The fruit also contains other polyphenols: flavonoids, kaempferol, ellagic acid and gallic acid (Rehman *et al.*, 2005).

According to Ayurveda, aamla fruit is sour (*amla*) and astringent (*kashaya*) in taste (*rasa*), with sweet (*madhura*), bitter (*tikta*) and pungent (*katu*) secondary tastes (*anurasas*) and all parts are used in ayurveda / unani medicines. Balances both Pitta and *vata* by virtue of its sweet taste. It may be used as a *rasayana* (rejuvenative) to promote longevity, and traditionally to enhance digestion (*dipanapachana*), treat constipation (*anuloma*), reduce fever (*jvaraghna*), purify the blood (*raktaprasadana*), reduce cough (*kasahara*), alleviate asthma (*svasahara*), strengthen the heart (*hrdaya*), benefit the eyes (*chakshushya*), stimulate hair growth (*romasanjana*), enliven the body (*jivaniya*), and enhance intellect (*medhya*) (Rehman *et al*., 2005). Indian gooseberry is a common constituent and most notably is the primary ingredient in an ancient herbal *rasayana* called *Chyawanprash*. *Emblica officinalis* tea may ameliorate diabetic neuropathy.

Aloe vera (*Aloe barbadensis*)

The botanical name of aloe vera is *Aloe barbadensis miller*. It belongs to *Asphodelaceae* (*Liliaceae*) family, and is shrubby, or absorescent, perennial, xerophytic, succulent, pea-green color plant. It has long triangular, fleshy leaves that have spikes along the edges. The fresh parenchymal gel from the centre of the leaf is clear, this part is sometimes dried to form *aloe vera* concentrate or diluted with water to treate aloe juice products. The sticky latex liquid is derived from the yellowish green pericyclic tubules that line the leaf (rind): this is the part that yields laxative anthraquinones (Schulz *et al*, 1997). Aloes are indigenous to South Africa and South America, but are now cultivated worldwide except in tundra deserts and rainy forests.

Chemical Composition: Dried aloe contains 73.07% carbohydrates, 4.73% protein, 0.27% fat and trace amounts of tannins (0.155g/100g), oxalate (0.68‘3g/100g) and Phytate (0.54g/100g). Aloe vera contains phyto-chemicals like saponins (5.651g/100g), flavanoids (3.246g/100g), alkaloids (2.471g/100g) and phenols (0.232g/100g) phenols, which is an indicative of cosmetic and medicinal value of Aloe barbadensis. It is also rich in minerals like Na, K, P, and Mg (Adesuyi *et al*., 2012).

Uses: Aloe vera may be effective in treatment of wounds and it also promotes the rate of wound healing. Topical application of aloe vera may also be effective for genital herpes and psoriasis. Aloe vera extracts may be useful in the treatment of diabetes and elevated blood lipids in humans (Boudreau and Beland, 2006) which is due to the presence of compounds such as mannans, anthraquinones and lectins.

Table 1: Commonly used plants, its active components and functions (Frankic, 2009)

Sr. No.	Plant	Used parts	Major active component	Functions
Aromatic spices				
1.	Nutmeg	Seed	Sabinene	Digestion stimulant, antidiarrhoeic
2.	Cinnamon	Bark	Cimetaldehyde	Appetite and digestion stimulant, antiseptic
3.	Cloves	Cloves	Eugenol	Appetite and digestion stimulant, antiseptic
4.	Carda-mom	Seed	Cineol	Appetite and digestion stimulant
5.	Coriander	Leaves, Seed	Linalol	Digestion stimulant
6.	Cumin	Seed	Cuminaldehyde	Digestive, Carminative, Galactagogue
7	Anise	Fruit	Anethol	Digestion Stimulant, Galactagogue
8	Celery	Fruit, Leaves	Phtalides	Appetite and digestion stimulant
9.	Parsley	Leaves	Apiol	Appetite and digestion stimulant, antiseptic
10.	Fenu-greek	Seed	Trigonelline	Appetite stimulant
Pungent spices				
11.	Capsicum	Fruit	Capsaicin	Digestion stimulant
12.	Pepperr	Fruit	Piperine	Digestion stimulant
13.	Horsrad-ish	Root	Allylizotiocianat	Appetite stimulant
14.	Mustard	Seed	Allylizotiocianat	Digestion stimulant
15.	Ginger	Rizome	Zingerone	Gastric stimulant
16	Garlic	Bulb	Allicin	Digestion stimulant, antiseptic
Herbs				
17	Rosemary	Leaves	Cineol	Digestion stimulant, antiseptic, antioxidant
18	Thyme	Whole plant	Thymol	Digestion stimulant, antiseptic, antioxidant
19	Sage	Leaves	Cineol	Digestion stimulant, antiseptic, carminatif
20	Laurel	Leaves	Cineol	Appetite and digestion stimulant, antiseptic
21	Mint	Leaves	Menthol	Appetite and digestion stimulant, antiseptic

Aloe vera extracts have been used as immunostimulant that aids in fighting cancers in cats and dogs (King *et al.,* 1995). Extracts of aloe vera might have anti-bacterial and anti-fungal activities which possibly could help to treat minor skin infections such as boils, benign skin cysts and may inhibit growth of fungi causing tinea. Juice from the pulp is useful for treating jaundice, menustrual disorders, scalp disorders, skin diseases, burns and haemorrhoids. Moghadassi and Verma (2011) has reviewed that it is useful for skin damaged from X-rays. On other hand concentration of glucose in gelatin results in high osmotic pressure that protects skin from live bacteria. Aloe vera includes antrokinone chemicals that are known for anti-virus, anti-bacterial and anti-cancer properties.

Effect of Phytobiotics in Livestock and Poultry

Phytobiotics have beneficial effects on production, repruction and health status in livestock and poultry. Beneficial effects of herbs or botanicals in farm animals may arise from activation of feed intake and secretion of digestive secretions, immune stimulation, anti-bacterial, coccidiostatic, anthelmintic, antiviral or anti-inflammatory activity and antioxidant properties. Often the desired activity of herbs is not constant. Conflicting results may arise from the natural variability of the composition of plant secondary metabolites. Variety and environmental growth conditions, harvesting time and state of maturity, method and duration of conservation and storing, extraction method of the plants, as well as possible synergistic or antagonistic effects, anti-nutritional factors or microbial contamination are factors which may substantially affect the results of *in vivo* experiments.

Poultry

Phytobiotics have been used to imrove production and health status of poultry by various research workers. Different effects of phytobiotics can be summarized as follows:

Body weight gain

Improved body weight gain was observed in broilers by inclusion of turmeric powder @ 0.75-1% in diet (Al-Kaissie *et al.*, 2011). Higher gain in body weights were observed by addition of turmeric @ 2 and 3 g/kg feed in broilers (Samarasinghe *et.al.*, 2003). Higher body weight was a consequence of increased feed consumption. Significant increase in body weights of broilers were observed with supplementation of *Ocimum sanctum* leaves (Funde, 2005; Lanjewar *et.al.*, 2008). Gupta and Charan (2007) included tulsi powder ranging from 0-600 mg/kg diet and maximum weight gain in broilers was observed with 200 mg/kg of diet. A combination of amla, Tulsi and turmeric in broiler diets has resulted significant higher body weight gains in broilers (Reddy, 2010).

Feed intake

Increased feed consumption in broilers was observed by supplementation of diet with powdered rhizome of *Curcuma longa* @ 1g/kg (Kumari *et al.*, 1994). Apparently lower feed intake was observed with individual supplementation of turmeric and Tulsi leaf powder in broiler diet (Lanjewar *et al.,* 2008). A combination of Amla, Tulsi and Turmeric (0.25%, 0.5%) in broilers diet has significantly increased feed intake in broilers ((Reddy, 2010).

Feed efficiency

Cabuk *et al.* (2006) measured production parameters of broilers which were supplemented by a mixture of oregano, laurel, sage, anis and citrus essential oils. The mixture of essential oils significantly improved feed conversion, which was attributed to more effective availability of nutrients due to the changes in intestinal ecosystem. Pande (2000) observed significant ($P<0.01$) improvement in feed efficiency of broilers with the supplementation of polyherbal preparations containing tulsi as an integral part of Composition. Turmeric powder inclusion (0.75%-1%) has improved feed efficiency in broilers (Ahmadi, 2010; Al-Kaissie *et al.*, 2011). Supplementation of turmeric powder at the rate of 0.5 per cent level resulted in better feed efficinecy of 1.2 and 2.0% in starter and finisher phase, respectively (Durrani *et al.*, 2006). Significant feed conversion ratio was observed by inclusion of *Aloe vera* and *Curcuma longa* in broilers (Mehala and Moorthy, 2008). Broilers fed with Amla, Tulsi and Turmeric either alone or in combination @ 0.25% and0.5% levels resulted in better feed efficiency ((Reddy, 2010).

Mortality pattern

No mortality upto 35 days of age in broilers was observed when turmeric was supplemented at 0.5 percent level in broiler diet (Durrani *et al*, 2006), and less mortality was observed @ 0.1% inclusion (Kumar *et al.,* 2005). Supplementation of amla and turmeric powder @ 5g/kg of feed in broiler diet resulted no mortality in broilers (Singh *et al.*, 2007) and less mortality by inclusion of amla, tulsi and turmeric @ 0.25 and 0.5% (Reddy, 2010). Hundred per cent livability was observed with inclusion of Aloe vera and Curcuma longa and their combination in broiler diets (Mehala and Moorthy, 2008).

Economics of phytobiotics supplementation

Inclusion of turmeric powder @ 0.5% in ration of broilers has substantially decreased the feed cost per unit live body weight gains ranging from 13.5 to 6.2 compared to control rations (Eevuri and Putturu, 2012). A combination of

Aloe vera and *Curcuma longa* inclusion in broiler diets has resulted significant difference in the feed cost per unit live body weight in the broiler weights up to six weeks of age (Mehala and Moorthy, 2008). Reddy (2010) reported a reduction of 4% feed cost per unit live weight gain in broilers with feed supplemented with combination of amla, tulsi and turmeric @ 0.25%.

Carcass traits

A reduction of fat percentage at 1% (Al-Sultan, 2003) and 1.2% (Samarasinghe *et.al.*, 2003) over body weights were reported by inclusion of turmeric powder in broiler ration. There was increase in the liver weight (Kurkure *et.al.*, 2002), spleen weight (Al-Sultan, 2003) and whole giblets weight (Durrani *et al.*, 2006) with supplementation of turmeric powder, while tulsi leaf product supplementation has no insignificant effect. A combination of amla and turmeric powder @5g/Kg feed in broiler diet has improved the dressing percentage (Singh *et al.*, 2007).

Serum biochemical constituents

Supplementation of broiler diets with tulsi leaf powder had a significant increase in serum HDL, cholesterol in laying hens was reported by Deshpande (2006). The levels of liver enzymes (ALT & ALP) were substantially reduced by feeding broilers with diet supplemented with turmeric powder and Tulsi leaf powder (Prakash *et al.,* 2009). Supplementation of tulsi leaf powder in feeds containing aflatoxins had significantly reduced AST, ALT, and ALP enzyme activities (Sapcota and Upadhyaya, 2009).

Significant reduction of total serum cholesterol, serum LDL cholesterol and serum triglycerides was observed in broilers fed with tulsi leaf powder supplementation (Lanjewar *et. al.,* 2008). Lower SGOT levels in serum was obtained in broilers fed with tulsi supplementation, whereas no significant difference in SGPT, uric acid, and creatinine (Gupta and Charan, 2007). Supplementation of broiler diet with amla has resulted in lower cholesterol, higher SAP, higher SGPT and normal SGOT in broilers (Vidhyarthi *et. al.,* 2008)

Immune response

Feeding of broilers on diet supplemented with Curcuma *longa* (1g/kg) has increased humoral response against RD vaccine tested by HA and HI tests (Kumari *et al.*, 1994). Emadi and Kermanshahi (2007) reported increased immunoglobulins IgA, IgM, IgG and decrease in ratio of monocytes in broilers fed with turmeric rhizome powder. Combination of amla and tulsi in broiler diet improved the antibody titres against Newcastle disease (Pande and Vijaykumar, 1994). Feeding of broilers with diet supplemented with amla, tulsi and turmeric powder either alone or in combination resulted in high HI titre values to ND vaccination (Reddy, 2010).

Coccidia prevention

Resistance of coccidia to currently used coccidiostatics to treat coccidiosis represents a serious problem in poultry industry. The use of plant extracts to treat coccidiosis is not a new approach. When searching for the best natural extract to treat coccidiosis, we have to take into account that the extract needs to be at least partially soluble in lipids to penetrate the cellular membrane, because coccidia are located inside the cells. Two plants, *Dichroa febrifuga* and *Sophora flavescens* are rich in alkaloids which are effective in treating coccidiosis (Youn and Noh, 2001). As infections with *Emeria tenela* include also lipid peroxidation in the intestine, herbs and spices with strong antioxidant potency may represent a good supportive treatment. In one of the latest studies Naidoo *et al.* (2008) studied the capacity of four plants which would be appropriate to treat coccidiosis: leaves of *Combretum woodii*, leaves and stem of *Artemisia afra*, a whole plant and seeds of *Vitis vinifera*. Extracts of all chosen plants improved the feed conversion to the same extent as coccidiostatic toltrazuril.

Pigs

In the pig production, most problems can be expected in the time of weaning. Weaning can be accompanied by infections, especially with enterotoxic *Escherichia coli*. The use of herbs and spices in piglet nutrition can reduce the incidence of infections. Results from Roselli *et al.* (2007) showed that alicin from garlic protects intestinal cells from increased permeability of membrane in pigs infected with *E. coli*. Garlic also contains active substances which suppress the action of fungi and viruses (Zigger, 2001) and improve the feed intake and daily weight gain of piglets (Janz *et al.*, 2007). Cinnamaldehyde, an active component of cinnamon, possesses antibacterial properties. Zigger (2001) observed larger feed intake and live weight gain of weaned pigs fed feed supplemented with garlic and cinnamon extracts. The mortality due to intestinal disorders dropped from 3.9 to 1.2%. Namkung *et al.* (2004) found that a mixture of cinnamon, thyme and oregano extracts inhibited the growth of coliform bacteria.

Combination of carvacrol, cinnamaldehyde and capsicum oleoresin beneficially effected gastrointestinal ecosystem and gastric emptying of weaned pigs (Manzanilla *et al.*, 2004). The same mixture was tested for its antioxidative properties in laboratory and found that the mixture effectively protected pig's blood lymphocytes against oxidative DNA damage at the concentration of 271.2 mg/ kg of feed. Its effect was comparable to that of 90.4 mg/kg of vitamin E. The antioxidant capacity of propylene glycol extracts effectively prevented oxidative DNA damage in peripheral lymphocytes (measured as % DNA in the tail of the comet and OTM (Olive tail moment), but did not prevent lipid peroxidation, measured by 8-OHdG (8-hydroxydeoxyguanosine). Although most studies concerning the effect of herbs and spices in pig production have been conducted on piglets, Allan *et al.* (2005) carried out an experiment on swine. Swine were fed 1000 ppm of dried oregano leaves and flowers enriched with 500 g/kg of oregano essential oil. Observed beneficial effects of oregano supplementation were: lower mortality rate, less culling during lactation period, shorter service interval, more live born and less stillborn piglets.

Ruminants

Herbs and spices have been introduced also to ruminant nutrition. Microbial ecosystem in the rumen is composed from complex anaerobic microbial population of bacteria, fungi, protozoa, methanogeneous arachea and bacteriphage. Numerous metabolites produced in rumen during microbial fermentation affect the basic digestive and metabolic functions and productivity of the host. Researchers have been searching for new possibilities to modulate the microbial fermentation in the rumen. The main goal of manipulating the rumen fermentation is to increase the effectiveness of digestion and metabolism of nutrients, to increase the productivity of the animals and to suppress the undesirable processes as methanogenesis. In intensive farming systems the feed additives, including antibiotics, were used to increase the production of milk, meat and wool. The ban on antibiotic use increases the production costs what triggered the need to search for antibiotic alternatives also in ruminant nutrition.

There are numerous studies showing beneficial effects of herbs and spices on feed intake, immune functions and health, rumen fermentation and productivity of calves, dairy cows, heifers and also beef cattle (Kraszewski *et al.*, 2002; Greathead, 2003; Wawrzynczak *et al.*2000; Cardozo *et al.* 2006). There are some data of the positive effect of plant supplements in nutrition of sheep and goats (Butter *et al.*, 1999). Kudke *et al.* (1999) supplemented calves with powder of *Azadirachta indica* tree. Supplemented calves had higher weight gain than unsupplemented ones. The unsupplemented group had much higher incidence of parasite infections. Gladine *et al.* (2007) tested the antioxidant effect of marigold, grape, rosemary and citrus extracts in sheep. Lipid peroxidation was induced by continuous infusion of linseed oil into the duodenum. The extracts were applied directly into rumen through the rumen cannula. The results showed that all tested plant extracts kept their antioxidant capacity *in vivo* in sheep. The most bio-efficient in limiting lipid peroxidation was marigold extract. Cardozo *et al.* (2006) studied the effect of alfalfa extract, anise, capsicum, and a mixture of cinnamaldehyde and eugenol on ruminal fermentation in beef heifers. The results indicated that tested concentrations of cinnamaldehyde and eugenol mixture, anise oil and capsicum oil may be used as modifiers of rumen fermentation in beef production systems. Same authors tested six natural plant extracts (garlic, cinnamon, anise, yucca, oregano and capsicum extract) and three secondary plant metabolites (cinnamaldehyde, eugenol, anethole) at five doses and two different pH (7.0 and 5.5) to determine their effect on *in vitro* microbial fermentation using ruminal fluid of heifers (Cardozo *et al.*, 2005). Results demonstrated that the effect of herbs and spices on ruminal fermentation in beef cattle may differ depending on ruminal pH. At pH 5.5, garlic, capsicum, yucca and cinnamaldehyde altered ruminal fermentation in favour of propionate, which is more energetically efficient.

Results obtained in the research of Benchaar *et al.*(2007) showed limited effects of 750 mg/day of essential oil mixture (thymol, eugenol, vanillin, guaiacol and limonene) on nutrient utilization, ruminal fermentation, and milk performance of cows fed diets containing alfalfa or corn silage as a sole forage source. Polish researchers showed that 2% of mixture of *Urtica dioica*, *Pradix teraxci*, *Agrimonia eupatoria*, *Fructus carvi* and *Matrica Chamonilla* improves the quality of milk (Kraszewski *et al.*, 2002). Tannins, the secondary plant metabolites found in stem, wood, leaves, fruits and seeds of many plant species can positively affect the protein digestion in ruminants. Tannins bind to proteins and form complexes which pass through the rumen undegraded. These proteins which pass the microbial degradation in the rumen are then successfully utilized by the animal and provide the proteins necessary especially in the special physiological states (like early lactation) and in the cases when feed is not of the best quality (Waghorn *et al.*, 1990). Extracts from herbs and spices help to prevent and alleviate different kinds of health problems. They are effective in treatment of endometritis (inflammation of the endometrium) in cows. Esparza-Borges and Ortiz-Márquez (1996) evaluated the effect of extracts of garlic (*Allium sativum*, L), eucalypt (*Eucalyptus globulus*, Labill.) and *Gnaphalium conoideum* on acute endometritis of Holstein cows. The most effective of all extracts was the garlic extract, however, also eucalypt worked beneficially. Tannins also prevent bloat of the rumen (Butter *et al.*, 1999) and possess antihelmitic properties (Barry and McNabb, 1999).

Summary

With the trend towards more "natural" animal production systems, anti-microbial agents are being banned. In this aspect, herbs and spices are not just appetite and digestion stimulants, but can, with impact on other physiological functions, help to sustain good health and welfare of the animals and improve their performance. Several phytochemicals like essential oils or dietary fiber can contribute to a balanced microflora (eubiosis), an optimal precondition for an effective protection against pathogenic micro-organisms and an intact immune system. Herbs and botanicals contain many different antioxidants with a high potential for the protection of nutrients against oxidation in the digestive tract, in metabolism as well as in the products. Herbs and Spices are used in the nutrition of poultry and they affect the gain in body weight, increase the feed intake, improve feed conversion ratio, reduced mortality, better carcass quality, better serum biochemical constituents, inproved immune response and reduces the cost of production. Current studies show promising results regarding the use of phytochemicals as growth and production promoters. There is still a need to clarify the phytochemical composition and the mechanisms of action for many herbs, spices and their extracts and furthermore, to assess the appropriate dose that should be safely used in specific circumstances and animal species.

References

Abbas, R.Z., Iqbal, Z., Khan, M.N., Zafar, M.A and Zia, M.A. 2010. Anticoccidial activity of curcuma longa L. in broilers. *Brazilian Archives of Biology and Technology*. 53: 63-67.

Adesuyi, A.O., Awosanya, O.A., Adaramola, F.B. and Omeonu, A.I. 2012. Nutritional and Phytochemical Screening of Aloe barbadensis. *Current Research of Journal Biological Sciences*. 4(1): 4-9.

Aghsaghali, A.M. 2012. Importance of medical herbs in animal feeding: A review. *Annals of Biological Research*. 3(2): 918-923.

Ahmadi, F. 2010. Effect of turmeric powder on performance, oxidative stress state and some of blood parameters in fed on diets containing aflatoxin. *Global Veterinaria*. 5: 312-317.

Al-Kassie, G.A.M., Mohseen, A.M. and Abd-Al-Jaleel, R.A. 2011. Modification of productive performance and physiological aspects of broilers on the adition of a mixture of cumin and turmeric to the diet. *Research Opinions in Animal & Veterinary Sciences*. 1: 31-34.

Allan P and Bilkei G. 2005. Oregano improves reproductive performance of sows. *Theriogenology*.63: 716–721

Al-Sultan, S.I. 2003. The effect of Curcuma longa (turmeric) on overall performance of broiler chickens. *International Journal of Poultry Science*. 2:354-53.

Barry T.N and McNabb, W.C. 1999. The implications of considered tannins on the nutritive value and temperature forage fed to ruminants. *British Journal of Nutrition*. 81: 263–272

Benchaar, C., Petit, H.V., Berthiaume, R., Ouellet, D.R., Chiquette, J and Chouinard, P.Y. 2007. Effects of essential oils on digestion, ruminal fermentation, rumen microbial populations, milk production, and milk composition in dairy cows fed alfalfa silage or corn silage. *Journal of Dairy Science*. 90: 886–897

Biswas, N.P. and Biswas, A.K. 2005. Evaluation of some leaf dusts as grain protectant against rice weevil Sitophilus oryzae (Linn.). *Environment and Ecology*. 23(3):485–88.

Boudreau, M.D. and Beland, F.A. 2006. An evaluation of the biological and toxicological properties of Aloe barbadensis (miller), Aloe vera. *J. Environ Sci. Health Environ Carcinog Ecotoxicol Rev.*, 24(1): 103-54.

Butter, N.L., Dawson J.M and Buttery P.J. 1999. Effect of dietary tannins of ruminants. In: Secondary plant products. Caygill J.C., Mueller-Harvey I. (eds.). Nottingham, Nottingham University Press: 51–70.

Bye, R. and Linares, E. 1999. Medicinal plant diversity in Mexico and its potential for animal health sciences. In: Proc. Alltech's 15th Annual Symp. on Biotechnology in the Feed Industry. (Ed.T.P. Lyons and K.A. Jacques). pp. 265-294.

Çabuk M., Bozkurt, M., Alçiçek, A., Akbaþ, Y and Küçükyýlmaz, K. 2006. Effect of a herbal essential oil mixture on growth and internal organ weight of broilers from young and old breeder flocks. *South African Journal of Animal Science*. 36:135–141.

Cardozo, P.W., Calsamiglia, S., Ferret, A and Kamel, C. 2005. Screening for the effects of natural plant extracts at different pH on *in vitro* rumen microbial fermentation of a high-concentrate diet for beef cattle. *Journal of Animal Science.* 83: 2572–2579.

Cardozo, P.W., Calsamiglia, S., Ferret, A and Kamel, C. 2006. Effects of alfalfa extract, anise, capsicum, and a mixture of cinnamaldehyde and eugenol on ruminal fermentation and protein degradation in beef heifers fed a high-concentrate diet. *Journal of Animal Science.* 84: 2801–2808

Castillo, M., Martín-Orúe, S.M., Roca, M., Manzanilla, E.G., Badiola, I., Perez, J.F and Gasa J. 2006. The response of gastrointestinal microbiota to avilamycin, butyrate, and plant extracts in early-weaned pigs. *Journal of Animal Science.* 84: 2725–2734

Cowan, M.M.1999. Plant products as antimicrobial agents. *Clinical Microbiology Reviews*. 12: 564-582.

Deshpande, R.R. 2006. Effect of dietary supplementation of Tulsi leaf powder (Ocimum sanctum) on egg yolk cholesterol and serum lipid profile in commercial layers. M.V.Sc. thesis submitted to Maharashtra Animal and Fishery Sciences University, Nagpur, India.

Durrani, F.R., Mohammad, Ismail., Asad, Sultan., Suhail, S.M., Naila, Chand and Durrani, Z. 2006. Effect of different levels of feed added turmeric (Curcuma longa) on the performance of broiler chicks. *Journal of Agricultural Biology Science.* 1:9-11.

Eevuri,T. R and Putturu, R. 2013. Use of certain herbal preparations in boiler feeds- A review. *Vet. World*. 6(3): 172-179.

Emadi, M. and Kermanshahi, H. 2007. Effect of turmeric rhizome powder on immunity responses of broiler chickens. *Medwell Journal of Animal and Veterinary Advances*. 6(7): 833-36.

Esparza-Borges, H and Ortiz-Márquez, A. 1996. Therapeutic efficacy of plant extracts in the treatment of bovine endometritis. *Acta Horticulturae* (ISHS),426:39–46

Frankic, T., Voljc, M., Salobir, J and Rezar,V. 2009. Use of herbs and spices and their Extracts in animal nutrition. *Acta Agri. Slovan*. 94 (2) : 95-102

Funde, S.T. 2005. Effect of *Ocimum sanctum*, *Emblica officinalis* and Levamisole on immune response in immunosuppressed broilers. Department of Veterinary medicine. Thesis submitted to Maharastra Animal and Fishery Science University, Nagpur.

Gladine, C., Rock, E., Morand, C., Bauchart D and Durand, D. 2007. Bioavailability and antioxidant capacity of plant extracts rich in polyphenols, given as a single acute dose, in sheep made highly susceptible to lipoperoxidation. *British Journal of Nutrition*. 98: 691–701

Gowda, N.K.S., Ledoux, D.R., Goerge, E.R., Bermudez, A.J. and Chen, Y.C. 2009. Antioxidant efficacy of curcuminoids from turmeric (Curcuma longa L.) powder in broiler chickens fed diets containing aflatoxin B1. *British Journal of Nutrition*. 102: 1629-1634.

Gowda, N.K.S., Ledoux, D.R., Rottinghaus, G.E., Bermudez, A.J and Chen, Y.C. 2008. Efficacy of turmeric (*curcuma longa*), containing a known level of curcumin, and a hydrated sodium calcium aluminosilicate to ameliorate the adverse effects of aflatoxin in broiler chicks. *Poultry Science.* 87: 1125-1130

Greathead, H. 2003. Plants and plants extracts for improving animal productivity. *Proceedings of the Nutrition Society.* 62: 279–290

Gupta, G. and Charan, S. 2007. Exploring the potentials of Ocimum sanctum (Shyama tulsi) as a feed supplement for its growth promoter activity in broiler chickens. *Indian Journal of Poultry Science.* 42(2):140-43.

HMPC (Committee on Herbal Medicinal Products). 2009. Assessment Report on *Curcuma Longa* L. Rhizoma. European Medicines Agency.

Jacobsen, C., Let, M.B., Nielsen, N.S and Meyer, A. S. 2008. Antioxidant strategies for preventing oxidative flavour deterioration of foods enriched with n-3 polyunsaturated lipids: a comparative evaluation. *Trends in Food Science & Technology.* 19: 76–93.

Janz, J.A.M., Morel, P.C.H., Wilkinson, B.H.P and Purchas, R.W. 2007. Preliminary investigation of the effects of low-level dietary inclusion of fragrant essential oils and oleoresins on pig performance and pork quality. *Meat Science.* 75: 350–355.

Jugl-Chizzola, M., Ungerhofer, E., Gabler, C., Hagmüller, W., Chizzola, R., Zitterl-Eglseer, K and Franz, C. 2006. Testing of the palatability of *Thymus vulgaris* L. and *Origanum vulgare* L. As flavouring feed addititve for weaner pigs on the basis of a choice experiment. *Berliner und Münchener Tierärztliche Wochenschrift.* 119: 238–243.

Kim, H.J., Yokozawat, Kimhy, Tohda, C., Rao, T.P. and Juneja, L.R. 2005. Influence of Amla (Emblica Officinalis Gaertnl) on hypercholesterolemia and lipid peroxidation in cholesterol-fed rats. *J. Nutr. Sci. Vitaminol.* 51:413-418.

King, G.K., Yates, K.M. and Greenlee, P.G. 1995. The effect of Acemannan Immunostimulant in combination with surgery and radiation therapy on spontaneous canine and feline fibrosarcomas. *Journal of the American Animal Hospital Association.* 31 (5): 439–47.

Kirtikar, K.R. and Basu, B.D. 1996. Indian *Medicinal Plants. Vol. IV.* International Book Distributors, Dehradun. pp. 2423-25.

Kraszewski J., Wawrzynczak S and Wawrzynski M. 2002. Effect of herb feeding on cow performance, milk nutritive value and technological suitability of milk for processing. *Annals of Animal Science.* 2(1): 147–158.

Kudke, R.J., Kalaskar S.R and Nimbalkar, R.V. 1999. Neem leaves as feed supplement for livestock. *Pushudhan.* 14: 12.

Kumar, M., Choudhary, R.S and Vaishnav, J . K. 2005. Effect of supplemental prebiotic, probiotic and turmeric in diet on the performance of broiler chicks during summer. *Indian Journal of Poultry Science.* 40:137-141.

Kumari, P., Gupta, M.K., Ranjan, R., Singh, K.K and Yadava, R. 1994. Curcuma longa as feed additive in broiler birds and its patho-physiological effects. *National Toxicology Program Technical Report Series.* 435:1-18.

Kurkure, N.V., Kalorey, D.R. and Ali, M.H. 2002. Herbal Nutraceuticals: An alternative to antibiotic growth promoters. *Poultry Fortune.* 31-32.

Lalla, J.K., Hamrapurkar, P.D. and Mamania, H.M. 2001.Triphala churna from raw materials to finished products. *Indian Drugs.* 38:87-94.

Lanjewar, R.D., Zanzad, A.A., Ramteke, B.N. and Deshmukh, G.B. 2008, Effect of dietary supplementation of tulsi (O. *sanctum*) leaf powder on the growth performance and serum lipid profile in broilers. *Indian Journal of Animal Nutrition.* 25(4):395-97.

Lee, J. H. , Cho, S., Paikl, H. D., Choi, C. W., Nam, K. Thwang., S. G and Kim, S. K. 2014. Investigation on antibacterial and antioxidant activities, phenolic and flavonoid contents of some thai edible plants as an alternative for antibiotics. *Asian Australasian J. Anim. Sci.* 27(10): 1461-1468.

Manjunatha, H and Srinivasan, K. 2006. Protective effect of dietary curcumin and capsaicin on induced oxidation of low-density lipoprotein, iron-induced hepatotoxicity and carrageenan-induced inflammation in experimental rats. *The FEBS Journal*. 273: 4528–4537.

Manzanilla, E.G., Perez, J.F., Martin, M., Kamel, C., Baucells, F and Gasa J. 2004. Effect of plant extracts and formic acid on the intestinal equilibrium of early-weaned pigs. *Journal of Animal Science*. 82: 3210–3218

Mehala, C. and Moorthy, M. 2008. Production performance of broilers fed with *Aloe vera* and *Curcuma longa* (Turmeric). *International Journal of Poultry Science*. 7(9):852-56

Miquel, J., Bernd, A., Sempere, J.M., Diaz-Alperi, J and Ramirez, A. 2002.Curcuma antioxidants: pharmacological effects and prospects future clinical use. *A review Archives of Gerontology and Geriatrics*. 34:37-46.

Mishra P, Srivastava V, Verma D, Chauhan, O.P. and Rai, G.K. 2009. Physico-chemical properties of chekiya variety of Amla (*Emblica Officinalis*) and effect of different dehydration methods on quality of powder. *African Journal of Food Science*. 3(10):303-06.

Moghaddasi, S.M. and Verma, S.K. 2011. Aloe vera their chemicals composition and applications: A review. *Int J Biol Med Res*., 2(1): 466-471.

Mondal, S., Varma, S., Bamola, V. D., Naik, S. N., Mirdha, B. R., Padhi, M. M., Mehta, N and Mahapatra, S.C. 2011. Double-blinded randomized controlled trial for immunomodulatory effects of *Tulsi* (*Ocimum sanctum* Linn.) leaf extract on healthy volunteers. *Journal of Ethnopharmacology*.136 (3):452–56.

Naidoo, V., McGaw, L.J., Bisschop, S.P.R., Duncan, N and Eloff, J.N. 2008. The value of plant extracts with antioxidant activity in attenuating coccidiosis in broiler chickens. *Veterinary Parasitology*. 153: 214–219.

Namkung, H., Li, M., Gong, J., Yu, H., Cottrill, M and Lange, C.F.M. 2004. Impact of feeding blends of organic acids and herbal extracts on growth performance, gut microbiota and digestive function in newly weaned pigs. *Canadian Journal of Animal Science*. 84 (4): 697–704.

Pande, C.B and Vijaykumar. 1994. A study on immunomodulating action of Zeestress, Proc. Scientific symposium on recent advances in Veterinary Microbiology, B.C.K.V.V., Nadia, W.B, India.

Pande, C.B. 2000. Zeestress-A promising adaptogenic, antistress and immuno modulator – A review. *Pashudan*.15(12):4

Papatsiros, V.G., Tzika, E.D., Papaioannou, D.D., Kyriakis, S.C., Tassis, P.D and Kyriakis, C.S. 2009. Effect of *Origanum vulgaris* and *Allium sativum* extracts for the control of proliferative enteropathy in weaning pigs. *Polish J. Vet. Sci*. 12: 407-414.

Pasqua, R.D., Hoskins, N., Betts, G and Mauriello, G. 2006. Changes in membrane fatty acids composition of microbial cells induced by addiction of thymol, carvacrol, limonene, cinnamaldehyde, and eugenol in the growing media. *Journal of Agricultural and Food Chemistry*. 54: 2745–2749.

Penolazzi, L., Lampronti, I., Borgatti, M., Khan, M.T.H., Zennaro, M., Piva, R and Gambari, R. 2008. Induction of apoptosis of human primary osteoclasts treated with extracts from the medicinal plant Emblica officinalis. *BMC Complementary and Alternative Medicine*. 8:59.

Platel, K and Srinivasan, K. 2001. Studies on the influence of dietary spices on food transit time in experimental rats. *Nutrition Research*. 21: 1309–1314.

Prakash, A., Singh, S.P., Varma, R., Choudary, G.K and Ram, S. 2009. Ameliorative efficacy of tulsi in Lead toxicity in cockerels. *Indian Vet. J.,* 86: 344-46.

Prakash, P. and Gupta, N. 2005. Therapeutic uses of *Ocimum sanctum* Linn (Tulsi) with a note on Eugenol and its pharmacological actions. A short review. *Indian Journal of Physiology Pharmacology*. 49(2):125-131.

Rai, V., Mani, U.V and Iyer, U.M. 1997. Effect of Ocimum sanctum Leaf Powder on Blood Lipoproteins, Glycated Proteins and Total Amino Acids in Patients with Non-insulin-dependent Diabetes Mellitus. *Journal of Nutritional and Environmental Medicine.* 7(2):113–18.

Rehman, H.U., Yasin, K.A. and Choudhary, M.A. 2007. Studies on the chemical constituents of Phyllanthus emblic. *Natural Product Research.* 21(9):775–81.

Roselli, M., Britti, M.S., Le Huërou-Luron, I., Marfaing, H., Zhu W.Y and Mengheri, E. 2007. Effect of different plant extracts and natural substances (PENS) against membrane damage induced by enterotoxigenic Escherichia coli K88 in pig intestinal cells. *Toxicology in Vitro*. 21: 224–229.

Samarasinghe, K., Wenk, C., Silva, K.F.S.T. and Gunasekara, J.M.D.M. 2003. Turmeric (*Curcuma longa*) root powder and mannanoligosaccharides as alternatives to antibiotics in broiler chicken diet. *Asian-Australian Journal of Animal Sciences.* 16:1495-1500.

Sapcota, D. and Upadhyaya, T.N. 2009. Efficacy of dietary Ocimum sanctum against Aflatoxin B in Broilers. *Indian Veterinary Journal.* 86:1163-65.

Saxena, M.J. 2008. Herbs- a safe and scientific approach. *International Poultry Production*, 16 (2): 11–13.

Schulz, V., Hansel, R and Tyler, V. E. 1997. Rational Phytotherapy: A Physicians guide to herbal medicine. Berlin: Springer. pp. 306.

Sidhu, S., Pandhi, P., Malhotra, S., Vaiphei, K. and Khanduja, K.L. 2011. Beneficial Effects of Emblica officinalis in l-Arginine-Induced Acute Pancreatitis in Rats, *Journal of Medicinal Food.* 14(1-2):147-155.

Singh, N., Singh, J.P. and Singh, V. 2007. Effect of dietary supplementation of herbal formulation on dressing percentage and mortality in broiler chicks. *Indian Journal of Field Veterinarians.* 2:22-24.

Soudamini, K.K. and Kuttan, R. 1992. Inhibition of lipid peroxidation and cholesterol levels in mice by curcumin. *Indian Journal of Physiology Pharmacology.* 36:239-243.

Srinivasan, K. 2005. Spices as influencers of body metabolism: An overview of three decades of research. *Food Research International*. 38:77–86.

Suanarunsawat, T., Ayutthaya, W.D.A. and Songsak, T. 2011. Lipid-Lowering and Antioxidative Activities of Aqueous Extracts of Ocimum sanctum L. Leaves in Rats Fed with a High-Cholesterol Diet. *Oxidative Medicine and Cellular Longevity.* 1-9.

Suresh, D and Srinivasan, K. 2007. Studies on the *in vitro* absorption of spice principles – curcumin, capsaicin and piperine in rat intestines. *Food and Chemical Toxicology.* 45: 1437–1442.

Tirupathi, R. 2010. Effect of herbal preparations on the performance of broilers. M.V.Sc., Thesis Submitted to Sri Venkateswara Veterinary University, Tirupathi, India.

Vidhyarthi, V.K., Nring, K. and Sharma, V.B. 2008. Effect of herbal growth promoters on the performance and economics of rearing broiler chicken. *Indian Journal of Poultry Science.* 43(3): 297-300.

Waghorm G.C., Jones W.T., Shelton, I.D and McNabb, W.C. 1990. Considered taninnns and the nutritive value of herbage. Proceedings of the Newzeland Grassland Association. 51: 171–176.

Wang, R., LI, D and Bourne, S. 1998: Can 2000 years of herbal medicine history help us solve problems in the year 2000? *In*: Proc. of Alltech's 14th Annu. Symp., Biotechn. in the Feed Industry (T.P. Lyons, Ed.). Alltech Technical Publications, Nottingham University Press, Nicholasville, KY. pp. 273-292.

Wawrzynczak, S., Kraszewski, J., Wawrzynski, M and Kozlowski, J. 2000. Effect of herb mixture feeding on rearing performance of calves. *Annals of Animal Science.* 27 (3): 133–142.

Webster's Encyclopedic Cambridge Dictionary of the English Language. 1989. Gramercy Books, New York.

Wenk, C. 2003. Herbs and Botanicals as feed additives in Monogastric animals. *Asian-Aust. J. Anim.Sci.* 6(2): 282-289.

Windisch, W., Schedle, K., Plitzner, C and Kroismayer, A. 2008. Use of phytogenetic products as feed additives for swine and poultry. *Journal of Animal Science.*86: E140–E148.

Youn, H.J. and Noh, J.W. 2001. Screening of the anticoccidial effects of herb extracts against eimeria tenella. *Veterinary Parasitology.* 96: 257–263.

Zigger, D. 2001. Helathier pigs on diet with garlic and cinnamon. *Feedtech.* 5: 8/9: 17. http://www.allaboutfeed.net/allabouts/ id935-3370/application_in_pig_diets.html (1. 10. 2013).

25

Mycotoxin Binders as Feed Additive for Livestock and Poultry

M. T. Banday and Pankaj Kumar Singh

Introduction

Mycotoxins are substances produced by fungal infections which in some cases do protect plants or plant seeds against parasites (Schardl *et al.,* 1996). However, upon ingestion, inhalation or dermal contact, these mold producing substances are poisonous to vertebrates. Even with more than 100,000 species of fungi known, only some of them are active mycotoxin producers. The fungi species which are known to produce the most hazardous mycotoxins in agriculture and in animal production are *Fusarium, Aspergillus* and *Pencillium* spp. Mycotoxins vary greatly regarding their toxicological effects as they do not belong to a single class of chemical compounds (Jewers, 1987). The environmental conditions in which they are preferably produced also differ. Fusarium toxins, such as trichothecenes, Zearalenone and fumonisins are commonly produced on the field, whereas aflatoxins and ochratoxins are produced by *Aspergillus* and *Penicillium* spp. mainly after harvest and during poor storage conditions.

It is not easy to detect and to diagnose problems related to mycotoxins in animals. Effects of mycotoxins are diverse, varying from immune suppression to death in severe cases depending on toxin-related (type of mycotoxin consumed, level and duration of intake), animal related (species, sex, age, breed, general health, immune status, nutritional status) and environmental (Farm management, hygiene, temperature) factors. All these factors contribute a great deal to the final susceptibility of animals to the presence of mycotoxins. So, it is not surprising if animals exhibit mycotoxins even at apparent "low levels" of mycotoxins present in the feed. According to the united Nation's Food and Agriculture Organization (FAO) approximately 25% of world's grain supply is contaminated with mycotoxins. The economic loss due to this has been estimated to run into millions of rupees annually. The greatest economic impact of mycotoxin contamination is felt by crop and animal producers as well as food and feed producers.

Occurrence of Mycotoxins

Contamination of feed stuffs with mycotoxins is a global problem. However, in certain geographical areas some mycotoxins are encountered more often than the others.

Environmental condition	Mycotoxin contaminating feed stuff
Winter conditions with high moisture.	Vomitoxin, Zeralenone, Ochratoxin, Diacetoxyscirpenol (DAS)T_2 toxin, Fumonisins
Warm and humid conditions.	Aflatoxins, ochratoxin and Fumonisins

Classification of Mycotoxins

(A). Classification of mycotoxins based on etiological agents

Mycotoxins	Fungi
Asperigillus toxin	*Asperigillus flavus*
Aflatoxin B_1,B_2,G_1,G_2	*Asperigillus parasiticus*
Cyclopiazonic acid	*Asperigillus flavus*
Ochratoxins	*Asperigillus ochracius.*
Ochratoxins	*Pencillium viridicatium*
Citrinin	*Pencillium citrinium*
Monoacetoxyscirpenol (MAS)	*Fusarium tricinctumFusarium solani*
Beoxynivalenol (DON, vomitoxin)	*Fusarium graminearium*
Zearalenone	*Fusarium gramineariumFusarium roseum*
Fumonisins B_1, B_2	*Fusarium inoniliformeFusarium Proliferatium*
Ergopeptines	*Clavicep purpurea*
Ergovaline	*Acremonium coenophialum*

(B). Classification of Mycotoxins based on their biological effects

Hepatotoxins	Nephrotoxins	Neurotoxins	Carcinogens
Aflatoxin, Sporidesmin Rubratoxin B Sterigmatocystin Tricothecenes Ochratoxin A Phomposin A Cyclopiazonic acids	Ocratoxin A Citrinin Aflatoxin, Oosporin Cyclopiazonic acid Sterigmatocystin	Trichothecenes (Vomitoxin, Satratoxin) Salframine Peintrem A Ergot alkaloids Ochratoxin A	Aflatoxin Sterigmatocystin Luteoskyrin Patulin Penicilic acid, T2 Ochratoxin Citrinin
Immunosuppressants	**Teratogenes**	**Gemtoxins**	**Dermitoxins**
Aflatoxin T2 Ochratoxin Citrinin Oosporin	Aflatoxin T2 Ochratoxin Citrinin Oosporin	Zearalenone Ergot alkaloids Aflatoxins	Trichothecenes (T2 Toxin, nivalenol)

Species susceptibility to mycotoxins

Susceptibility to mycotoxins by different species of animals varies as follows:

Mycotoxins	Poultry	Dairy	Pig	Horse
Aflatoxin	++	++	++	++
Zeralenone	++	++	++	++
Vomitoxin	++	++	++	++
Ochratoxin	++	++	++	++
Fumonisin	++	++	++	++
T-2 toxin	++	++	++	++

Mycotoxins and their effect on poultry production

Feed refusal which is a rapid, direct response to the presence of mycotoxin in the feed can be a problem with some animals. Growth inhibition is another important toxic effect of mycotoxins in poultry feed. A diet high in protein largely prevents the inhibitory growth caused by Aflatoxin, but at a cost. The concentration of bile required for lipid absorption and digestion from the intestinal tract is decreased during aflatoxicosis. The practical consequence of the malabsorption is to decrease the efficiency of feed conversion into animal products and hence increase the cost of production. Impairment of the mechanisms of resistance to infectious agents is one of the more insidious effects of mycotoxins. Aflatoxin, ochratoxin and T-2 have all been shown to act upon the immune system, and so their presence in poultry diets could have disastrous consequences.

Egg production can also be affected by mycotoxins in the diet. Aflatoxin apparently inhibits egg production as the point of commitment of an ovum to maturation; ova already committed are not influenced apart from less yoke than normal being deposited. Decreased egg production, with thin, rubbery shells was observed during field outbreaks of ochratoxin. Aflatoxin has also been implicated in hatchability, impaired semen quality in White Leghorn strains, under pigmentation, altering the response of chickens to environmental extremes and increased sensitivity to sodium chloride. In addition to affecting pigmentation, ochratoxin decreases the tensile strength of the large intestine by decreasing its collagen content and thus causes the fragile large intestine to rupture during processing (Hamilton, *et al.*, 1982). The most obvious effect of mycotoxins on poultry production is mortality, as this can be readily diagnosed and quantified.

Changes in the organ weight and texure may be observed in carcass after slaughtering if animals were fed mycotoxin contaminated diets. Enlarged liver, spleen and kidneys, and reduction in size of bursa and thymus usually occur due to mycotoxicosis (Espada *et al.*, 1992; Jand *et al.*, 2005; Leeson *et al.*, 1995). Other changes like necrosis of lymphoid and hemapoitic tissues, liver necrosis (fumonisins), gizzard lesions (trichothecenes) and increase in weight of kidney and liver were also reported (Hoerr *et al.*, 1981; Huff. *et al.*, 1988; Lun *et al.*, 1986; Satheesh *et al.*, 2004).

Effects of different mycotoxins are discussed as follows

1. Aflatoxin

It was death of about 100,000 turkey poults in the United Kingdom in 1960 following the ingestion of poultry feed containing Brazilian groundnut cake which led to the discovery of a group of compounds now called the Aflatoxin. Chemical and microbiological investigations soon revealed that the toxic effects produced by the Brazilian groundnut cake had resulted from the presence of quantities of four secondary metabolites of the mould *Asperigillus flavus* in the diet. As these compounds fluoresced either green or blue in ultraviolet light, they were designated Aflatoxin B_1, Aflatoxin B_2, Aflatoxin G_1 and Aflatoxin G_2. Almost all agricultural commodities support the growth of the Aflatoxin-producing fungi *Aspergillus flavus* and *Aspergillus parasiticus*. Formation of flatoxins can occur during the pre-and post-harvest stages of food production as long as a suitable environment for mould growth is provided. Optimal conditions for aflatoxin production are a water activity in excess of 0.85 and a temperature of 27°C, conditions which are frequently encountered in the Mediterranean region. Different crops vary in their ability to support fungal colonization because of differences in the chemical composition of each commodity. The incidence and magnitude of aflatoxins contamination varies with seasonal and geographical factors and also with the conditions, under which the crop is grown, harvested, stored and transported (Jewers *et al.,* 1986).

Aflatoxin B_1 is probably the most potent naturally occurring hepatic carcinogen known, and it is for this reason that many countries have introduced legislation restricting the amount of this compound allowed in foods and feeds. The acute toxicity of aflatoxin B_1 in poultry varies from species to species. The LD_{50} single dose (mg/kg body weight) is 0.3 for ducklings, and 6.0-16.0 for chickens. The acute toxicity of aflatoxins is governed by the age, sex and strain of the animal, the route of administration, the composition of the diet and the environmental conditions.

During chronic aflatoxicosis in chickens, lower body weight gain and biochemical alterations are observed at the minimal level of 0.62 mg AFB_1/kg of feed (Hamilton, 1984). Aflatoxins, ochratoxins and trichothecenes-all these toxic compounds are known to decrease the resistance of animals rendering them more susceptible to diseases (Ghosh *et. al,* 1990; Sharma, 1991; Dwivedi and Burns, 1994; Leeson *et. al,* 1995). In general aflatoxins are the most immune suppressive (CAST, 2003). Aflatoxin B_1 intoxication in broilers imparts the cell-mediated immunity by decreasing albumin and globulin. (Ghosh *et. al.* 1990) which is in accordance with the knowledge that aflatoxin exerts an important role in the inhibition of protein synthesis. Ghosh *et al,* 1990) have reported that poultry diets containing 250-5010 μg/kg of aflatoxins, predisposed poultry to attack by bacteria and viruses. This is consistent with the known effect of Aflatoxin on the immune system as a result of action on the T-cells (Hoerr *et al,* 1981). Aflatoxin appears to have a dose-related effect on the lymphatic tissues of poultry. Thymic involution of poults, without an associated effect on the bursal component of lymph tissue, has been induced by moderate levels of Aflatoxin (0.5 ppm AFB_1). Impaired function of lymphocytes and phagocytes appears to be the major aflatoxin-induced deficiency in immunogenesis. Aflatoxin has been reported to affect the uptake of a number of essential components of the diet, such as coccodiostats, vitamins and amino acids. This can have a disastrous effect on poultry health and production.

2. Ochratoxin

The ochratoxins were discovered as a result of mouldy corn cultures being fed to ducklings, rats and mice (Hamilton *et al.,* 1982), and ochratoxin A was later found to be the major toxic component of *Aspergillus oebraccus* (Doupnic and Peckham, 1970; Peckham *et al.,* 1971). It has since been shown that ochratoxin is produced by numerous species of Aspergillus and Pencillium. In recent years, ochratoxicosis, the disease state induced by ochratoxin, has been reported in broilers, layers, and turkeys from different stated in the USA, with levels of ochratoxin in feed ranging from 0.3 to 16 ug/g (Wyatt, *et al.,* 1973).The minimal dietary growth dose for young broiler chicks is 2 μg/g for ochratoxin. This compares with 2.5 and μg/g for Aflatoxin and T-2 toxin respectively. Visible examination of the chicks showed no clinical symptoms due to ochratoxin, but the study of affected broiler chickens has shown severe hydration and emaciation, proventricular haemorrhages, and visceral gout with white urate deposits throughout the body cavity and internal organs. The main target organ is the kidney, with the liver also affected to a lesser degree. The kidneys of affected birds become enlarged, and the livers exhibit a tan colored appearance, have increased glycogen content, but the liver fat does not increase. Inhibition of liver phosphorylase and decreased mitochondrial respiration are thought to be responsible for the increased liver glycogen.

Aflatoxicosis and ochratoxicosis result in a rubbery condition of the bones apparently related to increased tibial diameters and perhaps poor mineralization of bone tissue in young broiler chicks. A variety of toxic manifestations has been reported for mature laying hens fed 1-4 g/g of ochratoxin. These include a decrease in egg production, food consumption and serum protein concentrations, and an increase in prothombin times.

3. Cyclopiazonic acid

Cyclopiazonic acid was first discovered as a result of a scientific investigation of the secondary metabolites of *Pencillium cyclopium* (*P. griseofulvum*). Subsequent investigations revealed that it is produced by several economically important fungi including *A. flavus* (Tung *et al.,* 1971.) Recent toxicological studies have shown that the principal target organs of this mycotoxin are the liver, kidney and digestive system, with liver involvement occurring primarily in rats. Feeding trials with chickens have produced mucosal necrosis and inflammation in the crop, proventriculus and gizzard. In these chickens the kidneys often were swollen or distended. Dalvi (1986) has reappraised the etiology of the classical Turkey X diseases syndrome and has suggested that the arching of the neck and the head drawn back (opisthotonus), and the legs stretched (extended) fully backwards are effects produced by cyclopiazonic acid but not by aflatoxins. Thus, he concluded that this mycotoxin co-occurred with the aflatoxins in the Brazilian groundnut cake responsible for Turkey X disease.

4. Citrinin

Citrinin is produced by several species of Pencillium and Aspergillus. Like ochratoxin, citrinin is a nephrotoxin. Reduced growth rates, decreased feed and water consumption have been observed as toxic manifestations of this mycotoxin. Decreased feed intake was observed with 62.5 μg/g of citrinin, whereas body weight and water intake were affected by 500 and 250 ug/g respectively in chickens. Two week old chicks have been found to increase their water intake by about two-fold in four hours when fed 500 ug/g of citrinin. Acute diarrhea follows. Removal of the citrinin from the diet results in a return to normal water consumption and a subsidence of the diarrhea in about six hours.

Nephrotoxicity and hepatoxicity in chickens occurs at dietary levels of 250 ug/g of citrinin with liver and kidney enlargements of 11 and 22% respectively. Serum sodium levels are also changed. Necropsy of affected birds revealed the presence of pale and swollen kidneys (Wyatt, *et al.,* 1975).

5. Tricothecenes

This is a group of toxic fungal metabolites based on the structure and which are produced by a number of species of the genus Fusarium. T-2 toxin has been the most extensively studied tricothecene in poultry, and it has been found that the primary effect of T-2 toxicosis in young broiler chicks is oral necrosis.

The ability of the three thricothecenes, diacetoxyscirpenol, croticin and T-2 toxin, to cause oral necrosis and to affect body weight gain has been studied in growing chicks. At a dietary inclusion rate of 5 μg/g of diacetoxyscirpenol, a body weight reduction of 24% resulted, whereas T-2 toxin produced an 11% reduction at the same incorporation rate. A more severe oral response was observed with croticin fed at 10 μg/g. it is generally regarded that the presence of oral lesions in poultry is the primary means of diagnosing tricothecene toxicosis in the field.

Dietary T-2 toxin has also been found to affect the nervous system by producing an abnormal positioning of the wings, hysteroid seizures, or an impaired righting reflex. In addition, it can induce abnormal feathering, drastically decrease feed intake without impairing feed efficiency, decrease egg production and cause thinning of egg shells destruction of the hematopoietic system (Wyatt, *et al.,* 1973).

6. Slaframine

Slaframine, 1-acetoxy-6-aminohydrondolizine, is a parasympathomimitic secretogogue produced by the mould Rhizoctonia leguminicola. A recent study of chronic toxicity in broiler chicks has shown that daily oral intubation in saline for 21 days to 240 chicks at 8.8 or 17.8 µg/g body weight does not affect weight, feed intake and utilization, pancreatic weight, liver weight, and small intestine weight. However, changes in digestive function were observed: the protein content of the digesta, the digesta lipase and specific activities, and the digesta trypsin-specific activities were lowered at the higher dose level (Elissaide *et al.,* 1994).

Table 1: Toxic effects of major mycotoxins and mechanisms of action

Toxin	Toxic effects	Cellular and molecular mechanisms of action
Aflatoxin	Hepatotoxicity Genotoxicity Oncogenicity Immunomodulation	Formation of DNA adducts Lipid peroxidation Bioactivation by cytochromes P450 Conjugation to GS-transferases
Ochratoxin A	Nephrotoxicity Genotoxicity Immunomodulation	Effect on protein synthesis. Inhibition of ATP production Detoxification by peptidases
Patulin	Neurotoxicity In vitro mutagenesis	Indirect enzyme Inhibition
Trichothecenes (Toxin T-2, DON, Aflatoxin Rubratoxin-B Phomposin A, Ochratoxin-A	Hematotoxicity Immunomodulation Skin toxicity	Induction of apoptosis in haemopoietic progenitor cells and immune cells. Effect on protein synthesis Abnormal changes to immunoglobulins
Zeralenone	Fertility and Reproduction	Binding to oestrogen receptors Bioactivation by reductases Conjugation to glucuronyltransferases
Fumonisin B1	Central nervous system damage Hepatotoxicity Genotoxicity Immunomodulation	Inhibition of ceramide synthesis Adverse effect on the sphinganin/ sphingosin ratio Adverse effects on the cell cycle.

Methods to limit the effects of mycotoxins

In order to avoid mycotoxins, several strategies have been investigated (Doyle *et al*., 1982; Park, 1993; Bauer, 1994; Ramos and Hernandez, 1997) which can be divided into pre-and postharvest technologies. The best procedure to prevent the effect of mycotoxins in the minimizing of the mycotoxin production itself

(Miedaner and Reinbrecht, 1999), e.g. by harvesting the grain at maturity and low moisture and storing it at cool and dry conditions which is difficult to perform in countries with a warm and humid climate. Furthermore, the growth of fungi and therefore the production of mycotoxins can be limited by the use of propionic acid or ammonium isobutyrate. Feed additives like antioxidants, sulphur-containing amino acids, vitamins, and trace elements can be useful as detoxicants (Nahm, 1995).

Decontamination methods

Various methods have been tried to decontaminate aflatoxin contaminated commodities (e.g. groundnut, cottonseed, palm kernel cake/meal and maize). These include physical methods (sorting irradiation techniques, heating), chemical methods (acids, bases, oxidizing agents, solvent extraction), biological methods (Virdi *et al.*, 1989). These methods are discussed as follows:

i) Physical methods

The physical methods are focused on the removal of mycotoxins by different adsorbents added to mycotoxin-contaminated diets (Ramos *et al.*, 1996) with the hope of being effective in the gastro-intestinal tract more in a prophylactic rather than in a therapeutic manner. At present, however, the utilization of mycotoxin-binding adsorbents is the most applied way of protecting animals against the harmful effects of decontaminated feed.

Methods such as grain cleaning by washing with water or sodium carbonate to reduce the contamination of maize with *Fusarium* toxins and manually sorting out contaminated grains by physical aspects or by flouresence to detect the presence of some mycotoxins have been used. Other toxin inactivating techniques have been proposed like high temperature, UV, X-rays or microwave irradiation and solvent extraction of toxins (Scott, 1998). Dilution of contaminated feed with safe feed is also done. The addition of binding agents able to fix mycotoxins may reduce the bioavailability of these compounds in animals and limit the risks associated with the presence of residues in animal products destined for human consumption. In the case of AFB_1, hydrated sodium calcium aluminosilicates (HSCAS) and phyllosilicates derived from natural zeolites have a high affinity, both *in vitro* and *in vivo*. However, many studies have shown these compounds to be unable to adsorb other mycotoxins. Cations are able to fix AFB_1 and ZEN (Piva and Galvano, 1999). Bentonites have a lamellar crystalline microstructure, the composition and adsorption properties of which depend on interchangeability of the cations positioned in the various layers. Bentonites have been shown to be effective for the adsorption of AFB_1 and T-2 but not for the adsorption of ZEN or nivalenol (Ramos *et al.,* 1996). Other clays such as kaolin, sepiolite and montmorilonite, bind AFB_1 less efficiently than HSCAS and betonites.

Activated carbon is obtained by pyrolysis and the activation of organic compounds. It has a more heterogeneous porous structure. Activated carbon is also able to bind mycotoxins (Galvano *et al.,* 1996) Resins such as cholestyramine and polyvinyl polyprrodoxynivalenol are also able to bind OTA and AFB_1(Piva and Galvano, 1999) these binders may have adverse nutritional effects. The large amounts that must be added to get a perceptible effect may also reduce the bioavailability of certain minerals or vitamins in the diet. Some of these binders are not biodegradable and may affect the efficacy of retention basins and digesters used for the treatment of animal effluents. Clays render wet floors slippery, increasing risk of accidents for both animals and human staff. They also present a risk of contamination by heavy metals and dioxin (Devegowda, 2000).

ii) Chemical methods

A variety of chemical agents such as acids, bases (ammonia caustic soda), oxidants (hydrogen peroxide, ozone), reducing agents (bisulphites), chlorinated agents and formaldehyde have been used to degrade mycotoxins in contaminated feeds, particularly aflatoxins (Scott, 1998). Chemically, some mycotoxins can be destroyed with calcium hydroxide monthylamine (Bauer, 1994), ozone (McKenzie *et al.,* 1997; Lemke *et al.*, 1999) or ammonia (Park, 1993). Ammonia gas appears to be the most promising approach as it is capable of reducing the Aflatoxin level in situ by more than 95% and is applicable to a variety of contaminated commodities using batch and continuous processing methods (Petterson, 2004). Extensive work has been carried out on the toxicology of ammoniated aflatoxin contaminated commodities, including studies in poultry. The average ammoniation costs vary between 5 and 20% of the value of the commodity (Coker, 1998). Main drawbacks of this kind of chemical detoxification are the ineffectiveness against other mycotoxins and the possible deterioration of the animal's health by excessive residual ammonia in the feed.

(iii) Microbiological methods

Biological methods are not yet used in practice though the number of corresponding patents increases continuously (Erber, 1996; Duvick and Rood, 2000). These methods include fermentation procedures with microorganisms. One example is the conversion of aflatoxin B_1 (particularly by *Flavobacterium auranticum*) to harmless degradable products. The conversations, however, are generally slow and incomplete (Sweeney and Dobson, 1998; Arici, 1999; Bata and Lasztity, 1999; Karlovsky, 1999).Certain strains of lactic acid bacteria, propinobacteria and bifidobacteria have cell wall structures that can bind mycotoxins (Yoon and Baeck, 1999) and limit their bioavailability in the animal body. Mycotoxins are then eliminated in the faeces without, significant detrimental effects on the animals or any risk for toxic residues to be found in edible animal products. Research is currently underway to develop new classes of natural organic mycotoxin binders. Glucomannans extracted from the external part of the cell wall of the yeast *Sacchromyces cerevisiae* are able to bind certain mycotoxins.

Use of Toxin Binders

The means to reduce the uptake of mycotoxins from contaminated feed is by the use of toxin binders. Mycotoxin-adsorbing agents are large molecular weight compounds that should be able to bind the mycotoxins in contaminated feed without dissociating in the gastrointestinal tract of the animal. The theory is that the binder decontaminates mycotoxins in the feed by binding them strongly enough to prevent toxic interactions with the consuming animal and to prevent mycotoxin absorption across the digestive tract. In this way the toxin-adsorbing agent complex passes through the animal and is eliminated via the faeces. This prevents or minimizes exposure of animals to mycotoxins. The aim of these toxin binders is to inhibit the uptake of mycotoxins by an animal *in vivo*. These adsorbant materials are intended to act like a "chemical sponge" and adsorb mycotoxins in the gastrointestinal tract, thus preventing the update and subsequent distribution to target organs. The addition of mycotoxin binders to contaminated diets has been considered the most promising dietary approach to reduce effects of mycotoxins (Galvano *et al.,* 2001). The theory is that the binder decontaminates mycotoxins in the feed by binding them strongly enough to prevent toxic interactions with the consuming animal and to prevent mycotoxin absorption across the digestive tract.

The use of mycotoxin binding agents is occasionally recommended to farmers in order to protect the animals against harmful effects of mycotoxins occurring in contaminated feeds.

Desirable Characteristics of a Mycotoxin Binder

The selected toxin binder should fulfill the following criteria

- Binding efficacy
- Binding stability
- Effective inclusion rate
- Least active on nutrients

Binding efficacy

This is an estimation of the binding activity of the additive to different mycotoxins. Different physical and chemical properties of the additive will influence the binding efficacy. Only a careful selection of the clay, minimum impurities, desirable pore size and ionic characters will yield good binding efficacy. Again different mycotoxins have varying affinity for different clays. The mycotoxins have predominantly negative charges as in aflatoxin or mixed charges as in other mycotoxins. The selection of clay shall match this charge distribution in an opposite way. This will enable the binder to bind mycotoxins and avoid its absorption. The pore size in the clay shall be moderate to fit mycotoxin molecule. If it is excess, there is every possibility that it will absorb water and along with that vital nutrients. Binding efficacy is expressed as percentage of binding of a mycotoxin, either in a single toxin environmental or in a multi-toxin environment.

Binding stability

The binding of mycotoxin takes place in gizzard, at an acidic PH. It does not happen in the feed outside the body. Hence this requires certain activation time after the feed is consumed by the bird. If the activation time is too long, the feed may pass through the gizzard without toxin binder interaction. Secondly, once the mycotoxins are bound by additive in the gizzard, the chime passes through the intestine. There the toxin-binder complex is subject to a more neutral or slightly alkaline environment. If a careful selection is not made about a binder, at this place, the bound toxins may be released and the bird absorbs toxins. A stable binding is that the binder binds mycotoxins in the gizzard, but will not release them even at neutral or alkaline PH, ensuring that the complex passes through the intestine and ensure safety to birds.

Effective inclusion rate

For the safety of the entire feed the toxin binder shall be distributed evenly in the feed. The binder does not have an attraction property; rather the vicinity of the toxins to binder during churning the gizzard is a determining factor for effective binding. This has to be achieved with a proper surfactant.

Least active on nutrients

Nutrients differ in their molecular size from that of mycotoxins. The selection of the binder based on a porosity which is suitable for mycotoxin eventually excludes nutrients is an ideal binder. These additives are ideally called toxisorbants (specific to mycotoxins) and needs to be differentiated from enterosorbents which behave indifferently to mycotoxins and vital nutrients. Some HSCAS will have a hydration property owing to their very large surface area; thereby they swell in the intestines. Such clays are clearly not suitable binders as they bind nutrients along with mycotoxins.

Types of mycotoxin binders

Mycotoxin-adsorbing agents are large molecular weight compounds that should be able to bind the mycotoxins in contaminated feed without dissociating in the gastrointestinal tract of the animal. In this way the toxin-adsorbing agent complex passes through the animal and is eliminated via the faeces. This prevents or minimizes exposure of animals to mycotoxins.

Potential mycotoxin-adsorbing agents include activated carbon, aluminosilicates (clay, bentonite, montmorillonite, zeolite, phyllosilicates, etc.), complex indigestible carbohydrates (cellulose, polysaccharides in the cell walls of yeast and bacteria such as glucomannans, peptidoglycans, and others), and synthetic polymers such as cholestryamine and polyvinylpyrrolidone and derivatives.

I. Aluminosilicates

Silicate minerals are the largest class of mycotoxin sequestering agents and most studies on the alleviation of mycotoxicosis by the use of adsorbing agents have focused on aluminosilicates. Within this group, there are 2 important subclasses: the phyllosilicate subclass and the tectosilicate subclass. Phyllosilicates include bentonites, montmorillonites, smectites, kaolinites, illites. The tectosilicates include zeolites.

a. Bentonites

Bentonites are originally created from the weathering of volcanic ash in situ (Ramos *et al.,* 1996). They belong to the phyllosilicate group and are adsorbing agents with a layered crystalline microstructure and variable composition. Bentonites are generally impure clay consisting mostly of montmorillonite. Due to their montmorillonite content, bentonites swell and form thixotropic gels (Diaz and Smith, 2005). It was demonstrated that several bentonites had the ability to bind AFB_1 in a buffer solution (Masimango *et al.*, 1978). Dvorak (1989) showed that 2 samples of bentonite were effective in sequestering AFB_1 in different liquid media such as water, saline solution, blood serum of pigs, stomach fluid of pigs, and bovine rumen fluid. The initial concentration of 3.6 mg or 18 mg AFB_1 per litre was reduced to 0.3-27% after exposure to 50 g of adsorbing agent. Ramos and Hernandez (1996) demonstrated the high capacity of bentonites to sequester aflatoxins in several different solutions. They hypothesised that bentonites would sequester aflatoxins *in vivo*, thus reducing aflatoxicosis.

Over the past two decades the use of bentonite to suppress aflatoxins has been demonstrated for many farm animals. In a series of experiments on growing swine, Lindemann *et al.* (1993) demonstrated that the addition of sodium bentonite (0.5%) to diets contaminated with 800 ppb AFB_1 improved average daily feed intake and increased average daily gain. Bentonite supplementation significantly improved concentrations of blood urea, total protein, albumin and activities of AST, ALP and GGT, which were significantly altered by AFB1 (Lindemann *et al.*, 1993). Positive effects of bentonite on growing pigs consuming aflatoxin contaminated diets without negatively affecting mineral metabolism were observed (Schell *et al.*, 1993). Nageswara and Copra (2001) examined the effects of 1% sodium bentonite and activated charcoal additions to feed that contained 100 µg AFB1/kg on the AFM1 content in goat's milk. Concentrations of AFM_1 excreted in the milk were reduced 65% by bentonite and 76% by activated charcoal (Nageswara and Chopra, 2001). Similarly, Diaz *et al.* (2004) found that 0.25% activated carbon had no effect in reducing AFM_1 in cow's milk, but several commercial sodium bentonite products effectively reduced AFM_1 in milk.

b. Montmorillonites

Montmorillonite is a layered silicate which adsorbs organic substances either on its external surfaces or within its interlaminar spaces (Ramos *et al.,* 1996). Modified montmorillonite nanocomposite (MMN) is a new absorptive additive. Developped with nano-modification techniques, MMN has a sizable surface area, higher porosity, and stronger cation exchange activities along with more active sites, which make its nanoparticle effect easy to exert, and as a result, its adsorption efficacy is greatly enhanced.

Efficacy of a modified montmorillonite nanocomposite to reduce the toxicity of aflatoxins in broiler chicks was assessed (Shi *et al.*, 2006). *In vitro* data demonstrated that montmorillonite nanocomposite had a high ability to adsorb aflatoxins from aqueous solution and it was supposed to be a good candidate for *in vivo* testing. The results of *in vivo* study with broilers indicated that aflatoxins could significantly affect overall animal health and performance (Shi *et al.*, 2006). Abdel-Wahhab, (2005) observed that addition of a montmorillonite (5 g/kg) to aflatoxin-exposed rats prevented the degenerative changes in hepatic and renal tissue thus offering protection against serum biochemical parameters affected by the toxin and it was postulated that the bentonite formed a complex with the toxin thus preventing the absorption of aflatoxin across the intestinal epithelium. Scheideler (1993) examined the effects of various commercial aluminosilicate products on aflatoxin toxicity to chicks and reported that these additives lessened the growth reduction caused by AFB_1. Desheng *et al.* (2005) measured *in vitro* 0.6 g AFB_1/kg adsorption to Ca^{2+} montmorillonite from water and observed that a 0.5% Ca^{2+}-montmorillonite addition to feed contaminated with 200 μg AFB_1/kg significantly reduced aflatoxicosis in chickens. They assessed that the concentrations of Ca, P, Cl, Fe, Zn in broiler bones were not affected by aflatoxin and montmorillonite, but the concentrations of Mn, Pb, and F were decreased by the clay (Desheng *et al.*, 2005).

c. Zeolite

Zeolites are a group of silicates consisting of interlocking tetrahedrons of SiO_4 and AlO_4 (Kabak *et al.*, 2006; Ramos and Hernandez, 1997). Zeolites are crystalline hydrated aluminosilicates of alkali and alkaline-earth cations characterized by an infinite three-dimensional structure. Zeolites have large pores that provide space for large cations such as sodium, potassium, calcium. They are characterized by their ability to lose and absorb water and exchange constituent cations without damage to the crystalline structure (Diaz and Smith, 2005; Papaioannou *et al.*, 2002). Clinoptilolite is a natural zeolite whose main application is the adsorption of heavy metals from aqueous solutions (Kleiner *et al.*, 2001). Zeolites are similar to molecular sieves as well to ion-excahnge resins and are suitable for the distinction of different molecules by size, shape and charge.

Several studies assessed *in vivo* response of zeolites in different animal models. Scheideler (1993) fed a natural zeolite to broilers consuming 2.5 μg/kg AFB_1. Zeolite added at 1% of the di*et al*leviated the growth depression and reduced the increase in liver lipid concentration caused by AFB_1. Zeolite did however decrease serum phosphorus and chloride concentrations independently of aflatoxin exposure (Scheideler, 1993). Sova *et al.* (1991) showed that zeolite inclusion @ 5% of the diet significantly reduced heterophilia and lymphopenia in broilers consuming an aflatoxin-contaminated diet. Harvey *et al.* (1993) fed growing broiler chickens a diet contaminated with 3.5 ppm total aflatoxins with or without the inclusion of zeolites incorporated into the diets @ 0.5%. They reported that zeolite reduced the toxicity of aflatoxin by 41%, as indicated by weight gains, liver weight, and serum biochemical measurements. This result is compared positively with its *in vitro* capacity of binding aflatoxins (Harvey *et al.*, 1993). Abdel-Wahhab and Nada (1998) found a significant reduction in AFM_1 output in urine of rats exposed to aflatoxin, and concluded that zeolite reduced intestinal absorption of the toxin and hepatic metabolism into AFM_1 and hence excretion in urine (Abdel-Wahhab and Nada, 1998).

d. HSCAS (Hydrated sodium calcium aluminosilicate)

HSCAS is perhaps the most studied mycotoxin-sequestering agent among the mineral clays (Galvano *et al.*, 2001); (Diaz and Smith, 2005; Kabak *et al.*, 2006). It is a naturally occurring and heat-processed calcium montmorillonite that is commonly used as an anticaking additive in animal feed (Wang *et al.*, 2008). Phillips *et al.* (1998) showed that HSCAS was able to form the most stable complex with AFB1 (with an adsorption of more than 80% of toxin present in the medium). The complex was demonstrated to be stable in water at pH 2, 7 and 10 and a temperature of 25 and 37°C. Further studies with molecular modelling assessed that aflatoxin binding to HSCAS might be attributed to the chemisorption of aflatoxins at surfaces within the interlamellar region of HSCAS, whereas the exterior surfaces of the clay were responsible for only minor adsorption of toxins (Phillips, 1999; Phillips *et al.*, 2008). Kubena *et al.* (1990) reported that the reaction between AFB_1 and HSCAS was fast and reached the equilibrium after 30 min. Calculations based on the equilibrium binding experiments indicated that approximately 200-230 nmol of aflatoxin could be maximally bound per milligram of HSCAS (Kubena *et al.*, 1990).

It was reported that incorporation of HSCAS in the diet of chicks at a level of 0.5% by weight significantly reduced the effects of purified AFB_1 (Phillips *et al.*, 1988). Since these early studies, HSCAS has been reported to reduce the effects of aflatoxins in a variety of young animals including rodents, chicks, turkey poults, ducklings, lambs, pigs, mink and trout (Phillips, 1999). Levels of AFM_1 in milk from lactating dairy cattle and goats were also diminished with the inclusion of HSCAS in the diet (Ellis *et al.*, 2000; Harvey *et al.*, 1991; Smith and Phillips, 1994). All these studies support the conclusion that HSCAS has a notable capacity for the aflatoxins at levels in the diet at, or below, 0.5% w/w (the level that is recommended for anti-caking activity in animal feeds).

2. Activated charcoal

Activated carbon (AC) is a non-soluble powder formed by pyrolysis of several organic compounds and manufactured by activation processes aimed at developing a highly porous structure (Galvano *et al.*, 2001). AC is known as one of the most effective and non-toxic group of sorbents and has been shown to be a tenacious adsorbing agent of a wide variety of drugs and toxic agents. It has been commonly used as a medical treatment for severe intoxications since the 19th century (Huwig *et al.*, 2001). The sequestrant properties of AC depend on many factors including pore size, surface area, structure of the mycotoxin and doses. Superactivated charcoal differs from AC in that the particle size is reduced, thereby increasing surface area. The specific surface area of AC indeed varies from 500 m^2/g to 3500 m^2/g for superactivated charcoals (Ramos *et al.*, 1996).

Hatch *et al.* (1982) used activated charcoal as an antidote against a lethal dose of AFB_1 in goats (3 mg AFB1/ kg BW). Animals consuming activated charcoal 8 hours after aflatoxin exposure had no visible liver lesions and had a lower percentage of hepatic damage (3%) compared to animals receiving no antidote (25%). In subsequent studies, the effects of activated charcoal have been variable (Hatch *et al.*, 1982). When chickens were fed a mixture of 6 mg AFB1/kg BW and 200 mg of activated charcoal/ kg BW, functional alterations to the liver were reduced or prevented during a 2-month study (Dalvi and Ademoyero, 1984). Dalvi, (1984) fed broiler chicks 10 ppm aflatoxin for 8 weeks. Improvement in feed consumption and weight gain was observed in the charcoal treated birds. The treatment was also able to reduce the effect of AFB_1 on the microsomal cytochrome P450 and the activities of benzphetamine demethylase and SGOT. Galvano *et al.* (1996) showed reduced aflatoxin residues in milk of cows consuming different sources of charcoal, but responses to charcoal did not exceed that seen with HSCAS. Experiment with turkey poults (Edrington *et al.*, 1996), responses to charcoal with broilers (Edrington, 1997), rats (Abdel-Wahhab, 1999) and mink (Bonna *et al.*, 1991) also suggest that charcoal may not as effective in binding aflatoxin as are clay based adsorbing agents.

The activated charcoal showed a good capacity to adsorb FB_1 from water solution and was proposed as a candidate to reduce fumonisin toxicity. However, it was not effective *in vivo* at least with respect to the alteration of sphingolipid metabolism. In particular, the biomarker of fumonisin exposure (kidney sphinganine/sphingosine ratio) in rats fed fumonisin-contaminated diet (4 ppm FB_1+FB_2) mixed with 2% activated charcoal was not significantly different from that of rats fed fumonisin-contaminated di*et al*one, suggesting that activated charcoal may not be effective at reducing the toxicity of fumonisins. Activated charcoal may be important in binding *in vitro* ZEA and/or DON (Avantaggiato *et al.*, 2005; Bueno *et al.*, 2005; Doll *et al.*, 2004); however, the relevant *in vivo* effectiveness needs to be assayed.

3. Organic polymers as binders

Some complex indigestible carbohydrates (cellulose, polysaccharides in the cell walls of yeast and bacteria such as glucomannans, and peptidoglycans, and others) and synthetic polymers such as cholestryamine and polyvinylpyrrolidone can adsorb mycotoxins.

a. Yeast cell walls

Besides its excellent nutritional value, yeast or yeast cell walls can also be used as adsorbents for mycotoxins (Grunkemeier, 1990; Bauer, 1994). The in vitro adsorption of ochratoxin by yeast (consisting of 40% sterilized yeast and 60% fermentation residue of yeasts used for beer production) is dependent on the pH being at maximum in acidic solutions (at pH 3: 8.6 mg/g, at pH 8: 1.2 mg/g. however, in trials with pigs employing a feed supplement of 5% of yeast, only a slight reduction of the ochratoxin A concentration in blood plasma, bile, and tissues was achieved. By the use only a yeast cell walls instead of whole cells, the adsorption of mycotoxins can be enhanced. The cell walls harboring polysaccharides (glucan, mannan), proteins, and lipids exhibit numerous different and easy accessible adsorption centres including different adsorption mechanisms, e.g. hydrogen bonding, ionic, or hydrophobic interaction. Therefore, it was possible to bind 2.7 mg zearalenone per gram of cell walls. The binding was rapid and reached quilibrium after only 10 min, which is superior to commercial available clay-based toxin binders (Volkl and Karlovsky, 1999).

In another context, it was shown that yeast killer toxins were adsorbed by the polysaccharides and not by the proteins or fatty acids of yeast cell walls (Radler and Schmitt, 1987) and that this adsorption was not unspecific because cellulose and glycogen were not able to bind killer toxins. Cell walls derived from the *Saccharomyces cerevisiae* yeast are also used as a dietary mycotoxin-adsorbing agent. Yeast cell walls consist almost entirely of proteins and carbohydrates. The carbohydrate fraction is composed primarily of glucose, mannose, and N-acetyglucosamine. Glucans and mannans, the two main sugars, are found in about equal concentrations in *Saccharomyces cerevisiae*. Yeast mannan chains of various sizes are exposed on the external surface and are linked to cell wall proteins (Evans and Dawson, 2000).

b. Bacteria

Lactic acid bacteria (LAB) are a group of gram-positive, acid-tolerant, generally non-sporulating bacteria that have common metabolic and physiological characteristics. These bacteria, usually found in decomposing plants and lactic products, produce lactic acid as the major metabolic end-product of carbohydrate fermentation. The main strains that comprise the LAB are *Lactobacillus*, *Leuconostoc*, *Pediococcus*, *Lactococcus*, and *Streptococcus* as well as the more peripheral *Aerococcus*, *Carnobacterium*, *Enterococcus*, *Oenococcus*, *Sporolactobacillus*. Strains of lactic acid bacteria such as *Lactobacillus rhamnosus* strain GG and *Lactobacillus rhamnosus* strain LC-705 are used to remove mycotoxins. *Streptococcus thermophilus* NG40Z and

C5 have also been tested for their ability to detoxify mycotoxins (El-Nezami *et al.,* 1998). The cell walls harboring polysaccharides, proteins and lipids exhibit numerous different and easy accessible adsorption centers. It has been suggested that cell wall peptidoglycans and polysaccharides are the two most important elements responsible for binding by lactic acid bacteria (Kabak *et al.,* 2006).

4. Micronized fibers

Undigestible dietary fiber has adsorbance potential for mycotoxins. Dietary fibers can be obtained from different plant materials such as cereals (wheat, barley, oat), pea hulls, apple, bamboo, etc. They are constituted mainly of cellulose, hemicelluloses and lignin and can be obtained in ultrafine (<100μ) or less fine (>100μ) fractions (Aoudia *et al*., 2009). Alfalfa fiber has reduced the effects of zearalenone (James and Smith, 1982; Stangroom and Smith, 1984) in rats and swine and T-2 toxin in rats (Carson and Smith, 1983).

5. Synthetic Polymers

Cholestyramine is an anion exchange resin which is used for the binding of bile acids in the gastro-intestinal tract and for the reduction of low density lipoproteins and cholesterol. The in vitro binding capacity of this resin for ochratoxin A and zearalenone was 9.6 mg/g (Bauer, 1994) and more than 0.3 mg/g (Ramos *et al*., 1996), respectively, but in vivo, cholestyramine had only a very small effect on the reduction of the ochratoxin concentration in blood, bile and tissues. Cholestyramine has proven to adsorb zearalenone (Ramos *et al*., 1996; Doll *et al*., 2004) and fumonisins (Solfrizzo *et al*., 2001). In rats consuming ochratoxin, cholestyramine reduced plasma ochratoxin and increased fecal ochratoxin excretion (Kerkadi *et al*., 1998).

Summary

Mycotoxins are secondary metabolites of molds that have adverse effects on livestock and poultry. Factors influencing the presence of mycotoxins in feeds include environmental conditions related to storage. The diseases caused by exposure to mycotoxins are known as mycotoxicoses. Mycotoxins have various acute and chronic ill effects on livestock and poultry depending on species and susceptibility of an animal within a species. The economic impact of mycotoxins include loss of life, increased health care costs, reduced livestock production and disposal of contaminated feeds. The addition of mycotoxin binders to contaminated diets has been considered the most promising dietary approach to reduce effects of mycotoxins. The toxin binders decontaminate mycotoxins in the feed by binding them strongly enough to prevent toxic interactions with the consuming animal and to prevent mycotoxin absorption across the digestive tract. Mycotoxin adsorbents offer an attractive short-term solution to the challenge of mycotoxin-contaminated animal feeds. The only complete solution to the mycotoxin challenge will be the long-term of eliminating mycotoxins from the food and feed chains through improved quality control based on better analytical techniques coupled with genetic advances in plant resistance to fungal infestation.

References

Abdel-Wahhab, M.A., Hasan, A.M., Aly, S.E. and Mahrous, K.F. 2005. Adsorption of sterigmatocystin by montmorillonite and inhibition of its genotoxicity in the Nile tilapia fish (*Oreachromis nilaticus*). *Mutation Research*. 582: 20-27.

Abdel-Wahhab, M.A., Nada, S.A. and Amra, H.A. 1999. Effect of aluminosilicates and bentonite on aflatoxin-induced developmental toxicity in rat. *Journal of Applied Toxicology*. 19:199-204.

Aoudia, N., Callu, P., Grosjean, F. and Larondelle, Y. 2009. Effectiveness of mycotoxin sequestration activity of micronized wheat fibres on distribution of ochratoxin A in plasma, liver and kidney of piglets fed a naturally contaminated diet. *Food and Chemical Toxocology*, 47:1485-1489.

Arici, M. 1999. Degradation of mycotoxins by microorganis. *Ernahrung*. 23: 298-301.

Avantaggiato, G., Solfrizzo, M and Visconti, A. 2005. Recent advances on the use of adsorbent materials for detoxification of Fusarium mycotoxins. *Food Additives and Contaminants*. 22: 379-388.

Bata, A. and Laszitity, R. 1999. Detoxification of mycotoxin-contaminated food and feed by microorganisms. *Trends Food Sci. Technol*. 10: 223-228.

Bauer, J., 1994. Moglichkeiten zur Entgiftung mykotoxin haltiger Futtermittel. *Monatsh. Veterinarmed.* 49:175-181.

Bonna, R.J., Aulerich, R.J., Bursian, S.J., Poppenga, R.H., Braselton, W.E. and Watson, G.L.J. 1991. Efficacy of hydrated sodium calcium aluminosolicate and activated charcoal in reducitng the toxicity of dietary aflatoxin to mink. *Archives of Environmental Contamination and Toxicology*. 20: 441-447.

Bueno, D.J., Di Marco, L., Oliver, G and Bardon, A. 2005. In vitro binding of zearalenone to different adsorbents. *Journal of Food Protection*. 68: 613-615.

Carson, M.S., and Smith, T.K. 1983. Effect of feeding alfalfa and refined plant fibres on the toxicity and metabolism of T-2 toxin in rats. *Journal of Nutrition*. 113:304-313.

CAST Report. 2003. Mycotoxins: risk in plant animal, and human systems (Richard, J.L and Payne, G.A. eds.) council for Agricultural science and technology Task Force report No. 139, Ames, lowa, USA.

Coker, R.D. 1998. The chemical detoxification of Aflatoxin-contaminated animal feed. Nat. Toxicants Food. 284-298.

Dalvi, R. R. 1986. An overview of afflatoxicosis of poultry: Its characteristics, prevention and reduction. *Vet. Res. Commun*. 429- 443.

Desheng, Q., Fan, L., Yanhu, Y. and Niya, Z. 2005. Adsorption of aflatoxin B-1 on montmorillonite. *Poultry Science*. 84: 959-961.

Devegowda, G. 2000. Metre les mucotoxins sir la louche: d'ou vinent les glucomanans esterifies. *Feed Times*. 4:12-14.

Diaz, D.E., Hagler, W.M., Blackwelder, J.T., Eve, J.A., Hopkins, B.A., Anderson, K.L., Jones, F.T and Whitlow, L.W. 2004. Aflatoxin binders II: Reduction of aflatoxin M1 in milk by sequestering agents of cows consuming aflatoxin in feed. *Mycopathologia*. 1-8.

Diaz, G.J., Cortes, A and Roldan, L. 2005. Evaluation of the efficacy of four feed additives against the adverse effects of T-2 toxin in growing broiler chickens. *Journal of Applied Poultry Research*. 14: 226-231.

Diaz, D.E. and Smith, T.K. 2005. Mycotoxin Sequestering Agents: Practical Tools for the Neutralization of Mycotoxins. *The Mycotoxin Blue Book*. pp. 323-340.

Doll, S., Dänicke, S., Valenta, H. and Flachowsky, G. 2004. In vitro studies on the evaluation of mycotoxin detoxifying agents for their efficacy on deoxynivalenol and zearalenone. *Archives of Animal Nutrition*. 58: 311-324.

Doupnik, B and Peckham, J.C. 1970. Mycotoxicity of Aspergillus ochraceus to chikcs. *Appl. Microbiol*. 19: 594-597.

Doyle, M.P., Applebaum, R.S., Brackett, R.E. and Marth, E.H. 1982. Physical, chemical and biological degradation of mycotoxins in foods and agricultural commodities. *J. Food Prot*. 45: 946-971.

Duvick, J. and Rood, T.A. 2000. Zearalenone detoxification compositions and methods. US 6074838.

Dwivedi, P. and Burns, R.B 1984. Effect of ochratoxin A on immunoglobulin's in broiler chicks. *Res. Vet. Sci*. pp. 117-121.

Edrington, T.S., Kubena, L. F., Harvey, R. B and Rottinghaus, G. 1997. Influence of a superactivated charcoal on the toxic effects of aflatoxin or T-2 toxin in growing boilers. *Poultry Science*. 76:1205-1211.

Edrington, T.S., Sarr, A.B., Kubena, L.F., Harvey, R.B. and Phillips, T.D. 1996. Hydrated sodium calcium aluminosilicate (HSCAS), acidic HSCAS, and activated charcoal reduce urinary excretion of aflatoxin M_1 in turkey poults. Lack of effect by activated charcoal on aflatoxicosis. *Toxicological Letters*. 89: 115-122.

Elissalde, M. H., Ziprin, R. L., Hiff, W.E., Kubena, L.E and Harvey, R.B. 1994. Effect of ochratoxin A on salmonella-challenged broiler chicks. *Poultry Science*. 73: 1241-1248.

Ellis, R.W., Clements, M., Tibbetts, A. and Winfree, R. 2000. Reduction of the bioavailability of 20 ug/kg Aflatoxin in trout feed containing clay. *Aquaculture*. 183: 179-188.

El-Nezami, H., Kankaanpaa, P., Salminen, S., Ahokas, J. 1998. Ability of dairy strains of lactic acid bacteria to bind a common food carcinogen, Aflatoxin B_1. *Food and Chemical Toxicology*. 36:321-326.

Erber, E. 1996. Futtermittelzusatz zur Inaktivierung von Mykotoxinen, PCT Int. Appl. WO 9612414.

Espada, Y., Domingo, M., Gomez, J and Calvo, M.A. 1992. Pathological lesions following an experimental intoxication with aflatoxin B_1 in broiler chickens. *Res. Vet. Sci*. pp. 275-279.

Evans, J. and Dawson, K.A. 2000. The ability of Mycosorb to bind toxins present in endophtyeinfected tall fescue. *In*: Biotechnology in the Feed Industry: Proceeding from Alltech's 16th Annual Symposium (T.P. Lyons and K.A. Jacques, Eds), Nottingham Press University, Nottingham, UK, pp. 409-422.

Galvano, F., Galofaro, V and Galvano, G. 1996. Occurrence and stability of aflatoxin M_1 in milk and milk products: a world wide review. *Journal of Feed Protection*. 59:1079-1090.

Galvano, F., Piva, A., Ritieni, A., Galvano, G. 2001. Dietary strategies to counteract the effects of mycotoxins: A review. *Journal of Food Protection*. 64: 120-131.

Ghosh, R. C., Chauhan, H.V and Roy, S. 1990. Immunosuppression in broilers under experimental aflatoxicosis. *British Veterinary Journal*. pp. 457-462.

Hamilton, P.B. 1984. Determining safe levels of mycotoxins. *Journal of food Protection*. 45: 570-575.

Hamilton, P.B., Huff, W.E., Harris, J. R and Wyatt, R.D. 1982. Natural occurrences of ochratoxicosis in poultry. *Poultry Science*. pp. 1832-1841.

Harvey, R.B., Kubena, L.F., Ellisalde, M.H. and Phillips, T.D. 1993. Efficacy of zeolitic compounds on the toxicity of aflatoxin to growing broiler chickens. *Avian Disease*. 37: 67-73.

Harvey, R.B., Phillips, T.D., Ellis, J.A., Kubena, L.F., Huff, W.E. and Petersen, H.D. 1991. Effects of aflatoxin M_1 residues in milk by addition of hydrated sodium calcium aluminosilicate to aflatoxin contaminated-diets of dairy cows. *American Journal of Veterinary Research*. 52: 1556-1559.

Hoerr, F. J., Carlton, W.W. and Yagen, B. 1981. Mycotoxins caused by a single dose of T-2 toxin or diacetoxyscirpenol in broiler chickens. *Veterinary Pathology*. pp. 652-664.

Huff, W.E., Harvey, R.B., Kubena, L.F. and Rottinghaus, G.E. 1988. Toxic synergism between aflatoxin and T-2 toxin in broiler chickens. *Poultry Science*. pp. 1418-1423.

Huwig, A., Freimund S., Kappeli O and Dutler H. 2001. Mycotoxin detoxication of animal feed by different adsorbents. *Toxicology Letters*. 122: 179-188.

James, L.J., and Smith, T.K. 1982. Effect of dietary alfalfa on zearalenone toxicity and metabolism in rats and swine. J. Anim. Sci. 55:110-118.

Jand, S.K., Paviter Kaur and Sharma, N.S. 2005. Mycotoxicosis in poultry: a review. *Indian Journal of Animal Sciences*. pp. 465-476.

Jewers, K. 1987. Problems in relation to sampling of consignments for mycotoxin determination and interpretation of results. Proc. 2nd International Conference on Mycotoxins, 28 September - 2nd October, Bangkok, Thailand.

Jewers, K., Coker, R.D., Blunden, G., Jazwinski, Mary J and Sharkey, A.J. 1986. Problems involved in the determination of aflatoxin levels in bulk commodities. lnternat. *Biodeteriorafion*. 22:83-88.

Kabak, B., Dobson, A.D and Var, I. 2006. strategies to prevent mycotoxin contamination of food and animal feed. A review. *Crit Rev. Fod sci Nutr* (46(8): 593-619.

Karlovysky, P.1999. Biological detoxification of fungal toxins and its use in plant breeding, feed and food production. *Nat. Toxins*. 7: 1-23.

Kerkadi, A., C. Barriault, B. Tuchweber, A.A. Frohlich, R.R. Marquardt, G. Bouchardand, and Yousef, I.M. 1998. Dietary cholestyramine reduces ochratoxin A-induced nephrotoxicity in the rat by decreasing plasma levels and enhancing fecal excretion of the toxin. *J. Toxicol. Environ. Health*. 3:231-250.

Kleiner, I., Flegar-Mestric, Z., Zadro, R., Breljac, D., Stanovic Janda, S., Stojkovic, R., Marusic, M., Radacic, M., Boranic, M., 2001. The effect of the zeolite clinoptilolite on serum chemistry and hematopoiesis in mice. *Food and Chemical Toxicology*. 39: 717-727.

Kubena, L.F., Harvey, R.B., Huff, W.E., Corrier, D.E., Phillips, T.D., Rottinghaus, G., 1990. Efficacy of a hydrated sodium calcium aluminosilicate to reduce the toxicity of aflatoxin and T-2 toxin. *Poultry Science*. 69: 1078-1086.

Lemke, S.L., Mayura, K., Ottinger, S.E., McKenzie, K.S., Wang, N., Fickey, C., Kubena, L.F and Phillips, T.D. 1999. Assessment of the estrogenic effects of zearalenone after treatment with ozone utilizing the mouse uterine weight bioassay. *J. Toxicol. Environ. Health A*. 56: 283-295.

Lesson, S., Diaz, G. and Summers, J.D. 1995. Poultry Metabolic Disorders and Mycotoxins, University Books, Guelph, Ontario, Canada.

Lindemann, M.D., Blodgett, D.J., Kornegay, E.T. and Schurig, G.G., 1993. Potential ameliorators of aflatoxicosis in weanling/growing swine. *Journal of Animal Science*. 71: 171-178.

Lun, A.K., Young, L.G., Moran, E.T., Jr., Hunter, D.B and Rodriguez, J.P. 1986. Effects of feeding hens a high level of vomitoxin-contaminated corn on performance and tissue residues. *Poultry Science*. pp. 1095-1099.

Masimango, N., Remacle, J., Ramaut, J.L., 1978. The role of adsorption in the elimination of aflatoxin B1 from contaminated media. *Applied Microbiology and Biotechnology*. 6: 101-105.

McKenzie, K.S., Sarr, A.B., Mayura, K., Bailey, R.H., Miller, D.R., Roggers, T.D., Norred, W.P., Voss, K.A., Plattner, R.D., Kubena, L.F. and Phillips, T.D. 1997. Oxidative degradation and detoxification of mycotoxins using a novel source of ozone. *Food Chem. Toxicol.* 35: 807-820.

Miazzo, R., Peralta, M.F., Magnoli, C., Salvano, M., Ferrero, S., Chiacchiera, S.M., Carvalho, E.C.Q., Rosa, C.A.R and Dalcero, A. 2005. Efficacy of sodium betonite as a detoxifier of broiler feed contaminated with Aflatoxin and fumonisin. *Poultry Science.* 84: 1-8.

Miedaner, T. and Reinbrecht, C. 1999. Fusarein in Getreide- Beduutung von Pflanzenbau and Resistenzzuchtung zur verminderung von Ertragsverlusten und einer Kontamination mit Mykotoxinen. *Getreide. Mehl. Brot.* 35:135-140.

Nageswara, S.B. and Chopra, R.C., 2001. Influence of sodium bentonite and activated charcoal on aflatoxin M1 excretion in milk of goats. *Small Ruminant Research.* 41: 203-213.

Nahm, K.H. 1995. Possibilities for preventing mykotoxicosis in domestic fowl. *World Poult. Sci. J.* 51: 177-185.

Papaioannou, D.S., Kyriakis, C.S., Papasteriadis, A and Roumbies, N. 2002. A field study on the effect of in-feed inclusion of a natural zeolite (clinoptilolite) on health status and performance of sows/gilts and their litters. *Research in Veterinary Science.* 72: 51-59.

Park, D. L. 1993. Perspectives on mycotoxin decontamination procedures. *Food Addit. Contam.* 10: 49-60.

Peckham, J.C., Doupnik, B. and Jones, Jr., O.H. 1971. Acute toxicity of ochratoxins A and B in chicks. *Appl. Microbiol.* pp. 492-494.

Petterson, H. 2004. Controlling mycotoxins in animal feed. In: Mycotoxins in food, detection and control (Magan, N. and Oisen, M., eds.), Woodhead Publishing Limited. Cambridge, pp. 262-304.

Phillips, T.D., 1999. Dietary clay in the chemoprevention of aflatoxin induced disease. *Toxicological Science*, 52: 118-126.

Phillips, T.D., Afriyie-Gyawu, E., Williams, J., Huebner, H., Ankrah, N.-A., Ofori-Adjei, D., Jolly, P., Johnson, N., Taylor, J., Marroquin-Cardona, A., Xu, L., Tang, L. and Wang, J.S. 2008. Reducing human exposure to aflatoxin through the use of clay: A review. *Food Additives and Contaminants.* 25: 134-145.

Phillips, T.D., Grant, P.G and Sarr, A.B. 1995. Selective chemisorption and detoxication of aflatoxins by phyllosilicate clay. *Natural Toxins.* 3: 204-213.

Phillips, T.D., Kubena, L.F., Harvey, R.B., Taylor, D.R. and Heildebaugh, N.D., 1988. Hydrated sodium calcium aluminosilicate: a high affinity sorbent for aflatoxin. *Poultry Science.* 67: 243-247.

Piva, A and Galvano, F. 1999. Managing mycotoxins impact. Nutritional approaches to reduce the impact of mycotoxins in: Lyons T.P, Jackues, K.A (Eds), proceeding of Alltech's 15th Annual Symposium, Nottingham University Press, Nottingham, UK, pp. 381-399.

Ramos, A. J and Hernandez, E. 1997. Prevention of aflatoxicosis in farm animals by means of hydrated sodium calcium aluminosilicate addition to feedstuffs: A review. *Animal Feed Science and Technology.* 65: 197-206.

Ramos, A.J and Hernandez, E. 1996. In vitro Aflatoxin adsorption by means of a montmorillonite silicate. A study of adsorption isotherms. *Animal Feed Science and Technology.* 62: 263-269.

Ramos, A.J, Fink-Gremmels, J and Hernandez, E.1996. Prevention of toxic effects of mycotoxins by means of non-nutritive adsorbent compounds. *J. Food Prot.* 59: 631–641.

Rosa, C.A.R., Miazzo, R., Magnoli, C., Salvano, M., Chiacchiera, S.M., Ferrero, S., Saenz, M., Carvalho, E.C.Q and Dalcero, A., 2001. Evaluation of the efficacy of bnetonite from the south of Argentina to ameliorate the toxic effects of aflatoxin in broilers. *Poultry Science.* 80: 39-144.

Satheesh, K., Ahmad, M.N., Srilatha, C. and Rao, T.C.S. 2004. Effects of fumonisin-B1 toxicity on serum biochemistry in broiler chicken. *Indian Veterinary Journal.* pp. 1097-1099.

Schardl, J., Jensen, H and Latge, J., 1996. Epichloe festucae and related mutualistic symbionts on grasses. *Fungal Genetics and Biology*. 33:69-82.

Scheideler, S. 1993. Effects of various types of aluminosilicates and aflatoxin B1 on aflatoxin toxicity, chick performance, and mineral status. *Poultry Science*. 72: 282-288.

Schell, T.C., Lindemann, M.D., Kornegay, E.T and Blodgett, D.J. 1993. Effects of feeding aflatoxin-contaminated diets with and without clay to weanling and growing pigs on performance, liver function, and mineral metabolism. *Journal of Animal Science*. 71: 1209-1218.

Scott, P.M. 1998. Industrial and farm detoxification processes for mycotoxins. *Rev. Med. Vet.* 149: 989-998.

Sharma, R.P. 1991. Immunotoxic effects of mycotoxins. In: Mycotoxins and Phytoaiexins (Sharma, R.P. and Salunkhe, D.K., eds.), CRC Press. Boca Raton, Florida. pp. 81-99.

Shi, Y.H., Xu, Z.R., Feng, J.L and Wang, C.Z. 2006. Efficacy of modified montmorillonite nanocomposite to reduce the toxicity of aflatoxin in broiler chicks. *Animal Feed Science and Technology*. 129: 138-148.

Singh, G.S., Chauhan, H.V., Jha, G.J. and Singh, K.K. 1990. Immunosuppression due to chronic ochratoxicosis in broiler chicks. *J. Comp Pathol*. 399-410.

Solfrizzo, M., A. Visconti, G. Avantaggiato, A. Torres, and Chulze, S. 2001. In vitro and in vivo studies to assess the effectiveness of cholestyramine as a binding agent for fumonisins. *Mycopathologia*. 151:147-53.

Sova, Z.H., Pohunkova, H., Reisnerova, H., Slamova, A and Haisl, K. 1991. Hematological and histological response to the diets containing aflatoxin B1 and zeolite in broilers of domestic fowl. *Acta Veterinaria Hungarica*. 60: 31-40.

Stangroom, K.E., and Smith, T.K. 1984. Effect of whole and fractionated dietary alfalfa meal on zearalenone toxicosis in rats and swine. *Can. J. Physiol. Pharmacol*. 62:1219-1224.

Sweeney, M.J and Dobson, A.D.W. 1998. Review: mycotoxin production by Aspergillus, Fusarium and Pencillium species. *Int. J. Food Microbial*. 43: 141-158.

Tung, H.T., Smith, J.W. and Hamilton, P.B. 1971. Aflatoxicosis and brushing in the chicken. *Poultry Science*. 50: 796.

Virdi, J.S., Tiwari, R.P., Saxena, M., Khanna, V., Singh, G., Saini, S.S. and Vadehra, D.V. 1989. Effects of aflatoxin on the immune system of the chick. *J.Appl.Toxicol*. pp. 271-275.

Volkl, A. and Karlovsky, P. 1998. Zearalenone and feed additives: adsorption and desorption as a function of pH. University of Hohenheim, Stuttgart, Germany.

Wang, P., Afriyie-Gyawu, E., Tang, Y., Johnson, N., Xu, L., Tang, L., Huebner, H., Ankrah, N.A., Ofori-Adjei, D., Ellis, R.W., Jolly, P., Williams, J., Wang, J. S and Phillips, T.D. 2008. NovaSil clay intervention in Ghanaians at high risk for aflatoxicosis: II. Reduction in biomarkers of aflatoxin exposure in blood and urine. Food Additives and Contaminants, 25: 622-634.

Wyatt, R.D., Doerrr, J.A., Hamilton, P.B. and Burmeister, H.R. 1975. Egg production, shell thickness, and other physiological parameters of laying hens affected by T-2 toxin. *Appl. Microbiol*. pp. 641-645.

Wyatt, R.D., Hamilton, P.B. and Burmeister, H.R. 1973. Neural disturbances in chickens caused by dietary T-2 toxin. *Appl. Microbial*. pp. 757-761.

Yoon, Y and Baeck, Y.J. 1999. Aflatoxin binding and antimutagenic activities of Bifidobacterium bifidum HY strains and their genotypes. *Koeran Journal of Dairy Science*. 21: 291-298.

Zeitfracht Medien GmbH
Ferdinand-Jühlke-Straße 7
99095 Erfurt, Deutschland
produktsicherheit@kolibri360.de